INSIGHT
OF
ZODIAC SIGN

First Edition

by

SANJAY KUMAR GUPTA

@ Copyright Author-2017

Caution

ISBN-13: 978-1981135615

ISBN-10: 1981135618

Publisher: Createspace

ABOUT THE BOOK

Many people need to know about their financial status, growth in carrier and their future whether it is going to improve or is it going to get worse. Zodiac denotes an annual cycle of twelve signs and the apparent path of the Sun across the heavens. The term Zodiac and the names of the twelve signs are today mostly associated with horoscope astrology. Your zodiac signs give a summary about the basic characteristics of your personality, nature, area of interest, your strengths, weaknesses, personal life, professional life. Your date of birth determines which one is your zodiac sign. The dates assigned to the signs of the zodiac are given in front of each zodiac sign below. Different astrologers assign slightly different dates to these zodiac signs. A horoscope divided into twelve zodiac signs can predict a Kundali chart, which defines the pathways of various planets and stars. The zodiac signs that help in defining your horoscope include Aries, Taurus, Gemini, Cancer, Leo, Virgo, Libra, Scorpio, Sagittarius, Capricorn, Aquarius, and Pisces. Horoscope is a placement of planets and stars during the birth of any individual in an astrological chart which is calculated with the relevance of date, time and birth place of a person. This chart can also be calculated for an event, a question, countries, and occasions. Various symbols are used which represent planets, signs and geometric connections. This varies from person to person due to the continuous transition of planets. Horoscope is the first step to make astrological calculations. Horoscope can be used to predict future, past and present of a person. Other than this, various aspects of life including career, love, marriage, relationships etc can also be predicted by a horoscope. In Indian culture, every important work of life is done by drawing calculations from the horoscope. An accurate judgement of a horoscope is very important and is the main objective of studying astrology. Prediction of the right horoscope is very important that an astrologer should know complete calculations along with the use of new computing methods.

Astrology is not knowing your future, but planning your future by averting the misshapenness by action at the right time and wearing Gems, wearing Yantras, chanting Mantras and Prayers. It is important to realize that success comes only with the right actions at the right timing. JP Morgan made a fortune using astrology for "Right Timing". This knowledge is made available to you though this book. By using the book, your life will be more prosperous than ever before. It is important to work "Smartly" but not hard. This Book gives you the followings:

1.   More rewarding life.

2.   Pictures of your career and love life.

3.   Ever dream of becoming a Star.

4.   Financial fortune in life.

5.   The "Best you can be every day".

6.   Start of a professional practice.

7.   Predict Horoscope of individual of choice.

Money controls the way we live our lives, and this amazing readings of Zodiac signs will give you the insight, you need to take control of your financial future. It could be one of the most important readings that you will get through this Book that can truly change your life. This Book will isolate time to capture the situation and reveal its significance. At last, you have discovered a direct channel that will allow you an insight into your own destiny. The technically advanced matters allow you to deliver your reading to you quickly and effectively. Not only will your reading be incredibly accurate but also you will have it available to read and analyze at your own pace. As you continue to consult this book for answers to important issues in your life, an accurate record of your responses will be chronologically archived to allow you to a greater understanding of yourself as you relate to the complex world around you. This means that you can make money by using "Right Time" methods by reading your Horoscope to gain knowledge on when and how to do things. Gain Wisdom through knowledge of Zodiac Signs influences by understanding when and not to act. Gain Power and Status using your own cycles that empower you toward success. Avoid Problems before they happen by knowing the right time to act. Be Comfortable with gifts provided by the universe, as you deserve them. This is my exhaustive study, collections and the presentation of the subject, which matter much and even much more than the presentation of the subject matter is long years of experience and association with the astrology work all over India and abroad. This is so much strife and struggle in the present time as it was never before. This book is easily approachable and compact. It is full of all information at one place to be referred easily and quickly by anybody whether busy in any profession.

I am confident that the readers and experts will consider my effort. I am confident that this book will help to all in achieving his object and success in every field of astrology. The need for development has been felt for quite some time back that an authoritative book is written on astrology work which may contain all the aspect of astrology with illustrations so that complete information is conveyed in a simple language. I have tried to make clear about what is the correct Zodiac Sign astrology work. These are all correct and true facts & figures incorporated here in a single book for the use by the common men. This gives the authenticity to the book. This book is a tool for the Jyotish Students, for the Beginners, for the somewhat Advanced Students and for the Professionals too. I am sure you are going to give the recognition and importance of this book within a short time and that will be a positive proof of my efforts and impartiality.

ABOUT THE AUTHOR

 **Author: Sanjay Kumar Gupta**
He is civil Engineer from Mumbai University. He is a meritorious, genius and very intellectual person since childhood. He has deep interest in Astrology since childhood. He is deeply committed and involved in Astrological educational research and development work. Sincerity, deep study and hobby has put him into the use of many new techniques and methods for horoscope prediction based on Planet, Nakshatra and Zodiac Sign. He has studied many books and articles on the subject. He thought of the necessity of a consolidated book covering all the aspects, topics and subject matters pertaining to "Insight of Zodiac Sign" astrology at one place. This will result improving the general quality levels of the reader to a greater satisfaction. He has taken a lead in upgrading the process of awareness of various matters benefiting the fresh learner in astrology. Anybody can avail of practical knowledge on various topics related to astrology and can make predictions of individuality in detail of any person with the help of this Book. This book is a tool for the Jyotish Students, for the Beginners, for the somewhat Advanced Students and for the astrologers too. I am sure you are going to give the recognition and importance of this book within a short time and that will be a positive proof of my efforts and impartiality. He just believes the information presented in this book should be available to the public.

TABLE OF CONTENTS

CHAPTER 1

ZODIAC SIGN

PART-1
INTRODUCTION

The Zodiac is a band of group of stars or the positions of celestial bodies. The Zodiac is divided into twelve divisions of 30 degrees each called "Sign". Each segment is called a Sign or Rashi. The alternate Sign starting from Aries onwards are known as male and female on the other hand. The four Signs from Aries onwards indicate East, South, West and North, while the remaining Signs repeat in the same way. Gemini, Cancer, Capricorn Aries, Taurus and Sagittarius are night Signs. Leo, Libra, Scorpio, Aquarius, Pisces and Virgo are day Signs. If a Sign is aspect by its Lord, or by a planet friendly to its Lord, or by Mercury, or by Jupiter, it is said to be Strong Sign. Planets other than the above do not lend strength by aspect. The Rising sign, at a particular time, is the sign of the zodiac positioned on the eastern horizon on the cusp of the first house at birth and is called Lagna. The Lagna lord is considered more powerful in Kundali chart. Similarly, there are Badhaka Rashi (Sign) defined as per the Janma-Lagna. It works as the Badhaka Rashi and harms badly that house in which it is occupying. The ending portions, 30th degree of Cancer, Scorpio and Pisces are called Gandanta. It is said, that one born in Gandanta will not survive. He will either lose his mother, or he will end the dynasty, i.e. he is the last of his descent and will not have any children. If, however, he survives, he becomes a king with many elephants and horses. Zodiac signs having different characteristics act in different ways and are also known for its different nature. Every Zodiac Sign falls into one of four elements. There are four Elements, such as earth represents common sense; fire represents action, air represents thinking and communication skills, and water represents the ability to feel and intuitively know. Each Element is assigned to each sign depending to their orientation in the zodiac. Many astrologers consider the element of each of the planets when determining which of the elements may be more significant in a horoscope.

Aries, Leo, and Sagittarius are Fire. Taurus, Virgo, and Capricorn are Earth. Gemini, Libra, and Aquarius are Air. The lack of Air Element in a native birth chart indicates difficulty in the expression of that person. Communication of ideas and the ability to conceptualise may prove difficult. Cancer, Scorpio, and Pisces are Water.

The method for evaluating the strength of an element in a birth chart is to assign a value of 8; the element is considered "Preponderance" in that element. If we get less than 8 points with this approach in one element, it is considered "Absence" in that element. A preponderance of Fire Element indicates high spirits, great faith in self, enthusiasm, direct, honesty, intensely assertive, most daring, individualistic, active and self-expressive, good natured, fun loving, natural leader, having a good time than on material possessions, big egos. A preponderance of Earth Element indicates cautious, conventional, dependable but quite responsible, methodical, organizer, a builder, and a hard-worker. It provides the skills and attitude necessary to succeed readily in the world of business and never gamble or take unnecessary chances. The preponderance of Air Element suggests a strong emphasis on thought, ideas and intellectual and they communicate and express ideas with mental agility and become the impractical dreamers, constantly thinking, people-oriented, but more inclined toward the group than the individual. The preponderance of Water Element indicates close emotional relationships, romantic, sentimental, affectionate, secure bond with partner, communicate best in non-verbal ways; emotionally, psychically, or through forms as art, dance music, poetry and photography.

"Astrology" or "Jyotish" means the 'science of light' and is related with the Light and Magnetic Field emitted by the planet. 'Hindu Astrology' is founded by the Maharishi Aryabhatta, Parasara, Varaha Mihira, Jaimini, Garga, Kalidasa and Kalyan Varma. The origin of this science can go back as old as 4000 years. The astrology fully knows the individual's future as indicated by horoscope but can't certify the same as it is not 100 percent Mathematics or Science". Horoscope is like a snapshot of a particular place in time and space. For casting the natal horoscope of an individual the time of birth, date of birth and place of birth is needed. There are 12 houses and 12 Signs in a horoscope from which an astrologer can predict about various areas of the life of an individual. Many people need to know about their financial status, growth in carrier and their future whether it is going to improve or is it going to get worse. Zodiac denotes an annual cycle of twelve signs and the apparent path of the Sun across the heavens. The term Zodiac and the names of the twelve signs are today mostly associated with horoscope astrology. Your zodiac signs give a summary about the basic characteristics of your personality, nature, area of interest, your strengths, weaknesses, personal life, professional life. Your date of birth determines which one is your zodiac sign. This Book enables the astrologer to know that what the future has in store for the native. For the calculation of the timing of various events indicated in the horoscope the knowledge of impact of major period/sub period and transit is used. The dates assigned to the signs of the zodiac are given in front of each zodiac sign below. Different astrologers assign slightly different dates to these zodiac signs. A horoscope divided into twelve zodiac signs can predict a Kundali chart, which defines the pathways of various planets and stars. The zodiac signs that help in defining your horoscope include Aries, Taurus, Gemini, Cancer, Leo, Virgo, Libra, Scorpio, Sagittarius, Capricorn, Aquarius, and Pisces. Horoscope is a placement of planets and stars during the birth of any individual in an astrological chart which is calculated with the relevance of date, time and birth place of a person. This chart can also be calculated for an event, a question, countries, and occasions.

PART-2
GLOSSARY OF
ASTROLOGY

Affinity: It is a mutual attraction between the planets.

Affliction: Affliction of a planet is formed by, (1) its placement in the 6th, 8th or 12th houses or (2) association with the ruler of the 6th, 8th or 12th lords; or (3) association with natural malefic; or (4) association of Badhaka planet; or (5) its Combustion or placement between two natural malefic. Affliction of a planet is such a bad condition in which his energies or powers are considered to be zero and give an adverse effect.

Air signs: Air signs are the signs of Gemini, Libra, and Aquarius.

Angle (Kendra): The Ascendant (first house), Descendant (fourth house), Mid-heaven Seventh house) and I Mum Collie (tenth house) are called Kendra (Angle). This refers to the cusps of the first, seventh, tenth, and fourth houses, respectively.

Ascendant: The Ascendant or Lagna or Rising Sign is the Sign in which an individual is born. Ascendant is the rising sign or the sign on the cusp of the first house of the natal chart. It is the sign and degree of that sign that is rising on the eastern horizon at the moment of birth, with respect to the place of birth.

Aspects: Aspects are an important part of modern astrology. As the planets move in their elongated orbits around the Sun, they form various angular relationships with one another, using the Sun or Earth as the centre. These are called aspects, the most popular aspects result from dividing the circle by numbers like 1, 2, 3, 4, resulting in aspects such as the conjunction (0 degrees), opposition (180 degrees), trine (120 degrees), square (90 degrees), and so forth. When two planets form an aspect with one another, their energies and natures are said to combine and work in harmony or discord. When two planets are exactly on opposite sides, they are in opposition.

Association: It is a relationship between two or more planets by their position in a house.

Astrologer: The person who practices astrology is called Astrologer.

Birth Time: This is the moment of first breath of a new born individual.

Cardinal Signs: Cardinal signs are the signs of Aries, Cancer, Libra, and Capricorn, and are related to the change of the seasons.

Chart: It is a figure or sketch consisting of 12 houses, in which the position of the planets and the Signs are given.

Conjunction: Two planets situated together in a house or occupying position close to each other within a certain orb or reaching nearer are called in Conjunction.

Critical Degrees: Critical degrees are 0, 13, and 26 degrees of cardinal signs (Aries, Cancer, Libra, and Capricorn); 8-9 and 21-22 degrees of fixed signs (Taurus, Leo, Scorpio, and Aquarius); and 4 and 17 degrees of mutable signs (Gemini, Virgo, Sagittarius, and Pisces). Many astrologers consider 0 and 29 degrees of any sign critical degrees as well. The classic critical degrees are considered sensitive, emphasized, and often strengthening, points. The 0 and 29 degrees are more crisis-oriented, especially in predictive work.

Cusp: Cusp is the beginning of a house in the chart or a cusp of a sign is the degree when one sign ends and the other begin. There is a lot of talk about being "born on the cusp". A sun sign is either one sign or the other, regardless of whether it is almost another sign.

Cycle: It is a complete revolution or rotation made by a planet around the Sun.

Debilitation (Khala) Sign: When a planet occupies its Sign of Fall for Neecha Bhanga, the planet is in Debilitation and is considered weak or Neecha.

Degrees of Maximum Exaltation & Debilitation: Maximum degree of Exaltation or Debilitation is the defined degree of planet position in a Sign.

Derivative House: This is a house related to another house in the Chart, which signify the events of individual. Example:

The 3rd house is the house of the brothers and sisters and the fourth house is the mother's house. Accordingly, the third house from the fourth house, i. e. the 7th house in a natal chart will describe the signification of the brothers and sisters of the native mother.

Descendant: The point opposite the Ascendant, i.e. the seventh house is called Descendent.

Dispositors: The lord of the sign is called Dispositors of a planet positioned in that Sign. The Dispositors of a planet is the soul essence of that planet, like prime minister. It dictates the planet to reacts in the way he likes. Example: Venus is the Dispositors of Saturn as because of ruler ship of Libra and Saturn has to act as per choice of Venus.

Domicile (Domes): A domicile (domes) of a planet is a Sign of ruler ship and is regarded as a home. Example: The domiciles of Mercury are Gemini and Virgo as because Mercury is the ruler of them. They are domicile or gaudier of Mercury, where he rejoices.

Earth: Earth signs are the signs of Taurus, Virgo, and Capricorn.

Elements: Elements is the signs divided into 4 groups called elements, like Fire, Earth, Air, and Water.

Emphatic: Emphatic aspects are those that emphasize or align two planetary energies with one another, like conjunction (0 degrees), where two planets at the same point in the same zodiac sign and are said to be in conjunctions. Their natures are fused or blended into one. Opposition (180 degrees) aspect is that two planets are at opposite sides of the zodiac. The energies are in alignment with each other. They can pull together or apart, depending upon the nature of the planets involved.

Exaltation: The planet in a particular Sign is called the planet in Exaltation Sign. Planet is dignified during his Exaltation.

Face: The "faces" arise from a subdivision of zodiac Sign into six equal parts. This is an obsolete term meaning the division of each sign into six equal parts of 5° each. The faces derive from the ancient Egyptian decants.

Feminine signs: Feminine signs are the signs of Taurus, Cancer, Virgo, Scorpio, Capricorn, and Pisces. The remaining

6 signs of the zodiac are masculine signs. Feminine signs are also referred to as "negative" or "receptive" signs.

Fire signs: Fire signs are the signs of Aries, Leo, and Sagittarius.

Fixed signs: Fixed signs are the signs of Taurus, Leo, Scorpio, and Aquarius.

Ghati: The Ghati is a measure of the Time. One day = 24 hours = 60 Ghati. 2 ½ Ghati =1 hour; one Vighati = 24 seconds; and 60 Vighati = 1 Ghati.

Hard aspects: Hard aspects are generally considered the conjunction, opposition, and square. In midpoint work, the hard aspects also include the semi-square aspect. These represent challenge, obstacles, and substance. They provide meat and potatoes in our life. Too many can block or obstruct the life flow, yet too few can cause life to be weak or thin.

Horizon: It is the visible juncture of Earth and the sky and represented in a horoscope.

Horoscope: It is the Janma Kundali, which depicts the positions of different signs and planets at the time of birth. It also represents the Rising Sign at the place of birth and the location of planets in various signs.

House Cusp: This is the zodiacal degree at which a house begins.

Janma Rashi: Janma Rashi is the Sign occupied by Moon at the time of birth of the native.

Latitude: This is the celestial angular distance measured north or south of the plane of the ecliptic. This is the distance of a planet from the Equator.

Local Mean Time: This is the actual time in a given location based upon the Sun's position at the Mid Heaven (noon) of the place. It is abbreviated as L M T.

Longitude: It is the distance in degrees or in arc on Earth from 0° Aries eastward to any given point that intersects the ecliptic, such as, 10° Taurus is expressed as longitude 40°.

Luminaries: The Sun and the Moon are Luminaries.

Masculine Signs: Aries, Gemini, Leo, Libra, Sagittarius and Aquarius are considered Masculine Signs.

Masculine signs: Masculine signs are the signs of Aries, Gemini, Leo, Libra, Sagittarius, and Aquarius. The remaining

6 signs of the zodiac are feminine signs. Masculine signs are also referred to as "positive" signs.

Midheaven: Midheaven is the sign, and degree of that sign, on the cusp of the tenth house of the natal chart. It is the highest point in the zodiac at the moment of birth, and in relationship to the place of birth. Its abbreviation is 'MC'.

Moolatrikona: The position of the planets in the particular Sign is considered the Moolatrikona of planet. Planet is dignified during his Moolatrikona. The Moolatrikona sign is usually a part of its own house with the exception of the Moon, where it behaves almost as favourably and gets more strength.

Moon sign: The Moon sign is the zone of the zodiac in which the Moon is positioned when a person is born. This is also called Rashi too.

Mutable signs: Mutable signs are the signs of Gemini, Virgo, Sagittarius, and Pisces.

Natal chart: Natal chart is the horoscope drawn for a person's birth. It is also known as a birth chart. The natal chart is a map of one's life, similar to taking a picture of the planets at the time of one's birth. The universe stopped at that moment in time. It reveals what the universe has to say about who a person is and what he or she may become. The chart wheel is a map of the space surrounding a person at the time of birth. The wheel is divided in 12 sections called houses. Planets in the heavens are placed on the chart wheel in the houses that correspond to where they actually are in the sky at birth. The horoscope cast at the birth time of the individual, showing the position of Signs and Planet with respect to house, is called a Natal Chart or Nativity.

Native: It refers to a person (male or female) for whom a horoscope is cast and studied.

Navamsa Chart: It is a nine Divisional Chart of a Sign with 3° 20' segments each.

Opposite: It is point at a distance of 180° in the Zodiac or the 7th house apart.

Orb: Orb is determining whether one planet forms an aspect to another, astrologers allow an "orb" of influence, which is a specific number of degrees. Aspects between planets gradually form, become exact and separate. When an aspect is exact, it has its greatest impact. Yet, the effect of most aspects can be felt for some time before and after the moment when it is exact. The range within which an aspect is in operation is called its 'orb of influence', or simply its orb. An orb of one or two degrees of arc on either side of the exact aspect is considered a close or tight orb, while an orb of 10 degrees is loose. This is the degrees of Longitude in the Zodiac.

Own Sign: It is a sign ruled by that planet, which own the sign.

Part of Fortune: This is the Arabian Part most commonly used by western astrologers. It is the degree, which is calculated by subtracting the Sun's longitude from the sum of the Ascendant and Moon's Longitude. The degree occupied by the Part of Fortune symbolizes good fortune. It is also called Fortuna and Pars Fortune. Wherever the part of fortune is found in the natal chart is the place where a person is thought to possess natural talent.

Personal planets: Personal planets are the inner planets and luminaries include Sun, Moon, Mercury, Venus, and Mars. They have a personal and direct effect on the native's personality.

Planet: Sun, Moon, Mars, Mercury, Jupiter, Venus, Saturn, Rahu, Ketu, Uranus, Neptune and Pluto are the twelve heavenly bodies which appear to move in the Zodiac and influence the human body and are called Planets.

Planetary rulers: Each sign is ruled by a planet, called planetary rulers.

Prediction: Knowing about natives past, present and future with the help of horoscope is prediction. But, in one accident many lives are taken, does that indicate similarity in everyone's horoscope? No, because the horoscope of the place or a vehicle in which passengers are travelling, supersedes the horoscopes of the natives. Hence, in a calamity everyone's horoscope does not necessarily indicate death. However, this point needs further research and views from the readers

Quadrants: Quadrants are 4 "quadrants" in a chart, and each starts at the cusp of the first, fourth, seventh, and tenth houses.

Retrograde motion (Vakra/Saktha): When observed from Earth, it appears the apparent backward motion of a planet or moving in reverse direction than its natural direction of travel and is called Retrograde Motion. A planet is considered "retrograde" when it appears to be moving backwards. Both "retrograde" and "direct" are terms used in astrology to describe the direction of planetary movement with relation to the Earth. Note that the planets do not actually move backwards. However, they appear (from our perspective on Earth) to back up for periods of time. The Sun and the Moon never retrograde.

Rising Planet: The planets which are positioned in the Rising Sign or Ascendant in the natal chart are Rising Planets.

Rising Sign (Lagna): The earth is rotating once a day around its axis. One of the 12 zodiacal signs is entering the 1st house every two hours. The Sign entering into 1st house at the birth time is known as Rising Sign or Lagna.

Ruling Planet: The planet which rules the Ascendant or Lagna Sign is called the Ruling Planet.

Soft aspects: Soft aspects are generally considered the sextile and trine aspects, and are also referred to as "easy" and "flowing" aspects. The soft aspects bring ease, clarity, and vision to our lives. We can see, grasp, and understand what is happening. Too few of the soft aspects means we don't know what we are doing or what is happening in our lives, while too many soft aspects make for a life that runs cool, is overly mental, and lacking in substance.

Solar Chart: The horoscope Chart with the Sun in the Ascendant is called the Solar Chart.

Trikas (Badhakasthana): The Houses 6, 8 and 12 are called Trikas or Badhakasthana. These are the Badhaka houses and considered the evil houses of suffering.

Trikona (Trine): The Houses 1, 5 and 9 are called Trikona or Trine.

Triplicity: It is a group of three signs belonging to the same element: fire (Aries, Leo, Sagittarius); earth (Taurus, Virgo, Capricorn); air (Gemini, Libra, Aquarius); and water (Cancer, Scorpio, Pisces). The members of a triplicity lie 120° apart in the zodiac, forming a triangular and harmonious relationship with each other and form an equilateral triangle in a horoscopes diagram.

Upachaya: The 3rd, 6th, 10th and 11th houses from the Lagna are called the Upachaya.

Vishnu House: The houses 1st, 5th & 9th stand for Vishnu and hence called Vishnu House.

Waning: The phase of the Moon during which the visible portion of the Moon decreases, is called Waning.

Water signs: Water signs are the signs of Cancer, Scorpio, and Pisces.

Waxing: The phase of the Moon during which the visible portion of Moon grows larger, is called Waxing.

Winning Planet: When a planet is in association with an enemy planet and its degrees are more than that of enemy planet, then that planet is said to be a winning planet.

Zodiac: It is literally the circle of stars. Zodiac is defined a band of the heaven approximately 14° wide, centred on the Ecliptic, against which the Sun and other planets are seen to move, as seen from the Earth. Zodiac is a circle of 360 degrees, divided into 12 equal sectors of 30 degrees each that are the astrology signs.

PART-3
ASTROLOGY SYNBOL

Sign	Sign Picture	Sign Symbol	Lord Planet Symbol
Aries			Mars
Taurus			Venus
Gemini			Mercury
Cancer			Moon

Leo			Sun
Virgo			Mercury
Libra			Venus
Scorpio			Mars
Sagittarius			Jupiter
Capricorn			Saturn

Aquarius			Saturn
Pisces			Jupiter

PART-4
KUNDALI CHART

South Indian Kundali – Chart:

In South Indian Style Chart, the position of the signs is always fixed and the position of the Ascendant is always changing. The houses are counted in a clockwise direction. The upper top left but one Rectangular box, being denoted by the digit 1 is always Aries. The next Rectangular box right to it, being denoted by the digit 2, is always Taurus and so on as written in the Chart. The digit 1 through 12 indicates the position of the Signs fixed in clockwise direction. It is always fixed in the South Indian style of Chart. The Ascendant (Lagna) falls in one of the Sign depending on rising Sign and is called Lagna or Ascendant of the Chart. The counting of the Houses is always done in clockwise direction from the Ascendant (Lagna) as first house through twelfth house. The planets occupy the House according to their longitudinal position.

12 (Pieces) Venus	1 (Aries)	2 (Taurus)	3 (Gemini) Ascendant, Mars
11 (Aquarius) Mercury, Ketu	Lagna Chart-1		4 (Cancer) Moon
10 (Capricorn) Sun			5 (Leo) Rahu
9 (Sagittarius) Jupiter, Saturn	8 (Scorpio)	7 (Libra)	6 (Virgo)

North Indian Kundali – Chart:

In the North Indian style Chart, the Houses are always fixed and the rising Sign falls in the 1st House, which is called Lagna or Ascendant, which is at the top in the centre. The Lagna (Ascendant) as well as the other houses are always fixed and are counted in anti-clockwise direction from the 1st House or the Lagna or the Rising Sign. The planets occupy the House according to their longitudinal position. The numbers shown in this format tell us which sign is in the Lagna and other Houses as shown in the Chart.

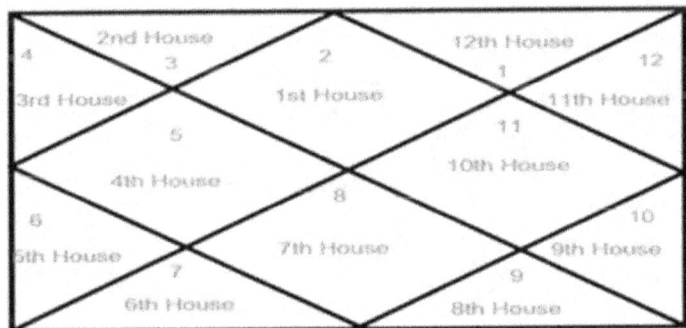

East Indian Kundali – Chart:

In the East Indian (Bengali) style Chart, the Houses are always fixed and the rising Sign falls in the 1st House, which is called Lagna or Ascendant, which is at the top in the centre. The Lagna (Ascendant) as well as the other houses are always fixed and are counted in anti-clockwise direction from the 1st House or the Lagna or the Rising Sign. The planets occupy the House according to their longitudinal position. The numbers shown in this format tell us which sign is in the Lagna and other Houses as shown in the Chart. In Bengali style zodiac is again fixed and ascendant & planets move anti clock wise along the zodiac unlike South Indian System. In western style the ascendant is fixed again and placed on the left hand side whereas

Western style Kundali – Chart:

It is a circular Chart divided in twelve parts in which Sign and Planets position are given as shown below.

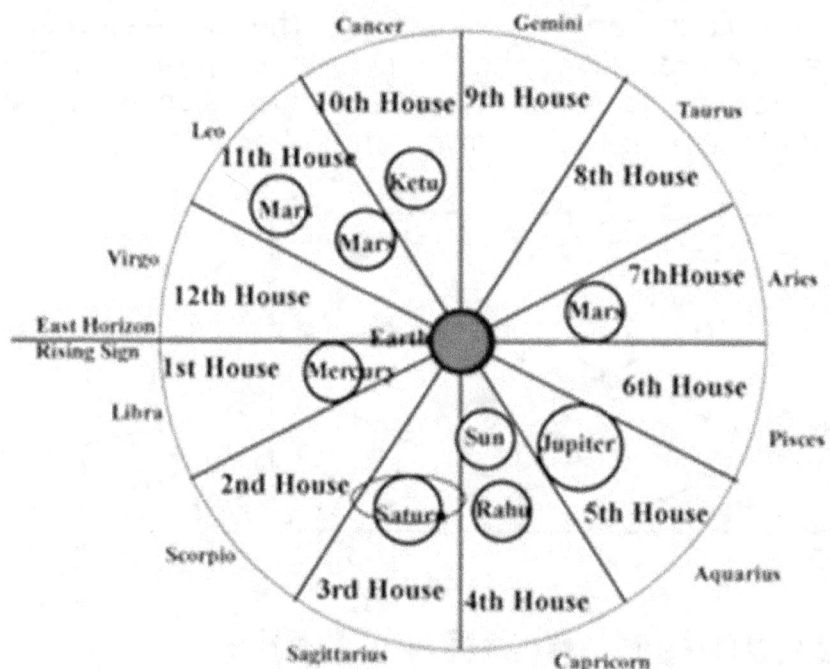

PART-5
CLASSIFICATION OF KUNDALI CHART

Natal Chart: Natal chart is the horoscope drawn for a person's birth. It is also known as a birth chart. The natal chart is a map of one's life, similar to taking a picture of the planets at the time of one's birth. The universe stopped at that moment in time. It reveals what the universe has to say about who a person is and what he or she may become. The chart wheel is a map of the space surrounding a person at the time of birth. The wheel is divided in 12 sections called houses. Planets in the heavens are placed on the chart wheel in the houses that correspond to where they actually are in the sky at birth. The horoscope cast at the birth time of the individual, showing the position of Signs and Planet with respect to house, is called a Natal Chart or Nativity.

Chalit Chart: Chalit Chart shows the actual Zodiac Sign and the actual planet position in a particular house. But Chalit Chart shall not be used for determining the aspects or knowing the sign in which a planet is posited or knowing the strength of a planet. Chalit Chart is the actual calculation of the 12 houses of the Lagna Chart and therefore, actual position of planet by making the house division for corrects predictions, because the planets show their behaviour according to the house they occupy.

Transit Chart: The Transit Chart or Progression Chart is prepared by positioning the rising Sign in 1st House and planets in the particular fouling Sign at that particular time. Then, this Chat is compared to find the aspect of transiting planet on Natal chart planet to find daily, weekly or monthly forecast of the horoscope when the native wish to find the best time for important event happening such as a change of job, the best time for marriage or having children etc. In short, the Transit Charts are a study aid for planetary influences in the individual life during a particular period.

Moon-Sign (Rashi) Chart: In Moon-Sign (Rashi) Chart, the Sign with Moon placed in the first House as Ascendant and rests other Signs with their Planets follow the Ascendant.

Tithi Parivesha Chart: The Tithi Parivesha Chart manifests concretely the Vimshotari Maha Dasa and Antar Dasa sequence. In this chart, the lord of the Vimshotari Maha Dasa is placed in the first House as Ascendant. In relation to this, the inner workings of the Antar Dasa are revealed with the help of the Antar Dasa lord planet positioned in different Houses. The Antar Dasa Lord brings the timing of the events that come about as a result in life. Tithi Parivesha Chart distils all this along with the transit of the Moon in the Rashi Chakra (Rashi-Chart).

Divisional Charts: In Divisional Charts, each Sign is divided into various divisions, as per the requirements of the type of the Chart. The Divisional Charts are used for study of (1) strength of the planets, (2) important aspects of life. Each Chart gives a clear history of one of aspects of life of the native, such as (1) The physique of the native is known by Lagna Chart, (2) wealth by Hora Chart, (3) happiness through co-born/sibling by Drekana Chart, (4) fortunes from Chaturthamsa Chart, (5) sons and grandsons from Saptamsa Chart, (6) spouse and planet strength from Navamsa Chart, (7) power and position from Dasamsa Chart, (8) parents from Dvadasamsa Chart, (9) pleasure and adversities through conveyances from Shodasamsa, (10) worship from Vimsamsa, (11) learning from Chaturvimshamsa, (12) strength and weakness from Saptavimshamsa, (13) evil effects from Trimsamsa, (14) auspicious and inauspicious effects from Khavedamsa, (15) all indications from Akshavedamsa and (16) Shashtiamsa charts. The Divisional Charts are discussed in different volume of the book.

Part One

PART-6
Sign Characteristics

Table 1: Sign Characteristics

Ascendant	Benefic Planets	Malefic Planets	Most Malefic	Neutral Planets
Aries	Mars, Sun, Jupiter,	Venus, Saturn	Mercury	Moon
Taurus	Venus, Sun, Mars, Mercury, Saturn	Moon	Jupiter	--
Gemini	Mercury, Venus, Saturn	Sun, Jupiter	Mars	Moon
Cancer	Moon, Mars, Jupiter	Mercury, Venus	Saturn	Sun
Leo	Sun, Mars, Jupiter	Moon, Mercury, Venus	Saturn	--
Virgo	Mercury, Venus, Saturn	Sun, Moon, Jupiter	Mars	--
Libra	Venus, Mercury, Saturn	Sun, Moon	Jupiter	Mars
Scorpio	Mars, Sun, Moon, Jupiter	Venus	Mercury	Saturn
Sagittarius	Jupiter, Sun, Mars	Moon, Mercury,	Venus,	--

		Saturn		
Capricorn	Saturn, Mercury, Venus	Moon, Mars, Jupiter	Sun	--
Aquarius	Saturn, Sun, Mars, Venus	Moon, Mercury	Jupiter	--
Pisces	Jupiter, Moon, Mars	Sun, Mercury, Saturn	Venus	--

Table 2: Sign Characteristics

Lagna (Ascendant) Sign	Death inflictor (Maraka Graha)	Raja Yoga Karaka Graha	Neutral Planets	Badhaka Rashi (Signs)
Aries (Mesha)	Mercury, Venus	Jupiter	Venus	Aquarius
Taurus (Vrishabha)	Jupiter, Venus, Moon	Saturn	Sun	Scorpio
Gemini (Mithuna)	Mar, Sun, Moon	Saturn	Moon	Leo
Cancer (Karka)	Venus, Saturn, Mercury	Mars	Venus	Taurus
Leo (Simha)	Saturn, Venus, Moon	Mars	Mercury	Aquarius
Virgo (Kanya)	Jupiter, Moon, Sun	Venus	Moon	Scorpio
Libra (Tula)	Jupiter, Sun	Mercury, Rahu	Mars	Leo
Scorpio (Vrischika)	Mercury, Venus, Saturn	Moon	Saturn	Taurus

Sagittarius (Dhanu)	Venus, Moon, Mercury	Sun, Ketu	Mercury	Aquarius
Capricorn (Makara)	Mars, Jupiter	Mercury	Mars	Scorpio
Aquarius (Kumbha)	Sun, Jupiter, Moon	Venus	Sun	Leo
Pisces (Meena)	Saturn, Venus, Sun, Mercury	Mars	Mercury	Taurus

Table 3: Sign Characteristics

Lagna Sign	Person Nature (Guna)	Cause of Death	Person's Nature	Varna
Aries (Mesha)	Rajasic	High Fever	Violent	Kshatriya (Warrior)
Taurus Vrishabha	Rajasic	Fire, Weapon	Auspicious	Sudra (Service Person)
Gemini (Mithuna)	Rajasic	Cataract, Asthma, Mental Deviation Loss of Appetite	Violent	Vysya (Trader)
Cancer (Karka)	Sathwic	Cholera	Auspicious	Brahmana (Intellectual)
Leo (Simha)	Sathwic	Wild Beast, Fever, Boils, Enemies	Violent	Kshatriya (Warrior)
Virgo (Kanya)	Thtamasic	Women, Venereal	Auspicious	Sudra (Service

		Disease, Fall from height		Person)
Libra (Tula)	Sathwic	Brain Fever, Typhoid	Violent	Vysya (Trader)
Scorpio (Vrischika)	Sathwic	Jaundice	Auspicious	Brahmana (Intellectual)
Sagittarius (Dhanu)	Sathwic	Tree, Water, Wood, Weapon	Violent	Kshatriya (Warrior)
Capricorn (Makar)	Thamasic	Stomach Ache, Loss of Appetite	Auspicious	Sudra (Service Person)
Aquarius (Kumbha)	Thamasic	Cough, Fever, Consumption	Violent	Vysya (Trader)
Pisces (Meena)	Sathwic	Drowning	Auspicious	Brahmana (Intellectual)

Table 4: Sign Characteristics

Lagna Sign	Sign Lord	Affecting Disease	Affected Part of Body
Aries (Mesha)	Mars	Bile (Pitta)	Head, Face, & Brain
Taurus (Vrishabha)	Venus	Cold (Sleshma)	Neck, Throat, & Gland
Gemini (Mithuna)	Mercury Rahu	Gas (Vata)	Shoulder, Lungs, Hand, Blood, & Hand's Bone
Cancer (Karka)	Moon	Cold	Chest, Breast, Stomach,

			Shoulder's Bones,
Leo (Simha)	Sun	Bile	Back, Waist, Heart, Spinal Bones
Virgo (Kanya)	Mercury	Gas	Lever, Back Bones, Tili, Gurda
Libra (Tula)	Venus	Cold	Skins
Scorpio (Vrischika)	Mars Pluto	Bile	Penis, Thighs, Nectar
Sagittarius (Dhanu)	Jupiter Ketu	Gas	Waist, Veins
Capricorn (Makar)	Saturn	Gas	Knees, Bone's joints
Aquarius (Kumbha)	Saturn Uranus	Gas	Feet, Digestive Organs
Pisces (Meena)	Jupiter	Gas	Ankles, Palms

Table 5: Sign Characteristics

Lagna Sign	Aspect on other Signs	Sign Age	Relation
Aries (Mesha)	Leo, Scorpio Aquarius	28 ½ YEARS	Chandra Sadga
Taurus (Vrishabha)	Libra, Cancer Capricorn	18 YEARS	Chandra Sadga
Gemini (Mithuna)	Sagittarius, Virgo Pisces	33 ½ YEARS	Chandra Sadga
Cancer (Karka)	Taurus, Libra Capricorn	40 YEARS	Chandra Sadga
Leo	Aries,	28 ½	Surya Sadga

(Simha)	Scorpio Aquarius	YEARS	
Virgo (Kanya)	Gemini, Sagittarius Pisces	18 YEARS	Surya Sadga
Libra (Tula)	Taurus, Cancer Capricorn	33 ½ YEARS	Surya Sadga
Scorpio (Vrischika)	Aries, Aquarius Leo	40 YEARS	Surya Sadga
Sagittarius (Dhanu)	Gemini, Virgo Pisces	28 ½ YEARS	Surya Sadga
Capricorn (Makar)	Aspect on other Signs	18 YEARS	Surya Sadga
Aquarius (Kumbha)	Leo, Scorpio Aquarius	33 ½ YEARS	Chandra Sadga
Pisces (Meena)	Libra, Cancer Capricorn	40 YEARS	Chandra Sadga

Table 6: Sign Characteristics

Lagna Sign	Sign Gender	Mode	Sign Element (Tatva)	Affected Part of Body
Aries (Mesha)	Male	Odd	Fire	Head, Face, Brain
Taurus (Vrishabha)	Female	Even	Earth	Throat, Gland Right Eye
Gemini (Mithuna)	Male	Odd Dual	Air	Neck, Nose, Lungs, Blood, Hand's Bone Ear,
Cancer (Karka)	Female	Even	Water	Chest, Breast, Stomach,

				Bones Shoulder's,
Leo (Simha)	Male	Odd	Fire	Upper Stomach, Back, Waist, Heart, Spinal Bones
Virgo (Kanya)	Female	Even Dual	Earth	Digestive Organs, Kidney, Lever, Back Bones,
Libra (Tula)	Male	Odd	Air	Skins
Scorpio (Vrischika)	Female	Even	Water	Uterus, Ovary, Penis, Nectar
Sagittarius (Dhanu)	Male	Odd Dual	Fire	Thighs, Waist, Veins
Capricorn (Makar)	Female	Even	Earth	Knees, Feet, Digestive Organs
Aquarius (Kumbha)	Male	Odd	Air	Calf Mussels, Feet, Digestive Organs
Pisces (Meena)	Female	Even Dual	Water	Left Eye, Ankles, Palms

Table 7: Sign Characteristics

Planet	Moola-Trikona Sign	Moola-Trikona. Degree	Detriment Sign
Sun	Leo	0 – 20	Aquarius
Moon	Taurus	4 – 30	Capricorn
Mars	Aries	0 – 12	Libra Taurus
Mercury	Virgo	16 – 20	Sagittarius Pisces

Jupiter	Sagittarius	0 - 10	Gemini Virgo
Venus	Libra	0 – 15	Scorpio
Saturn	Aquarius	0 - 20	Cancer Leo
Rahu	Virgo	--	--
Ketu	Pisces	--	--

Table 8: Sign Characteristics

Planet	Exaltation Sign	Debilitation Sign	Max. Exaltn. Debilitn. Degree	Detriment Sign
Sun	Aries	Libra	10	Aquarius
Moon	Taurus	Scorpio	3	Capricorn
Mars	Capricorn	Cancer	28	Libra Taurus
Mercury	Virgo	Pisces	15	Sagittarius Pisces
Jupiter	Cancer	Capricorn	5	Gemini Virgo
Venus	Pisces	Virgo	27	Scorpio
Saturn	Libra	Aries	20	Cancer Leo
Rahu	Taurus Gemini	Scorpio Sagittarius	--	--
Ketu	Scorpio Sagittarius	Taurus Gemini	--	--

Table 9: Sign Characteristics

Lagna Sign	Element	Mode of Expression	Positive Quality	Negative Quality
Aries	Fire	Cardinal	Vital	Impulsive
Taurus	Earth	Fixed	Stable	Stubborn
Gemini	Air	Mutable (Dual Signs)	Adaptable	Cursory

Cancer	Water	Cardinal	Protective	Jealous
Leo	Fire	Fixed	Authority	Autocratic
Virgo	Earth	Mutable (Dual Signs)	Detailed	Critical
Libra	Air	Cardinal	Diplomatic	Vacillating
Scorpio	Water	Fixed	Resurgent	Ruthless
Sagitt.	Fire	Mutable (Dual Signs)	Discerning	Moralistic
Capri.	Earth	Cardinal	Principled	Miserly
Aquarius	Air	Fixed	Liberal	Eccentric
Pisces	Water	Mutable (Dual Signs)	Charity	Anxiety

Table 10: Sign Characteristics

Lagna Sign	Person's Characteristics	Person's Profession
Aries (Mesha)	Hasty, impulsive, restless, short-tempered	Govt. job, surgeon, mechanics, industrialists, athletes, Police, Military Service, Fire Service, Sports, Engineering, arm manufacturing, trade union leader
Taurus (Vrishabh)	Slow in movement, inclined to ease and luxury, faithful & obedient	Musician, singer, actors, banking, tailors, property dealing, Jewellery business, money lending, commission agent, financial institutions, handicrafts, fancy articles, five star hotels, drama, cinema, music, poet, story writer.
Gemini (Mithuna)	Good speakers,	media and journalism, accountants, translators,

	witty and humorous, inquiring and curious, fond of knowledge, fun seeking,	writers, Information and broad casting, space department, education department, book publishing , mathematics department, auditors, law and order councillor, ambassador.
Cancer (Karka)	Emotional , forgiving, sensitive	Export and Import, naval and marine, fishing, nursing, interior design, food, petroleum, historians, shipping, transport department, agriculture, hotel business.
Leo (Simha)	Dominative, behaves like ruler	Govt. Job, Politics, Administrator, Social Services, Charitable institutions, Engineering, Industry, religion, investing, diplomacy.
Virgo (Kanya)	Intelligent, good speaker, tactful	Auditing, Accounting, Business, Teacher, writer, retail shops, computing, astrology, media, doctors, healing.
Libra (Tula)	Good talker, judicious in dealings	Shop, commission agents, bank, Life insurance, law department, hotel business, bar and Restaurant, Dancing Hall, Beauty parlour, Music, Dance , Cinema, judges, artists, cosmetics, fashion, receptionists, advertising, interior decorating, prostitutes.
Scorpio (Vrischika)	Peevish, straight	Iron Industries, Engineering, and

	forward, likes to hide or run away from people and crowds	Instrument Manufacturing, raw materials, priest, astrology, mantra and tantra, occult practices, chemicals, drugs, liquids, insurance, doctors, nurses, police, occult.
Sagittarius (Dhanu)	Honest, easy going, even-tempered, kind hearted, gambles	Forest department, law, religion, banking and finance, entrepreneurs, athletes, law department, temple, financial institutions, education department, ordnance depot, military training, social service, charitable institutions.
Capricorn (Makara)	Witty, and changeable, good organizer, cautious, secretive, ambitious, preserving, pragmatic	Hotels, food products, engineer, doctor, business, building work, Granite stone and sand business, Labourer like porters, coolies, drivers, shoe polishing, shoe makers, plumber and mining.
Aquarius (Kumbh)	Studious, philosopher, honest, benevolent	advisors, consultants, philosophers, astrologers, engineers, computer, Psychology, Religion, Teaching, Research and Development, Administration, Service in Space Dept., Defence, Fire, Jail, Bomb manufacturing, tourist guide, central excise CBI Dept.
Pisces (Meena)	Lazy, emotional,	doctors, captain, hospital, prisons Education

	timid, honest, talkative, intuitive , psychic, fond of good food and company	Department, Religious Institutions, Medicine, Financial, Law Department, External Affairs, Bank, Navy, shipping, temple worker, priest

Table 11: Sign Characteristics

Name of Sign	Effects	Direction	Progeny Nature	Feature
Aries	Positive	Northern	Barren	Bestial
Taurus	Negative	Northern	Semi-Fruitful	Bestial
Gemini	Positive	Northern	Barren	Human
Cancer	Negative	Northern	Fruitful	Bestial
Leo	Positive	Northern	Barren	Bestial
Virgo	Negative	Northern	Barren	Human
Libra	Positive	Southern	Semi-Fruitful	Human
Scorpio	Negative	Southern	Fruitful	Bestial
Sagittarius	Positive	Southern	Semi-Fruitful	Half Bestial & Human
Capricorn	Negative	Southern	Semi-Fruitful	Bestial
Aquarius	Positive	Southern	Semi-Fruitful	Human
Pisces	Negative	Southern	Fruitful	Half Bestial & Human

CHAPTER-2 ZODIAC SIGN ASTROLOGY

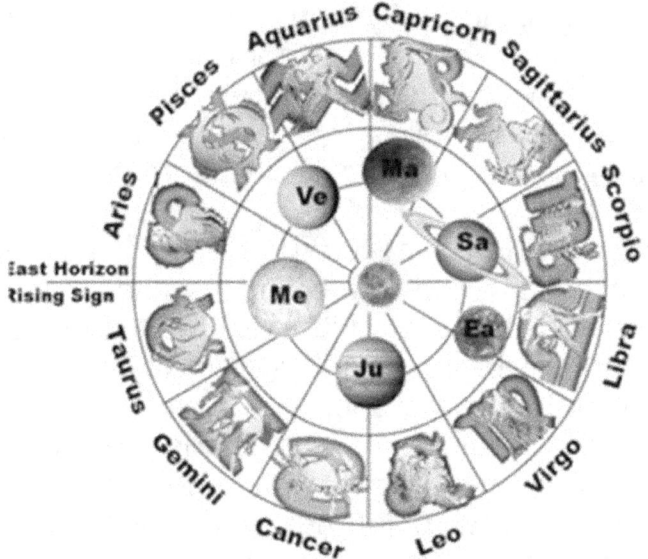

The Zodiac is divided into twelve divisions of 30 degrees each called "Sign".   Each segment is called a Sign (Rashi). The Zodiac is a band of group of stars or the positions of celestial bodies. Zodiac has twelve signs having different characteristics. They act in different ways and are also known for its different nature. There are no incompatible zodiac signs in astrology, which means that any two signs are more or less compatible. Two people whose zodiac signs are highly compatible will get along very easily because they are on the same wavelength. But, people whose zodiac signs are less compatible will need to be more patient and tactful in order to achieve a happy and harmonious relationship. As we all know, zodiac signs belong to four elements:
Fire: Aries, Leo, Sagittarius

Earth: Taurus, Virgo, Capricorn
Air: Gemini, Libra, Aquarius
Water: Cancer, Scorpio, Pisces
Signs that have the same element are naturally compatible because they understand each other best, and in addition, Air is highly compatible with Fire, and Water is highly compatible with Earth. The strongest attraction is expected in opposing signs and their potential is always great. Synastry is a branch of astrology where two natal charts are compared in order to determine the quality of the love connections between zodiac signs. Synastry or a relationship horoscope can be a useful tool for partners who want to know the strengths and weaknesses in their relationship. Comparing signs can also help in gaining a better understanding of the partner, which will result in a better relationship.

Gender (Male/Female) of Sign: The alternate Sign starting from Aries onwards are known as male and female on the other hand. The Gender of the Sign will help the Astrologer in assessing one's children, brothers and sisters in terms of Males and Females, from the horoscope.

Directions of Signs: The four Signs from Aries onwards indicate East, South, West and North, while the remaining Signs repeat in the same way. A journey undertaken by a person towards the direction indicated by the Lagna or the Moon at the commencement of journey yields fruitful results.

Night and Day Signs: Gemini, Cancer, Capricorn Aries, Taurus and Sagittarius are night Signs. Leo, Libra, Scorpio, Aquarius, Pisces and Virgo are day Signs.

Strength of Signs: If a Sign is aspect by its Lord, or by a planet friendly to its Lord, or by Mercury, or by Jupiter, it is said to be Strong Sign. Planets other than the above do not lend strength by aspect.

Lagna Sign: The Rising sign, at a particular time, is the sign of the zodiac positioned on the eastern horizon on the cusp of the first house at birth and is called Lagna. The Lagna lord is Atmakarka and is considered more powerful in Lagna.

Badhaka Signs: Similarly, there are Badhaka Rashi defined as per the Janma-Lagna. It works as the Badhaka Rashi and harms badly that house in which it is occupying.

Gandanta of Sign: The ending portions, 30th degree of Cancer, Scorpio and Pisces are called Gandanta. It is said, that one born in Gandanta will not survive. He will either lose his mother, or he will end the dynasty, i.e. he is the last of his descent and will not have any children. If, however, he survives, he becomes a king with many elephants and horses.

PART-1
ZODIAC SIGN ARIES
HOROSCOPE

The Flying Ram Guided by the story of the Golden Fleece, an Aries is ready to be the hero of the day, fly away and carry many endangered, powerless people on their back. The power of the ram is carried on his back, for he is the gold itself, shiny and attractive to those ready for betrayal. The story of glory that isn't easy to carry is in these two horns, and if this animal doesn't get shorn, allowing change and giving someone a warm sweater, they won't have much to receive from the world. Each Aries has a task to share their position, power, gold, or physical strength with other people willingly, or the energy will be stopped in its natural flow, fear will take over, and the process of giving and receiving will hold balance at zero. As the first sign in the zodiac, the presence of Aries always marks the beginning of something energetic and turbulent. They are continuously looking for dynamic, speed and competition, always being the first in everything - from work to social gatherings. Thanks to its ruling planet Mars and the fact it belongs to the element of Fire, Aries is one of the most active zodiac signs. It is in their nature to take action, sometimes before they think about it well. The Sun in such high dignity gives them excellent organizational skills, so you'll rarely meet an Aries who isn't capable of finishing several things at once, often before lunch break! Their challenges show when they get impatient, aggressive and vent anger pointing it to other people. Strong personalities born under this sign have a task to fight for their goals, embracing togetherness and teamwork through this incarnation.

Aries ruling: Aries rules the head and leads with the head, often literally walking head first, leaning forwards for speed and focus. Its representatives are naturally brave and rarely afraid of trial and risk. They possess youthful strength and energy, regardless of their age and quickly perform any given tasks.

Strengths: Courageous, determined, confident, enthusiastic, optimistic, honest, and passionate

Weaknesses: Impatient, moody, short-tempered, impulsive, and aggressive

Aries likes: Comfortable clothes, taking on leadership roles, physical challenges, individual sports

Aries dislikes: Inactivity, delays, work that does not use one's talents

Element: Fire

Quality: Cardinal

Color: Red

Day: Tuesday

Ruler: Mars

Lucky Numbers: 1, 8, 17

Aries most compatible: Libra, Leo

Aries Love and Sex: Aries is a fire sign with the need to take initiative when it comes to romance. When they fall in love, they will express their feelings to the person they are in love with, without even giving it a considerable thought. The compatibility of an Aries with other signs of the zodiac is very complex. Aries in love may shower their loved one with affection, sometimes even an excess of it, forgetting to check the information they get in return. They are very passionate, energetic and love adventures. An Aries is a passionate lover, sometimes even an addict to pleasures of the flesh and sexual encounters. With their opposing sign being Libra, the sign of relating, tact, and diplomacy, it is the furthest point from their natural personality. This can present a problem in their romantic experience, for they don't seem to have enough patience and focus on their partner, as much as they do on the passionate approach they always nurture. They have to embrace all matters of Venus, with all of its love, tenderness, joy, peaceful satisfaction, and foreplay. Still, their partner should keep in mind that they need the adrenaline and excitement every day, and their relationship can only be strong and long lasting if their primal needs are met.

Aries Friends: Social life of an Aries representative is always moving, warm, and filled with new encounters. They are tolerant of people they come in contact with, respectful of different personalities and the openness they can provoke with simple presence. Their circle of friends needs a wide range of strange individuals, mostly in order for them to feel like they have enough different views on personal matters they don't know how to resolve. Since people born in the sign of Aries easily enter communication, direct and honest in their approach, they will make an incredible number of connections and acquaintances in their lifetime. Still, they often cut many of them short for dishonesty and unclear intentions. Long-term friendships in their lives will come with those who are just as energetic and brave to share their insides at any time.

Aries Family: Independent and ambitious, an Aries often knows where they want to go at a young age, separating from their family a bit early. Even as children they can be hard to control, and if they don't receive enough love and patience from their parents, all of their intimate bonds later in life could suffer. A lot of anger comes from the sign of Aries if too many restrictions come their way, and only when they come from liberal families will they nurture their bonds with an easy flow. Even when this isn't the case, they will take on family obligations when they need to be taken care of, never refusing more work as if their pool of energy is infinite.

Aries Career and Money: This is an area of life in which an Aries shines brightest. Their working environment is the perfect place for their ambition and creativity to show, with them fighting to be as good as possible. A natural born leader, Aries will prefer to issue orders rather than receive them. Their speed of mind and vast energy to move helps them to always be one step ahead of everyone else. All they need to do in order to succeed is follow their chosen path and not give up on professional plans guided away by emotions. When faced with a challenge, an Aries will quickly assess the situation and come to a solution. Competition does not bother them and instead encourages them to shine even brighter. They can have great careers in sports and challenging environments, and enjoy their chosen path as managers, policemen, soldiers, etc. Even though Aries representatives can be wise and save some money for a rainy day, this is not often the case for the joy of spending it and taking risks is even greater. They live in the present and aren't that focused on the future, and this can make them irrational and hasty when it comes to financial decisions. Still, they seem to always find a way to earn money and compensate for what they have spent, in a natural flow of energy that needs to come back when invested wisely.

Aries Man: Independence is the key to understanding an Aries, for they don't like to take orders from others. In order to seduce an Aries man, you need to learn to play the game by his rules. This man often finds the chase for the subject of his desire more thrilling than the catch, and his conquering nature makes him often chase after partners he can't have. To get his attention, one must play hard to get, as if sending a message that he needs to fight for a prize, and winning the one he truly wants to be with. This is a man in love with a good challenge and in a rush to become their partner's "knight in shining armour", so he needs to be left to be one from time to time. His life partner might have to yell back in a fight, building strong boundaries and earning his respect. On a bad day an Aries can be self-centred, arrogant and stubborn, but he is also courageous, adventurous, and passionate. A relationship with this man can be fun and exciting, but it easily gets someone hurt if their partner doesn't recognize the energy needed for their relationship to last.

Aries Woman: Aries women are fearless and natural leaders. They are energetic, charismatic, dynamic, and in love with challenges and adventures. If you want to attract the attention of an Aries woman, you must let her seduce you and appeal to her independent nature. A woman born under the Aries zodiac sign is extremely passionate and sexual, which makes her irresistible to the opposite sex. She is constantly on the move and will never allow herself to be overrun by a man, at the same time craving for love but trying to hold on to control. To attract a woman born in this sign, one has to take action but not give the impression that control has been taken over. She needs to be free to show initiative and fight for affection of her loved one, expecting the same in return. Once she falls in love, she is extremely faithful, and at times overly jealous. Dating her means giving her all the attention she needs, giving her time and constant effort to prove there is love behind the act. Confident and domineering, she doesn't just need someone to follow, but someone to be equally energetic and strong. A relationship with an Aries woman can be interesting, full of adventures and excitement, but only if one is ready to take on a less dominant role from time to time.

Aries   Compatible
Signs: Sagittarius, Leo, Aquarius, Gemini, Libra
Famous Aries:
A.P.J. Abdul Kalam.

Bob Woodward
Charlie Chaplin,
Diana Ross,
Elton John,
Eric Clapton,
Hrithik Roshan.
Kareena Kapoor.
Leonardo DaVinci,
Mahatma Gandhi.
Marlon Brando,
Priyanka Chopra.
Quentin Tarantino,
Ranbir Kapoor.
Robert Downey Jr.,
Russel Crowe,
Shah Rukh Khan.
Sundar Pichai.
Thomas Jefferson,
William Shatner,

PART-2
ZODIAC SIGN TAURUS HOROSCOPE

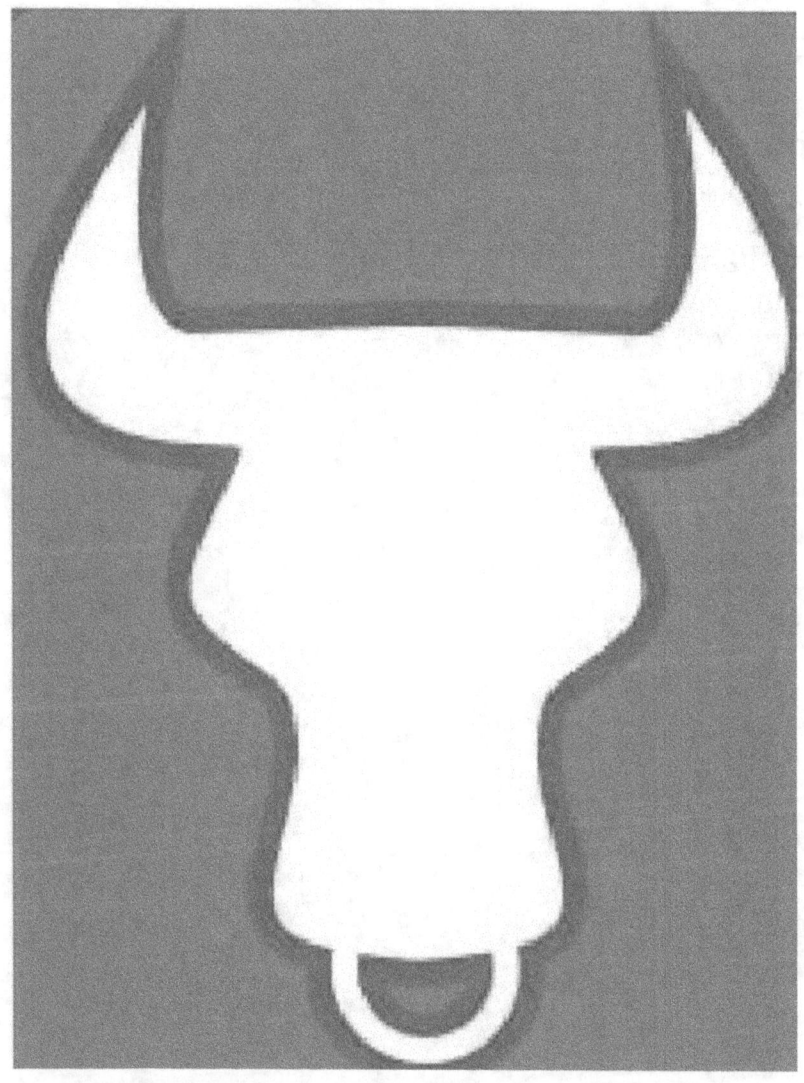

The Wandering Bull being the one who betrayed their best friend, goddess Hera herself, this is an unfortunate being that has to wander the Earth in order to find freedom. As if something was always poking them behind their back, reminding them of happiness that once was, stinging and pushing forwards, they close up in their own worlds, lonely and separated from their core. To find love, a Taurus has to travel the world, change perspective or make a shift in their entire belief system and their system of values. Practical and well-grounded, Taurus is the sign that harvests the fruits of labour. They feel the need to always be surrounded by love and beauty, turned to the material world, hedonism, and physical pleasures. People born with their Sun in Taurus are sensual and tactile, considering touch and taste the most important of all senses. Stable and conservative, this is one of the most reliable signs of the zodiac, ready to endure and stick to their choices until they reach the point of personal satisfaction. Taurus is an Earth sign, just like Virgo and Capricorn, and has the ability to see things from a grounded, practical and realistic perspective. They find it easy to make money and stay on same projects for years, or until they are completed. What we often see as stubbornness can be interpreted as commitment, and their ability to complete tasks whatever it takes is uncanny. This makes them excellent employees, great long-term friends and partners, always being there for people they love. Earthly note makes them overprotective, conservative, or materialistic at times, with views of the world founded on their love of money and wealth. The ruler of Taurus is Venus, the planet of love, attraction, beauty, satisfaction, creativity and gratitude. This tender nature will make Taurus an excellent cook, gardener, lover, and artist. They are loyal and don't like sudden changes, criticism or the chase of guilt people are often prone to, being somewhat dependable on other people and emotions they seem to be unable to let go of. Still, no matter their potential emotional challenge, these individuals have the ability to bring a practical voice of reason in any chaotic and unhealthy situation.

Element: Earth
Quality: Fixed
Color: Green, Pink
Day: Friday, Monday

Ruler: Venus

Lucky Numbers: 2, 6, 9, 12, 24

Taurus most compatible Zodiac Sign: Scorpio, Cancer

Strengths: Reliable, patient, practical, devoted, responsible, stable

Weaknesses: Stubborn, possessive, uncompromising

Taurus likes: Gardening, cooking, music, romance, high quality clothes, working with hands

Taurus dislikes: Sudden changes, complications, insecurity of any kind, synthetic fabrics

Taurus Love and Sex: One always has to be prepared to have patience for a Taurus lover. They are extremely sensual, touch, smell and all pleasurable senses being extremely important to them, but they also need time to create a safe environment and relax in their sexual encounters. When they create enough intimacy with a loved one, they become a bit gooey, sometimes even needy, and have to keep their emotions in check, holding on to practical reasoning, while embracing change and initiative of their partner at all times. For long-term relationships they often choose people from the same social environment that are able to respond to their intellectual needs, but also the expectations of their family and close friends. Holding on to traditional values and the practical side to life, this is a sign that rarely chooses a partner who won't fulfill basic expectations of their upbringing, often showing and receiving attention through gifts and material things. If they stick to the moral code too stiffly and refuse all taboos and adventurous approaches, they could end up swimming in frustration and anger issues they don't know how to resolve, often manifesting through the person standing in front of them. Taurus compatibility with other signs can be complicated. This is a sign of physical pleasure, hedonism and the flow of emotion that isn't reserved for just anyone.

Taurus Friends: People born in this sign are loyal and always willing to lend a hand of friendship, although they can be closed up for the outer world before they build trust for new social contacts they make. Many of their friendships begin in childhood with a tendency to last them a lifetime. Once they make a clear intimate connection to another person, they will do anything they can to nurture the relationship and make it functional even in the hard times.

Taurus Family: Home and matters of the family are very important to every Taurus. This is a person who loves kids and appreciates time spent with people who love them, respecting family routines, customs, and present in all events and gatherings. They will enjoy hosting house parties for both their family and friends, and don't mind cooking a meal for a room full of people if they just have fun in return.

Taurus Career and Money: Taurus representatives usually love money and will work hard in order to earn it. They are reliable, hardworking, patient and thorough, as an employee or someone in a position of power. When focused on a specific project, they will firmly stick to it, no matter what happens in the world around them. Stability is the key to understand their working routine. The search for material pleasures and rewards is an actual need to build their own sense of value and achieve a satisfying luxurious, yet practical way of life. Their job is observed as a means to make it possible. Taurus is a Sun sign well organized with their finances, and all of their bills will be paid without delay. They care for their pension, taking responsibility and saving some money for a rainy day, able to make due with a really small and a really big salary just the same. Occupations that fit them are agriculture, banking, art, and anything that involves culinary skills.

Taurus Man: If you are in search for a strong, loyal and generous man, Taurus is the person you are looking for. He is trustworthy, patient and tender when in love, always in search for a returned emotion. He will not pick on subtle hinds and suggestive looks from those who flirt with him, being a bit slow on the uptake as if waiting for someone to ask them out. He dislikes artificiality of any kind, and values conversations filled with genuine statements, especially when it comes to compliments and love declarations. A Taurus man needs time to build trust and anyone on a chase for his heart needs to take the time earning it. As a person of very few words, he will seem impossible to penetrate at times, as if nothing can touch him. An invitation for a delicious home-cooked mean is always a safe bet when dating this man, as well as choosing a place that is comfortable and cosy, rather than popular or modern. Turned to nature and common thinking, he will see sex as something that comes when the time is right, rarely puts any pressure on his partner and feels like it is something to be enjoyed, not so much something to crave for. A part of his fixed, static character is the potential inability to forgive betrayal, and he needs to feel truly safe to settle down with one partner for good.

Taurus Woman: If you want to seduce a woman born with her Sun in Taurus, you will need to appeal to her sense of romance. Taurus women want to be courted and slowly seduced, even when they have already decided to enter a relationship with someone. They need things to move slowly, and will rarely jump into a sexual bond quickly and without thinking long and hard about her choices. A Taurus woman longs for true love and security. It is very unlikely that she will give into her desires and instincts quickly, and if someone wishes to have her heart, they will have to spend a lot of time and energy into the game of winning her over, making her feel comfortable. Once she falls in love she becomes affectionate, intimate, close and loyal, standing by her partner for as long as he is faithful to her. She has an eye for beautiful things and appreciates simplicity of fine things in life, so the way to approach her is through enjoyable shared moments, respect for privacy, fine food and a gentle touch. This is a woman who doesn't like to feel rushed when dating and needs to have her time. Once she feels comfortable and secure with someone, she will happily and quickly give her heart without holding back.

Taurus compatible Signs: Cancer, Virgo, Capricorn, Pisces

Al Pacino,
Andre Agassi,
Barbara Streisand,
Bono,
Carol Burnett,
Dennis Hopper,
Famous Taurus:
Farida Jalal

Immanuel Kant
James Brown,
John Wilkes Booth,
Karl Marx, Cher,
Madhuri Dixit
Moushmi Chatterjee
Puja Bedi
Rabindra Nath Tagore
Sachin Tendulkar
Salvador Dali,
Sri Sri RaviShankar

PART-3
ZODIAC SIGN GEMINI HOROSCOPE

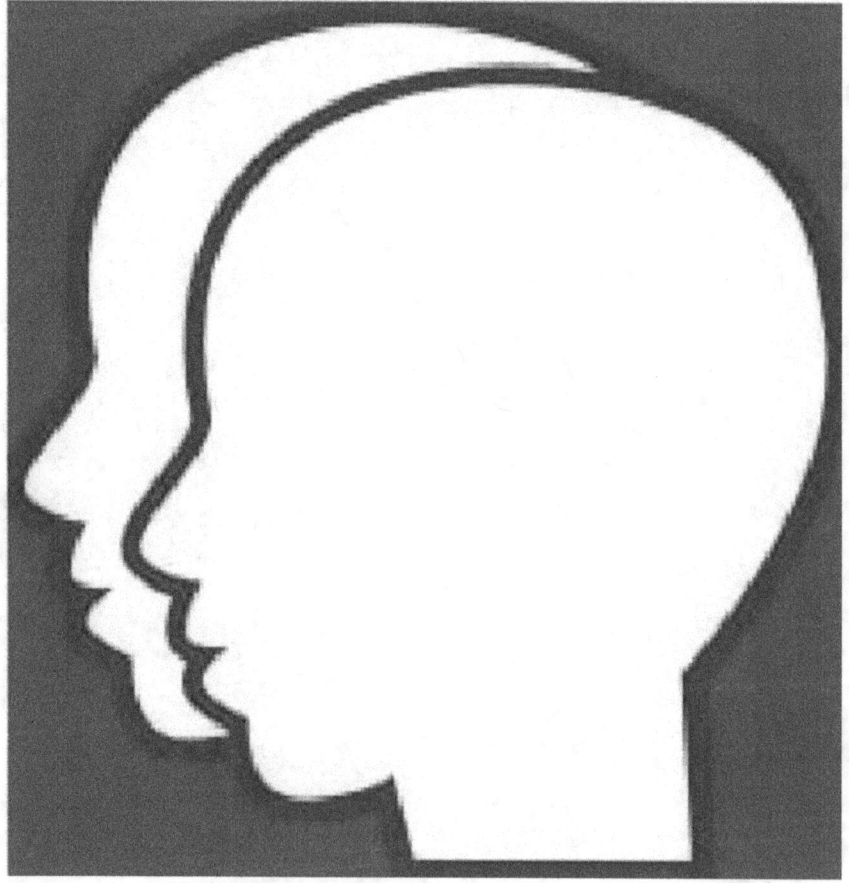

The Caring Twins There is so much childish innocence in the nature of Gemini, telling their tale of brotherhood, love between best friends and relatives who are entirely different by character, circumstances, physical appearance or upbringing. They are in this world to mend differences and make them feel right, ready to give their life for a brother or a friend. Gemini Love and Sex Fun and always ready for an intellectual challenge, Gemini sees love first through communication and verbal contact, and find it as important as physical contact with their partner. When these two combine, obstacles all seem to fade. Inquisitive and always ready to flirt, a Gemini could spend a lot of time with different lovers until they find the right one who is able to match their intellect and energy. They need excitement, variety and passion, and when they find the right person, a lover, a friend and someone to talk to combined into one, they will be faithful and determined to always treasure their heart. Expressive and quick-witted, Gemini represents two different personalities in one and you will never be sure which one you will face. They are sociable, communicative and ready for fun, with a tendency to suddenly get serious, thoughtful and restless. They are fascinated with the world itself, extremely curious, with a constant feeling that there is not enough time to experience everything they want to see. The sign of Gemini belongs to the element of Air, accompanying Libra and Aquarius, and this connects it to all aspects of the mind. It is ruled by Mercury, the planet that represents communication, writing, and movement. People born under this Sun sign often have a feeling that their other half is missing, so they are forever seeking new friends, mentors, colleagues and people to talk to. Gemini's changeable and open mind makes them excellent artists, especially writers and journalists, and their skills and flexibility make them shine in trade, driving and team sports. This is a versatile, inquisitive, fun loving sign, born with a wish to experience everything there is out there, in the world. This makes their character inspiring, and never boring.

Element: Air
Quality: Mutable
Color: Light-Green, Yellow
Day: Wednesday
Ruler: Mercury
Lucky Numbers: 5, 7, 14, 23
Gemini most compatible Zodiac Sign: Sagittarius, Aquarius
Strengths: Gentle, affectionate, curious, adaptable, ability to learn quickly and exchange ideas

Weaknesses: Nervous, inconsistent, and indecisive

Gemini likes: Music, books, magazines, chats with nearly anyone, and short trips around the town

Gemini dislikes: Being alone, being confined, repetition and routine

Gemini Love and Sex: Fun and always ready for an intellectual challenge, Gemini sees love first through communication and verbal contact, and find it as important as physical contact with their partner. When these two combine, obstacles all seem to fade. Inquisitive and always ready to flirt, a Gemini could spend a lot of time with different lovers until they find the right one who is able to match their intellect and energy. They need excitement, variety and passion, and when they find the right person, a lover, a friend and someone to talk to combined into one, they will be faithful and determined to always treasure their heart. The biggest challenge for any Gemini's love life is to find an emotion that lasts, especially as they get older and realize that they are already in a repetitive mode of superficial or disappointing bonds. Their personality doesn't allow much depth, for they are on a mission to spread information, not to dig into them and find mistakes, holes, or resolutions. They look at life from a relative perspective of movement, being the one to circle the Sun, moving forwards and backwards from Earth's point of view, never certain of their own direction. Gemini might make sharp turns, leaving those who love them behind, but there are partners that could follow in their pace, ready to build a loving foundation through time.

Gemini Friends: Those born with their Sun in Gemini are very social and love to spend time with friends and family, especially its younger members. A Gemini has an abundance of social contacts and loves to chat, search for understanding, always looking for strong willed people to communicate with. Without a clear flow of words spoken, they will quickly lose interest in the entire theme of any conversation, and need to stay on the move, feeling inspired and pushed forwards by the information shared.

Gemini Family: Family is very important to a Gemini, especially their children once they build a strong emotional bond with them. Lack of stability they show to their partners with their expectations extremely high, don't reflect on family as much, and they seem to have a more modest and calm approach to those he shared a home with. Although responsibilities carried by their family life can stand to be a challenge for their nature, they will find a magical way to be in two places at once, getting everything done just as they are supposed to.

Gemini Career and Money: In constant need of intellectual stimulation, the most suitable job for a Gemini has to be challenging to their brain. They are skilful, inventive and often very smart, with a need for a dynamic working environment and a lot of social contacts met in the office. The best careers they can choose are those of traders, inventors, writers, orators, preachers and lawyers, but any career that gives them the opportunity to communicate freely while keeping them on the move and busy at all times, is an excellent choice. As if they were created for multitasking, problem solving and bringing new ideas to life, they need a workplace that won't keep them stuck in a routinely, repetitive tasks that don't allow them to shine. Deciding between practicality and pleasure can be a difficult choice for Gemini. Even though they believe that money is just a necessary evil, most of them will not spend much time thinking where to earn it or how they spent it. They need strong grounding to keep their finances in check and organized, giving them a sense of confidence and security they often don't even know they need.

Gemini Man: A Gemini man is enthusiastic and full of life, never disappointing with dull moments. He is adventurous and humorous, and this makes him the perfect partner if a person is energetic and dynamic, in search for some laughter and fun. Gemini men are naturally chatty and flirtatious, and you can meet them at public gatherings, conferences, and traffic jams. Their personality is marked by dualism, making them inconsistent but clever, and amazingly attractive to others. This man is impossible to cling to, and need a partner who gives him enough freedom and space, followed by mental stimulation and variety. To win his heart, one has to be fun, stimulating, adventurous, laughing at his jokes and ready to learn from him day after day. As if opposed to his eloquent nature, this isn't a man ready to discuss emotions that much, and will prefer if they are shown rather than spoken of. Sex with a Gemini man can be a wonderful experience, but if his partner is unwilling to experiment, he will get bored. As in all things in life, he needs new experiences, verbal contact, and freedom of expression when it comes to sexual relationships.

Gemini Woman: If you want to attract a Gemini woman, you will have to be able to keep up with her dual nature. She can be passionate and gentle one moment, and aloof and distant the next. This is a result of her natural born tendency to stay safe and on a distance from other people, prepared to run off into a carefree love story that waits for her just around the corner. This is an enthusiastic, witty, intellectual and soft spoken woman, while at the same time extremely open-minded and always ready to meet someone new. Although a Gemini woman is usually not very shy, getting in a serious and committed relationship will take time and a lot of patience. However, once she finds a man who can satisfy her sexual and intellectual desires, this woman will be the one to suggest starting a family, marriage, and growing old together, although this might happen in an unusual way. She is impressed by partners who teach her new things and have insights that she sees and ingenious. Her sex life is a story to be told, but only to those who are ready to listen, usually the one specific person she finally managed to build true intimacy with.

Gemini   compatible
Signs: Aries, Leo, Libra, Aquarius, Sagittarius
Allen Ginsberg,
Ameesha Patel
Bob Hope,
Boy George,
Ekta Kapoor
Famous Gemini:
Henry Kissinger,
Joe Montana
Karan Johar
Mithun Chakraborty

Nicole Kidman,
Norman Vincent Peale,
Paul Gauguin,
Prince,
Ralph Waldo Emerson,
Sandra Bernhard,
Shilpa Shetty
Shweta Shetty
Sonakshi Singha
Sonam Kapoor
Steffi Graf,
Stevie Nicks,
William Butler Yeats,

PART-4
ZODIAC SIGN CANCER
HOROSCOPE

The Brave Crab Sent to this Earth by something they believe in, only to mess with someone bigger than they are, this isn't an animal aware of their strength. Patriotism can make them endanger their own wellbeing, fighting for someone else's cause, as if others can become their higher power. The Crab knows where they're going, but this is often in a wrong direction, at least until they learn their lessons and start relying solely on themselves. Deeply intuitive and sentimental, Cancer can be one of the most challenging zodiac signs to get to know. They are very emotional and sensitive, and care deeply about matters of the family and their home. Cancer is sympathetic and attached to people they keep close. Those born with their Sun in Cancer are very loyal and able to empathize with other people's pain and suffering. The sign of Cancer belongs to the element of Water, just like Scorpio and Pisces. Guided by emotion and their heart, they could have a hard time blending into the world around them. Being ruled by the Moon, phases of the lunar cycle deepen their internal mysteries and create fleeting emotional patterns that are beyond their control. As children, they don't have enough coping and defensive mechanisms for the outer world, and have to be approached with care and understanding, for that is what they give in return. Lack of patience or even love will manifest through mood swings later in life, and even selfishness, self-pity or manipulation. They are quick to help others, just as they are quick to avoid conflict, and rarely benefit from close combat of any kind, always choosing to hit someone stronger, bigger, or more powerful than they imagined. When at peace with their life choices, Cancer representatives will be happy and content to be surrounded by a loving family and harmony in their home.

Element: Water
Quality: Cardinal
Color: White
Day: Monday, Thursday
Ruler: Moon
Lucky Numbers: 2, 3, 15, 20
Cancer most compatible Zodiac Sign: Capricorn, Taurus
Strengths: Tenacious, highly imaginative, loyal, emotional, sympathetic, and persuasive
Weaknesses: Moody, pessimistic, suspicious, manipulative, insecure
Cancer likes: Art, home-based hobbies, relaxing near or in water, helping loved ones, a good meal with friends
Cancer dislikes: Strangers, any criticism of Mom, revealing of personal life

Cancer Love and Sex: Cancer is a very emotional sign, and feelings are the most important thing in their relationships. Gentle and caring, they will show their sensibility to the world without even thinking they might get hurt. For partners, they always choose a person who is able to understand them through non-verbal, silent contact, and a shared daily routine, and their affection won't last long with superficial, flaky or unreliable partners. The lack of initiative these individuals suffer from won't make it easy for them to build a sex life they wish for, if they don't find a partner who is able to make them feel calm, protected, and free to express. This is a dedicated sign, ready to make many unhealthy compromises only to keep their image of a family going, and could choose partners who are in a way selfish of abusive. Shared responsibility and a life together with their partner makes them feel secure and ready for the next step in life, no matter if it is a child, a new job, or simply a cleanup in the field of friendships and relationships that became obsolete or hurtful. In love with children, parenthood, marriage and traditional values, they can still be misguided by people they admire and trust into changing their honest approach to a modern one that doesn't fit their true personality. Compassion and understanding that a Cancer chose to send your way shouldn't ever be taken for granted, or they will show you just how bad of a match you two are in the long run.

Cancer Friends: When it comes to friendships, Cancer representatives will gladly connect with new social contacts, but are extremely sensitive of people not approved by their closest surrounding. Filled with respect for people they communicate to easily, they see all contacts through their emotional prism rather than simple curiosity or status. Most of all they enjoy socializing at home, where intimate atmosphere can be made and deep understanding shared in circumstances under their control. Intuitive and compassionate, they are sometimes impossible to understand from an extremely rational point of view.

Cancer Family: Cancer is the sign of family and these individuals care about family bonds and their home more than any other sign of the zodiac. Deeply sentimental, they tend to diligently preserve family memories, keeping them intact for years. When their personal lives are fulfilled, they make wonderful, caring parents that seem to know how they children feel even when they are miles apart.

Cancer Career and Money: When a job needs to get done, a Cancer will roll their sleeves up and finish it successfully. If they are left alone to work, they usually perform better than when surrounded by other people, loyal to their employer and focused on the task. They will have great careers as nurses, housekeepers, gardeners, politicians and decorators.

For Cancer representatives, security and money are of great importance and stand for the real reason they work as much as they do. They easily earn money and aren't used to spending it all in one day. It is their goal to save, invest, and watch their investments grow daily. Resourceful and good at managing time and finances, this is a sign that is often in charge of all money in the household, keeping their partner or other family members under control.

Cancer Man: A Cancer man is conservative just enough to know that initiative is important, but often fails to show it before he feels safe to do so. His partners need to make the first move, but still doing it subtly to let him still feel like he is leading the way. This is a complex individual, very sensitive, shy and overly protective of his loved ones. When he prefers women, he will subconsciously search for the ideal wife and mother. A Cancer man is an emotional person, who loves to take care of other people. He wants to feel needed, and protective, receiving a lot of attention from his partner through kind words and subtle concerns and compliments to make his day. Although he can be moody, pessimistic and clingy, he is a creative and generous partner in search for someone to share a life with.

Cancer Woman: Cancer personality can be quite complicated, but deep inside they are home-loving and conservative people. A Cancer woman is vulnerable, emotional, and not likely to quickly fall in love. Once her trust is earned, she will be passionate and loyal. In order to seduce her, one has to be proactive and make the first move, respecting her need to be treated like a lady. She is not the right choice for someone in search for a one-night stand, and needs more from her partner than just casual encounters. Romantic and ready to love, this woman needs a romantic partner who believes in love, while also in tune with her unspoken feelings. Despite her cautious nature, a Cancer woman is deeply erotic and when feeling secure to show her true personality and emotions, they will be expressed through an incredible sex life. To have lasting relationship with a Cancer woman, she needs someone faithful, respectful and honest, for doesn't forget betrayal and becomes very rigid and unpredictable when hurt.

Cancer Compatible Signs: Taurus, Virgo, Scorpio, Pisces, Capricorn

Famous Cancers:
Amrish Puri
Dalai Lama,
George M. Cohen,
Gerald Ford,
Karisma Kapoor
Katrina Kaif
Naseeruddin Shah
Priyanka Chopra
Rajendra Kumar
Sanjeev Kumar
Vivian Dsena

PART-5
ZODIAC SIGN LEO
HOROSCOPE

The Lion in the Cave is the story of the Lion, which always speaks of bravery. This is an animal fearless and impossible to challenge, hurt or destroy, their only weaknesses being fear and aggression towards those they confront. Living in a cave, a Lion always needs to have one, nesting and finding comfort in hard times. However, they should never stay there for long. With their head high, they have to face others with dignity and respect, never raising a voice, a hand, or a weapon, bravely walking through the forest they rule. People born under the sign of Leo are natural born leaders. They are dramatic, creative, self-confident, dominant and extremely difficult to resist, able to achieve anything they want to in any area of life they commit to. There is a specific strength to a Leo and their "king of the jungle" status. Leo often has many friends for they are generous and loyal. Self-confident and attractive, this is a Sun sign capable of uniting different groups of people and leading them as one towards a shared cause, and their healthy sense of humour makes collaboration with other people even easier. Leo belongs to the element of Fire, just like Aries and Sagittarius. This makes them warm hearted, in love with life, trying to laugh and have a good time. Able to use their mind to solve even the most difficult problems, they will easily take initiative in resolving various complicated situations. Ruled by the Sun, Leo worships this fiery entity in the sky, quite literally as well as metaphorically. They are in search for self-awareness and in constant growth of ego. Aware of their desires and personality, they can easily ask for everything they need, but could just as easily unconsciously neglect the needs of other people in their chase for personal gain or status. When a Leo representative becomes too fond and attached to their achievements and the way other people see them, they become an easy target, ready to be taken down.

Leo Love and Sex: This Fire sign is passionate and sincere and its representatives show their feelings with ease and clarity. When in love, they are fun, loyal, respectful and very generous towards their loved one. They will take the role of a leader in any relationship, and strongly rely on their need for independency and initiative. This can be tiring for their partner at times, especially if they start imposing their will and organizing things that aren't theirs to organize in the first place. Each Leo needs a partner who is self-aware, reasonable and on the same intellectual level as them. Their partner also has to feel free to express and fight for themselves, or too much light from their Leo's Sun might burn their own personality down. Sex life of each Leo is an adventure, fun and very energetic. This is someone who has a clear understanding of boundaries between sex and love, but might fail to see how important intimacy and emotional connection is to the quality of their sex life. Every Leo needs a partner to fight through their awareness and reach their sensitive, subconscious core, in order to find true satisfaction in a meaningful relationship.

Element: Fire

Quality: Fixed

Color: Gold, Yellow, Orange

Day: Sunday

Ruler: Sun

Lucky Numbers: 1, 3, 10, 19

Leo most compatible Zodiac Sign: Aquarius, Gemini

Strengths: Creative, passionate, generous, warm-hearted, cheerful, humorous

Weaknesses: Arrogant, stubborn, self-centred, lazy, and inflexible

Leo likes: Theatre, taking holidays, being admired, expensive things, bright colours, fun with friends

Leo dislikes: Being ignored, facing difficult reality, not being treated like a king or queen

Leo Friends: Leo is generous, faithful and a truly loyal friend, born with a certain dignity and commitment to individual values. Born with a need to help others, they will do so even if it takes a lot time and energy. Strong and reliable, this individual has the ability to appeal to almost everyone and has the energy to host celebrations and different events with people that bring out the best in them. They are rarely alone, for interactions with others give them the sense of self-esteem and awareness they need, but could have trouble finding friends able to keep pace and follow the high energy they carry everywhere they go.

Leo Family: Family matters won't be the first thing Leo will think about when they wake up in the morning or lie to bed at night. Turned to themselves for the most part, they tend to become independent as soon as possible. Still, a Leo will do anything to protect their loved ones, proud of their ancestry and roots in good and bad times.

Leo Career and Money: Leos are highly energetic and tend to always be busy, no matter the need for their employment.

They are ambitious, creative and optimistic and once they dedicate to their work, they will do everything just right. The best possible situation they can find themselves in is to be their own bosses or manage others with as little control from their superiors as possible. Jobs that allow open expression of artistic talent, such as acting and entertainment, are ideal for a Leo. Management, education and politics are also a good fit, as well as anything that puts them in a leadership position which naturally suits them. Leos love to be surrounded by modern and trendy things, and although money comes easy to them, they spend it less responsibly than some other signs of the zodiac. Extremely generous, they could provide many friends with financial help, supporting them through bad times. Although this doesn't always prove to be wise, it always makes them feel good.

Leo Man: A Leo man wants to be treated like a king in their intimate relationship and this is not their narcissistic characteristic, but a true inner need that all people with deep self-respect have to feel. Plans with him are always big and dramatic, and showering with admiration, devotion and attention come really natural both ways. This is a man who gives many gifts when they are in love, often expensive and posing as a statement of his effort. Any partner that wants to stay with him has to prove that they are worthy of royal treatment and ready to give enough of it back. A Leo man will love compliments, and although he appears confident, he needs a lot of praise to start feeling safe around their loved one too. However romantic and passionate, this man will rarely choose a woman that doesn't "go well" with his appearance, or doesn't make him look good in the eyes of specific groups in the outer world. He is known to easily take the roll of an eternal bachelor, always on the hunt and celebrating love and life. He will put himself in the centre of attention, and his partner could compete with a number of admirers, but their relationship is not in danger for as long as he is adored the way he loves to be. When treated right, he will stick around forever.

Leo Woman: Leo women are very warm-hearted and driven by the desire to be loved and admired, and as all people born under the Leo zodiac sign, love to be in the limelight.

To seduce her, one has to treat her well, respect her, compliment her and see her fit to live a luxurious lifestyle she deserves. She appreciates romantic partners and will expect to be the centre of someone's world, giving the person she loves the same royal treatment. Dating a Leo woman requires acceptance of her flaws and admiration for her qualities. She doesn't like competing for love and wants to have clarity on her role in her partner's life. A woman born under the Sun sign of Leo will always enjoy a visit to a theatre, an art museum or a fancy restaurant. She wants to be showered with flowers and gifts, but as grandiose displays of affection rather than an expensive routine of her partner. She can be a bit domineering and needs to stay in control of her own life. If her partner holds her as valued and queen-like as she is, there is infinite warmth, care, and attention in her heart to respond.

Leo compatible Signs: Aries, Gemini, Libra, Sagittarius

Famous Leos:

Amelia Earhart,

Gautam Rode

Huma Qureshi

Madhurima Tuli

Princess Anne,

Sanjay Dutt

Sara Khan

Vikram Singh Chauhan

Whitney Houston,

PART-6
ZODIAC SIGN
VIRGO HOROSCOPE

The Disappointed Goddess Seeking goodness in humankind is the story of Virgo, and disappointment seems to be inevitable from their point of view. The first time they came from their cloud and jumped onto planet Earth, it felt like their mission is to use their existence for good, discovering ways of justice and purity in other people. Once they fail to find it too many times, Virgos will pull away, get lost, turn to substance abuse, or simply separate from other people to sit on the bench, criticize and judge.

Virgos are always paying attention to the smallest details and their deep sense of humanity makes them one of the most careful signs of the zodiac. Their methodical approach to life ensures that nothing is left to chance, and although they are often tender, their heart might be closed for the outer world. This is a sign often misunderstood, not because they lack the ability to express, but because they won't accept their feelings as valid, true, or even relevant when opposed to reason. The symbolism behind the name speaks well of their nature, born with a feeling they are experiencing everything for the first time. Virgo is an Earth sign, fitting perfectly between Taurus and Capricorn. This will lead to a strong character, but one that prefers conservative, well-organized things and a lot of practicality in their everyday life. These individuals have an organized life, and even when they let go to chaos, their goals and dreams still have strictly defined borders in their mind. Constantly worried that they missed a detail that will be impossible to fix, they can get stuck in details, becoming overly critical and concerned about matters that nobody else seems to care much about. Since Mercury is the ruling planet of this sign, its representatives have a well-developed sense of speech and writing, as well as all other forms of communication. Many Virgos may choose to pursue a career as writers, journalists, and typists, but their need to serve others makes them feel good as caregivers, on a clear mission to help.

Element: Earth
Quality: Mutable
Color: Grey, Beige, Pale-Yellow
Day: Wednesday
Ruler: Mercury
Lucky Numbers: 5, 14, 15, 23, 32
Virgo most compatible Zodiac Sign: Pisces, Cancer
Strengths: Loyal, analytical, kind, hardworking, practical
Weaknesses: Shyness, worry, overly critical of self and others, all work and no play
Virgo likes: Animals, healthy food, books, nature, and cleanliness
Virgo dislikes: Rudeness, asking for help, taking centre stage
Virgo Love and Sex: The sign of Virgo leads Venus to its tragic fall and speaks of one's inability to feel worthy, beautiful, or lovable. Compatibility of Virgo with other zodiac

signs is mostly based on the ability of their partner to give them all the love they need to start feeling safe and open up enough to show their soft, vulnerable heart. They will rarely have direct statements of love, but intimacy brings out all of the beauty of their emotional self-expression. A Virgo will prefer a stable relationship than having fun, casual lovers, except if they become one, using their charm and superficial communication to win hearts without ever investing their own. Methodical and intellectually dominant, each Virgo seems to have an equation in their mind that their partner has to follow. They will rarely have many sexual experiences with different people, for they need to feel important to someone and find real physical pleasure in order to give their whole self to someone. The sign of Virgo is easily attached to the symbolism of a virgin, but the truth is their quality is mutable, and their need for change often overcomes their self-imposed restrictions and moral boundaries when it comes to sex.

Trust needs to be built with Virgo, slowly, steadily and patiently, and each partner they have in life has a chance to be nurtured and cared for, but only if they give enough to deserve special treatment of Virgo.

Virgo Friends: Virgos are excellent advisors, always knowing how to solve a problem. This can make them helpful and extremely useful to have around, but also brings out their need to search the problem in everything and everyone around them. They will care for people they build a solid relationship with, treasuring them for years and nurturing them in every possible way. An intimate friendship with a Virgo is always earned by good deeds.

Virgo Family: People born with their Sun in Virgo are very dedicated to their family and attentive to elderly and sick people. They understand tradition and the importance of responsibility, proud of their upbringing and everything that made their mind be as dominant as it is.

Virgo Career and Money: Virgos are practical, analytical and hard-working, always knowing exactly where to look for the core of any problem. Their methodology makes them shine at jobs that require good organization, dealing with paperwork, problem solving and working with their minds and their hands. When they focus, perfection is to be expected from their work, for no other sign has such an eye for details as Virgo. In love with books and artistic expression, they make good critics, while their need to help humankind serves them best if they decide to become doctors, nurses or psychologists.

Virgo stands for all practical and used things, and it is in the nature of these individuals to save money and always put something on the side. They will see irrational spending as a bad habit or a matter of being spoiled, and always hold on to practical solutions that don't cost much. Unfortunately, this approach can sometimes make them a bit cheap and too concerned about everything they might lack tomorrow. They need to learn to indulge in some hedonism too.

Virgo Man: To seduce a Virgo man, one must respect his need for cleanliness and order. In most cases he enters romance slowly, carefully, and likes to take his time getting to know a person before starting anything serious. When he receives information on what to expect, his partner has to be sure to deliver nothing less. A Virgo man might put up a cool front, but don't let him fool you. He has deep and sensual needs, and only if his partner is patient enough, able to withstand his tendency to overanalyze everything, he will eventually warm up. It takes obvious and hard work to sweep a Virgo man off his feet. He needs a partner to inspire, remind him of his own talents, and will often find such a person in platonic and completely irrational spheres while settling for less her, on planet Earth. In search for someone honest, patient and tidy, he is always ready to settle down with the right person for a very long time.

Virgo Woman: A Virgo personality is a mix of intelligence, attention to detail, common sense, and commitment, and a woman born with her in this sign is very smart, modest, and capable. Trying to seduce this woman can seem intimidating at first, for she is likely to put on a facade of indifference. However, she is not cold but practical, realistic and cautious when it comes to starting new relationships. She will never simply give herself to someone without assessing their character and emotions shared with her from the beginning. She doesn't, under no circumstances, fall under a category of spoiled, materialistic women. As all Earth signs she will enjoy the material world and see any gift she gets as a blessing, but still truly enjoy things only when they are extremely practical and easy to use, digest, or when in need of fixing. Attracted to intelligent but distant men, a Virgo woman has a strong capacity to love, but chooses to wait for sharing emotions until she is ready. Private and defensive, this is someone who needs her defence mechanisms respected but still broken by the right partner.

Virgo   compatible
Signs: Taurus, Cancer, Scorpio, Capricorn, Pisces
Famous Virgos:
Elvis Costello,
H.G. Wells,
Joan Jett,
Lily Tomlin,
Michael Jackson,
Mother Teresa,
Raquel Welch,
Richard Gere,
Rocky Marciano,

PART-7
ZODIAC SIGN LIBRA
HOROSCOPE

The Measure of Our Souls The shortest myth of them all seems to present a good analogy to the shortest constellation in the sky, you might even say that it is non-existent, presented by the pliers of Scorpio. Libra is one dot of balance in the sea of different extremes, manifested only through the fifteenth degree of this magnificent sign, an object among animals and people. There is something awfully insecure about Libra, as if they were unsure which plate to burden next, aware that things pass and teach us to be careful around other people. Whatever we do in our lifetimes, only serves to point the way for our Souls towards that "higher power" to finally measure our existence. Telling

us where we went wrong or what we did right, Libras unconsciously teach us that true liberation hides in lightness. People born under the sign of Libra are peaceful, fair, and they hate being alone. Partnership is very important for them, as their mirror and someone giving them the ability to be the mirror themselves. These individuals are fascinated by balance and symmetry, they are in a constant chase for justice and equality, realizing through life that the only thing that should be truly important to themselves in their own inner core of personality. This is someone ready to do nearly anything to avoid conflict, keeping the peace whenever possible. The sign of Libra is an Air sign, set between Gemini and Aquarius, giving these individuals constant mental stimuli, strong intellect and a keen mind. They will be inspired by good books, insurmountable discussions and people who have a lot to say. Each Libra representative has to be careful when talking to other people, for when they are forced to decide about something that is coming their way, or to choose sides, they suddenly realize that they might be in the wrong place and surrounded by wrong people. No partner should make them forget that they have their own opinion. Planet ruling the sign of Libra is Venus, making these people great lovers but also fond of expensive, material things. Their lives need to be enriched by music, art, and beautiful places they get a Element: Air
Quality: Cardinal
Color: Pink, Green
Day: Friday
Ruler: Venus
Lucky Numbers: 4, 6, 13, 15, 24
Libra most compatible Zodiac Sign: Aries, Sagittarius
Strengths: Cooperative, diplomatic, gracious, fair-minded, social
Weaknesses: Indecisive, avoids confrontations, will carry a grudge, self-pity
Libra likes: Harmony, gentleness, sharing with others, the outdoors
Libra dislikes: Violence, injustice, loudmouths, and conformity

Libra Love and Sex: Finding a compatible partner will be the main priority in the life of people born with their Sun in Libra. Once they start a romantic relationship, maintaining peace and harmony become the most important thing and their primary goal. Their charming personality and their dedication to each relationship makes their compatibility with others satisfying, but that fallen Sun they have to heal often creates trouble in their emotional world. Libra is the sign of marriage, making its representatives open for traditional pathways of love. Even though the element of Air gives them a lot of flexibility, they will still feel the strong pull towards tradition and their desires will eventually turn to love put on paper, well-organized and serving a purpose to create a certain image for the outer world. In a way, each Libra is in search of a partner who has the ability to set clear boundaries, as if expecting to be protected by them but without their pride being endangered in the process. This is a sign deeply connected to sexuality for Scorpio rises where it ends. They search for deep, meaningful relationship and although they don't have trouble relating with people they aren't really close to, the only true satisfaction in their love life comes from complete surrender of body and soul. It is the gravity of Libra to share their entire life with someone, with a challenge to be independent and aware of their core personality at the same time. When a Libra has made up their mind on being with someone, they have already chosen well, but it will help to know what stands in their way of achieving happiness or pushes them forwards.

Libra Friends: Libra representatives are highly social and put their friends in the limelight, but sometimes raise their expectation bars too high, and choose friendships that make them feel superior to the person standing in front of them. Their nature makes them indecisive which is why they might show a lack of. Still, this won't make them any less invested in their relationships when someone else takes the baton and shows interest in them. Tactful and calm, they can communicate through any problem if they want to, and will often help others understand the other side of their personal conflicts and trouble with other people.

Libra Family: Born into a family that gave them a certain weakness of the Sun, Libra can often transfer guilt between family members without even being aware of doing so. In constant search for harmony these individuals have a tendency to agree with their parents and siblings only to avoid conflict, being the one to pull back when a challenge comes their way. They need to nurture their personality and often turn to solitude only to discover their own point of view among many. If they are well built and worked on their inner sense of power, they discover ease in being a good parent and role model, ready to share everything they know with their children.

Libra Career and Money: For each Libra, the key to a happy life is in a fine balance, meaning they will not commit to work without setting apart enough time for their private life and their loved ones, and if they do, they will feel like they need to set free from it. They can be loved leaders even though they sometimes lack the initiative needed to organize people who work for them, and will work hard to deserve privileges that come their way. In search for truth and justice, they are good lawyers and judges, and can also be successful as diplomats, designers and composers if they have nurtured their artistic side from childhood. They will work well in a group, and can be convincing and gifted speakers. Financial aspect of their lives is often under control, which probably wouldn't be the case if they had an easier time deciding what they want to buy. As soon as they start questioning their financial choices, chances are they won't even spend any money at all, simply because it was hard to make a decision of any kind. They balance between saving and spending pretty well and even though they enjoy fashion and fine clothes, they rarely let their desires for spending get the best of them.

Libra Man: Libra men appreciate all that is beautiful and search for a partner to inspire them with their appearance. This might sound superficial, but the fact is they need mental and visual stimuli to make decision processes easier and push them into a serious relationship to begin with. Once they have decided to be with someone, they usually make serious, long-term bonds, enduring with ease through the hard times knowing they have already made the perfect choice to begin with. A Libra man wants to discuss everything with his partner, from daily matters to big shared endeavours in life. This is a man in search for a partner with strength of will and confidence, someone to guide the way when he feels lost or insecure. Once he finds the right person, he will do anything to make them happy, turning their attention solely to their partner and often forgetting himself in the process. This man is deeply romantic in his core and in search for true love to last him a lifetime.

Libra Woman: To seduce a Libra woman one has to be a good conversationalist and listener. She enjoys being taught about new things and enjoys talking about herself and her personal interests, just as much as she likes sinking deep into her partner's life. She is charming, intelligent, and finds solutions to problems that arise along the way with certain ease. Her partner needs to keep her interested and on her toes at first, making her question her own initiative and choices, while also straightforward and surprising enough. Being ruled by Venus a Libra woman has a natural tendency to have certain mood swings, but this won't make her any less just in her ways. Once she falls in love and shares a home with a partner, she will take care of them, make them look good, and keep their social life organized and well-adjusted to social norms they live in.

Libra compatible Signs: Gemini, Leo, Sagittarius, Aquarius

Famous Libran:

Lil Wayne
Hilary Duff
Halsey
Zach Galifianakis
Julie Andrews
Brie Larson
Lena Headey
Gwen Stefani
Alicia Vikander
Susan Sarandon
Dakota Johnson
Kate Winslet

PART-8
SCORPIO ZODIAC
SIGN HOROSCOPE

Scorpio-born is passionate and assertive people. They are determined and decisive, and will research until they find out the truth. Scorpio is a great leader, always aware of the situation and also features prominently in resourcefulness. Scorpio is a Water sign and lives to experience and express emotions. Although emotions are very important for Scorpio, they manifest them differently than other water signs. In any case, you can be sure that the Scorpio will keep your secrets, whatever they may be. Pluto is the planet of transformation and regeneration, and also the ruler of this zodiac sign. Scorpios are known by their calm and cool behaviour, and by their mysterious appearance. People often say that Scorpio-born is fierce, probably because they understand very well the rules of the universe. Some Scorpio-born can look older than they actually are. They are excellent leaders because they are very dedicated to what they do. Scorpios hate dishonesty and they can be very jealous and suspicious, so they need to learn how to adapt more easily to different human behaviours. Scorpios are brave and therefore they have a lot of friends. Honesty and fairness are the two qualities that make Scorpio a great friend. People born under the Scorpio sign are very dedicated and loyal, when it comes to working. They are quick-witted and intelligent, so they would feel better to be in the company of witty and fun loving people. They are full of surprises and will give you everything you need, but if you let them down once - there's no return. Scorpios are very emotional, when they are in pain, it is simply impossible to make them feel better. They are very dedicated and they take good care of their family.

Element: Water

Quality: Fixed

Color: Scarlet, Red, Rust

Day: Tuesday

Ruler: Pluto, Mars

Lucky Numbers: 8, 11, 18, 22

Scorpio most compatible Zodiac Sign: Taurus, Cancer

Strengths: Resourceful, brave, passionate, stubborn, a true friend

Weaknesses: Distrusting, jealous, secretive, violent

Scorpio likes: Truth, facts, being right, longtime friends, teasing, a grand passion

Scorpio dislikes: Dishonesty, revealing secrets, passive people

Scorpio Love and Sex: Scorpio is the most sensual sign of the zodiac. Scorpios are extremely passionate and intimacy is very important to them. They want intelligent and honest partners. Once Scorpios fall in love, they are very dedicated and faithful. However, they enter into a relationship very carefully, because sometimes they need a lot of time to build trust and respect for partners.

Scorpio Friends and Family

Scorpio Career and Money: Scorpios are fantastic in management, solving and creating. When a Scorpio sets a goal, there is no giving up. Scorpios are great in solving tasks that require a scientific and thorough approach. Their ability to focus with determination makes them very capable managers. They never mix business with friendship. Jobs such as a scientist, physician, researcher, sailor, detective, cop, business manager and psychologist are appropriate for this powerful zodiac sign. Scorpio respects other people, so expects to be respected in return. Scorpios are disciplined enough to stick to the budget, but they are also not afraid of hard work to bring themselves in a better financial position. However, they are not inclined to spend much. Money means security and a sense of control for them, which means that they are good at saving money and make decisions carefully before investing in something.

Scorpio Man: Learning how to attract the Scorpio man isn't easy. Scorpio men are tedious, confident, intense, sexual and very competitive. Some of the negative Scorpio traits is the fact that they are highly obsessive, compulsive and jealous people. In order to seduce the Scorpio man, you will have to make sure to keep an air of mystery around you. Games are something that appeals to men born under the Scorpio astrology sign, which means that you will have to work to keep them interested. With a Scorpio man, it's all about the challenge and about capturing something, so don't make it easy and play hard to get. Being honest and affectionate with this sign is also very important. They are attracted to confident and flirtatious women. However, there must be more than physical attraction to get him to the point where he will allow you to seduce him. He also needs an emotional attraction, because one of the most important Scorpio characteristics is the fact that he is the most intensely feeling sign of the zodiac. He longs for a sexual experience that goes beyond physical limitations, so if you want to seduce him, just bare your soul. Never try to control the Scorpio man, because he needs to be in control at all times.

Scorpio Woman: The Scorpio personality is both complex and fascinating. Scorpio women are secretive, sexy, magnetic, but they also appear aloof and calm. The Scorpio woman has a great capacity for kindness and a desire to do well in the world. If you want to seduce her, you have to be patient and willing to let her take the reins of the relationship. Be a good listener and pay full attention when she is speaking. Dating with a Scorpio woman can be really entertaining, but do not let her wild side fool you into thinking that she will be an easy conquest. Don't expect a sexual encounter with the Scorpio woman on the first date. Earning her love requires a lot of time and patience, but once she falls in love, she will give everything to the relationship. The woman born under the Scorpio zodiac sign can be very possessive but also completely devoted partners. Be honest and avoid getting into arguments with her, because she is not the type to forgive and forget easily. The woman born under the Scorpio star sign is curious, so in order to keep her happy, you will need to find new ways to satisfy her curiosity. If you can allow her to retain her pride, she will respect you for it and will fall in love with you quickly.

Scorpio compatible Signs: Cancer, Virgo, Capricorn, Pisces

Famous Scorpios:

Hillary Clinton,

Ike Turner,

Jody Foster,

Katherine Hepburn,

King Hussein,

PART-9
ZODIAC
SIGN SAGITTARIUS
HOROSCOPE

Like the other fire signs, Sagittarius needs to be constantly in touch with the world to experience as much as possible. The ruling planet of Sagittarius is Jupiter, the largest planet of the zodiac. Their enthusiasm has no bounds, and therefore people born under the Sagittarius sign possess a great sense of humour and an intense curiosity. Curious and energetic, Sagittarius is one of the biggest travellers among all zodiac signs. Their open mind and philosophical view motivates them to wander around the world in search of the meaning of life. Sagittarius is extrovert, optimistic and enthusiastic, and likes changes. Sagittarius-born is able to transform their thoughts into concrete actions and they will do anything to achieve their goals. Freedom is their greatest treasure, because only then they can freely travel and explore different cultures and philosophies. Because of their honesty, Sagittarius-born is often impatient and tactless when they need to say or do something, so it's important to learn to express themselves in a tolerant and socially acceptable way.

Element: Fire
Quality: Mutable
Color: Blue
Day: Thursday
Ruler: Jupiter
Lucky Numbers: 3, 7, 9, 12, 21
Sagittarius most compatible Zodiac Sign: Gemini, Aries
Strengths: Generous, idealistic, great sense of humour
Weaknesses: Promises more than can deliver, very impatient, will say anything no matter how undiplomatic
Sagittarius likes: Freedom, travel, philosophy, being outdoors
Sagittarius dislikes: Clingy people, being constrained, off-the-wall theories, details
Sagittarius Love and Sex: People born under the sign of Sagittarius are very playful and humorous, which means that they will enjoy having fun with their partners. Partners who are equally open, will certainly suit the passionate, expressive Sagittarius who is willing to try almost anything. For this sign there is always a thin line between love and sex. Their love for change and diversity can bring a lot of different faces in their bedroom. But when they are truly in love, they are very loyal, faithful and dedicated. They want their partners to be intellectual, sensitive and expressive.
Sagittarius Friends and Family: Sagittarius is very fun and always surrounded by friends. Sagittarius-born love to laugh

and enjoy the diversity of life and culture, so they will easily acquire many friends around the world. They are generous and not one of those who lecture. When it comes to family, Sagittarius is dedicated and willing to do almost anything.

Sagittarius Career and Money: When Sagittarius-born visualizes something in their minds, the will do everything they can to achieve this. They always know what to say in a given situation and they are great salespeople. Sagittarius favours different tasks and dynamic atmosphere. Jobs such as a travel agent, photographer, researcher, artist, ambassador, importer and exporter suit this free-spirited person. The fun-loving Sagittarius enjoys making and spending money. Considered to be the happiest sign of the zodiac, Sagittarius does not care much where it will earn the following money. They take risks and are very optimistic. They believe that the universe will provide everything they need.

Sagittarius Man: Sagittarius men are fun-loving people and eternal travellers, who are interested in religion, philosophy and the meaning of everything. The man born under the Sagittarius astrology sign loves adventures and sees all the possibilities in life. He wants to explore each and every one of them to determine where the truth is. Some of the best Sagittarius traits are his frankness, courage, and optimism. He is a restless wanderer, so the best you can do to keep his attention is to share in his quest, appreciate his wisdom and respect his opinions. Some of the negative Sagittarius characteristics include his carelessness and impatience. The Sagittarius man can also be tactless, superficial, and over-confident at times.

He needs freedom and doesn't like clingy women. If you want to seduce him, you will have to learn when to hold on and when to let go. The Sagittarius man is a logical thinker and an enthusiastic listener, who will listen carefully to everything you have to say, before processing the information and coming to his own conclusions.

Sagittarius Woman: Sagittarius women are wild, independent, fun, friendly and outgoing. They enjoy expressing themselves in a sexual manner and they are determined to live life to the fullest. A Sagittarius personality is vibrant, inquisitive, and exciting. The woman born under the Sagittarius zodiac sign is an honest woman, who always speaks her mind and values freedom and independence. If you want to seduce the Sagittarius woman, you should ask her out on a date outdoors. She loves adventures and long conversations. Dating a woman born under the Sagittarius star sign requires an adventurous spirit, because she sees everything as a challenge and can't stand boredom. However, although she is wild, do not expect her to fall in love with you immediately. Once the Sagittarius woman does fall in love, she will be a loyal and caring partner. If you give her a reason to think a relationship with you is going to be difficult, she will simply walk out the door. The Sagittarius woman usually has multiple love affairs throughout her life, due to the fact that she will not settle for a relationship that makes her unhappy.

In order to date the Sagittarius woman, you will need to possess a free spirit and love of travel. She sees life as one big adventure and expects her partner to be adventurous, spontaneous and romantic.

Sagittarius compatible Signs: Aries, Leo, Libra, Aquarius

Famous Sagittarians:

Andrew Johnson,

Bruce Lee,

Dick Van Dyke,

Elvis Presley,

Fiorello LaGuardia,

Harpo Marx

Harry Chapin,

Keith Richards,

Kim Bassinger,

Lucky Luciano,

Steven Spielberg,

PART-10
ZODIAC
SIGN CAPRICORN
HOROSCOPE

The Goat of Fear A goat with the tail of a fish is created to face fear and create panic. It is the sign of decisions made to be protected from monsters in our minds, lives, and immediate physical surrounding. Always ready to transform into something that scares those scary things off, Capricorn speaks of each natural chain reaction of fear, where one scary thing leads to many others, rising up as defensive mechanisms that only make things worse. Immersed in their secrecy, they face the world just as they are – brave enough to never run away, but constantly afraid of their inner monsters. Belonging to the element of Earth, like Taurus and Virgo, this is the last sign in the trio of practicality and grounding. Not only do they focus on the material world, but they have the ability to use the most out of it. Unfortunately, this element also makes them stiff and sometimes too stubborn to move from one perspective or point in a relationship. They have a hard time accepting differences of other people that are too far from their character, and out of fear might try to impose their traditional values aggressively. Capricorn is a sign that represents time and responsibility, and its representatives are traditional and often very serious by nature. These individuals possess an inner state of independence that enables significant progress both in their personal and professional lives. They are masters of self-control and have the ability to lead the way, make solid and realistic plans, and manage many people who work for them at any time. They will learn from their mistakes and get to the top based solely on their experience and expertise. Saturn is the ruling planet of Capricorn, and this planet represents restrictions of all kinds. Its influence makes these people practical and responsible, but also cold, distant and unforgiving, prone to the feeling of guilt and turned to the past. They need to learn to forgive in order to make their own life lighter and more positive.

Element: Earth
Quality: Cardinal
Color: Brown, Black
Day: Saturday
Ruler: Saturn
Lucky Numbers: 4, 8, 13, 22
Capricorn most compatible Zodiac Sign: Taurus, Cancer
Strengths: Responsible, disciplined, self-control, good managers

Weaknesses: Know-it-all, unforgiving, condescending, expecting the worst

Capricorn likes: Family, tradition, music, understated status, quality craftsmanship

Capricorn dislikes: Almost everything at some point

Capricorn Love and Sex: It is not easy to win over the attention and the heart of a Capricorn, but once their walls break and their heart melts they stay committed for a lifetime. Their Their relationships with other signs can be challenging due to their difficult character, but any shared feeling that comes from such a deep emotional place is a reward for their partner's efforts. Shown sensitivity comes through acts rather than words, and years are often needed for them to open enough to chat about their actual emotional problems. Turned to their personal goals, whatever they might be, Capricorns can lack compassion and emotion when relating to their loved ones. The certain ease of a "normal" life is something they will give with full devotion, and their partner will be able to rely on them, use them as a stepping stone for any personal endeavour, and have a lasting bond with a constant tendency of growth. Still, this isn't someone willing to compromise much, and seems to have the need to create a problem only to resolve it or feel bad that it was never resolved before. Capricorn might be a stiff Earth sign, set in their way, but this makes them a perfect match for certain signs of the zodiac.

Capricorn Friends: Capricorn is intelligent, stable and reliable, and this makes its representatives loyal and extremely good friends, standing in one's life as pillars on their way to their dreams. They need to be surrounded by people who don't ask too many nosey questions, know where boundaries are set, but also warm, open-hearted and loyal enough to follow their lead. They will not collect too many friends in this lifetime, but turn to those who make them feel at peace, intelligent and honest at all times.

Capricorn Family: This is a sign with full understanding for family traditions. Capricorns feel connected to every single thing from their past and their childhood, and loves bringing out these memories whenever a season of holidays or birthdays is near. This is a sign of a typical conflict one has over dominance in their household, with their father being and extremely important figure in the way this person built their self-image over the years. As parents they tend to be strict but fair, readily taking on responsibilities that come with a child.

Capricorn Career and Money: Capricorns will set high standards for themselves, but their honesty, dedication and perseverance will lead them to their goals. They value loyalty and hard work over all other things, and keep associates with these qualities close even when they might be intellectually inferior. Concentrated and resourceful, this is someone who gets the job done, doesn't mind long hours, and commits to the final product completely. They shine in jobs that include management, finance, programming and calculations. Deeply rooted in tradition, the state, and the system they live in, a Capricorn needs all of their paperwork in perfect order, their documents clean, and their file impeccable. Money will be truly valued in lives of these individuals, and they won't have much trouble managing it and saving some for a rainy day, for as long as their debts don't swallow their actual abilities. Hard workers with a higher cause, they will do anything they can to set free from a loan or a mortgage of any time, but also know that true success will only come in the long run.

Capricorn Man: Capricorn men are determined and ambitious people, who want to reach the top to get the rewards. He prefers reality over uncatchable dreams, but isn't afraid to set some of their more realistic dreams in motion. His need for control is strong, and he could be judgmental towards his partners, expecting them to be something they really aren't. His nature is wrapped up around accomplishment and responsibility and he often doesn't set romantic relationships on the top of his priority list. A Capricorn man wants to take charge and be the one to make the rules from the beginning. He is in search for a practical, grounded partner, and almost always ends up with an emotional one who has a hard time controlling their heart. When starting a relationship, he will think about ways to respect the norm but also show his feelings, expecting the person in front of him to feel comfortable and attractive enough no matter the amount of affection he gives.

Capricorn Woman: Capricorn women are ambitious, persistent, responsible and reliable. She only wants to find someone to make her smile, and can't wait to open up and feel the real pull of emotion that makes her warm up to the possibilities that lie in the future. It will take some time for her to lower her guard and feel safe and comfortable enough to show just how sensitive and caring she can be when she is in love. She wants her partners responsible, calm, and hard-working, and needs to know that she is taken care of if something bad happens in the future. A Capricorn woman needs to feel comfortable with people she dates and needs time to decide what she wants out of each relationship. Born in a Sun sign that exalt Mars, her instincts and initiative are strong, and this makes her a passionate lover always in charge of her own life no matter the outer circumstances.

Capricorn compatible Signs: Taurus, Virgo, Scorpio, Pisces

Famous Capricorns:

David Bowie,
J.R. Tolkien,
Joseph Stalin,
Louis Pasteur,
Martin Luther King,
Pat Benatar,
Richard Nixon,
Rod Serling,

PART-11
ZODIAC
SIGN AQUARIUS
HOROSCOPE

Aquarius-born is shy and quiet, but on the other hand they can be eccentric and energetic. However, in both cases, they are deep thinkers and highly intellectual people who love helping others. They are able to see without prejudice, on both sides, which makes them people who can easily solve problems. Although they can easily adapt to the energy that surrounds them, Aquarius-born have a deep need to be some time alone and away from everything, in order to restore power. People born under the Aquarius sign look world as a place full of possibilities. Aquarius is an air sign, and as such, uses his mind at every opportunity. If there is no mental stimulation, they are bored and lack a motivation to achieve the best result. The ruling planet of Aquarius, Uranus has a timid, abrupt and sometimes aggressive nature, but it also gives Aquarius visionary quality. They are capable of perceiving the future and they know exactly what they want to be doing five or ten years from now. Uranus also gave them the power of quick and easy transformation, so they are known as thinkers, progressives and humanists. They feel good in a group or a community, so they constantly strive to be surrounded by other people. The biggest problem for Aquarius-born is the feeling that they are limited or constrained. Because of the desire for freedom and equality for all, they will always strive to ensure freedom of speech and movement. Aquarius-born have a reputation for being cold and insensitive persons, but this is just their defence mechanism against premature intimacy. They need to learn to trust others and express their emotions in a healthy way.

Element: Air
Quality: Fixed
Color: Light-Blue, Silver
Day: Saturday
Ruler: Uranus, Saturn
Lucky Numbers: 4, 7, 11, 22, 29
Aquarius most compatible Zodiac Sign: Leo, Sagittarius
Strengths: Progressive, original, independent, humanitarian
Weaknesses: Runs from emotional expression, temperamental, uncompromising, aloof
Aquarius likes: Fun with friends, helping others, fighting for causes, intellectual conversation, a good listener
Aquarius dislikes: Limitations, broken promises, being lonely, dull or boring situations, people who disagree with them
Aquarius Love and Sex: Intellectual stimulation is by far the greatest aphrodisiac for Aquarius. There's nothing that can attract an Aquarius more than an interesting conversation with a person. Openness, communication, imagination and willingness to risk are the qualities that fit well in the perspective of life of this zodiac sign. Their compatibility with other signs can be complex, Integrity and honesty is

essential for anyone who wants a long-term relationship with this dynamic person. In love, they are loyal, committed and not at all possessive - they give independence to their partners and consider them as equals.

Aquarius Friends and Family: Although Aquarius-born is communicative; they need time to get close to people. Considering that they are highly sensitive people, closeness to them means vulnerability. Their immediacy behaviour combined with their strong views makes them a challenge to meet. Aquarius will do anything for a loved one to the point of self-sacrifice if necessary. Their friends should possess these three qualities: creativity, intellect and integrity. When it comes to family, their expectations are nothing less. Although they have a sense of duty to relatives, they will not maintain close ties if the same expectations as in friendship are not fulfilled.

Aquarius Career and Money: Aquarius-born brings enthusiasm to the job and has a remarkable ability of exploitation of their imagination for business purposes. Career which enables a development and demonstration of the concept will suit this zodiac sign. Their high intellect combined with their willingness to share their talents, inspires many who work in their environment. Aquarius is a visionary type who likes to engage in activities that aim to make humanity better. When it comes to money, this zodiac sign has a talent to maintain a balance between spending and saving money. Most people born under the sign of Aquarius are well adapted to their feel for style and they are not afraid to show it. It is not uncommon to see an Aquarius boldly dressed in bright colour suits. Careers such as acting, writing, teaching, photography or piloting, are suitable for this sign. The best environment for them is one that gives them the freedom to solve the problem without strict guidelines. Aquarius is an unconventional type and if given the opportunity to express their talent, can achieve remarkable success.

Aquarius Man: You should learn about all the positive and negative Aquarius traits if you want to seduce man born under this zodiac sign. Aquarius men are unpredictable, intelligent, social, independent and excellent communicators.

Some of the negative Aquarius characteristics include unreliability, stubbornness, indecision, and inflexibility. If you want to seduce the man born under the Aquarius astrology sign, you will have to be cool about it. If you ever come on too strong emotionally for this man, he will never even think about taking you to bed. The Aquarius man can often live inside his own mind, so he needs a companion to talk about the progressive thoughts he creates. So, if you want to seduce him, you should be friends first. Most of their romances tend to start out as friendships which gradually evolve into something more serious. This means that the Aquarius men will never enter into a romantic relationship with someone who isn't already a friend. Be patient and keep things on a platonic basis until you're ready to make your first big move. Respect his need for freedom, his individuality, and his desire to make a difference in the world.

Aquarius Woman: Aquarius personality is independent, mysterious, free-spirited and eccentric. Aquarius women have a unique sense of humour and a practical outlook in life. However, inconsistency is a constant problem for women born under the Aquarius zodiac sign. The Aquarius woman longs for romance and good conversation, although she may seem like a cold and aloof person. She is an excellent sex partner, but only if she is convinced that you are interested in more than just a one night stand. If you want to seduce a woman born under the Aquarius star sign, you will have to appeal to the multiple different sides of her personality. Aquarius women are drawn to people who stand out from the crowd, so if you want to attract her attention it's important to come across as a bit different from everyone else she knows. The Aquarius woman is a highly imaginative sex partner, who wants to try out new things on a regular basis. However, make sure to let her know that she is more than just a sex partner to you. The worst mistake you can make when trying to seduce the Aquarius woman is to be too demanding and pushy. Give her lots of freedom, because she is a very independent woman and won't tolerate any kind of control. In order to attract her attention, show her that you're the cool and calm type. The Aquarius woman feels uncomfortable with people who openly express their feelings, so avoid emotionally charged issues and appeal to her intellect instead.

Aquarius compatible Signs: Aries, Gemini, Libra, Sagittarius

Famous Aquarius:

Anton Chekov,

Charles Darwin,

Charles Dickens,

Chris Rock,

Franklin D. Roosevelt

Galileo,

Garth Brooks,

Geena Davis,

Jennifer Aniston,

Justin Timberlake,

Norman Rockwell,

Oprah Winfrey,

Peter Gabriel,

Phil Collins,

Prince Andrew,

Vanessa Redgrave,

Virginia Woolf,

Wolfgang Amadeus Mozart,

Yoko Ono,

PART-12
ZODIAC SIGN PISCES
HOROSCOPE

Pisces are very friendly, so they often find themselves in a company of very different people. Pisces are selfless; they are always willing to help others, without hoping to get anything back. Pisces is a Water sign and as such this zodiac sign is characterized by empathy and expressed emotional capacity. Their ruling planet is Neptune, so Pisces are more intuitive than others and have an artistic talent. Neptune is connected to music, so Pisces reveal music preferences in the earliest stages of life. They are generous, compassionate and extremely faithful and caring. People born under the Pisces sign have an intuitive understanding of the life cycle and thus achieve the best emotional relationship with other beings. Pisces-born is known by their wisdom, but under the influence of Uranus, Pisces sometimes can take the role of a martyr, in order to catch the attention. Pisces are never judgmental and always forgiving. They are also known to be most tolerant of all the zodiac signs.

Element: Water
Quality: Mutable
Color: Mauve, Lilac, Purple, Violet, Sea green
Day: Thursday
Ruler: Neptune, Jupiter
Lucky Numbers: 3, 9, 12, 15, 18, 24
Pisces most compatible Zodiac Sign: Virgo, Taurus
Strengths: Compassionate, artistic, intuitive, gentle, wise, musical
Weaknesses: Fearful, overly trusting, sad, desire to escape reality, can be a victim or a martyr
Pisces likes: Being alone, sleeping, music, romance, visual media, swimming, spiritual themes
Pisces dislikes: Know-it-all, being criticized, the past coming back to haunt, cruelty of any kind
Pisces Love and Sex: Deep in their hearts, Pisces-born is incorrigible romantics. They are very loyal, gentle and unconditionally generous to their partners. Pisces are passionate lovers who have a need to feel a real connection with their partners. Short-term relationships and adventures are not peculiar to this zodiac sign. In love and relationship, they are blindly loyal and very caring.
Pisces Friends and Family: Gentle and caring, Pisces can be the best friends that may exist. In fact, they often put the needs of their friends in front of their needs. They are loyal, devoted, and compassionate and whenever there is some problem in the family or among friends, they will do their best to resolve it. Deeply intuitive, Pisces can sense if something is wrong, even before it happens. Pisces are expressive and they will not hesitate to express their feelings to the people around them. They expect others to be open to

them as they are. Communication with loved ones is very important for them.

Pisces Career and Money: Intuitive and often dreamy, Pisces feel best in a position where their creative skills will come to the fore, even better if it's for charity. Occupations that fit Pisces are: attorney, architect, veterinarian, musician, social worker and game designer.

Inspired by the need to make changes in the lives of others, they are willing to help even if that means to go beyond the boundaries. This zodiac sign is compassionate, hard-working, dedicated and reliable. Pisces-born can be great at solving problems. For the most part, Pisces don't give money too much thought. They are usually more focused on their dreams and goals, but they will try to make enough money to achieve their goals. In this area, there can be two sides of the Pisces - on one hand; they will spend a lot of money with little thought, while on the other hand they can become quite stingy. Yet, in the end, there will always be enough money for a normal life.

Pisces Man: Romance rules the world of the Pisces men. The Pisces lives to please and love. The best way to seduce a Pisces man is to open up to him completely. Some of the best Pisces traits are his sensitivity, compassion, and kindness. He is a gentle person who will figure out what you want and then serve it up as often as possible. He is always looking for ways to help others and knows exactly what you want almost before you do. His desire to please leaves him susceptible to manipulation and lies. He will use his wild imagination to please you. The Pisces man loves to laugh, so if he finds you funny and easy to be around, you are on a good way to seduce him. He seems calm on the outside, but on the inside, you will find a different person, as the Pisces man battles between strong emotions. Encourage him to open up and release those feelings. One of the greatest Pisces characteristics is the ability to tune into others emotions. So, if you are dating a Pisces man, you can look forward to an emotionally fulfilling relationship.

Pisces Woman: Pisces women are known as kind, imaginative, compassionate, selfless and extremely sensitive individuals. If you want to attract the woman born under the

Pisces zodiac signs, you need to be romantic and to have a good sense of humour. It is also important to be a good listener. The Pisces personality is compassionate and full of unconditional love. Once you have captured her attention, she will be quick to open up to you. A sex with the Pisces woman will be explosive and you will never be bored with her in the bedroom. The woman born under the Pisces star sign enjoys lively discussions about spiritual things and the supernatural. She is intuitive and will quickly figure out if all you are looking for is just sex. She wants to be treated with respect and you will never be able to seduce her fully during the first few dates. If you're honest and if you open up to her, she will immediately feel more connected with you. The Pisces woman is very sensitive by nature, so she is not quick to forgive and forget. If her heart has been wounded in the past, she will have a hard time opening herself back up to the thought of a new romantic relationship.

Pisces compatible Signs: Taurus, Cancer, Scorpio, Capricorn
Famous Pisces:
Albert Einstein,
Alexander Graham Bell,
Andrew Lloyd Webber,
Dr. Seuss,
Elizabeth Barrett Browning,
Elizabeth Taylor,
George Frideric Handel,
George Washington
James Taylor,
Jerry Lewis,
Kurt Cobain,
Michelangelo Buonarotti,
Renoir, Cindy Crawford,

CHAPTER-3
SUN SIGN

Sun sign astrology is the form of a simplified system of astrology which considers only the position of the Sun at birth, which is said to be placed within one of the twelve zodiac signs. This sign is the star sign of the person born in that twelfth month of the year. The Moon has the fastest apparent movement of all the heavenly bodies; so it is often used as the main indicator of daily trends for astrology forecasts. The Sun sign astrology was not invented until 1930. The astrologer R. H. Naylor was the first astrologer to make accurate predictions based on "Sun Sign". The date of birth determines the Sun-Sign and gives special attributes as mentioned below:

Table: Sun Sign with Date of Birth

Sun Sign	From	To	Element
Aries	March 21st	April 19th	Fire
Taurus	April 20th	May 20th	Earth
Gemini	May 21st	June 20th	Air
Cancer	June 21st	July 22nd	Water
Leo	July 23rd	August 22nd	Fire
Virgo	August 23rd	September 22nd	Earth
Libra	September 23rd	October 22nd	Air
Scorpio	October 23rd	November 21st	Water
Sagittarius	November 22nd	December 21st	Fire
Capricorn	December 22nd	January 19th	Earth
Aquarius	January 20th	February 18th	Air
Pisces	February 19th	March 20th	Water

PART-1
ARIES HOROSOPE

(March 21st - April 19th):

Who works from morning to evening, and never likes to be outdone?

Who is outspoken, alert, and ambitious and walks almost like a run?

Element: Fire

Quality: Cardinal

Ruler: Mars

Lucky stone: Coral & Pearl. The glittering Coral will gives all the courage, makes him/her rich and gives a comfortable future. Topaz and Moonstones will be auspicious too.

Lucky Number: The number 1, 9 & 14 can bring luck in life.

Colour: Red, Peacock blue

Lucky Colour: Revel in the magic of peacock blue or Shades of Red will be lucky.

Day: Friday, Tuesday and Saturday. Dig gold on Tuesday.

Lucky Day: Dig gold on Tuesday.

Lucky Flower: Sweet Pea.

General Insights: Aries are energetic and turbulent. They are continuously looking for dynamic, speed and competition. They are always first in everything - from work to social gatherings. People are meant to emphasize the search for answers to personal and metaphysical questions. This is the biggest feature of this incarnation. It is in their nature to take action, sometimes before they think about it well. They have excellent organizational skills, so you'll rarely meet an Aries who doesn't like to finish more things at once. The challenges are increased when they are impatient, aggressive and vent anger on others. Aries rules the head and leads with the head, often literally walking head first, leaning forwards for speed and focus. They are naturally brave and rarely afraid of trial and risk. They possess youthful strength and energy, regardless of age and they perform tasks in record time. By aligning with themselves they could achieve the best results.

Personal Quality: The Arian is vital, impulsive, born leader, adventurous, brave, fearless, highly dominating, and full of energy, aggressive, argumentative, good athletes and soldiers and makes enthusiastic lovers. He/She is outspoken, alert and quick to act and speak. He/She is always willing to help the persons in need. He/She is not a follower. He/She is large hearted and speak straight forward for what he/she feel about the person. He/She is childishly egocentric, extremely demanding and liable to throw tantrums if denied. He/She is quick to anger and known for his impatience, and is prone to be arrogant. Under planetary afflictions he is subject to brain fever, dizziness, nosebleed, neuralgia, inflammation of the cerebral hemispheres, and diseases of face.

Positive Quality: He/She is courageous, determined, confident, enthusiastic, optimistic, honest, passionate, generous, a lover of justice, and wishes to earn by own efforts and never looks at others wealth. He/She gives time, effort, money and sympathies to others. He/She likes to be challenged and enjoys solving any obstacle. He/She has both moral and physical courage.

Negative Quality: He/She is impatient, moody, short-tempered, impulsive, and aggressive, but not very tactful in communicating and will never bend, has strong feeling of admirations and is not diplomatic and is sharp tongue and shows anxiety. He/She has a spending nature to maintain the image. He/She gets nervous when things are not moving his/her way. He/She is quick-tempered, violent, impatient, egotistical and intolerant.

Physical Appearance: He/She is angular, slim in early life, although may fill out later. He/She has sharp elbows and knees.

Relationships: He/She is possessive, jealous, faithful and idealistic, passionate and incurably romantic. He/She tends towards joyous sex and close relationships throughout the lives. His/Her life partner will be Leo and Libra. Aquarius and Sagittarius might be very helpful to him/her in business.

Aries Career: His/Her income and social status will rise at the age of 48-52 and will get promotion at the age of 30, 36 and 45. He/She will deal with 'futures' on the money market or hacking through the Amazonian forest. He/She can be Dentist, Director, E M T, Entertainer, Entrepreneur, Landlord, Lawyer, and Make-up artist, Optometry, Producer, Sports person and Stockbroker.

Aries Health: He/She is always in a tearing hurry and often has a fast metabolism, which keeps the weight down. He/She is prone to stress and suffer from tension headaches.

Ideal Partner: He/She vibes best with one of Gemini, Leo, Sagittarius, Aquarius

Compatible Signs for Aries: Gemini, Leo, Libra, Sagittarius and Aquarius.

Incompatible Zodiac Signs: Taurus, Cancer, Virgo, Scorpio, Capricorn, Pisces

Greatest Overall Compatibility: Leo, Libra, Sagittarius.

Best for Compatibility: Libra.

Aries Liking: Comfortable clothes, taking on leadership roles, physical challenges, individual sports.

Aries Disliking: Inactivity, delays, work that does not use one's talents.

BUSINESS ASTROLOGY FOR ARIES

Aries are born leaders and love being in charge. However they lack in art of diplomatic dealing. For a successful business they should work upon to handle people and situations tactfully. Business booster Gem is Garnet.

Aries Compatibility with Aries

Aries to Aries Sexual Intimacy and Compatibility: With their self-respect aiming high and strong personalities, it is easy for these two to take off their clothes and enjoy one another. Their biggest problem could be their possible selfishness. Since sexual harmony is probably the most important segment of the relationship to their sign ruled by Mars, this might lead to fights and the exchange of many sharp words.

The worst possible scenario is if one of them thinks extensively about the other's satisfaction, while the other has no awareness of the needs of their giving partner. Since

Mars is a planet primarily connected to sex as a means to continuation of the species, it is quite often that two Aries partners lack the ability to satisfy each other in a sensual, Venus way. Mars will stand for satisfaction in the physical sense, but no emotion. Hence, there might be a lack of true human interaction in this segment of relationship. Their thoughts must be turned to their partner with no exception if they plan for their sexual relationship to work (65%).

Aries to Aries Trust: They start acting defensive and get angry a lot. Now you can imagine how two Aries would look together if they weren't honest? Pretty much like children at the playground fighting about nothing at all, while keeping some sort of unnatural closeness to neutralize the feelings of guilt. With that said, we can claim with certainty that two Aries in a loving relationship share trust as the same goal. Not only trust in what the other person is saying, but more importantly, trust to open up and say what's on your mind without presence of fear of their partner's reaction. In many cases, Aries partner is not full of love and support due to their lack of tact and impulsive nature. This is the main reason why two Aries in a relationship don't always work. 55%

Aries to Aries Communication and Intellectuality: They are like handsome, naked people, banging their heads against a brick wall. This is the best possible image to make you understand the mindset of Aries, as they are brainstorming. Intelligence has nothing to do with the image, as they can be extremely intelligent due to the possibility of the position of Mercury in their sign, but some kind of strange stubbornness is what can make them senselessly stupid. The usual Aries brain has a sort of need to always prove something to others. Now this would be great if the debate had some calm, rational thinking as a base, but even when their arguments are valid, they can't seem to stay calm and distant for long enough to make their opinion be heard. It is difficult for two of them to find peace in communicating, unless they share most of the same opinions. Mars, their ruler, is in its low energy set and is a planet of aggression. When you try to pin previous description to your Aries couple, you may not find it valid. But if you look again, you might see all types of hidden aggression between them, especially if they are tired. It is as if they are used to it and now they simply function in this way. 85%

Aries to Aries Emotions: They aren't very emotional. There is no better person than another Aries to understand the Aries. As a Fire sign, Aries is warm and passionate but it can be a bit difficult to understand their soft side because of their seemingly "masculine" and soldier-like nature. Two Aries can share deep emotional understanding as if they speak the same language.

Unfortunately, Sun is exalted in Aries and often too warm for their balanced functioning. If a second Sun would enter our solar system, can you imagine what kind of damage its gravity would do? This said, you can understand that it is easier to be with a person ruled by a planet when you yourself are a star. It is not as if this emotional relationship is impossible, but it can sometimes be difficult for both parties to remain independent, strong and true to their self. 75%

Aries to Aries Values: The thing Aries value most is a person's ability to be straightforward and clear, so two Aries will understand each other perfectly on this matter. If they would agree on specifics, that is questionable. Because of their primal nature, they like to see themselves as fighters for justice. This gives them a typical "Robin Hood" perspective, so it is possible to say that they also value truth, honour and respect. This would be out of context and more of an "in general" description of their values. Values any other sign could also have, but not be that passionate about them. 95%

Aries to Aries Shared Activities: Mars makes people active, so let's say they can share everything physical, from walks, sports and sex, to mountaineering. Activities to be shared are easily found in this relationship because of their similar energy type and potential. 95%

Conclusion: When two Aries come together, it is imperative for at least one of them to have mastered the art of staying calm. If this is achieved by one of them, not through passive aggression but through rational thought, their relationship can be truly rewarding. As two warm and passionate people, they can share many adventurous moments that raise their energy levels sky high. If, however, none of them has this rational, grown-up ability, it is only possible to prolong their relationship based on superficial activities and sex, of course. Since the sign of Aries takes Saturn, the wise ruler of time, patience and responsibility to its detriment, one of these partners will have to learn their lesson and take responsibility for the future of their relationship if they are to last in time. 75%

Aries Compatibility with Taurus

Aries to Taurus Sexual Intimacy and Compatibility: The Aries ruled by Mars and Taurus by are very sexual people. Both planets are in connection with physical relations, but their biggest difference is in their final goal when it comes to sex. Aries is guided by a simple instinct and Taurus is all about satisfaction. They enjoy themselves through the entire sexual experience. To satisfy Taurus, you need to be emotionally involved, gentle and passionate at the same time, and willing to put some time and effort into the art of sex. Aries representatives usually get satisfied with having sexual relations at all. This goes for both male and female representatives of the sign. For their mutual satisfaction it is imperative for Aries to develop an atypical sense of touch and work on their sensuality in order to keep their Taurus happy. Let us also remember that Taurus is a fixed sign, pretty much set in their ways, and when it comes to sexual satisfaction they will rarely compromise and settle for less than perfect. 65%

Aries to Taurus Trust: Both of these signs have the ability to form a stable relationship filled with honesty. Neither of them is flaky or runs away from a good challenge. This can contribute to a positive attitude and open agreements on honesty when they are together. They both have a need to search for their one true love, as Mars and Venus always do. This can lead to infidelity and typical love triangle issues, due to lack of emotion from Aries or a lack of self-worth by Taurus partner. Still, if they communicate well from the beginning, they will usually find how important mutual trust is to both of them and try hard not to jeopardize it. 85%

Aries to Taurus Communication and Intellectuality: Aries and Taurus both have horns. They both are stubborn, but they are not even stubborn in a similar way to share some understanding. Aries grabs their convictions and simply doesn't let go. They will kick and scream (literally) until they convince Taurus that they are right about something (consider it the smallest thing in the entire Universe). When Taurus notices this Aries' behaviour, they dig in. They don't move, ever. They don't even possess the sense of sound. You can almost expect a deep voiced "moo" as they get more and more irritated. Their shoulders go up, their eyebrows make an "M" and they simply stand there, annoying their Aries partner even more. Who could be this inhuman to just stand there and not listen to a word their loved one says in that high pitched tone? Taurus can. Not because of the anger, but because they are in fact too sensitive to deal with this kind of behaviour. Taurus never looks too sensitive. Their Venus role is grounded and strong, but this is a sign in which the Moon is exalted, Uranus falls and Mars is in detriment. You can imagine how this person can react to shouting and aggression of any kind. Their intellect is not an issue at all. If they can find their way through those hard headed conflicts, it is all the same if they were intelligent or stupid because they must love each other very much. The cure for this condition is in the middle, of course. Taurus needs to set strong boundaries and act securely from the safe zone they've created and Aries needs to take a step back and lower their voice, just a little bit. This would be a good place to start. 40%

Aries to Taurus Emotions: They are both highly emotional, but they don't show it in the same way. It is safe to assume that as much as they may love each other, it will be difficult for both of them to know they are loved. Aries shows their emotions loudly and openly, in a way that is sort of rough and inpatient. They don't give much time for the other person to give an emotion back, and act as Fire, their element, without much sense for anyone. Taurus may find this superficial, too intense or even phony, as they don't recognize this type of behaviour as love. Taurus shows their emotions in a silent, slow process of giving. They will show love through cooking, touching and gentle words. The problem is in the fact that Aries finds this boring, stiff or even untrue. To make each another feel loved, they will both have to learn to show affection to their partner in a way that differs a lot from their natural one. This can be a small or a large obstacle and the outcome depends only on their readiness to listen to the needs of their loved one. 60%

Aries to Taurus Values: This is an area in which they match quite well. Although they seem completely different, their main objectives are pretty much the same. They both value material security, since Aries is ruled by Mars, a planet connected to the fear of existential crisis, and Taurus is an Earth sign, material in their core and very inclined to the financial world. They both cherish character and strength, physical and verbal, and need someone who will not disappoint them as the first impression fades. 90%

Aries to Taurus Shared Activities: Aries and Taurus really want completely different things. While Aries is active, ready to run, train and needs to use the energy through any kind of physical activity, Taurus has the need to rest and gather energy almost all the time. Even if they go for coffee together, there is a big chance Aries partner will finish theirs and be bored in about 20 minutes, while Taurus partner would sip their coffee enjoying it, than order some cake. The only activity they would both truly enjoy is a walk through the park and any slow outdoor activity to restore contact with nature. 40%

Conclusion: This is a relationship full of personal challenges and individual depth. If they want to succeed as a couple, many internal issues in both must be solved. Only if they both accomplish peace in their lives, have just enough education, just enough other relationships and acquired just enough humour, they might be able to put aside their differences and listen to each other well enough. It is not that hard, except when you are used to using your horns. 63%

Aries Compatibility with Gemini

Aries to Gemini Sexual Intimacy and Compatibility: When Aries and Gemini engage in sexual activities, they don't know where they could end up. With Aries' libido and Gemini's ideas, they might be a bit too creative and harshly judged by their environment. It is a good thing that they both don't care that much about other's opinions anyway. In its healthy image, this is mostly a combination of passion, energy and curiosity. In a not so healthy one, their sexual relationship can be full of nasty words and verbal aggression. The good thing is that neither is too sensitive and easily hurt, so this can be exciting and unique for both of their experiences. Since Aries is a warrior by nature, Gemini's approach to sex might be too playful for their taste, but this is usually only until they open up to the everlasting game provided by Gemini partner. Their main goal is to stay as uninhibited as possible, so the Air sign of Gemini can give oxygen to the Fire of Aries. 90%

Aries to Gemini Trust: Lack of trust is probably the biggest problem in this relationship. Aries is possibly very jealous. Gemini is always changing the face they wear for the world. Most Gemini representatives aren't even aware of their basic individuality, convinced that they change personality overnight. Although this is not exactly true, the sense that Aries can get from them is not exactly a recipe for trust. Because of this, Aries might get angry, Gemini distracted and distant, to the point where Aries starts looking for another partner even if the relationship has not ended and Gemini doesn't even care anymore. 40%

Aries to Gemini Communication and Intellectuality: Aries is not the master of the art of conversation. Conversation is Gemini's main life theme. Even if they talk less than a typical Gemini usually does, their inner dialogue must be rich. Both partners should approach the relationship as if Gemini person was there to teach Aries how to have a good conversation. Since Gemini loves to be in a role of a teacher and loves to be in a relationship in which their partner learns something from them, this should be a good approach for both of them. That is if Aries' ego allows this "submissive" role. Still, we are all aware that there are some Gemini representatives that simply talk too much about nothing significant at all. This would be a reason for Aries to lose their temper and think of their partner as superficial and even stupid. This loss of respect is truly bad for their own ego, since the decision to be with this partner was theirs in the first place. They should remember that there is always somebody in the world who wouldn't think of this specific Gemini as stupid and that everyone deserves to be with someone who doesn't see them in this light. When this disrespect happens in their relationship, Aries should consider letting their Gemini go and look for someone who suits them better. 85%

Aries to Gemini Emotions: Emotion is a tricky territory for them. Aries partner has warm, passionate emotions and a problem to express them. Gemini often doesn't go very deep beneath the surface to look for someone's hidden qualities and isn't really that emotional by nature. So this is a mix of an emotional partner who can't communicate how they feel, and a rational one who talks about everything else. The good thing is that Aries does not lack the fierceness to turn Gemini's attention to them and make them listen. When they manage to connect in a constructive dialogue, it is possible for them to explain to each other where they stand and how they feel, and this way set a good foundation for future emotional exchange. 60%

Aries to Gemini Values: Aries values a person's ability to be clear and concise, and Gemini's need to talk around everything, it seems pretty obvious that this is not a perfect match. Now think of Gemini. It is kind of hard to think of any of their values except for the fact that they value everything interesting, and this is a kind of understatement since they find almost everything interesting. Well this is not exactly true. Gemini partners value knowledge and someone's literal abilities, as well as a fine rational mind. This is something Aries can fulfil to a certain point, in case they don't react on impulse to everything Gemini says. It is not that difficult for these partners to respond to each other's needs, but if they don't share similar education, interests and strength of character, they might see each other unworthy of their affection. 75%

Aries to Gemini Shared Activities: Gemini is all for activity however insane certain activities and Aries will feel liberated in this relationship. It is hard to say who will lead and who will follow. Aries always leads with that atomic energy, while Gemini always comes up with new ideas and initiative. They motivate and challenge each other all the time and they both never say "no". Choices of activities Aries come up with must be truly aggressive and ridiculous for Gemini not to engage. If Gemini thinks of something that Aries would maybe like to refuse, their ego wouldn't let them and they would jump in anyway, just to prove that they can. This much excitement should be followed by enough rest and time spent at home. 50%

Conclusion: The overall relations of this couple would be good, exciting and challenging, a relationship where both partners can learn a lot and be active in a healthy way. The main problem with their romantic involvement is the lack of trust, especially if Aries partner gets too attached to Gemini, always fighting for their freedom. The need for conversation with a lot of essence is bigger than any positive or any negative aspects of their relationship and both of them should always have this in mind. In general, there is a big chance these two will end up together, because their shared love of adventure is bigger than most of their troubles. 74%

Aries Compatibility with Cancer

Aries to Cancer Sexual Intimacy and Compatibility: Cancer is extremely asexual. The problem with sexual relations with an Aries is that Aries partners are usually not that gentle to begin with. Cancer and the Moon is a psychological challenge for all of us to understand that they are sexual beings. This would be fine if the members of this universal Cancer family weren't convinced in their asexual nature as well. Their emotional characteristics allow only for sexual relationships with meaning and enough tenderness. Only when they meet the right person to set them free, they come to learn about the other aspects of their sexuality. They need to learn to show emotion. For them intimacy is something built, not implied. If they manage to reconcile these huge differences at the beginning of their relationship and if none of them is forced to do anything they are not ready for, their attraction to each other should do the trick and their sexual relationship could become truly sensual and exciting for both of them. 70%

Aries to Cancer Trust: The trust is something different for this couple. Usually the problem they encounter is a trust issue when it comes to intimacy. Aries has a different view on intimacy. In the eyes of their Cancer partner they can seem pushy and even aggressive with an attitude that doesn't lead to anything close to relaxed. As much as Cancer would like to understand the straightforward nature of Aries, it will be extremely difficult to see it as anything other than beastly. There is also a problem with the way they show and recognize emotions. It may be hard for an Aries partner to understand that they are loved if someone only asks annoying questions, tries to tie them down and doesn't want to have sex. On the other hand, Cancer will probably feel violated in every way, unless Aries partner slows down and has an atypical show of gentle emotions. Usually any sort of mistrust is a consequence of the lack of ability to believe in each other's feelings for one another, for they don't really recognize them well. 50%

Aries to Cancer Communication and Intellectuality: Both of them have the same tendency to act on an impulse and cut the conversation short before they even got to the point of it. It is not their intention to react in this way, but they push each other's buttons and it is very hard for them to stay focused and solve the issue they talked about. Their interests differ too much, so even when they are trying to have a peaceful conversation about something impersonal, it is still a battle to keep the attention to the subject in question, whoever initiated the talk. Their only shared characteristic is the cardinal quality of both signs, which gives them a good understanding on each other's "ad hoc" personalities. This will make it easier for the couple to recover from all of the possible conflicts and misunderstandings. Still, in the eyes of a Cancer partner, this type of relationship doesn't have a purpose and they might find themselves fighting in a way they don't feel comfortable with. As their signs are ruled by Mars and the Moon, it is an archetypal story of hurt and emotional pain, so their intentions have to be truly pure. They have to treat each other in a gentle, thought-out way, measuring every word they say. This can be exhausting for both of them, unless they fully accept the fact that they don't need to change their personality, only the way they express it and make a game out of it. 20%

Aries to Cancer Emotions: Aries and Cancer are both deeply emotional, although Aries is often described as if they had an emotional disability. They are warm, passionate and have high expectations of their partner when it comes to scratching beneath the surface. Their boundaries may be too strict as they fear their own sensitivity and sometimes act like heartless soldiers. Cancer wears their emotions as a winter coat and hides them only when feeling ashamed to show them. They accept their emotional nature as a given and work toward realizing a personal world full of respect for their soft side. We often say that Cancer wants to have a family and raise children, but this is not due to their need to reproduce or stay in the house all day long, but because they need a safe haven for their emotional side and enough people to share their compassionate nature with. Even though these approaches to their emotions seem different, they understand each other's depth and in most cases respect each other in this area of existence, in case Aries leaves their impatient nature out of their relationship. The problem appears when they are supposed to understand how they feel about each other, as feelings are not easily shown when dealing with partner's personality they don't fully understand. 70%

Aries to Cancer Values: Aries values a lot of significance to someone's state of energy, focus and consistency, Cancer values the ability to stay rational and stable, qualities they have a difficulty achieving, or being in a state of emotional balance. Their values aren't even connected, except for the fact they both have the idea that some sort of future balance, that can be quite hard to achieve, would make them better. 40%

Aries to Cancer Shared Activities: Aries wants to devote to their physical body, sports and all the ways to keep their creative energy high, Cancer wants to sleep, dance and eat all they long. Mostly they share sexual activities and the time for rest, since Cancer probably has no intention of following that insane Aries pace.
30%

Conclusion: This relationship can be painful for both partners and needs a lot of work put into it in order to work. It requires both of the partners to adapt and make changes in their behaviour, while tip toeing around each other most of the time. It is not an easy road, but the rewards are such inner understanding of passion, full of emotion and the ability to create something truly unique. If they succeed, they will probably never be satisfied with a different partner. 47%

Aries Compatibility with Leo

Aries to Leo Sexual Intimacy and Compatibility: This is a warm and passionate connection. They have similar sexual preferences and they definitely take each other seriously, whatever the level of their relationship. As they both have extremely strong personalities, they could fight and make up all the time, but enjoy it in a way some Water signs might find crazy. They have a sexual connection that cannot be interrupted, changed or faded through time, since they are both individual sources of energy, waiting for someone to follow. Still, if one of them has issues with their ego, they could slump into an energy drain system, where they insult each other and destroy each other's confidence and libido. This is a very rare possibility, but it is always there when two signs that present an extraordinary soil for the Sun come together. 90%

Aries to Leo Trust: There are many trust issues in every Aries to Leo couple. Their strong convictions and the need for loyalty mend these problems in most cases. They both think highly of themselves and see how their surroundings react to their partner. Leo is a strong willed, attractive masculine sign, and no matter if male or female, they will have this magnetic aura around them most of the time. This will make Aries strangely jealous and possessive, ready to fight for what belongs to them. Exactly why this problem culminates, for no one can ever possess Leo, the King of the zodiac. Still, their mutual understanding for the passionate nature they share and the determination of both partners to solve any problem that stands in their way, might just make them stay together for years, building security and trust every day. 60%

Aries to Leo Communication and Intellectuality: At first, their conversations are unbelievably energized, full of admiration and respect. As they get closer, it can be expected for them to put in an emotional note, but not a very gentle one for their emotions burn like the Fire element they belong to. This is a certain promise of a lot of fights, loud statements and interruptions. Interestingly, as two very passionate people, they usually get over their fights quickly and don't care much about specific words spoken in the heat of the moment. The moment they cool down, their relationship will easily go back to normal and their sexual life will blossom every time they fight. They are interested in similar things. Aries is interested in Leo, while Leo is interested in everything great about Leo. This is a very nice distribution of interests, since Leo will cherish all the things Aries will say and do for them, and give it all back multiple times greater. This is, of course, the scenario of two healthy individuals in these roles. If one of them has greater psychological issues, their communications will turn to senseless talk about one of them, and their relationship will become a battlefield for the ego where they both constantly need to prove something to each other. 90%

Aries to Leo Emotions: Their emotional natures are very similar. Sun is connected to love, pure and simple, not the nurturing kind but the creative, warm, passionate and playful love. They understand each other's emotional state perfectly, even when one of them would like to flee from the passionate need of the other to give resistance to something small and seemingly irrelevant. Emotional compatibility of this magnitude is a cure for all other imperfections they might encounter, while evolving in time and spicing their sexual relations with even more warmth and emotion then the beginning of their relationship might have promised. 99%

Aries to Leo Values: Their values are almost as if these were the words that describe them. Anything concise needs strong Mars energy, the ruler of Aries, while Leo brings clarity to all. Their only problem is the need they both have to be the leader and the brave one in the relationship. It is not as if they both can't be brave, but they have a tendency to compare to one another and search for their role as a lead. This can come between them and manifest as a typical battle of the sexes (in case they are of the opposite sex) or as a fight for dominance of any kind, consuming the quality of their relationship only because of their need to be the one with dominant values in general. 95%

Aries to Leo Shared Activities: They have enough energy to do everything together, but the problem is they don't actually need the same daily routine. Aries loves to be active, walk, exercise and always feels excited when about to do something for the first time. Leo is a sign of fixed quality, a drama queen of signs, and the one that wants to show off and go where they can be seen. Leo's energy is easily focused to coffee shops and places where they can rest enjoy and be the centre of everyone's attention. This is a waste of time in the eyes of an Aries, always ready for something different and exciting. For as long as Aries has the will to initiate their shared activities and compromises for a show off from time to time, they can find things to do together and enjoy them. But if Aries partner gets tired and irritated by Leo's laziness, there is a big chance they will simply separate their activities all together. 65%

Conclusion: The relationship of Aries and Leo is passionate and turbulent, but they don't seem to mind an occasional fight and a sharp word. When they fall in love deeply, they are almost impossible to separate as they stubbornly hold on to the idea of their future together. Although they are not two of the most romantic believers in love, they are passionate in their beliefs and when they find love, they will fight for it until there is literally nothing left of their relationship. It is meaningless to advise gentle behaviour or looking for peace, because the entire world of their relationship is based on the element of Fire they share. It is pointless to look for peace, when the opposite of peace is what attracts them in the first place. For as long as they love each other and stay faithful and true, they will be tied up in a relationship they need to fight for every day. Their main objective is to find a way to enjoy the fight and have fun. 83%

Aries Compatibility with Virgo

Aries to Virgo Sexual Intimacy and Compatibility: Aries comes in as a brute with no manners or tact to sweep them of their feet with a passionate nature that looks superficial and completely unattractive. Aries may look at Virgo and think of Virgin Mary, her chastity and what we would call a total absence of sex. It is hard to say if Aries and Virgo would present the clumsiest or simply the worst couple when it comes to sex. There is nothing more asexual for Aries than a person without an obvious sexual identity. To express their sexuality or feel sexual at all, Virgo needs patience, verbal stimulation and a lot of foreplay. The real question is – how did these two get attracted to each other in the first place? Their intimate life can be good only in case Aries accepts to wait and communicate about things they don't find important at all, or if Virgo was so disappointed in their previous relationships that they turned into a sexual predator, open for an interesting turn in their intimate life. 10%

Aries to Virgo Trust: In most relationships, Virgo is faithful, but hates being lied to. In case they overcome their sexual difficulties and stay together against odds, their problems with trust shouldn't be significant. Aries usually has the need to be honourable and straightforward, except in rare cases when they cannot contain their sexual appetites. They have a need to be honest and ask for honesty in return. 70%

Aries to Virgo Communication and Intellectuality: This could lead to endless, pointless fights, because Aries will never change their nature, or their priorities, while Virgo will seem like a crazy person screaming, with gloves on and a huge bottle of antiseptic liquid in their hand. The good thing is that before they get to this stage of the relationship, they will probably find each other extremely repellent and break up instead. These two are very annoying to each other that they might annoy everyone around them. When you think of a partner who brings out the best in you, Aries and Virgo are the worst possible match. The downside of an Aries partner is their impulsive nature, readiness to fight and the tendency to lose their mind over something that might not be that big of a deal. The possible downside of Virgo is hysteria and continuous, never ending talks, when they are not understood. It is a known fact that Virgo likes things clean. Well Aries is like an animal in their cage, especially if they are crazy enough to decide to live together. To good that could come out of this strange bond is their intellectual cooperation, in case they share the same interests or work. They will awake each other's intelligence, challenge each other's mind and probably think of entirely different, but constructive solutions for problems that might occur. 30%

Aries to Virgo Emotions: Aries would think before acting, not a usual thing they would do. If they knew each other as friends, going through their emotional experiences with other people prior to them becoming a couple, they could know each other well enough to make their relationship work. When Aries goes from friendship into a sexual relationship, they tend to be much more considerate and gentle. When it comes to emotions, we could say that their emotional compatibility is better than their sexual one. Still, as Virgo is primarily an intellectual sign, a sign where Venus falls and the lack of emotion is evident, and Aries usually mixes up love and sexual attraction, it is hard to achieve a quality emotional connection between them. Their best chance for love would be the silent observation by Virgo partner for some time before they get together, because this would give a rational advantage in knowing the person they are starting a relationship with. Without sexual involvement Aries is more tolerant and a better listener, so friendship will provide more substance to their romantic relationship. 20%

Aries to Virgo Values: Their relationship could be based on their joint business though. This would give more meaning to their conversations and everyday life. They both value hard work and ambition, as well as clear and sharply deduced information. This is what makes them great as colleagues, but this is not exactly the most important set of values a happy couple would share. Other things they value don't coincide that much. Aries is all for bravery and an attitude while Virgo thinks of these as stupid, unless they are a part of tradition or have historic significance. Virgo values intelligence while Aries thinks success has nothing to do with it and sees it as a possible reason for loneliness and sorrow. Still, these would rather be the reasons to tease each other and have a nice laugh, than they would have the capacity to tear their relationship apart. 50%

Aries to Virgo Shared Activities: This is a couple that could go for a run because it's healthy, spend time in the nature because it's healthy, think about their bodies together because it's healthy, have regular intimate relations, once a week, because it's healthy and to sum up, anything that's healthy would be easy to incorporate in their relationship. Also, it is a very good thing that Aries doesn't care much about "empty time", such as watching television or opening solitaire, because Virgo would rather read or clean than subject to these activities for the "permanent damage of the brain". It is a good thing that Virgo is a sign of mutable quality and always concerned about their health, or they would never think about following Aries to their activities. Still, there is always a chance that Virgo will use their health to get out of these activities and spend some time alone. The problem in this area of their relationship is connected to activities that don't leave much sense of dignity or make careful Virgo feel scared, or the activities too boring for an Aries personality. 70%

Conclusion: It's a good thing that the relationship between an Aries and a Virgo is never boring. Although in most cases they are not really meant to last, it can still be a fun experience if none of them takes their potential for a shared future too seriously. In case they take the best out of their relationship, giving it enough freedom and unpredictability,

Virgo would incorporate some of Aries' energy, while Aries would allow Virgo to teach them how to organize their thoughts and communicate calmly. This way they might come to the point where their relationship could actually last and the outcome depends on their ability to relax and have fun together. 42%

Aries Compatibility with Libra

Aries to Libra Sexual Intimacy and Compatibility: Aries is a sign of Saturn's debilitation and Libra exalts it, so their main issue is the lack of emotion and poor boundaries when it comes to sex. Saturn can cool things a bit too much and be a challenge to overcome in their attempts to get sincerely close. Aries and Libra are both signs of masculine nature, and so they are a primal opposition. When they engage in intimate relations, it is expected for all their libido and possible problems with sexual expression to surface. Attraction they feel toward each other is great, but their signs combined present passive-aggressive behaviour in general and as a couple they could have a tendency to hurt each other in intimate relations. When they connect through real emotion and respect each other's boundaries, they have a potential for a very good sex life, as Aries gives initiative and energy to indecisive Libra, lifting their libido and Libra awakens the fineness of Aries, teaching them how to be selfless lovers and enjoy thinking about the satisfaction of their partner. 80%

Aries to Libra Trust: Since Aries doesn't put much time or thought into their actions, the lack of conversation about every single detail from their personal life could easily arise suspicion in the mind of a Libra. Trust is not their forte and problems with it could torture them for years. Libra partner has a problem with insecurity in general and needs to show their worth through relationships with different people. They love to be loved and seem to be hungry for the approval of those around them. Aries finds this stupid but easily gets jealous and threaten their mutual sense of stability and belief in other person's choices. Due to Libra's lack of confidence, it is also possible that they will doubt everything their partner does. The most important thing here is for Libra to work on their self-esteem and keep their focus on their own life instead of trying to blend into the life of their partner. 40%

Aries to Libra Communication and Intellectuality: They have conflicts, but Libra tries to flee from most of the time, their communication usually serves to feed the hungry Sun of a Libra partner, or Aries' hungry Saturn. Their opposition covers the points of debilitation and exaltation of Saturn and Sun, and this is mainly shown in their communication and everyday functioning. This means that their role in each other's life is quite simple. Aries needs to boost their Libra partner's spirits all the time, showing them how capable and brave they can be, while Libra takes on the responsibilities of their Aries partner and shows them how to reach a certain goal. All of this can be quite tiresome at times, especially if one of them has a problem with this unconditional role play, or doesn't recognize the effort of their partner. Mostly they will talk about their daily activities and events since they don't share many interests. While talking about different activities and people, they find a common language as Aries helps Libra not to obsess about others and Libra helps Aries to understand different views than their own. Their communication might be great if they were in the same profession or at least share a workplace, because that would cover the basic interests they share and give them more space to find the middle between their opinions. 55%

Aries to Libra Emotions: Libra has good understanding of the nature of Aries. They don't understand their actions and their way of display of emotion, but the core of emotion and sensitive personality is easy to reach from their perspective. As crazy as this may seem with lack of qualities their relationship might suffer, this is a couple that understands each other very well when it comes to emotions. Aries can awake Libra's ability to show them because of their own openness. This is something every Libra needs, as they have trouble letting their guard down. Libra, on the other hand, has enough depth to look inside Aries personality instead of superficially examining their behaviour. It is safe to say that this is a couple that could solve any issue with love they have for each other and although their troubles could be great, this is possibly such a deep emotional connection that all problems fade next to it. 99%

Aries to Libra Values: Libra knows the way for Aries to reach their goals by discovering new values in relationship with them. In general, their individual values are different in so many ways, but it is exactly the purpose of their relationship to question them and set them straight. Aries values direct, energized approach and outspoken people. Our values set the direction that leads us to our goal for personal development. Aries has a goal in the sign of Capricorn for this is the sign in their tenth house. Capricorn is ruled by Saturn that exalts in Libra. In the practical sense, this means that Libra helps Aries achieve their goals, while following necessary values. This is an interesting observation because the sign of Aries is the sign of Saturn's debilitation and doesn't seem to understand the set of values or exact steps that would lead them to their goal. Libra values tact, fineness and prestige. While Aries gives their best to live in the now, Libra examines the past to set distant targets in the future. They have a lot to learn from each other, but if they do, they might just set their mutual values somewhere in the middle. 70%

Aries to Libra Shared Activities: They don't maintain their relationship, but still it might work, but only if Libra partner lets go of their idea that they need to include their partner in everything they do. Aries could help Libra by supporting their independence in any possible way, while accepting involvement in a part of activities Libra cares about. This is the couple that finds it very difficult to coordinate their activities. They want to do opposite things most of the time, and the only activity they always agree on sharing is sexual activity. Although this is a pillar for a good relationship and everything else they cannot share might seem irrelevant for some time, they need to find a way to do something else they both enjoy. 30%

Conclusion: Libra is the sign of relationships in general. Any problem they might have with each other is something to be worked on, because it shows what their personal problem with any relationship is. When they are madly attracted to each other and fall in love, there is almost nothing that could separate them, no matter the differences. However difficult it might be to reconcile these two natures, remember that this is a primal opposition that represents partners by signification. Aries and Libra are the couple of the zodiac, as much as any other opposing signs, for they are each other's seventh house, house of relationships. Wouldn't we all like to find the middle ground with our loved one? They need to work on their bond, that's a fact, but their relationship is a promise of a perfect fit of two souls meant to be together. 62%

Aries Compatibility with Scorpio

Aries to Scorpio Sexual Intimacy and Compatibility: It is a good thing that Aries rarely belongs to this category, for it is a sign where all conservative and rigid opinions have fallen with Saturn. If Aries and Scorpio find an understanding inside their sexual relationship, they will probably become the atomic bomb of all sexual experiences you can think of. Still, it is hard for them to find their shared language. Pluto is known for its destructive qualities, usually related to sexual repression and it can intensify all things, sex primarily. So they are basically a combination of everything we don't want to deal with when it comes to sex, taboos and instinctive sexual behaviour. This is a contact that lacks pleasures and tenderness of Venus. Aries and Scorpio are signs with an unbreakable bond. Aries is our first breath, Scorpio is the last. They are two sides of the same coin, both ruled by Mars, a planet of instincts, necessities of the body and sexuality as one of these. When they are in a sexual relationship, it can be difficult to set all of the aggression aside. Not only are they both ruled by Mars, but Scorpio is ruled by Pluto, too. Both signs are the opposite of one's ruled by Venus and represent positions where Venus is in detriment. We could say that this means "lack of love", but it is not quite that simple. Since Scorpio is a Water sign, it is connected to our deepest, darkest ability to love. Scorpios need to feel emotion in their sexual experiences, but due to suppressive nature of our society, can live out some weird sexual scenarios that may seem "sick and twisted" to more conservative zodiac signs. They are, in fact, completely different. Aries likes things "straight" and simple. Scorpio, on the other hand, has a slight need to manipulate, play a game of seduction and takes sexual relations very seriously. They always want to transcend all of their previous sexual experiences and find someone they can merge their Soul with, to possess and adore until they die. Aries is much more simple and masculine when it comes to sex. It is a physical need that needs to be met. They usually have to build emotion inside a sexual relationship as they get to know their partner. This relationship's real possibility exists only if they share the need to satisfy one another and treat each other with enough tenderness. 50%

Aries to Scorpio Trust: They are both jealous and possessive by nature. Aries likes to win and be the best lover and partner anyone has ever had. Scorpio wants to be the only one that was ever loved by their Aries partner. As opposed to sexual compatibility, this issue is easy for them. "If you lie, you die." Not literally, of course, but a small lie could easily end their relationship. If they have doubts about each other's actions, it is very likely they won't last very long. 90%

Aries to Scorpio Communication and Intellectuality: Scorpio is too dark and difficult for Aries. Aries is too shallow from Scorpio's perspective. Aries will probably tap their foot impatiently while Scorpio goes on and on about all those deep and meaningful things. From the perspective of Aries, this is something nobody should think about, let alone talk about all of the time. This wouldn't be expressed as boredom (although this is always an option with Aries), but more as a need to act and stop obsessing about everything. What they both enjoy though is their shared ability to give a lot of information in only a sentence or two, but this could interfere with their communication even more, since they might say everything they need to in a couple of minutes and have nothing to talk about afterwards. 20%

Aries to Scorpio Emotions: There is archetypal "battle" of Mars with the Moon. The rejection of one's emotional Self and too much roughness in order to survive. It is really easy for Scorpio to get hurt here. It seems like they jumped into this relationship only for this reason, so they can repay some sort of a karmic debt. Aries will probably never know or understand what happened in Scorpio's emotional world because they simply didn't sense anything. They don't have a strong affection to emotion in general and they are both trying hard to be strong and unemotional. Since there is no one here to keep the emotional balance between them, it will be very easy for them to openly "cut" one another, possibly many times, before one of them decides to cut their bond entirely. 1%

Aries to Scorpio Values: They both value bravery and things that are concrete and clear. Still, they part ways in their further processing of these. While Aries considers something is done with as soon as it's cleared, Scorpio will dig for reasons why it would be unclear, or was unclear in the first place. So when together, they would both feel the need to clear things up, but Scorpio will obsess about them even when issues are solved and find new details that need to be cleared up, again and again. They need to be productive and fully independent, or they will drive their Aries partner crazy. When it comes to bravery, Aries thinks of bravery as a knight's tale, something to show when you are wearing your sword, while Scorpio thinks it is brave to sink into the darkness of the mind, go to the underground, the underworld or challenge the devil himself. This is exactly where the difference in their deep levels of the nature of Mars comes to light. Although everything seems the same, nothing is even remotely close to being similar at all, as soon as you scratch beneath the surface. 40%

Aries to Scorpio Shared Activities: You could say that their main shared activity is sex. Everything else is secondary anyway. 99%

Conclusion: They are most aggressive image of Fire and Water element. Fire evaporates Water, just like Aries shatters Scorpio's feelings. Water damps down Fire, just like Scorpio wears Aries out. They seem to bring out the worst in each other and this is nobody's fault, it is just hard to reconcile so much focused energy that moves in two different directions. Their relationship is like the process of nuclear fusion and often just too much to handle. 48%

Aries Compatibility with Sagittarius

Aries to Sagittarius Sexual Intimacy and Compatibility: Sagittarius is passionate about their cheerful personality. Sagittarius really cares about their opinions, convictions and moral value. They can spend their entire life analyzing these to see if they are wrong or right and search for the universal truth. When it comes to their optimism and good mood, they passionately protect them from anything too serious or hard. When Aries and Sagittarius engage in sexual relations it can be quite funny. Sagittarius partner has this innate ability to make a joke out of almost anything. The seriousness of an Aries when sex is in question is something that gives Sagittarius a strong impulse to make a joke. These are two Fire signs, both very passionate, each one in their own way. Aries is passionate when it comes to action, new things and of course – naked people and specific sexual positions. If they let someone taint them, it would shake their conviction that they should always smile and find a reason to be happy. Although Aries can be a bit vain about their sexual abilities and performance, in most cases Sagittarius is able to break this wall of strict, sexual tension and lead them to a more relaxed zone where they can relax and experiment. 95%

Aries to Sagittarius Trust: Aries feels they can share anything with their Sagittarius partner. The problem could appear if they have different views on the seriousness and depth of their relationship. If this is the case, usually a Sagittarius partner sees Aries as a short term, not that important partner. This is why they could easily cheat on them and probably wouldn't even call it cheating. Aries and Sagittarius are both aware of the excessive need for honesty in their life. Usually they don't have to talk much to understand each other and can easily spot when the other one is lying. This makes it extremely hard to create a situation of mistrust, especially because of the feeling of security Sagittarius partner gives to Aries, by taking everything in with dignity and serenity. In return, Aries partner that values their relationship more, would jump into their possessive nature with even more ease and never trust their Sagittarius partner again. 70%

Aries to Sagittarius Communication and Intellectuality: Aries gives initiative and focus and Sagittarius gives vision and faith. These signs are ruled by Mars and Jupiter, which means that they could have some disagreements on their convictions. In case these are not convictions they think of as their personality's foundation, this shouldn't be a huge problem. Still, it is possible for their set of beliefs to differ too much for them to even understand each other. When this happens, they fight whenever and wherever they can, since none of them has the ability to let their convictions go. This is a wonderful bond that is often seen in friendships that last for years. Their mutual understanding can be so deep, that even if they lack physical attraction, they would gladly substitute it with a life spent in this kind of intellectual relationship. They motivate and push each other wherever they might like to go. When they are together, they make each other feel as if nothing is impossible. Aries because they want to win, and Sagittarius because convictions are their forte and something they have surely thought about a lot. 90%

Aries to Sagittarius Emotions: This is a love that could last for a very long time, for as long as their respect their personal needs, individuality and the distance they possibly need from each other every once in a while. Although they are not considered very emotional, it is a mistake to assign emotionality only to the element of Water. This is an element which works from the heart and you can feel it in your chest. In search for an explanation of emotional nature of Fire signs, you should just imagine that warm feeling in your belly and that would be the best possible description. When they fall in love with each other, deeply and sincerely, it is almost possible for their passers-by to warm up in the middle of winter. These are extremely warm signs, due to their corresponding element of Fire, open for any kind of activity just to share time together and feel that wonderful emotion in their stomach. Their emotions are active, warm and on the move. Always changeable but creative and there to move them anywhere they want to go. 90%

Aries to Sagittarius Values: Aries has the idea of honour and heroic "sweep off feet" logic. In time, they both must have realized that Sagittarius gives this idea a new step up and brings it into a world of royalty. Not only does Sagittarius value honourable and heroic people, too, but they value honourable people with blue blood that give money and food to the poor, every day. Aries partner values things that are brought up to a higher level by their Sagittarius. In time they will both understand that Aries grows through this relationship and widens their entire system of values. Their main difference is in the fact that Aries values things concise and clear, while Sagittarius will easily disperse and go around the point for days. This can be met through their mutual value of truth, so honesty can be their cure for anything. 70%

Aries to Sagittarius Shared Activities: Their activities can be shared and fun, when they get together. This has nothing to do with their needs and tendencies, but with the potential of their entire relationship. They can go for coffee and they would have fun, but they could also go bungee jumping together and have even more fun. Aries is a sign in which Saturn falls. This means that they easily get tired or just bored and they always need new and exciting stimulations. Sagittarius is a sign of mutable quality, ready to change whatever needs changing in order to feel good. It is all the same to them. They are fully capable of respecting each other's personality, so even if their wishes for certain activities differ, this would be easily dealt with. 99%

Conclusion: This is definitely a couple with lots of potential. They might have to stand up to their environment and defend their feelings from others, but this won't shake them too much, for neither of them thinks that much about the opinion of others anyway. If they manage to mend their philosophical differences and respect each other's different opinions, they could become one of the warmest relationships in the zodiac. Their main relationship advice would be to always tell the truth to each other and not go crazy about their healthy differences. Their differences are exactly the thing that could make their sexual life more exciting. 87%

Aries Compatibility with Capricorn

Aries to Capricorn Sexual Intimacy and Compatibility: Saturn puts too much pressure on Mars and takes a lot of its energy. Their relationship will result in lack of sexual desire, the mutual feeling of incompetence or even impotence of one or both included parties. When this sort of relationship happens, it is in most cases triggered by some deep unconscious need to be held back and restricted when it comes to sexuality. As with everything that comes through the sign of Capricorn, with time Aries partner could achieve some sort of balanced state in which they are sexually satisfied and their instinctive needs are met. Unfortunately, Capricorn partner will lose their energy and the need to participate in this sort of sexual behaviour by that time. This is a very difficult combination of signs when it comes to sexual compatibility. Rulers of Aries and Capricorn are Mars and Saturn. These planets are considered archetypal or karmic enemies. When it comes to sexuality, it is mostly signified by Mars and its contact with Saturn may result in all sorts of physical and objective obstacles on the way to a healthy sex life. This will ultimately lead to their separation, for there is nothing light or easy with these two, especially when it comes to intimate matters. Because of the unconscious type of their relationship, they could be insanely attracted to each other, but in most cases their differences will keep them at a safe distance. At their best, Capricorn will support Aries' libido and control their passion to burn as slowly as possible. At the same time, Aries would consider their partner a teacher and learn about their body and the way to satisfy them. Still, this is a balance that is extremely hard to achieve when the clash of these two hard personalities happens. 5%

Aries to Capricorn Trust: They have deep misunderstandings in other areas of their relationship, but they will rarely betray each other's trust. This is something they will easily take for granted though, for when they are together, they seem to lose awareness of the things between them that should be treasured. Since they are both in extremes and "all or nothing" types of people, it will be easy for them to trust each other. One of them should have the sense to remind the other from time to time about the qualities that their bond includes. 99%

Aries to Capricorn Communication and Intellectuality: Capricorn partner does not allow their impulsive and from their perspective even stupid Aries partner to have their own opinions and value them as useful or practical. Although they will certainly respect their initiative and energy level, the rest of Aries behaviour is simply unacceptable in most matters. Since Capricorn has their feet on the ground and is capable to measure the situation rationally, they will hold on to their calm opinion about their partner's lack of tact and endurance or even of their idiotism. They should keep their conversation in touch with carrier goals, achievements at work and physical activity, for Capricorn exalts Mars, the ruler of Aries. Other than that, they don't have much to talk about. You can imagine how annoying this can be to an Aries, especially if you take into consideration their passionate need to set strong boundaries and be respected in every way possible. On the other hand, Aries will simply have no patience for their Capricorn partner. They will seem boring and as if they only want people in their life to be useful. This will be attributed to their selfishness and lack of emotion and heart. Both of them will be wrong in a way, for they would need to understand what they both could become if the right person or motivator came along. The problem is they could remain stuck in this pointless ego battle, until they both get so tired that they will hardly think of another relationship ever again. 20%

Aries to Capricorn Emotions: They will both probably have an image of the other person as someone they could become after some effort is put into their growth. The problem is that no one here wants to change. Their emotions could easily be connected to this first image they've had, however unrealistic it might be. So this is exactly what we would call the lack of understanding, because these two could be crazy in love with each other's false image and crazy persistent to change one another. In this special case it would be best not to call this paragraph "emotions", but "understanding". Their problem is in the way they understand each other to begin with. Their problem really isn't the lack of emotion to each other (although you could easily contribute their issues to emotional disability considering their Mars/Saturn nature), but the lack of understanding and acceptance of each other. They are often too stubborn and narrow minded to see that there is something different than what fits into their boundaries of possible personal characteristics. 1%

Aries to Capricorn Values: The problem they have is in their unrealistic expectations founded on the fact that they share some values. If Aries values someone's ability to endure and push themselves over all possible personal barriers, that doesn't mean they are ready to become this person, or have the control to be one. They both value independence, clarity and honesty, and in general their system of values is not what brings problems to their relationship. Mostly they are in sync when it comes to serious view of people they are surrounded with. Similarly, Capricorn might value speed and focus of some people, but this doesn't mean they will jeopardize their depth or attention to detail just to be faster, or leave deep psychological needs unattended only to stay focused at some goal. 60%

Aries to Capricorn Shared Activities: Aries doesn't understand is that studiousness always has its value, as much as Capricorn doesn't understand the physical need of their body to rest and act on impulse from time to time. They can easily find activities they both like when they include physical movement. They both need to pay attention to their bodies and everyday habits, so it would be good for them to have the same physical activity every day at the same time together. This way they would motivate each other. Aries will lift up Capricorn's energy and Capricorn would give endurance to their shared efforts. Capricorn sometimes just doesn't understand why Aries wants to run at 6 in the morning. Aries doesn't understand why Capricorn would spend all night doing something extremely boring, just to be thorough when it has no value in this particular matter. This could do wonders for their entire relationship, for they could understand each other's natures much better through this sort of basic activity. 40%

Conclusion: Their only chance of success is unconditional respect and the wideness of their views and expectations. This is not an easy relationship. None of the partners has any trace of lightness and blissful ignorance. This is why their relationship might seem like a competition to ruin the relationship in the best possible way. It is hard to say who will get out of it a winner, for they will both feel lousy most of the time and be relieved that they finally separated. If they stubbornly decide they love each other too much to let each other go, both of them would probably bang their heads against a wall for years to come. They could truly complement each other, but only in a scenario where they would look for good in one another and highlight each other's qualities. Unfortunately, the malefic nature of their rulers rarely allows for them to be this positive and acceptance oriented. If they got together, and whatever their story is, they should think about the things they could learn from each other instead of looking for each other's shortcomings, and always stay out of each other's business. 38%

Aries Compatibility with Aquarius

Aries to Aquarius Sexual Intimacy and Compatibility: This is a relationship that could bring out their worst nature and simply emphasize that they are a sign ruled by Mars, a cold, unemotional sexual hunter. While this can be really exciting to both of them, it will not be very fulfilling, because they both need to feel loved. There is an excess of masculinity and energy that could lead to very turbulent relations. Their roles are easy to understand with Aries giving energy and stamina to their Aquarius partner, and Aquarius giving crazy ideas and widening horizons of their Aries. Sexual contact between signs of Aries and Aquarius can be really stressful or extremely exciting. Usually it is both. Their signs go well together in general and they support each other easily, since they both have a lot of energy to follow one another. Still, when it comes to their sexual and intimate relations, they could lack emotion. Aries is a passionate sign with lot of warm, creative emotions. This is very fun at the beginning of their relationship, but after a while, it might get tiresome for there are not enough ideas to cover the emotional emptiness they could encounter. 65%

Aries to Aquarius Trust: Aries is ruled by Mars and needs to be the only one in the world that their partner ever lays eyes on. This could turn them into an angry, possessive person who obsesses about the movements of their partner. When we are discussing matters of trust between them that don't include other people, it is safe to say that they don't have a problem. They both simply don't understand why they would lie when there are so many interesting truths to discover. Trust is an important issue for Aries and Aquarius can understand that. This doesn't mean they will be faithful to their Aries partner forever, but they would think it is fair to keep an open relationship and tell them about their indiscretions. They need to be free to speak their mind and accept that they will never avoid conflict, but that it can be used in a constructive way to better understand each other and strengthen their relationship. 85%

Aries to Aquarius Communication and Intellectuality: Aries could find an idol in their Aquarius partner and full-heartedly enter any dialogue because they are excited about what they might discover and how their perspective would change. On the other hand, Aquarius enjoys this role in their partner's life due to their ego issues with the Sun positioned at this sign. They will share their thoughts with their partner, trying to be as interesting as possible. Aquarius is motivated by their Aries partner and enjoys making tiny jokes at their expense. Their conversations can be so exciting that many people would like to jump in. Aries is often kind of serious and asks for their boundaries to be respected. Aquarius partner will recognize this, laugh and shake their entire world. It is unimaginable to Aries, always moving straight, for someone to have such an open mind, going back and front, having new revelations every day and never losing energy for new, different topics. It is important for Aries not to take things personally when it comes to Aquarius humour and they might have a lot of fun together. Because of their strong natures, filled with energy, they could fight most of the time. In most cases, Aquarius will not stand for ridiculous conflicts and will build a brick wall somewhere between them if needed. Still, they usually tear it down at the end of the day, for they cherish each other the way they are after all. 90%

Aries to Aquarius Emotions: Aries and their partner are cold, distant and have no intention of opening their heart for them. Aquarius sees things differently and tries to stay rational at all times. In order for Aquarius to awaken their emotional nature, it usually takes a partner with enough flexibility and patience to get there. We wouldn't exactly say that Aries is patient, so you can imagine the problem that could appear. When Aries starts asking for the show of emotion, the true problem surfaces, for Aquarius might have shown how they feel the entire time, but no one would guess what they were showing. 40%

Aries to Aquarius Values: Although they can share a great conversation, their values go their separate ways as soon as they touch the subject of freedom. They both value freedom by first impulse. But in time, Aries realizes that they don't really value freedom that much when they see it at work. In fact, they would often change everything in their lives only to take away the freedom from their Aquarius partner. This is not a conscious need, but Aries can be like a spoiled child wanting things (and people) all for themselves. So with Aquarius changing direction as the wind and never changing their nature, Aries can find themselves truly unhappy for they want someone to share everything with, not only what the wind carries in. 30%

Aries to Aquarius Shared Activities: Unless Aquarius suggests something truly unacceptable to their Aries partner, they will have an abundance of possibilities when it comes to their shared activities. With so much energy, their only mistake would be to stay at home and not share a chance to get all that energy out of their systems. 99%

Conclusion: It is difficult for Aries to grow through togetherness and learn about their emotional nature. This is something they will never be satisfied with. This is a couple that lacks tenderness. They are not two brutes who let their relationship fade as soon as their passion does, but the distant examining look of Aquarius can take out the emotion out of it. Aries partner needs to be relaxed by their significant other, so they can melt down and show their true, warm emotional nature. In this relationship, they would have a distant partner that basically supports their primal, instinctive nature. Still, every relationship with Aquarius can surprise us as much as any individual Aquarius could. With them as a partner, there is always room for an enlightening scenario that leaves all things to free will. In case they decide to share their lives together, they should have a screaming room they could individually visit once in a while. This would probably do the trick. And about that lack of emotion, they could just put in a lot of physical tenderness to begin with and let things go from there. 68%

Aries Compatibility with Pisces

Aries to Pisces Sexual Intimacy and Compatibility: Their connection is like a "little death" making room for all that is new, untamed and inexperienced. It is hard for them to bond, as much as it is hard for all of us to transcend, go beyond our physical body and be one with the Universe. With that said, it is understandable how difficult it is for their sexual natures to accept one another. Aries stands for instinctive sex. The sign of Pisces stands for orgasm. Although Aries cares about their orgasm, they will not make an art out of it. Pisces would rather satisfy themselves than be with someone who doesn't understand the art of orgasms. Aries and Pisces are two signs that really have trouble connecting. The beginning of all things lies at 0° of Aries and their end at 29° of Pisces. When they end up together, it can be torture for both, because they just don't understand what each of them needs. Aries would even have some success in understanding the need for tenderness and physical touch, but what Pisces want is like an unreachable wonderland that no one needs. In fact, they simply don't understand what it is they need. Aries looks like an inexperienced child to their Pisces partner, and although this can open the door for Pisces to enter this relationship, it does not feel that good when they realize that this is not about to change. If they are both open enough to find their intimate language, their sex life has to be weird and kinky if they want to succeed. Pisces will feel suffocated in anything ordinary and less satisfying than what they know they deserve, while Aries is usually not very interested in sharing emotions all night long and waking up in the afternoon. 20%

Aries to Pisces Trust: Pisces will stay in their little world for as long as they can, only to avoid being hurt and lied to. With Aries holding their head high, their attractive, straightforward attitude and their libido, it is not easy for sensitive Pisces not to pick up those signals emitted all around. This will immediately give effect to the degree of their confidence. Aries will see their partner's world as a phony, unclear image that there is no need for, and find their Pisces partner shady and unworthy of their trust. 1%

Aries to Pisces Communication and Intellectuality: They don't complete each other, but the effect they have on each other can be like the correct medicine. Aries has a tendency not to look behind, question the past, or be too sharp and fast for their relationships. They could also have an ego with a shotgun, waiting for any potential partner to pass by and kill their desire to even think about dating an Aries, let alone be serious about a relationship with them. Aries and Pisces could find many things to talk about if they open up for each other's support and advice. Although they are interested in entirely different things most of the time, they are still connected as neighbouring signs and have a way of leaning on each other. Through their relationship they need to learn about their own weaknesses and how to mend them to be complete. Pisces are sensitive enough to explain to Aries how they should soften up but keep their boundaries strong. Pisces represent a dream land of Aries and they are able to show them that they could actually have a mission and a higher purpose, instead of just chasing through life. In return, Aries partner will help their Pisces partner find their grounding. They will not be that gentle about it, that is guaranteed, but could be realistic just enough to show Pisces how important it is to have initiative and build something you dream about in the real world. If they start their intimate relationship on these foundations, they could easily discover their middle ground for other segments of their relationship. In case they are not so open to change and are not in search of someone to help them create, they will hardly share many topics they both find interesting. 70%

Aries to Pisces Emotions: Aries to Pisces emotional worlds are like two different planets that rule their signs, Mars and Neptune. While Mars, the ruler of Aries, is covered in rust, a red colour desert with volcanoes, canyons and weather, Neptune is a blue gas giant, cold, whipped by winds and much farther from the Sun. This is exactly how their emotions differ. Those that Aries cherish most are well defined, strong, protected, and colour in a colour of passion. Pisces on the other hand, have a windy and changeable emotional world, colour blue like the colour of sadness and vision, and are easily cooled down as soon as they feel disappointment. 5%

Aries to Pisces Values: The core of these values is different for the two of them. It is strange how they both value honesty and have such trust barriers when they get together. When they get involved, trust becomes something like a sole purpose of their entire relationship. They will also both like fairytale heroes and value the usual pride, chastity and bravery scenario. Aries representatives will value them because of that sense of strength, power and because of the role of that one and only hero, smarter and braver than everyone else. Pisces value them for their ideals, happy endings and those utopian relationships between those few worthy men and women. 35%

Aries to Pisces Shared Activities: Pisces partner knows that sports are healthy, but they need to attach them to something from their infinite world. They could share a walk in the forest, or engage in water sports. Other activities that Aries would gladly take on are not "spiritual" enough for Pisces. Pisces always need to have "a second perspective" and this can seem crazy to their Aries partner. So water sports are fine, because of all the secrets of the water, the view of the ocean, being underwater and contemplating on the purpose of life, or a dive in the pool. Walk in the forest can be beautiful because they can hear the birds, trees saying "hello" and wait for two owls to rest on their shoulders. In the world of Aries, things are really so much simpler and if they want to enjoy something, they will simply go and enjoy. In the same manner, they would run when they run, practice when they practice and watch the ocean when they watch the ocean.

40%

Conclusion: Pisces is ruled by Neptune, in charge of our entire aura and our permeability for outside stimuli. This is a relationship disturbed mostly by the lack of trust and the ability of both parties to open up to their partner. Aries is ruled by Mars, the planet that rules our first chakra, responsible for our ability to set good boundaries. Since they are both responsible for our border with the outside world, it is hard to say which partner should loosen up and make it possible for them to come close. Their only chance of a happy ending is if Aries partner dives in and their Pisces partner wakes up. 29%

PART-2
TAURUS HOROSCOPE

(April 20th - May 20th):

Who loves good things and smiles through life except when crossed?

Who is stolid, tenacious and determined and thinks he knows the most?

Element: Earth

Quality: Fixed

Color: Blue, Pink, Green

Ruler: Venus

Color: Blue, Pink, Green

Lucky stone: Diamond, Emerald & Blue Sapphire. Wearing Diamond or Emerald or Blue Sapphire can bring wealth and makes a better person and can give him/her strength.

Lucky Number: 2, 4, 6, 11, 20, 29, 37, 47, 56. Number 2, 3 & 8 are best for good fortune.

Lucky Colour: Lotus pink or Shades of verdant green will do a world of good.

Day: Friday, Monday

Lucky Day: Monday is the day of new beginnings

Flower: Daisy

General Insights: Powerful and reliable, Taurus is the first when it comes to harvesting the fruits of his labour. They love everything that is good and beautiful, and they are often surrounded by material pleasures. People born under the Taurus sign are very sensual and tactile. Touch is extremely important for them, both in business and in romance. Stable and conservative, Taurus is among the most reliable signs of the zodiac. Stubbornness is a trait that is forcing him to expel things to the end, in order to comply with the standards. As an earth sign, Taurus can be overprotective of their loved ones. They are great in making money and they will stick to their projects until it is successfully completed. Bulls are often known for their stubbornness, but it can also be interpreted as a complete commitment to the execution of tasks. This makes them excellent workers and great friends, because they are always there, no matter what. Their ruling planet Venus represents love, attraction, beauty and creativity. Thus, Taurus can be an excellent cook, entertainer and artist. He is loyal and doesn't like sudden and unwanted changes. Taurus is the most dependable sign of the zodiac. Although some may have very conservative views of the world or can be too fond of money and wealth, they have the ability to bring practical voice of reason in any chaotic and unhealthy situation.

Personal Quality: The Taurus is stable, stubborn, generous, highly reliable, practical, ambitious, and good in the position of managers and achieves almost everything in life. He/She makes friendships very rarely but, once made, he/she is faithful. He/She is reliable, responsible, affectionate and loyal. He/She is easy to get along with and good team player. He/She can reach to the desired height with hard work, devotion and patience. He/She is practical, reserved and is possessing tremendous willpower and self-discipline. He/She is incredibly and uncompromisingly loyal. His/Her economic position will be good from the age of 35-46 and becomes a rich man in the society. His/Her early part of life is very struggling. His/Her children are generally intelligent and bright and he/she has a pleasant married life.

Positive Quality: He/She is reliable, patient, practical, devoted, responsible, stable, and helpful and does a lot of things for family and considers it as a sacred duty. He/She has ability to concentrate and never leaves anything unfinished and does not believe in shortcut of anything. He/She is warm, loving, gentle and charming most of the time. He/She is honest, reliable and loving.

Negative Quality: He/She very seldom changes the mind once made-up and does not bother at all for the result. He/She is stubborn, possessive, uncompromising, expressed in dullness, stubbornness and resistance to change. He/She is very suspicious and is afraid of getting deceived. He/She believes to the person so easily that anybody can cheat him/her. He/She cannot forget or forgive people so easily.

Physical Appearance: His/Her stature varies from short to medium to stocky. His/Her eyes are bright and soulful and he/she carries himself gracefully.

Relationships: He/She is intense and passionate. He/She demands perfection from mate and is exacting. This makes him/her ardent and fascinating lovers. He/She makes charming company and is loyal and devoted.

Career: He/She has a wide spread of potential careers, right from banking to the fine arts. He/She will hold on to one job for the rest of his/her working lives. Here are some occupations that a he/she might consider such as Advertising director, Antique dealer, Business person, Cashier, Clothing designer, Financial advisor, Florist, Patron of the arts, Perfumer, Real estate agent, Singer, Venture capitalist and Woodworker.

Health: Traditionally, he/she is endowed with a vigorous constitution and splendid health. If there is a weak point, it is usually his/her throat or neck. He/She is hopelessly addicted to food and alcohol.

Ideal Partner: He/She vibes best with one of Taurus.

Compatible Zodiac Signs: Scorpio, Cancer, Virgo, Capricorn and   Pisces

Incompatible Zodiac Signs: Gemini, Leo, Libra, Sagittarius, Aries Aquarius,

Greatest Overall Compatibility: Virgo, Capricorn

Best for Marriage and Partnerships: Scorpio

Taurus likes: Gardening, cooking, music, romance, high quality clothes, working with hands

Taurus dislikes: Sudden changes, complications, insecurity of any kind, synthetic fabrics

BUSINESS ASTROLOGY FOR TAURUS

These natives really work hard to attain their goals. But most of the time they rely on conservative methods. To take their business to new heights they should welcome new techniques and approaches.

Taurus Compatibility with Aries

Taurus & Aries Sexual Intimacy and Compatibility: Aries in its rough form is guided by a simple instinct, the need for continuation of the species and the transfer of genetic material to the next generation. On the other hand, Taurus is all about satisfaction. They don't even consider orgasm that big of a deal if they are enjoying themselves through the entire sexual experience. To satisfy Taurus, you need to be emotionally involved, gentle and passionate at the same time, and willing to put some time and effort into the art of sex. The fact that Aries is ruled by Mars and Taurus by Venus immediately shows us how sexual these signs are. Both planets are in connection with physical relations, but their biggest difference is in their final goal when it comes to sex. Aries representatives usually get satisfied with having sexual relations at all. This goes for both male and female representatives of the sign. For their mutual satisfaction it is imperative for Aries to develop an atypical sense of touch and work on their sensuality in order to keep their Taurus happy. Let us also remember that Taurus is a fixed sign, pretty much set in their ways, and when it comes to sexual satisfaction they will rarely compromise and settle for less than perfect. 65%

Taurus & Aries Trust: They both have a need to search for their one true love, as Mars and Venus always do. This can lead to infidelity and typical love triangle issues, due to lack of emotion from Aries or a lack of self-worth by Taurus partner. Both of these signs have the ability to form a stable relationship filled with honesty. Neither of them is flaky or runs away from a good challenge. This can contribute to a positive attitude and open agreements on honesty when they are together. Still, if they communicate well from the beginning, they will usually find how important mutual trust is to both of them and try hard not to jeopardize it. 85%

Taurus & Aries Communication and Intellectuality: They will kick and scream (literally) until they convince Taurus that they are right about something (consider it the smallest thing in the entire Universe). When Taurus notices this Aries' behaviour, they dig in. It is quite obvious that Aries and Taurus both have horns. This is an image you should definitely keep in mind while analyzing their communication skills. Not only are they both stubborn, but they are not even stubborn in a similar way to share some understanding. Aries grabs their convictions and simply doesn't let go. They don't move. They don't even possess the sense of sound. You can almost expect a deep voiced "moo" as they get more and more irritated. Their shoulders go up, their eyebrows make an "M" and they simply stand there, annoying their Aries partner even more. Who could be this inhuman to just stand there and not listen to a word their loved one says in that high pitched tone? Taurus can. Not because of the anger, but because they are in fact too sensitive to deal with this kind of behaviour. Taurus never looks too sensitive. Their Venus role is grounded and strong, but this is a sign in which the Moon is exalted, Uranus falls and Mars is in detriment. You can imagine how this person can react to shouting and aggression of any kind. Their intellect is not an issue at all. If they can find their way through those hard headed conflicts, it is all the same if they were intelligent or stupid because they must love each other very much. The cure for this condition is in the middle, of course. Taurus needs to set strong boundaries and act securely from the safe zone they've created and Aries needs to take a step back and lower their voice, just a little bit. This would be a good place to start. 40%

Taurus & Aries Emotions: Although some of the rituals can be taught in time, this is not a solution if they don't feel enough closeness. Aries shows their emotions loudly and openly, in a way that is sort of rough and inpatient. They don't give much time for the other person to give an emotion back, and act as Fire, their element, without much sense for anyone. Taurus may find this superficial, too intense or even phony, as they don't recognize this type of behaviour as love. Taurus shows their emotions in a silent, slow process of giving. These are both highly emotional signs, but they don't show it in the same way. It is safe to assume that as much as they may love each other, it will be difficult for both of them to know they are loved. They will show love through cooking, touching and gentle words. The problem is in the fact that Aries finds this boring, stiff or even untrue. To make each another feel loved, they will both have to learn to show affection to their partner in a way that differs a lot from their natural one. This can be a small or a large obstacle and the outcome depends only on their readiness to listen to the needs of their loved one. 60%

Taurus & Aries Values: They both value material security, since Aries is ruled by Mars, a planet connected to the fear of existential crisis, and Taurus is an Earth sign, material in their core and very inclined to the financial world. This is an area in which they match quite well. Although they seem completely different, their main objectives are pretty much the same. They both cherish character and strength, physical and verbal, and need someone who will not disappoint them as the first impression fades. 90%

Taurus & Aries Shared Activities: Even if they go for coffee together, there is a big chance Aries partner will finish theirs and be bored in about 20 minutes, while Taurus partner would sip their coffee enjoying it, than order some cake. Aries and Taurus really want completely different things. While Aries is active, ready to run, train and needs to use the energy through any kind of physical activity, Taurus has the need to rest and gather energy almost all the time. The only activity they would both truly enjoy is a walk through the park and any slow outdoor activity to restore contact with nature. 40%

Conclusion: This is a relationship is full of personal challenges and individual depth. If they want to succeed as a couple, many internal issues in both must be solved. Only if they both accomplish peace in their lives, have just enough education, just enough other relationships and acquired just enough humour, they might be able to put aside their differences and listen to each other well enough. It is not that hard, except when you are used to using your horns. 63%

Taurus Compatibility with Taurus

Taurus & Taurus Sexual Intimacy and Compatibility: If they don't lack in primal sexual drive, they will probably build a strong and gentle sexual relationship, in which both of them will have their needs met. The problem they might encounter is the possibility that none of them will have enough initiative. When two Taurus partners come together, the world seizes to exist as they both knew it before. They discover a pleasure of sex they have never had the opportunity to experience, for they perfectly understand each other's need for touch and the stimulation of all senses. The sign of Taurus is the most sensual sign of the entire zodiac, that is, when they are not too lazy to discover their sexuality. Still, they are usually stubborn enough to overcome this small obstacle. Since they share sexual fantasies and ideas about intimacy, with enough openness and communication they should be able to overcome any obstacle they stumble upon. That is, if they don't stubbornly wait for the other person to make the first move. This could make them both wait forever. 95%

Taurus & Taurus Trust: They can both understand the importance of honesty, but are often too afraid to open up and let someone sink into their true emotional world. This can lead to ridiculous assumptions about each other's behaviour and questioning of every word said by both of them. The problem with their trust issues is not in trust itself, but more in their inability to change. If either one of them has a history of unhealthy relationships in which they were disappointed and let down, it will be very difficult for them to restore trust with their current partner. They have the best possible chance to form a relationship based on trust if they haven't had that much experience prior to them coming together. This will allow enough flexibility to set a good foundation and respect each other's need for privacy without mistaking it for cheating. 75%

Taurus & Taurus Communication and Intellectuality: It is not easy for Taurus to accept change, so when one of them is at the doorstep of a big shift, they could easily end up in a fight. The partner that is stuck in one place has difficulty to understand why things couldn't stay the same, forever. It will be very hard to explain to them how much pressure the person making a change must have felt in the first place to understand that change is needed. Although these partners share so many interests and have a similar way of functioning, when they stick to their convictions they rarely decide to let them go even for a second. After all, they both have horns and a tendency to cross them as soon as one of them jumps out of the regular flow of things. They can be stubborn and closed up for any sort of emotional contact, especially when they get angry or someone hurts them. This is the main reason their communication is not that good at times, for they both have trouble opening up and living in the moment, without the fear of being hurt. The problem here is not the fear itself, other signs can feel it too, but the lack of awareness that this fear even exists. They should practice intimacy and tell each other embarrassing things and emotional issues they wouldn't share with anyone else, in order to build enough trust and make room for a meaningful dialogue. 50%

Taurus & Taurus Emotions: This is a never ending circle, as a system of Earth being circled by the Moon, again and again, month after month. There is a fixed, unchangeable nature to this motion and this is something the sign of Taurus lives with and senses every day. The sign of Taurus is an Earth sign in which the Moon, the ruler all of our emotions, is exalted. The sign itself is ruled by Venus, informing us of its balance and a contribution to the material manifestation of all emotions coming from the Moon. When two Taurus partners come together and in case none of them has their heart closed up, there is a deep emotional understanding they can share. This can be very rewarding for both of them, for they can both feel the needs of their partner and be able to take care of each other, while enjoying the fact that this time they are also taken care of. 99%

Taurus & Taurus Values: Be it the financial value of any object in their surroundings or the value of being loved, the awareness of it is something they consider imperative for their partner to have. It is a funny thing to talk about the system of values when we speak of Taurus. This is a sign that represents all value and withholds in itself the price of everything in this world. What a Taurus partner values is primarily someone's respect toward the value itself. Two Taurus in a relationship value values together and a conflict can arise only if they assign different values to different things. Still, in most cases, they will give enough space to one another to set an individual list of priorities and find a compromise if some things are more valuable to one of them, than to the other. 95%

Taurus & Taurus Shared Activities: Not only will they eat and sleep together, but they will also feel so much joy in the fact that they finally have someone to do these things with, without the sense of guilt. The main problem here is that they could easily neglect the needs of their physical bodies for a healthy lifestyle and activity. There is a lot of self-control needed in order for them to stay together and not get overweight or simply too lazy. 99%

Conclusion: The relationship between two Taurus representatives is something to cherish and hold on to, only if they are not both too stubborn in their intent to wait for the other person to make the first move. Because of their emotional and sensual nature, they can be very attentive to each other's needs and take care of one another when necessary. Their problem usually shows only through the double set of horns, making them sink too deep into their differences with no apparent reason. If they could open up to each other, and to their mutual need for change, this is a relationship both of them would find extraordinary. 86%

Taurus Compatibility with Gemini

Taurus & Gemini Sexual Intimacy and Compatibility: It is not as if they don't have the need to be touched, we all do, but they have to know they are loved and accepted in so many different ways and touch is just one of them. After all, they are one of the Air signs and in their world thoughts have to be preoccupied, while strong communication is something that gives the possibility of a good sexual life. Taurus is a sensual Earth sign with a deep need for physical touch and the joy of all senses of the body. Gemini needs intellectual stimulation and doesn't care that much about spending time in someone's arms. While Taurus could stay at home, in bed, all they long, cuddling with their sweetheart and ordering food, Gemini would like to get out and be intimate at all not so intimate places. Their sex life could become the source of most of their problems, as soon as Gemini gets bored or Taurus annoyed by the lack of emotional essence. 5%

Taurus & Gemini Trust: If they don't start their relationship on a clear and truthful foundation, where the first thing they learn about each other would be the level of commitment each of them wants, a true problem with trust will easily arise. Gemini partner will start thinking of excuses to get out of any obligation imposed on them by Taurus, only to avoid hurting their feelings. Trust can be a real issue with this couple. Gemini is not all that trustworthy when someone tries to tie them down. Taurus, in most cases, lives for the day when they will be tied down with someone simultaneously. In response, Taurus will sense something is wrong and start obsessing about their partner's behaviour and the things they say. This can come to the point of absolute distrust between them, especially if Taurus gets really angry and "vindictive" in their usual passive and stubborn manner. Gemini's intentions are often misinterpreted here and this can lead to a number of following situations that will hurt them both. It is very important for them to discuss into detail what they both wish for while their relationship grows. This way, they could prevent misunderstanding that could lead to an unrepeatable lack of trust. 10%

Taurus & Gemini Communication and Intellectuality: It is very easy for an Air sign such as Gemini to forget to have lunch or sleep for a couple of hours per night. There are so many interesting things in the world that they just don't want to miss. Taurus partner could cook a healthy meal, take care of their finances and insist on a daily schedule that would give their Gemini a good energetic base to invest in their ideas. Taurus is a sign that precedes Gemini. In the astrological sense, this tells us that the sign of Gemini wouldn't exist if Taurus wasn't there. Basically, this means that Gemini partner wouldn't do much if their physical needs weren't taken care of. This is where their relationship has a strong connection, for Gemini partner might need someone like Taurus to take care of their body and its needs. In general, their interests are not that similar, but they can find a way to communicate, for none of them lacks gentleness and a way with words. If Gemini partner decides to slow down a little bit and Taurus opens up, they could even find out it is possible for them to have fun together. After all, they are ruled by Venus and Mercury, two inner planets that are, when combined, in charge of fun, sweet talk and the art of conversation. 60%

Taurus & Gemini Emotions: In most cases, their best chance of a loving relationship is in the love Taurus feels. We wouldn't say that Gemini is insensitive or unemotional, but they certainly have different approach to their emotional nature than the Earth sign of Taurus, that exalts the Moon. It is not very likely they will share their emotions with ease and enjoy each other the way they might with some other zodiac representatives. Still, there is a gentle side to Taurus that can melt down even the coldest of hearts. When Taurus falls in love with a Gemini, they will do anything to understand their nature. There is nothing that a gentle nature of Taurus can't understand, however different from their personal primal character it might be. When their Gemini feels this deep and stable understanding, they could respond in a warm and childish way, learning that they can be free even when in love with someone like Taurus. 35%

Taurus & Gemini Values:

While Taurus values the Earth, the material world, their emotions and what is stable in their life, Gemini values the Universe, the world of ideas, their rational mind and change. This is where their differences and element natures strongly diverge. They should really try hard to accept the true value of both their worlds in order to work out their differences. 1%

Taurus & Gemini Shared Activities: This will drive their Gemini crazy. It is best for both of them to walk, a lot, for this can keep Gemini grounded, while Taurus always needs movement not to end up in a static, inert, horizontal state. They could find activities to enjoy together, but not at the same pace. Gemini likes things fast, while exciting, and Taurus would like to examine everything from the beginning to the end, set the value on each activity and thoroughly decide if they would want to repeat it or not. They could connect their passionate natures through some kind of art, especially if they managed to find a way to create something together. With Gemini's ideas and Taurus' practical sense, combined with the need for beauty, this should be a true work of art. 25%

Conclusion: Although their chances to reconcile their differences are slim, if Taurus partner puts their whole heart into it, they might manage to become the most relevant part of their Gemini's life as their base and their reliability in everything they do. The relationship between Taurus and Gemini doesn't give much promise to begin with. Still, the fixed quality of the sign of Taurus can give them enough endurance and persistence to last in their intent to be with a Gemini, long enough for them to really get to know each other well. In case they accept each other completely, Taurus will give Gemini their connection to planet Earth, to their body and their daily routine, giving them the base for health and normal functioning. In return, Gemini will give their Taurus wings and, better yet, teach them how to fly. 23%

Taurus Compatibility with Cancer

Taurus & Cancer Sexual Intimacy and Compatibility: They would probably never have the urge to have sex just for the sake of it, but this doesn't in any case mean they are asexual. Taurus is a sign of physical pleasure. Ruled by Venus, the planet of feminine sexuality, Taurus needs to approach their sexual experiences with the same studiousness with which they would approach any other thing in their life. Someone might think that Taurus and Cancer are two of the most asexual signs in the entire zodiac. This is an instinctive assumption based on the fact that both signs don't care for Mars very much, meaning they don't care for instinctive sex. They need to see, touch, smell, feel everything on their partner's body and enjoy making them satisfied. With Cancer's need for closeness and the lack of ability to make their sexual life light and carefree, Taurus seems to have the perfect touch to relax them and build trust within their sexual relationship. With the lack of Mars comes the lack of initiative, and this could be their problem when it comes to sex. In case they both don't have a sexual drive stronger than their love for food, they could end up in an asexual relationship, in which they would lie around the house, cook, eat and gain weight all day long. 95%

Taurus & Cancer Trust: When Taurus and Cancer fall in love with each other, they base their entire relationship on the feel of their partner. There is not much that can be hidden from this sensitive, "sixth" sense these two can share when they connect. It would take a lot to break their trust and this would certainly mean the end of their relationship. In most situations, none of them has the need to betray their partner, for their goal is the same – love, family and home. 99%

Taurus & Cancer Communication and Intellectuality: Whatever their current interests, they will communicate it in a slow, sensitive manner, leading a conversation to a point of deep mutual understanding. Still, Taurus can be truly stubborn at times. It doesn't really matter if they are right or not, for they simply close up for any further conversation as soon as one of their true convictions is touched. Cancer can't really do that much when this happens. They can try and be even more gentle and compassionate. They share lots of different interests and will easily talk about their relationship. Neither of these signs talks much, but they are perfectly capable of understanding each other's silence and give each word a lot of significance. Their most common topics would include love, home and children, except when they feel unready or when they are too young. It is important to understand that these are not the only issues on their minds, though. In case they are close to enlightened, this will certainly work. If not, they will get really emotional and discover that their Taurus partner in fact drives them crazy. As a result, Taurus will see their partner as a real lunatic, waving their hands for no reason and showing no rational behaviour whatsoever. 80%

Taurus & Cancer Emotions: While Cancer feels, senses and takes care of their Taurus partner in the emotional realm, Taurus will give love back through physical tenderness, material security and the gentle touch of practical sense that Cancer needs. When this cycle happens several times, their love seems like a chain reaction that will never stop growing. Taurus and Cancer are the rulers of the entire warm, earthly emotional world. It is not just due to their sensitivity, but the combination of their emotional expression is something almost unimaginable. If they meet in supporting circumstances, when they don't need to fight for each other or the possibility of their love, every emotion should simply build up on the previous one and things between them should run smoothly. But if they stumble upon an obstacle of any kind, before their love for each other develops, they will probably be discouraged and never discover what they might have felt if only they fought for one another. If they do fall in love, they will not lack the energy to fight for their relationship, no matter the obstacles. 99%

Taurus & Cancer Values: However, there can be a certain divergence in their view of the material world. Cancer is a Water sign, much more focused on the value of emotion, while Taurus will be turned to financial security. This usually reflects the fear Taurus feels when it comes to their material existence being in question. Most of all, they both value life and peace. Since the Moon is the ruler of the sign of Cancer and exalted in Taurus, they both deeply value all things the Moon represents, such as, family, compassion, understanding and bliss. In the eyes of Cancer this may seem superficial for they have a tendency to think of material reality in an idealistic way. If they have a family together, these issues should settle, because the love of Cancer partner to their children combined with their desire to give them everything they need, easily shifts their perspective and teaches them about the true value of money. 80%

Taurus & Cancer Shared Activities: We could say that they might share every single activity any of them thinks of, but this is not that often called an "activity". Most of the time and especially if they both have demanding jobs, they will simply share the activity of sleeping, eating or doing nothing. This is not a question of laziness, but more of an exaggerated need for the pleasure of rest. When shared, it seems to multiply and grow beyond both of their rational minds. 90%

Conclusion: Taurus and Cancer present the gentlest couple of the zodiac. When they fall in love, they will rarely find the reason to separate, because of their shared emotional goals for love, understanding, family and the feeling of home. This is the relationship that seems like a perpetum mobile of love, in case both partners don't already have too much emotional baggage that makes them unable to give and receive this depth of emotion. Even if they do, with no obstacles on the way, they will likely learn to forgive and forget as the flow of their relationship takes them to what they always desired. 91%

Taurus Compatibility with Leo

Taurus & Leo Sexual Intimacy and Compatibility: It is in the nature of both signs to spend time in a horizontal position and it might be hard for them to agree on whom is to be on top. When motivated, they can both be excellent lovers that put a lot of energy into their sexual activities, but with one another, their sex life will most likely become a battle for personal satisfaction and rest. The sexual relationship between a Taurus and a Leo can be in a way exhausting for both of them. This is mostly due to the fact that they can both be lazy. While Taurus likes to lie down and enjoy being loved, Leo likes to lie down and be served and taken care of. Their best possibility of a healthy sex life would be the one where both partners have already built their sexual identity and know how to satisfy themselves. In this case, sensual Taurus would take care of their Leo partner, while passionate Leo would bring excitement into their relationship. In this scenario they would both take care of their own personal needs, aware that they need to commit to their partner's satisfaction in order for a relationship to work. In general, they are a feminine and a masculine sign, and share a similar need for personal satisfaction. If they don't end up in a clinch in which they both have expectations and won't move until they are met, they could have a very rewarding sex life. After all, they are just two different sides of love, joy and life in colour. 50%

Taurus & Leo Trust: Their main problem could be the lack of will on any side to change behavioural patterns that might arise. As two fixed signs, they are most likely to stand on their own two feet when it comes to telling the truth. They understand that honesty is the base of any relationship that might last, and if they fall in love, none of them will want to jeopardize their future together. If, however, one of them has a habit to lie or cheat that they have developed in their previous relationships, they will probably continue the same behaviour in this one. It is of outmost importance that both of them develop their personalities and moral boundaries independently in order for them to be functional together. In case one of them believes that the other one could change, time could consume them through a sense of distrust that builds with disappointment because change doesn't occur. 60%

Taurus & Leo Communication and Intellectuality: It is hard to say who will be more annoying to whom. While Taurus holds on to their practical perspective, Leo holds on to their ego, and a conflict with no solution is born. Their fixed natures will make them hold on to their "side" however stupid it might be, with no actual intent of reconciliation. They are in luck because they are ruled by Venus and the Sun, both warm and with a tendency to be close to one another, because when it comes to their interests and their intellectual understanding, they could drive each other out of their minds. They both need someone with a not so rigid approach if they want to find the middle ground. Taurus will find a mutual language with Leo through their usual, materializing role. Any creative impulse of Leo could be followed by the realization plan thought out by their Taurus partner, if only they shared enough emotions to have patience for each other. Their creative strength is the strength of a Venus in combination with Sun, so we could say with certainty that they would create something in image of universal love. 5%

Taurus & Leo Emotions: They are both a personification of love, each one of them in their own way. When they get together, they will rarely feel this love for one another. Maybe we could view this as their mission to give love to the less fortunate zodiacal signs, or maybe their emotional nature has to give more and receive less. Taurus is a deeply emotional sign, in case they don't close up and live in their own little, safe material world. Leo is a passionate sign that represents love as a power of creation and all we feel gravity toward. Whatever the reason in most cases they simply don't fall in love with each other. There is a great possibility that they will simply stay in their own worlds, with no prospects of merging even in something close to a friendship. This is not because they don't like each other or feel some sort of hostility, but because they are like two islands in two different oceans. Each of them has their own nature, their own world with all its beauties, and they need someone closer to this world they hold on to. None of them has the role of a floating island in search for someone to merge with. If emotions are shared, they could be huge, but there would still be the issue of showing and recognizing them before we imagine a fairytale. 20%

Taurus & Leo Values: There is a peace to Taurus that Leo doesn't want to understand, for it seems like a boring place to visit. Leo values peace, too, but for them it is hidden in a different, much more joyful place or in public, such as peace between entire countries and continents. Taurus and Leo have different views on value itself. While Taurus values financial security and material beauty, Leo values everything shiny, bravery and someone's inner fire. To Taurus, Leo may seem like someone to strike a pose and have no depth at all, and although depth is not one of their primary values, it is still a very important one. They find nothing interesting in people without essence and neither does Leo, thinking of Taurus and how they don't open up to share anything deep. 1%

Taurus & Leo Shared Activities: Leo is guided by their nature, and a lion does sleep for 20 hours a day and plays for the rest of it. Even the fiercest of lions wait for their lioness to serve their food. The problem will occur when Taurus sets up a romantic image and Leo falls asleep, but this is still a scenario that could work. Their outside activities can be fun for both of them if they go out to fancy restaurants. Shared activities are very easy for both of them to find. Either they will lie down, sleep, eat and cuddle, or they will separate and do things without each other. Taurus is inert and loves to spend time on their couch while the rain outside falls and they hear nothing but the sound of a fireplace. This is where Leo can be seen, and act as a gallant person that deserves the best, while Taurus could enjoy really good food. Other than these, they probably won't have that many activities to share, but if they are not stubborn, they could enjoy everything that is not too demanding and physical together. 40%

Conclusion: The relationship of Taurus and Leo could be aggressively challenging, if not for their warm natures ruled by Venus and the Sun. They are both signs of fixed quality with entirely different natures. If they gather enough patience before they enter their relationship, they have a chance to become your archetypal couple of a girl and a boy. When their masculine and feminine principles are in balance, they can use them to mend their sexual, intellectual and financial circumstances and really enjoy each other. 29%

Taurus Compatibility with Virgo

Taurus & Virgo Sexual Intimacy and Compatibility: The gift of Taurus is their ability to relax their sexual partner by giving them enough attention and obsessing about their satisfaction. To their Virgo partner this seems almost unreal, for they would expect something rough and scary when it comes to sex. This is an ideal combination of partners for first sexual experiences, because Virgo can enter the world of sexuality in the gentlest way possible. The sexual relationship between a Taurus and a Virgo can be quite touching. Virgo partner is usually ashamed to show their sexuality, or their body for that matter. This is where Taurus gets in the picture as a hero setting their Virgo free. The problem here can arise because of the nature of Virgo and their need to go into detail and analyze everything. Not only can they damage the spontaneity of their sex life, but they could also affect their Taurus' self-esteem by finding little flaws on their body and in their actions. Virgo is a sign of virginity and is a place where Venus, representing all satisfaction, falls. The fear of being hurt is sometimes too big to handle and with Virgo's view of Venus they rarely understand the side of sexuality that is in relation to satisfaction and tenderness. Taurus is a sign ruled by Venus and their understanding of sex is quite different. They seem to have a mission to explain what tenderness is to those around them, and find someone like Virgo a perfect student for their teachings. They will gladly explain to their partner what the beautiful side of sex is, only if Virgo is ready to listen. 85%

Taurus & Virgo Trust: It is not easy to open up to such an enormous field of possibility when you feel so small. Taurus is much more relaxed and gives so much importance to the beauty of sex, so if Virgo doesn't feel adequate with their Taurus partner, it will not be easy for them to believe in their honesty or faithfulness. Virgo doesn't trust anyone with ease. This is due to the fact that Pisces are their opposing sign, and they see every partner in their life as a glimpse of the unknown. This mistrust will really hurt their Taurus partner, for they can't understand what they did to deserve it and they will probably blame it on Virgo's changeable nature, thinking that they are not that honest, either. 75%

Taurus & Virgo Communication and Intellectuality: It is often said that Taurus can be really stubborn and difficult to talk to, but it is almost certain that a Virgo will use their mutable quality to find different approaches in order to explain their point of view. Intellectual strength of Virgo is exactly what Taurus needs to build a better understanding of the world. As two Earth signs, they can both stick to their convictions and be too rigid not to accept another's point of view, but in most cases, the intellect of Virgo and the tenderness of Taurus can help them find a language they both understand whatever the situation. 90%

Taurus & Virgo Emotions: Because of the lack of trust and disappointment Virgo is almost always ready for, Taurus needs to stay put and never let them down in order to build the trust and let their feelings for each other evolve. The patience Taurus can have when they fall in love is what makes them such a good fit for Virgo. Since Virgo will not recognize their feelings right away, they will need time to set a strong emotional foundation. If they are not both too stiff and too afraid to get hurt, they can build a strong and deep emotional relationship with mutual respect intact. It is really important not to stay at a safe distance for too long, because they could easily build a relationship with no emotion and stay in it, unsatisfied, for years, even though they might have had potential to fall madly in love. 85%

Taurus & Virgo Values: With Taurus' ruler in fall at the sign of Virgo, their Earth to Earth understanding is a bit damaged. Since Venus represents all value, Virgo could show what Taurus would recognize as a lack of understanding in general. They don't exactly value the same things, but they will be okay for as long as the feminine side of Taurus isn't disrespected. However, they will value the nature of Earth element, stable, secure and slow, and this should give them enough time to mend the differences and find middle ground. 50%

Taurus & Virgo Shared Activities: If Taurus approaches their usual activities in a way to respect Virgo's occasional obsession with their health, they could think of a number of things to do together, and complement each other very well. Taurus can truly seem lazy to their Virgo partner, especially when they are on a satisfaction spree and don't leave the house except if they are on their way to a nice restaurant. The intimacy of their nature scares Virgo to the point of agony and they will quickly need a change of scenery not to feel like they are standing in one place for eternity. The preparation of healthy snacks would be just one of possible suggestions to satisfy the needs they both have. Virgo is a sign ruled by Mercury, and although it belongs to the element of Earth, they need to move. Taurus can be really static and it is important that they make a decision to follow their partner or they really won't have much of a future together. It is a good thing that Taurus is usually guided by inertia, so when they get used to movement, this will become a permanent state for both of them. Although you wouldn't connect these signs to travelling, when together, they could feel and follow the urge to travel the world. 55%

Conclusion: Virgo needs to be flexible enough to value their Taurus and give them the intellectual view on things they might idealize. In general, Taurus is there to teach Virgo about love, tenderness and sexuality. Their relationship could be a match made in heaven, only if they are not too scared of being hurt and too distrustful. If they do give in to each other and fall madly in love, they could be the combination of a clear heart, represented by Taurus, and a clear mind, represented by Virgo. They don't need any more for each other. 73%

Taurus Compatibility with Libra

Taurus & Libra Sexual Intimacy and Compatibility: While Taurus loves to be comfortable and relies on their sense of touch and taste, Libra will want everything pretty and rely on their eyesight and the sense of smell. They do connect in a way, but in most cases they have this different approach to Venus as a planet of sexual pleasure. The main difference between these signs is in their exalting planet. The relationship between Taurus and Libra has a special kick to it, since both signs are ruled by Venus, but represent its completely different characteristics. Taurus exalts the Moon and Libra exalts Saturn. It is like they adore opposite things and while Taurus will care for emotions and tenderness in a sexual relationship, Libra will rely on its depth and good timing. It will not be easy for them to understand what the other person wants and they could both end up seeming needy to one another – Taurus to Libra because of their emotional neediness and Libra to Taurus because of their physical one. However different they might be, they are still two signs ruled by Venus and can be fairly attracted to each other. As a feminine and a masculine sign, they could mend their differences and try to learn about "the other side of Venus" instead of expecting the impossible from one another. They are both gentle lovers who like their relationships without stress and drama, so with enough patience they could be a really good fit. 35%

Taurus & Libra Trust: If Libra can't make this basic decision, the uncertainty would be a punch for the ego of Taurus and it will be very hard to recover after they realize they are not wanted with certainty. Even if they are Libra's first choice, there is still a matter of flirting with so many other people. Taurus' trust can be damaged with Libra's need to be liked by everyone, especially if they are not sure if they want to be with their Taurus partner in the first place. With really insecure Libra specimens, it is almost impossible to have a trusting relationship, for their need for acceptance can go a long way and even suck them into unfaithfulness. Their quality however, is in their outlook on justice, and they will rarely act on their insecurities, but still, who could be sure when the vibe is so unstable, especially when someone as stable as Taurus tries to blend in. 30%

Taurus & Libra Communication and Intellectuality: The main challenge here is their primary opinion on each other. As two sides of Venus, these signs represent a peasant girl (Taurus) and a city lady (Libra). When you combine this with the fact that Libra has a fallen Sun, their ego problem would easily make them feed on their refined, city image and they could criticize the "peasant girl" for her lack of style and her rusticity. They simply don't understand that the fact that Taurus likes things comfortable doesn't always reflect on their lack of style and can't accept the differences in their appearance. These two will drive each other crazy. On one hand you would have Taurus, never doubting their character, never moving and annoyingly unchangeable. On the other, you would have Libra, indecisive and never certain of what they want. It seems impossible for them not to jump on each other's nerves daily. You could say that this is quite superficial, but it is actually a really deep problem with insecurity. Libra is not criticizing Taurus because they are such a fine person, but because they are afraid they are not. Taurus will easily get insecure because of this critical view and dip into their guilt trip, almost as if they always searched for someone to wake the guilt in them, but without the ability to change and accept criticism as constructive. Even if this is not something any of them will say out loud, it can be felt in their relationship, even by those around them. 5%

Taurus & Libra Emotions: Although Taurus and Libra both are looking for someone to sweep them off their feet, they will rarely have this with each other. Taurus will usually decide not to give enough space for Libra to discover them, while Libra will spend too much time looking for faults. If they are attracted to each other just enough they could fall crazy in love, but in most cases, they are both too careful to end up in a loving relationship. As signs ruled by Venus, they are both complemented by signs ruled by Mars and normally look for a partner with initiative, to have a fast, exciting start and not get enough time to think things over. 25%

Taurus & Libra Values: Venus, their ruler, represents value itself, so we could say that they value same things because of the similarities shown through what their ruler likes or not, but in different ways. Of course they both want true, magical, mystical love with Venus exalted in Pisces, but Taurus values tenderness and touch on their way to get there, while Libra values responsibility and seriousness. However, their final goals are the same and they do value one's ability to love them, most of all. This could be their real point of connection if they do fall in love with one another. 40%

Taurus & Libra Shared Activities: Libra will rarely go for a walk in the mud, but they could make a tour downtown, where they could both be seen wearing their new outfits. It is not hard to find activities to share for this couple, for as long as they are not boring to one of them. Although Taurus doesn't really have to go to matinees and art shows with no soul, they could have a nice time at a cosy art gallery where art work in warm colours is exhibited. With their mutual love for beautiful things and love in general, they will find a nice way to spend their time together if they are open enough to make some changes to their usual routine. 65%

Conclusion: This relationship is a lesson both of them will never forget, especially if they manage to build enough understanding and tenderness between them. Look out Libras, for Taurus is here to wake your inner fears and bring them all to surface! Taurus should be careful, too, for their need to feel guilt could blossom with a Libra. The only way they could ever be happy would be to embrace what they don't want to deal with in their own inner worlds. If they do this, well you can imagine what a Venus complete would be like. 33%

Taurus Compatibility with Scorpio

Taurus & Scorpio Sexual Intimacy and Compatibility: Any sort of sexual frustration could lead to a pretty dark approach to their sex life. Scorpio has this depressive need to die naked and sweaty in the arms of a loved one, while Taurus has the need to be loved this much. It may even sound romantic, but carries with it all unsolved emotional issues as baggage into their sexual encounters. This doesn't mean their sex life will be bad. On the contrary, they will both find it fantastic, because they will fill it with all sorts of emotions, good or bad. As all opposing signs, Taurus and Scorpio can be madly attracted to each other, more so because of the sexual nature of their signs. We wouldn't primarily link Taurus to sexuality, but it does represent sensuality and is a sign that governs physical pleasure. Their relationship is a connection of deepest emotions and sexuality that no other couple in the zodiac is privileged to have. In the end, emotion will be the only thing that is left and sex will be a way to connect rather than a means to personal satisfaction. This can become an obsession and even an addiction, but who would give up on the opportunity of such deep intimacy? As signs of fixed quality, when they click, it is impossible to separate them, and no one would want to when you consider the possible vengeance of Scorpio. They represent the basic contact between sexual planets Venus and Mars, while being from the physical and emotional realms as an Earth and a Water sign. They are the signification of a deeply intimate relationship and a very rich sex life, for as long as Scorpio is tender enough and Taurus ready to experiment. 95%

Taurus & Scorpio Trust: There is a fine line between two possibilities in a relationship of Taurus and Scorpio. The first possibility would be the one in which Taurus partner is really closed up, unreachable and too quiet. Scorpio rarely trusts anyone but themselves unconditionally and in a relationship with Taurus they need to build the sense of security. We wouldn't exactly say that Scorpio is insecure, but their deep emotional nature makes them question everyone's motives in caution not to get hurt. This could wake the suspicious nature of Scorpio and their obsessive interrogations will damage their mutual trust even more then they lacked it in the first place. The second option would be for Taurus partner to be open just enough to share what Scorpio needs to hear. If they manage to find this fine balance, they shouldn't have a problem. As they get more and more intimate, Taurus will feel secure enough to share everything Scorpio wants to know and Scorpio will realize that their stable and unchangeable Taurus won't disappoint them. 80%

Taurus & Scorpio Communication and Intellectuality: While Scorpio would go in depth about all those things Taurus doesn't seem to care about, they could be very surprised to find that behind the tender and alive nature of Taurus, there is a deep understanding of everything natural going on, however dark it might seem. As all opposing signs, they seem completely different and as if they have nothing in common. Still, we should keep in mind that opposing signs complement each other perfectly and their communication should be exciting, challenging and something to enjoy if they are both confident enough. In return, Scorpio will show Taurus the value of life from their perspective. Taurus will find it incredible how Scorpio as a sign of death and destruction, can understand the depth of life and emotion better than any other sign in the zodiac. 75%

Taurus & Scorpio Emotions: This emotional connection is really something to deserve. Not only do these two represent the axis of Moon's special dignities, exalted in Taurus and fallen in Scorpio, but they also have Venus as a ruler on one hand and the intense element of Water on the other. When they fall in love, they become an image of eternal love. There is no better personification of Had, the god of the underworld in ancient Greece, and Persephone – an idea of immortal love that can never die. 99%

Taurus & Scorpio Values: They value life and love in a way that no other sign understands. The depth of their belief system goes as far as planet Earth's core and if they begin their relationship on the same page, this could be what binds them for years. Although their perspectives differ when it comes to material and emotional values, their core is the same and everything else can be adjusted. 99%

Taurus & Scorpio Shared Activities: They do however need new, exciting and breathtaking experiences from time to time, but they would be ok having them alone if their Taurus partner wasn't interested. Although Scorpio is a sign of change, this doesn't mean they are not very slow in their everyday routine. As a fixed sign, they are, as much as Taurus, static and inert. There is a lot of energy to Scorpio, nuclear energy lies within their sign, but when it comes to everyday life, they tend to repeat patterns and blend in what mostly other people find "normal". Of all possible activities, they will mostly share sexual ones and all experiences of physical pleasure. They will both enjoy discovering how far their sexual desire could lead and this will keep them busy most of the time. 85%

Conclusion: This has to be the focus of their relationship, for they can't seem to understand platonic and imaginative relationships when they get together. There is no such thing as a platonic experience of romance, when the whole point of romance is to get physical. It is very possible that they will build their sexual life to the point where no other partner could ever satisfy their needs. Taurus and Scorpio are both signs of deepest physical pleasure, each in their own way. This could lead to a possessive relationship with no way out, although they probably wouldn't want to get out even if they could. The entire experience can be too dark for the Taurus partner, especially if their practical sense is challenged by Scorpio's character. In case they are both independent and ready to blend with someone else, they could be the perfect connection between sexual and emotional, the one that we all wish for. 89%

Taurus Compatibility with Sagittarius

Taurus & Sagittarius Sexual Intimacy and Compatibility: Although this is a delusion, these signs are too far apart in their basic character to understand each other's sexuality. Sagittarius would probably think of Taurus as a person who eats and sleeps all day long. There is nothing sexual about it. It is interesting though, how two people ruled by two beneficent planets such as Venus and Jupiter can't seem to find sexual satisfaction. The fact is that they can. When Taurus thinks of sexuality, Sagittarius is probably the last person on their mind. With their childish attitude that changes with the weather, there seems to be no room for any sexual activity in their life. Although this is a rare scenario, they could actually use their attributes to enhance sexual pleasure Venus would offer. If they understood each other as two individuals who deserve respect, they could find the missing link for a very interesting and fulfilling sex life. Taurus would take care of their Sagittarius partner and keep them satisfied. In return they would get a cheerful soul who knows how to make their relationship exciting. There is so much to be learned about the "light side" of sexuality here, and this could be a fun experience if Taurus loosened up a little and Sagittarius slowed down. 25%

Taurus & Sagittarius Trust: Taurus finds this repulsive at best, and if a relationship with a Sagittarius partner begins and they start acting this way, there is a big chance they will get dumped. Trust between these partners isn't something to be questioned and analyzed. The sign of Sagittarius is considered something like a synonym for honesty. It is true that representatives of this sign have no idea how to make up a lie, let alone tell it. Still, when it comes to romantic relationships, they often suffer from a Don Juan syndrome and can't get enough attention from one partner. If they do trust each other they can have a wonderful, trusting relationship for a while, but there is still no guarantee on how long this could last. If they don't, it won't be mended whatever they try to do. Usually loss of trust here simply breaks up the relationship and they both go their separate ways with no regret. 5%

Taurus & Sagittarius Communication and Intellectuality: With the Moon exalted in Taurus and Jupiter exalted in Cancer (ruled by the Moon), there is a certain feel, a soul, a tenderness to share between them. Their approaches to life are different, their characters incomparable, but the joy they can feel toward some things is completely the same. If they don't find this shared feeling of joy, they could both learn what bad communication really is. There are so many beautiful things in the world, and so much to talk about when you think about them. Taurus and Sagittarius both have this joy about them that can be awaken by their relationship. In most cases they can talk about the weather and be fine, but when they have a problem, this turns into a ridiculous conversation that isn't really a conversation at all. Sagittarius will want to jump out of their skin while waiting for Taurus to finish the sentence, as much as Taurus will look at their Sagittarius partner as a source of all stupidity. Taurus is your countryside and Sagittarius is the world, so their problems could easily include disrespect because of their origin or their goals. Although they will rarely end up in a fight or use ugly words, it can sometimes be too obvious how much they don't care for each other's worlds and how far apart they really are. 50%

Taurus & Sagittarius Emotions: If Taurus was a bit more prone to temporary infatuations or platonic relationships, they could fall in love with a Sagittarius enough to overcome the differences between them. Sagittarius on the other hand, is often infatuated and temporarily in love. There is a great chance they will fall in love with a Taurus if they like their physical appearances, but they won't last in those feelings for long enough to gently lead Taurus to mutual love. They are both connected with the Moon in a way, so there are some feelings to be shared. However, Sagittarius doesn't normally react with much emotion to static, from their perspective boring Taurus nature. Their pace is off and they rarely get in sync with their emotions. Most relationships between a Taurus and a Sagittarius partner that manage to last, are those that started as a friendship and had a chance to develop emotionally for years without them actually being in a romantic relationship. 25%

Taurus & Sagittarius Values: They could support each other's utopian worlds a bit too passionately, and this could lead to one of them, or both, being in a delusion about what reality is about. There is too much love and happiness in the world if they start sharing opinions and this can become like a drug to both of them. The combination of signs of Taurus and Sagittarius is a "flower child" full of love, understanding for the world and ultimately humane. The practicality of Taurus will usually break this pattern and hit a counter-attack with their reality checks and material issues so they can both remember where their values part ways – to security and utter lack of it. 60%

Taurus & Sagittarius Shared Activities: Taurus might not actually share food with joy, but they will certainly like to share the activity of eating. It could be quite easy for them to find other things to do together, too. The problem will surface the moment Taurus wants to go home and spend an evening in their warm bed, while Sagittarius' fun has just begun. They don't share the same passion toward the same things, and although they might have fun being together, their priorities are not the same. The Fire energy of Sagittarius will be put off by Taurus' Earth personality and this will be tedious for both of them. 20%

Conclusion: With their inner beauty and the understanding they share in search of the truth to life, these two might seem as a perfect couple. However, every positive needs a negative to complete it, and when we really observe, we can notice that often a Taurus and a Sagittarius don't even get attracted to each other. Taurus needs earthly pleasures in their relationships and as a fixed, Earth sign it is the slowest of all signs. This is not exactly someone who can easily understand the fast, changeable and fiery Sagittarius. The best possible scenario for their relationship would be for them to get to know each other very well and build a friendship without expectations, for years. In the end, this could result in deep understanding that would provide them both with enough patience to actually start a relationship that has a future. If not, they can always hold on to beauty in the world. Imagine how wonderful their world of creation could be if they joined their forces of good. 31%

Taurus Compatibility with Capricorn

Taurus & Capricorn Sexual Intimacy and Compatibility: Capricorn won't feel the need to show off and Taurus will let go of their fear of getting hurt. The problem in their relationship can be hidden in their understanding of the

Moon, for Taurus exalts it, and Capricorn doesn't like it very much. They could have trouble connecting on an emotional level if Capricorn doesn't fall in love deeply enough or has trust issues. Taurus and Capricorn can both be quite rigid when it comes to sex. This is exactly what could make them a perfect couple. In combination with other signs of the zodiac it can be hard for them to open up and feel the need to experiment, even though Capricorn will do their best to show how ingenious they are when it comes to sex. When they get together and get to know each other intimately, they will learn what it means to relax. This will be multiplied by Taurus' need to be loved unconditionally that they show in an endless loop, scaring their Capricorn away. Their different approaches to the combination of sexual instincts and love are what could make a gap between them. Taurus has a problem with initiative and aggression, not understanding Mars that well, while Capricorn needs initiative, physical strength and supports Mars. In their sex life, this could lead to a lack of emotion from Capricorn partner, leading to the frustration of Taurus, scared away by their libido with no emotional foundation. This could go as far as impotence and a general lack of sexual desire in both partners, unless they hold on to intimate nature of their sexuality and approach each other as different individuals with certain needs. 85%

Taurus & Capricorn Trust: Capricorn just isn't into lying. They don't even judge it but find it unnecessary and stupid. Even when they do tell a lie, in most cases it is an experiment with other human beings to see if they can guess where the truth lies. When they are intimately involved, they like things between them and their partner clean and true. Taurus can easily sense this and will feel secure enough to not give in to their occasional need to hide things from their partner. Taurus is ruled by Venus, a planet exalted in the sign of Pisces, so they have this understanding of importance of secrecy when they are in love. With Capricorn, they can find a way to hide their intimacy from the rest of the world and stay true to their loved one for a very, very long time. 99%

Taurus & Capricorn Communication and Intellectuality: The deep understanding of the Moon is something Taurus is blessed with and Capricorn lacks in their core. The fear of emotion can easily become a daily routine of neglect toward their personal emotional needs. Although they have different natures, they understand each other very well and motivate each other to grow each of them in their needed direction. Their differences are exactly there to make them a perfect couple, because they complement each other in a more subtle way then their opposing signs. Taurus has a mission to teach Capricorn about the significance of tenderness one should always have for one self. In return, Capricorn will help Taurus deal with responsibility and teach them how to reach their goals with no distracting emotions. It is not always easy for them to understand each other, but with enough compassion and openness to feel for the other person, they can support each other in a way no other pair of signs can. After all, they do belong to the element of Earth, and can make magic in our material reality when they reconcile their differences. 85%

Taurus & Capricorn Emotions: This is usually something like a pattern to be broken when they do begin a relationship, for they have enough time and patience for one another. It is difficult to say with certainty they will find emotional fulfilment with each other because they are both very careful when it comes to love. From the perspective of Taurus, this might not be the best emotional contact they've ever had, but from Capricorn's perspective, things can't get much better than being loved by a Taurus partner. However, there is a dose of almost unbearable satisfaction Taurus will feel when their long-term digging reaches the emotional core of their Capricorn partner. When this contact is reached, they will rarely feel the need to separate from them again. To Capricorn this may seem as if someone literally touched their heart and they will probably never want to let their Taurus partner go. 90%

Taurus & Capricorn Values: With shared sense for value of the material world, these two can get really far together. While Taurus would create and motivate, Capricorn would lead the way to success and financial security. Whatever their goals, they could easily reach them together due to the fact they share the same material values to begin with. Still, they don't have such a peachy situation when it comes to their approach to emotions and family. They should observe different sides of their personalities as complementing instead of destructive and find a way to coexist giving value to each other's shadow. 90%

Taurus & Capricorn Shared Activities: Their high ambition can lead them to a state of low energy and Taurus is there to mend their tired Soul with fine food and time for joy. On the other hand, if the creative, motivating side of Taurus is awaken by the striving nature of their Capricorn partner, they will become everything but lazy and make room for both of them to be satisfied and happy with what they've accomplished. You could say that Taurus is lazy and Capricorn never stops working, but this is not exactly the case. If any sign in the zodiac needs rest, it would be Capricorn. In short, they can do anything together, for as long as they hold on to a fine balance of activity and rest. 85%

Conclusion: With the ability to complement each other in a gentle, slow way, they are the most boring couple on the outside, with most exciting inner activity that stays hidden from the rest of the world. Taurus and Capricorn can form a relationship so deep that their creative power in the material realm could seem unreachable for other signs of the zodiac. If Taurus motivates their Capricorn partner, and Capricorn shows the way of accomplishment to their Taurus partner, they could work together, raise children and share a life with more fun than they are both used to, or simply form an unbreakable bond. When their deep emotions intertwine, they are bound to each other for eternity. 89%

Taurus Compatibility with Aquarius

Taurus & Aquarius Sexual Intimacy and Compatibility: However, they could really help each other blossom if they opened up for the possibility of unusual sexual encounters. If the tenderness of Taurus is projected on their independent, distant Aquarius partner, their creative and motivating side would awake, giving energy and speed to the productive gentle side of Taurus. The slow, tender and smooth nature of Taurus will be ridiculously annoyed by the changeable and unusual nature of Aquarius. In most cases, they are not even attracted to each other and think of each other as boring or crazy, depending on the situation. Imagine the sex life they could have, different from each other, two outcasts, if they only shared enough respect and emotion. They will rarely get this far, for they seem to be looking for different things in a relationship to begin with. Taurus would like to have a secure, unbreakable partnership and Aquarius wants to be free of any attachment to this world, let alone emotional relationships. It is not easy for them to mend these differences or keep them out of their sex life, because they wouldn't feel like themselves in a relationship with disregard of their primal needs. 15%

Taurus & Aquarius Trust: Guilt and self-criticism is the most difficult trait of Taurus, and one Aquarius is free from, finding it obsolete. This strict Aquarius opinion will scare Taurus to the point where they feel it is impossible to tell how they feel. This will end in a circle of lies and mistrust that cannot be repaired. If Taurus wasn't so stressed out by their Aquarius partner, they might decide to be true and honest. Aquarius doesn't really understand the attitude Taurus has and least of all their fear of not being good enough. There seems to be no flexibility in an Aquarius partner, although they tend to show a nature that is so open for people's differences. In order to build the subtle trust, Taurus needs to be brave and stop thinking about the consequences of everything they say, while Aquarius needs to let go of their righteous attitude and be careful about the way their Taurus partner feels in their presence. 20%

Taurus & Aquarius Communication and Intellectuality: If Taurus shows understanding for their partner's need to fly, they could actually help them materialize what they have dreamed about. This doesn't happen often, for Aquarius rarely finds Taurus as a person to talk to, slow and boring with a "small town" attitude that inhibits the progress of our civilization. As a contact of Earth and Air elements, they can be so far apart that they can't find anything to talk about.

The sign of Taurus brings Uranus to its fall and all of those bright ideas Aquarius has, seem to go through the sieve of reality given by Taurus. This wouldn't be a problem per se, but sometimes the narrow-minded Taurus doesn't exactly see the true possibilities of the material world and can bring down their Aquarius partner to the point where they don't see how any of their dreams is possible. Their differences are hard to reconcile and when they fall in love, every little thing could become a huge problem and a reason for both of them to think about ending the relationship. If Taurus wants a white picket fence, Aquarius wants a condo on 67th floor. If Taurus yearns for compassion, Aquarius doesn't care about opinions of others. If Taurus wants to go by foot, Aquarius wants to buy a plane ticket. In general, they can find that they aren't exactly made for each other; unless they both have enough flexibility to understand the ultimate difference in others, and enough openness to do things they don't care about just to see if they like them in the end. 10%

Taurus & Aquarius Emotions: Aquarius is distant enough as it is, and without excitement some other signs might offer, they will not exactly feel the electricity of being in love with an unmovable Taurus. They are lucky that the sign of Aquarius lifts up spirits of Venus, or they wouldn't really stand a chance. Taurus will rarely fall in love with an Aquarius due to the fact that they don't recognize their ruler in a good context. However, they both might get tricked by the middle ground between them. If Taurus sees the stable, Saturn side of Aquarius, and Aquarius recognizes the inner child in their Taurus partner, they could discover that they do belong together, even though this goes against the odds. 15%

Taurus & Aquarius Values: While Taurus values material things and grounded behaviour, Aquarius values freedom in any shape and form. One of them wants to be tied down, and the other wants to fly. There is really not much they can do, but accept the differences of their goals and natures, for there is truth and good in both approaches to life. They can find certain things to value together, if they put their minds into creating them through Aquarius field of ideas and Taurus' practical realizations. 1%

Taurus & Aquarius Shared Activities: By "recourses" we don't necessarily speak of money, but the overall energy for action. Taurus will gladly visit a strange place they have never been to, but after this, they will want to come home and have a nice dinner. For a short time there could be a number of beautiful things they could do together. After the recourses of Taurus have been spent, there is not much else they will want to do together. Aquarius doesn't have this need and wants to always be on the move. They could be taken care of through the efforts of their Taurus partner, if they had enough patience to keep them well fed, dressed in clean clothes and took care of household activities. Still, this compromise is rare to find, because the emotional satisfaction Taurus partner will get in return is not enough. 5%

Conclusion: They are ruled by Venus and Uranus, both planets rotating in a direction opposite to the direction of other planets. They are two outcasts, different and standing out together, they understand that East can be where west is, and vice versa. Taurus and Aquarius are people from two different worlds. Still, there is a strange similarity and connection between their rulers and although very challenging, this is a relationship where both partners could fall in love with each other, over and over again, every single day. They understand diversity, change of direction and the excitement of love. However, they will rarely get to the point to understand each other because of their excessive need for peace (Taurus) and excitement (Aquarius). What a strange pair these signs are. With such an obvious opportunity for electric love, they go around it and search for something else. 11%

Taurus Compatibility with Pisces

Taurus & Pisces Sexual Intimacy and Compatibility: They have the ability to get lost in each other, make their dreams come true and satisfy each other by pure existence. When it comes to sex, Taurus can easily end up in a rut if their partner isn't inspiring or creative enough. They don't even care, for as long as their emotional needs are met and their physical body respected. Pisces on the other hand, get lost in sexual experiences, and can even find them toxic if their impressions on other people are unrealistic. Taurus and Pisces are both all about pleasure. Taurus represents the art of love making, tenderness and sensuality. The sign of Pisces is a culmination of a sexual encounter – orgasm. This is a place where Venus is exalted, magical, mysterious and unbelievably satisfying for Taurus' ruler. When they meet the right Taurus partner, they can be intrigued and relieved by their nature, for what they see is actually what they get. Because of the emotional nature of the sign of Pisces and their deep sense of purpose, Taurus will feel loved to the point of getting lost in the sexuality of their partner. They will both pay very little attention to their own pleasure because of all those feelings guiding them. This is almost always a giving relationship where both partners are equally satisfied when it comes to sex. 99%

Taurus & Pisces Trust: The beauty of their contact is in the fact that when together, they both lose their need to hide and let their emotions grow with ease. The sign of Pisces is a sign of mutable quality, and they can unexpectedly change, without a clear reason. If this happens, Taurus will know that the trust is lost, however their relationship seemed just a couple of minutes ago. Because of the Pisces' tendency to enter each relationship with an idealistic approach, there is a great chance they will open up to their Taurus partner as soon as they realize how stable and secure they seem. It is a deep sense of broken intimacy, lost in the Pisces' need for emotional exhilaration. Basically, when Pisces partner gets bored, they start to think of excuses and lies, before they even realize that their relationship is over. It is up to Taurus to understand the flakiness of their partner. When they do, they can either accept the situation and fight for love, or end the relationship and move on. 80%

Taurus & Pisces Communication and Intellectuality: Subtlety of Pisces is something truly inspiring for Taurus and they will feel the need to get to know every detail in their partner's behaviour. Both of these signs are not very talkative and Pisces even lead Mercury to its fall. This is why they really need to form a strong emotional bond and listen to each other through very little words. Taurus and Pisces probably won't have the need to talk much. Instead, they will understand each other through all types of nonverbal communication, curious about each other's next movement. That field of talent and creative energy that Pisces carries along goes well with Taurus' need for beauty. Unfortunately, they can get lost in the world of Pisces and really lose that grip they have on reality. At first, this will be like a drug, an addiction, something they have been waiting for their entire life. As time goes by, the feeling will not be this good, for they will lose touch with themselves and have a feeling they don't know who they are anymore. The most important thing for Taurus in a relationship with Pisces is to stand their ground and hold on to their common sense, practicality and their usual need to live in reality. 95%

Taurus & Pisces Emotions: Taurus will feel, for however long, like the centre of someone's world, loved and cherished to the point of unbearable beauty. If this feeling goes on, they could stay in a beautiful relationship for a very long time. Taurus and Pisces have a magical emotional connection. For as long as Pisces don't change their mind and swim off, their relationship should be filled with love and wonder. With Pisces exalting Venus, the ruler of Taurus, this is not only love but adoration. As soon as Pisces partner feels this beautiful emotion dying down, they will make a spontaneous manoeuvre to distance themselves from their Taurus partner. The funny thing is that in most cases Taurus won't be hurt at all. That simple feel of inadequacy will be enough for both of them to let distance take its course. Even though Taurus has a tendency to get emotionally bound to their partner, their potential separation from Pisces will be as coming back to reality more than a devastating event. 99%

Taurus & Pisces Values: Taurus is turned to a material reality and Pisces to an emotional one. Their values differ a lot, but the one they share is incomparable to others – love. No other sign of the zodiac can truly understand the way these two value love, especially when they are in love with each other. 85%

Taurus & Pisces Shared Activities: At first, they will enjoy the same things, but Pisces partner will get bored very soon if the scenery doesn't change. It is not like them to stay in one place for too long. The main problem these signs will run into is in the fact that Taurus is a fixed sign and Pisces mutable. This can lead to a lack of understanding when it comes to the way they want to spend their time together. When Taurus is found in a beautiful situation, they will want to stay in it forever, holding on to the first image even when the beauty of it fades. This will slow down all movement and could really annoy their Pisces partner. 70%

Conclusion: This is a relationship based on love and full of it while it lasts. They both crave romance and beauty in their lives, and will do anything that is needed to keep the beauty going between them. Taurus will give their Pisces partner a chance to connect to the real world, showing them how to ground their creativity, while Pisces will lift up Taurus and make them a bit softer and more flexible. They seem to be on a mission of convincing them that true love exists. When their relationship is over, they will both know it instantly and very often a conversation about a breakup would be redundant. If they savour their trust and nurture the beauty of love they share, their relationship can last and be as inspiring as a dream coming true. 88%

PART-3
GEMINI HOROSCOPE

(May 21st - June 20th):

Who oscillates, communicates and changes often and who is fond of life, fun and pleasure?

Who has a free soul and loves others attentions and exchanging of an intellectual nature?

Element: Air

Quality: Mutable

Colour: Green, Yellow

Day: Wednesday

Ruler: Mercury

Colour: Green, Yellow

Day: Wednesday

Lucky Gem: Blue Sapphire, Diamond & Emerald. He/She can count on the Emerald, which will bless him with all the intelligence he needs.

Lucky Number: Number 3, 7, 8, 9, 12 & 23 will bring him good news.

Lucky Colour: Sky Blue; Reach for the skies with sky blue.

Lucky Day: Thursday; all his dreams come true on Thursday.

Lucky Flower: Rose

General Insights: Gemini can be sociable, communicative and ready for fun, while on the other hand it can be very serious, thoughtful, restless and even indecisive. As an air sign, Gemini is concerned with all aspects of the mind. This zodiac sign is ruled by Mercury, which is a planet that represents communication, writing and teaching others. They get fascinated by almost everything in the world and they have a feeling as if there is not enough time to experience everything they want to see. This makes them excellent artists, writers and journalists. Gemini sign means that sometimes people born under this sign have a feeling that their other half is missing, so they are forever seeking for new friends, mentors and colleagues. Gemini is versatile, inquisitive, fun loving and wants to experience everything out there, so their company is never boring.

Personal Quality: The Gemini is adaptable, dual natured, affectionate, courteous, kind, generous, scientists, and talented, bright, witty, entertaining army personnel, thoughtful towards poor and sufferer, adaptable and adjusting nature. He/She has unpredictable temperament. He/She is very well with Aquarius. He/She has wealth in later half of life. He/She will completely change his/her mind like Chameleon.

Positive Quality: He/She is usually quite, gentle, affectionate, curious, adaptable, ability to learn quickly and exchange ideas, creative and has a strong self-confidence. He/She is often the centre of attraction in the gathering and is versatile, adaptable, and inquisitive and always moves along with the times.

Negative Quality: He/She is sharp-tongued, nervous, inconsistent, and indecisive, and sometimes boring. He/She cannot concentrate well at one point and hence does not finish the job. He/She becomes cynical, biting, moody and quickly angered. He/She is superficial, restless, nervous, lacks concentration and conniving. He/She does not keep their promises.

Gemini likes: Music, books, magazines, chats with nearly anyone, and short trips around the town.

Gemini dislikes: Being alone, being confined, repetition and routine

Expressive and quick-witted, Gemini represents two different sides of personality and you will never be sure with whom you will face.

Physical Appearance: He/She has small, narrow hands and feet and slim. He/She is generally tall. He/She is highly energetic and exudes oodles of charisma.

Relationships: He/She is emotionally undemonstrative but enjoys being in a lively family, and seeks a partner with strong opinions. He/She is keen to make relationships but tends to be too egocentric.

Career: He/She is particularly suited to media work, in sales pitches and can be writer, Broker, Commentator, Concierge, Correspondent, Debater, Impersonator, Journalist, Librarian,

Linguist, News commentator, Novelist, Orator, and Playwright.

Health: He/She has the ill effects of smoking since he/she has delicate lungs.

Ideal Partner: He/She is best with Aquarians. Virgo, Libra and Aquarius people may help him/her.

Compatible Zodiac Signs: Aries, Leo, Libra, Sagittarius, Aquarius.

Incompatible Zodiac Signs: Taurus, Cancer, Virgo, Scorpio, Capricorn, Pisces

Greatest Overall Compatibility: Sagittarius, Libra, Aquarius

Best for Marriage and Partnerships: Sagittarius

Business booster Gem is Citrine.

BUSINESS ASTROLOGY FOR GEMINI

Gemini can convince anybody about anything. However at times they turn loud mouth and often are unable to keep secrets. For a successful business they should learn patience and remain grounded. Business

Gemini Compatibility with Aries

Gemini & Aries Sexual Intimacy and Compatibility: It is a good thing that they both don't care that much about other's opinions anyway. In its healthy image, this is mostly a combination of passion, energy and curiosity. In a not so healthy one, their sexual relationship can be full of nasty words and verbal aggression. When Aries and Gemini engage in sexual activities, who knows where they could end up. With Aries' libido and Gemini's ideas, they might be a bit too creative and harshly judged by their environment. The good thing is that neither is too sensitive and easily hurt, so this can be exciting and unique for both of their experiences. Since Aries is a warrior by nature, Gemini's approach to sex might be too playful for their taste, but this is usually only until they open up to the everlasting game provided by Gemini partner. Their main goal is to stay as uninhibited as possible, so the Air sign of Gemini can give oxygen to the Fire of Aries. 90%

Gemini & Aries Trust: Most Gemini representatives aren't even aware of their basic individuality, convinced that they change personality overnight. Although this is not exactly true, the sense that Aries can get from them is not exactly a recipe for trust. Lack of trust is probably the biggest problem in this relationship. Aries is passionate, ruled by Mars and possibly very jealous. Gemini is ruled by Mercury, the zodiac's trickster, always changing the face they wear for the world. Because of this, Aries might get angry, Gemini distracted and distant, to the point where Aries starts looking for another partner even if the relationship has not ended and Gemini doesn't even care anymore. 40%

Gemini & Aries Communication and Intellectuality: Since Gemini loves to be in a role of a teacher and loves to be in a relationship in which their partner learns something from them, this should be a good approach for both of them. That is if Aries' ego allows this "submissive" role. Still, we are all aware that there are some Gemini representatives that simply talk too much about nothing significant at all. This would be a reason for Aries to lose their temper and think of their partner as superficial and even stupid. Aries is simply not the master of the art of conversation. Gemini is a sign ruled by Mercury and conversation is their main life theme. Even if they talk less than a typical Gemini usually does, their inner dialogue must be rich. Both partners should approach the relationship as if Gemini person was there to teach Aries how to have a good conversation. This loss of respect is truly bad for their own ego, since the decision to be with this partner was theirs in the first place. They should remember that there is always somebody in the world who wouldn't think of this specific Gemini as stupid and that everyone deserves to be with someone who doesn't see them in this light. When this disrespect happens in their relationship, Aries should consider letting their Gemini go and look for someone who suits them better. 85%

Gemini & Aries Emotions: Gemini often doesn't go very deep beneath the surface to look for someone's hidden qualities and isn't really that emotional by nature. So this is a mix of an emotional partner who can't communicate how they feel, and a rational one who talks about everything else. Emotional realm is a tricky territory for this couple. Aries partner has warm, passionate emotions and a problem to express them. The good thing is that Aries does not lack the fierceness to turn Gemini's attention to them and make them listen. When they manage to connect in a constructive dialogue, it is possible for them to explain to each other where they stand and how they feel, and this way set a good foundation for future emotional exchange. 60%

Gemini & Aries Values: It is kind of hard to think of any of their values except for the fact that they value everything interesting, and this is a kind of understatement since they find almost everything interesting. Well this is not exactly true. Gemini partners value knowledge and someone's literal abilities, as well as a fine rational mind. When you calculate the fact that Aries values a person's ability to be clear and concise, and Gemini's need to talk around everything, it seems pretty obvious that this is not a perfect match. Now think of Gemini. This is something Aries can fulfil to a certain point, in case they don't react on impulse to everything Gemini says. It is not that difficult for these partners to respond to each other's needs, but if they don't share similar education, interests and strength of character, they might see each other unworthy of their affection. 75%

Gemini & Aries Shared Activities: It is hard to say who will lead and who will follow, as Aries always leads with that atomic energy, while Gemini always comes up with new ideas and initiative. They motivate and challenge each other all the time and they both never say "no". Gemini is all for activity however insane certain activities might be, and Aries will feel liberated in this relationship. Choices of activities Aries come up with must be truly aggressive and ridiculous for Gemini not to engage. If Gemini thinks of something that Aries would maybe like to refuse, their ego wouldn't let them and they would jump in anyway, just to prove that they can. This much excitement should be followed by enough rest and time spent at home. 50%

Conclusion: The main problem with their romantic involvement is the lack of trust, especially if Aries partner gets too attached to Gemini, always fighting for their freedom. The overall impression of this couple would be good, exciting and challenging, a relationship where both partners can learn a lot and be active in a healthy way. The need for conversation with a lot of essence is bigger than any positive or any negative aspects of their relationship and both of them should always have this in mind. In general, there is a big chance these two will end up together, because their shared love of adventure is bigger than most of their troubles. 74%

Gemini Compatibility with Taurus

Gemini & Taurus Sexual Intimacy and Compatibility: After all, they are one of the Air signs and in their world thoughts have to be preoccupied, while strong communication is something that gives the possibility of a good sexual life. While Taurus could stay at home, in bed, all they long, cuddling with their sweetheart and ordering food, Gemini would like to get out and be intimate at all not so intimate places. Taurus is a sensual Earth sign with a deep need for physical touch and the joy of all senses of the body. Gemini needs intellectual stimulation and doesn't care that much about spending time in someone's arms. It is not as if they don't have the need to be touched, we all do, but they have to know they are loved and accepted in so many different ways and touch is just one of them. Their sex life could become the source of most of their problems, as soon as Gemini gets bored or Taurus annoyed by the lack of emotional essence. 5%

Gemini & Taurus Trust: If they don't start their relationship on a clear and truthful foundation, where the first thing they learn about each other would be the level of commitment each of them wants, a true problem with trust will easily arise. Gemini partner will start thinking of excuses to get out of any obligation imposed on them by Taurus, only to avoid hurting their feelings. In response, Taurus will sense something is wrong and start obsessing about their partner's behaviour and the things they say. This can come to the point of absolute distrust between them, especially if Taurus gets really angry and "vindictive" in their usual passive and stubborn manner. Gemini's intentions are often misinterpreted here and this can lead to a number of following situations that will hurt them both. Trust can be a real issue with this couple. Gemini is not all that trustworthy when someone tries to tie them down. Taurus, in most cases, lives for the day when they will be tied down with someone simultaneously. It is very important for them to discuss into detail what they both wish for while their relationship grows. This way, they could prevent misunderstanding that could lead to an unrepeatable lack of trust. 10%

Gemini & Taurus Communication and Intellectuality: This is where their relationship has a strong connection, for Gemini partner might need someone like Taurus to take care of their body and its needs. It is very easy for an Air sign such as Gemini to forget to have lunch or sleep for a couple of hours per night. There are so many interesting things in the world that they just don't want to miss. Taurus is a sign that precedes Gemini. In the astrological sense, this tells us that the sign of Gemini wouldn't exist if Taurus wasn't there. Basically, this means that Gemini partner wouldn't do much if their physical needs weren't taken care of. Taurus partner could cook a healthy meal, take care of their finances and insist on a daily schedule that would give their Gemini a good energetic base to invest in their ideas.

In general, their interests are not that similar, but they can find a way to communicate, for none of them lacks gentleness and a way with words. If Gemini partner decides to slow down a little bit and Taurus opens up, they could even find out it is possible for them to have fun together. After all, they are ruled by Venus and Mercury, two inner planets that are, when combined, in charge of fun, sweet talk and the art of conversation. 60%

Gemini & Taurus Emotions: In most cases, their best chance of a loving relationship is in the love Taurus feels. We wouldn't say that Gemini is insensitive or unemotional, but they certainly have different approach to their emotional nature than the Earth sign of Taurus, that exalts the Moon. Taurus is a sign that precedes Gemini. In the astrological, it is not very likely they will share their emotions with ease and enjoy each other the way they might with some other zodiac representatives. Still, there is a gentle side to Taurus that can melt down even the coldest of hearts. When Taurus falls in love with a Gemini, they will do anything to understand their nature. There is nothing that a gentle nature of Taurus can't understand, however different from their personal primal character it might be. When their Gemini feels this deep and stable understanding, they could respond in a warm and childish way, learning that they can be free even when in love with someone like Taurus. 35%

Gemini & Taurus Values: While Taurus values the Earth, the material world, their emotions and what is stable in their life, Gemini values the Universe, the world of ideas, their rational mind and change. This is where their differences and element natures strongly diverge. They should really try hard to accept the true value of both their worlds in order to work out their differences. 1%

Gemini & Taurus Shared Activities: Gemini likes things fast, while exciting, and Taurus would like to examine everything from the beginning to the end, set the value on each activity and thoroughly decide if they would want to repeat it or not. This will drive their Gemini crazy. It is best for both of them to walk, a lot, for this can keep Gemini grounded, while Taurus always needs movement not to end up in a static, inert, horizontal state. They could find activities to enjoy together, but not at the same pace. They could connect their passionate natures through some kind of art, especially if they managed to find a way to create something together. With Gemini's ideas and Taurus' practical sense, combined with the need for beauty, this should be a true work of art. 25%

Conclusion: Although their chances to reconcile their differences are slim, if Taurus partner puts their whole heart into it, they might manage to become the most relevant part of their Gemini's life as their base and their reliability in everything they do. In case they accept each other completely, Taurus will give Gemini their connection to planet Earth, to their body and their daily routine, giving them the base for health and normal functioning. The relationship between Taurus and Gemini doesn't give much promise to begin with. Still, the fixed quality of the sign of Taurus can give them enough endurance and persistence to last in their intent to be with a Gemini, long enough for them to really get to know each other well. In return, Gemini will give their Taurus wings and, better yet, teach them how to fly. 23%

Gemini Compatibility with Gemini

Gemini & Gemini Sexual Intimacy and Compatibility: Although they will most certainly have an abundance of information on sexual activity, only after they have had some experience can they become great lovers. It is very rare for a Gemini as an Air sign, to be practical and find the way to manifest what they've read or heard about in the realm of reality and physical body. When we think of two Gemini in a sexual relationship, it is okay if we laugh a little. The image that comes to mind could easily be the image of two people with split personalities, trying to have sex by banging their heads together and talking at the same time. Their biggest quality is the ability to learn. With their desire to become great lovers, they will absorb knowledge through each of their relationship like a sponge. Two Gemini together will share information and coordinate their previous experiences with one another. They will be more satisfied when they teach their partner something, than they will be by sex itself. With their open minds and creative wit, there is probably not a single place they wouldn't want to have sex with, nor a position they wouldn't want to try out. It is not like they are promiscuous, but find joy and excitement in the change of scenery, especially if there is relative movement involved, too. So imagine a train, an airplane restroom or any means of transportation in which it is possible to hide. However, their sexual life can become empty when the excitement has passed if none of them has enough depth to bring into the act of sex. They are not even aware of the focus and emotional connection they need, until they find the right partner. Usually this is not another Gemini. Their hearts should be dug up and their relation to sexuality changed, before they get together with one of their kind. Any other scenario probably won't keep them satisfied for very long. 80%

Gemini & Gemini Trust: They both know themselves, so it is easy to understand each other in all those flaky, superficial and changeable moods. Basically one of them is going to move in two minutes and the other one in three, so how can they trust each other to stay? If they knew their own next move, they might be able to build the trust with someone so similar to them. They might not trust each other, but they don't really care. This is not something that will bother them. On the contrary, it will give them the freedom to be themselves, but rarely keep them in a relationship for too long. 50%

Gemini & Gemini Communication and Intellectuality: They will jump in each other's sentences and use all possible communication tools, starting with the usual phone conversations and chat that will grow into dozens of chat variations, depending on the emoticon that they want to use. Communication between a Gemini and another Gemini never ends. When they get together, there is always something to share, an idea to be up for discussion and a distance to be crossed on foot. It is wonderful to watch them together as they find someone who understands and speaks in the same language. For as long as there is enough respect and listening between them, the intellectual side of their relationship will be intact. 99%

Gemini & Gemini Emotions: It is to be expected that the more open Gemini will build a deep emotional bond with their partner, although their feelings might not be returned. The sign of Gemini is not that emotional to begin with. The good thing is that they both know this and find a rational explanation of the benefit of their mutual lack of emotion. Still, there is a need in each one of them to reach for something they can't have. Since they rely on an intellectual connection with someone, they could find true emotional satisfaction in their communication, but this is easily shattered due to the fact that mental compatibility is not the same as emotional, let alone sexual. Often, they will deny this and hold on to what they have, until one of them gets swept of their feet by someone who wakes their emotions with silence. 70%

Gemini & Gemini Values: When we say "freedom" we think of Aquarius, but in fact Gemini values freedom as much, if not even more than their Aquarius friend. This is a value that two Gemini will passionately share. They don't like to be bored with tiresome details, obligations of a meaningful relationship, or their partner's need for compassion that is not to be asked for. The problem is in the fact that they think too much, and feel much less. If they could get out of their head for a couple of seconds, they might realize that their chest is crying out for intimacy, closeness and compassion. 99%

Gemini & Gemini Shared Activities: Even if both of them didn't really want to do something, they will both do it together out of curiosity. After they share an experience, they will put it through a mind filter, talk it out, and move on to the next one. This title says it all. They share activities. Among all of them, if one of them wants to do something, well basically anything, the other one will follow out of pure curiosity, and vice versa. There is really no stopping for these two and no one can follow them like they can follow each other. 99%

Conclusion: Because of their possibly superficial approach, it is best if they have already had some relationships with depth before they met each other. This could give them the quality to last together for longer than a week or two. In most cases, this is not a relationship they will want to stay in, although their mutual understanding is perfect. The relationship between two Gemini will give other signs of the zodiac an almost certain headache. They will go everywhere together, do everything together and talk about everything with one another, again and again, until one of them loses interest in the other. It is like they are too similar, and at the same time a relationship of too many personalities. If each of them isn't gathered into one person, they will need someone who is, to hold their balance and not let them dissipate. In case they have built up personalities and each of them understands their own inner core, they can probably live forever and never consume the energy their connection brings. 83%

Gemini Compatibility with Cancer

Gemini & Cancer Sexual Intimacy and Compatibility: This is not always the case, of course, but it is very rare for a Gemini partner to manage to relax their Cancer and make them join their sexual adventure. Still, there is a link between them, pretty strong for that matter. Cancer continues the sign of Gemini, and in a way their nature is a consequence of endless conversations and rational explanations. Gemini would go outside and have the weirdest sexual experiences, and Cancer would stay at home and wait for the night to be loved by their tender partner. If Gemini partner has enough patience to talk to their partner about same things over and over, in time they could build enough intimacy to have a good sex life. In order for this scenario to develop, Cancer needs to talk, too. It will not be enough to say a sentence or two and expect that their Gemini will understand how they feel. If they manage to communicate, Cancer could actually wake the depth of their Gemini, giving them enough support to express their more emotional sexual personality. If they want their relationship to work, their sex life needs to be somewhere in the middle, such as intimate enough and exciting enough, for both partners to be satisfied. 5%

Gemini & Cancer Trust: As a rational sign, they usually find no reason to lie or cheat on their partner, for as long as they don't feel threatened by intimacy they are not ready for, or tied down. The major task of Cancer in a relationship with Gemini is to let them be free. If they start living together, Gemini will have to stay out of their love nest and Cancer will spend a lot of time alone. Gemini would go outside and have the weirdest sexual Gemini is a sign ruled by Mercury, our little trickster. It is not easy to trust a Gemini partner, especially if one is trying to take away their freedom. This is a challenge, but both signs can have enough dedication to their personal values, and these partners should know better than to ruin their relationship with trivial lies. 25%

Gemini & Cancer Communication and Intellectuality: In their relation to a Cancer partner, they usually feel the need to open up a bit more and share things that they wouldn't with other people. There is a certain "motherly" glow around Cancer, male or female, that gives Gemini enough room to set their inner child free. This can actually be a wonderful relationship that lasts much longer than other Gemini relationships, because there is usually enough understanding in Cancer for their childlike partner. There is probably not a single sign in the zodiac that Gemini couldn't talk to, so their communication skills mark this category a bit higher in every case. That is, if other aspects of the relationship are satisfying enough. It is safe to say that Cancer and Gemini make good friends and this could give a push in the right direction toward their emotional and sexual understanding as well. 70%

Gemini & Cancer Emotions: Cancer can try to share their emotions too often for Gemini to have enough patience to understand them, and this can be devastating for their emotional relationship. On the other hand, Gemini doesn't really show emotions in a way that Cancer will easily recognize. The main negative possibility here would be in the fact that Gemini doesn't always listen. It is often said that Gemini is a superficial sign, but in fact, they like to stay afloat and keep away from difficult conversations because of their need to move forward. It is a good thing they are capable of finding emotion in anything and anyone, so they might have a chance to understand the emotional nature of their childish Gemini. 10%

Gemini & Cancer Values: While Gemini is a rational sign, giving value to all that comes out of their mind, Cancer is an emotional guru, giving value to things their heart beats for. When it comes to relationships, they are probably the most vulnerable to differences in this category, for differences here make their primary goals different. They have completely different systems in which they value things in life. Although they will both be motivated to find love, the way their partner acts in a relationship could be considered "wrong" due to its opposite nature from what each of them values. 1%

Gemini & Cancer Shared Activities: If they have enough of it to follow Gemini's chaotic schedule, they could find a lot of enjoyable things to do together. They have completely different systems in which they value Activities Gemini and Cancer could share mainly depend on the energy level of Cancer partner. Cancer is a sign that exalts Jupiter, and its representatives like to travel, as much as they like to feel at home due to the sign itself. Although Gemini partner doesn't really have to go that far, they could find middle ground in travelling to places they both want to visit. However, the main characteristic of Gemini is their curiosity. It will lead them in all directions, they will feel the need to try everything and find more new, exciting experiences to share with someone. Unfortunately Cancer doesn't really enjoy this that much. While the relationship is new, it will all seem exciting and breathtaking. As time passes, Cancer will want to settle down, imagining them buying furniture together and raising children. It is not easy for them to accept that this nature Gemini has isn't something that will be different tomorrow. There is no settling down with a Gemini partner. This is just not their mission in life. 15%

Conclusion: In order for their relationship to last, they both need to make some adjustments. Gemini will hardly ever change their routine for someone, especially when they find someone's way of life boring, so the best thing to do here is to give them their freedom. If Cancer falls in love deeply enough, they will understand what their Gemini partner needs, and won't hold them back even if they wished for them to be different. Gemini partner has to open their heart and listen to those few words that Cancer wants to share. Gemini and Cancer are next to each other in the zodiac, and they are likely to be next to each other in friendship. When it comes to emotional or sexual relationships, there seems to be too many things that set them apart. Even though they can speak about many things, when it comes to discussions of their relationship, their views on it are different. Gemini needs to keep it interesting and Cancer needs to be heard, as much as felt. If they give each other enough freedom and understanding, they could be like children in love for the first time. 21%

Gemini Compatibility with Leo

Gemini & Leo Sexual Intimacy and Compatibility: Gemini is childish when it comes to sex, and rarely connects deep emotion with sexuality. Leo could be the right partner to teach them how to make a real intimate connection if they are not preoccupied with themselves. Everything that Leo would like to show, Gemini would gladly examine. They are a very good fit when it comes to sex, for Gemini gives their relationship ideas and excitement, while Leo brings in energy, creativity and love. Their sex life can be stimulated by their intellect and communication, for they both rely on their conscious Self and their mind. If Leo feels right in intimate relations with their Gemini, as a fixed sign they will give them stability and a chance to last together for a very long time. They will both want to experiment, have sexual encounters outdoors and will enjoy being naked. This is a perfect relationship for both partners to overcome shame and any sort of fear regarding intimacy and sexuality. 90%

Gemini & Leo Trust: One of them is distracted by everything and the other focused only on their own needs. Because of this they could end up in a situation in which their relationship lacks trust and it might take some time for them to notice, because they will not question one another at the beginning. Because of their natures, they could get lost, each of them satisfying their own needs. This will lead to all sorts of situations they don't want to share with each other, even adultery. Both Gemini and Leo can be bad listeners when it comes to other people's needs. It is of outmost importance for them to share and listen to one another from the beginning, in order for both of them to be able to satisfy their partner's needs instead of letting them slip away. 45%

Gemini & Leo Communication and Intellectuality: This is where Gemini jumps in as a faithful follower, to admire their Leo partner and teach them a thing or two. If they find themselves in an emotional relationship with each other, they will probably talk about everything else but their feelings. Leo might have the need to, in general, but Gemini partner will easily seduce them to go in another direction. Gemini and Leo are both rational and focused on their mental activity. Leo does have a deep emotional background, but pays a lot of attention to words and intellectual strength of their partner. They both don't have much use for sweet talk and if Gemini tries to use the way they have with words, it could work only up to a point where they start sounding fake. There is not much room for over thinking, and they will both probably say the first thing that comes to their mind. This is a great way to build trust between them, if they don't judge each other in the process and share some emotions along the way. Still, there is an issue of Leo's "perfect" Sun. This position gives so much energy to Leo that they sometimes feel obligated to burn everyone around them by imposing their will. This is a strange need of Leo, since the sign supports Sun greatly, but it is there. They usually have this idea to change the world and make it a better place, and this can drive their Gemini partner crazy in case they don't have enough room for their own opinions. 95%

Gemini & Leo Emotions: Although the mutable quality of Gemini partner doesn't really go well with the fixed quality of any sign, with Leo it could be a perfect match because of their warm, supporting and respectful nature. If Leo has enough patience to wait for Gemini's emotions to surface, they could get more than they've bargained for. The beauty of their relationship is the consciousness of both of them, leading to verbal display of emotions, once they feel safe with one another. The warmth that Leo is prepared to give to their Gemini partner will hardly leave them indifferent, while the charm and childish nature of Gemini is going to spark wonderful emotions in Leo, day after day. If they both recognize love, this is a wonderful love story full of support, respect and always something new to discuss. 85%

Gemini & Leo Values: Most of all they both value intelligence and clarity. As two signs ruled by planets in charge for our mental, rational behaviour, they will meet each other's needs perfectly. Gemini values the independence of their partner and their own freedom, and this is exactly what Leo can give them. On the other hand, Leo always values their partner's inner child and this is exactly what they will get in their Gemini. 99%

Gemini & Leo Shared Activities: Although Gemini wouldn't really want to spend that much money, Leo doesn't mind paying for everything, for as long as they are not feeling used. Still, Leo can be very lazy. Gemini is always on the move, and has the need to do at least three different activities every day. When Leo has time to rest, they will probably want to watch TV and move from the left to the right side of the couch all day long. It is a good thing Gemini wants to go everywhere and do everything, or these two might have a big problem finding activities they want to share. This way, Leo can lift their activities to a "higher level" going to all places that Gemini would suggest, but in a fancy, expensive version. This is something that could build a gap between their worlds, but usually they have enough respect for each other's needs that they can separate activities and be very happy together. None of them is needy and they don't want to spend each moment of their lives with their partner, so it will give them both enough freedom – for Gemini to move and for Leo to rest. 80%

Conclusion: They both consider their day best spent in laughter, and if they share friends, they could seem like a perfect couple. Their main challenge is the difference in their approach to change and they both need to make room for small adjustments in their behaviour if they want their relationship to last. Leo will need to make room for more movement and understand what seems to be "flakiness" of their changeable Gemini partner, while Gemini will have to understand that Leo is in fact keeping them together for however long they are meant to last. Gemini and Leo can have so much fun that it could make the rest of the zodiac envious. Their mutual respect can usually overcome any boundaries, and they should keep having fun and building their relationship on a solid foundation of childish joy. 82%

Gemini Compatibility with Virgo

Gemini & Virgo Sexual Intimacy and Compatibility: Their sexual relationship is hardly promising, but they both have the need to communicate. If they find a language they both understand, they might agree on the way their sexual life is to progress. However, there is a big chance that endless discussions will not lead to their mutual understanding, leaving them distant and not interested to share any sexual experiences. If they fall in love, they will use enough tenderness and respect to make their sex life work, but it will still rarely be satisfying for both partners. They are both curious, but not in the same way, and have an extrovert vs. introvert conflict. While Gemini would often like to be free to get naked and run around the streets, Virgo would prefer if everyone kept their bodily fluids to themselves. Gemini and Virgo are both ruled by Mercury, not a very sexual planet at first glance. However, this affects them in different ways, for Gemini is a masculine sign, always ready to explore, while Virgo is a feminine sign, shy and sensitive. To top it all, they both need verbal stimulation, but the words each of them would like to hear often aren't what the other person knows how to say. 5%

Gemini & Virgo Trust: Gemini is not someone Virgo can trust. When Virgo starts doubting and analyzing everything, Gemini will simply fly away, in the best case scenario. If they have a touch through maleficent planets, such as Mars, this will not end there, but in endless fights and conflicts on trust and fidelity. What a combination! With Virgo's trust issues and Gemini's trickster nature, both of them are about to go crazy. The good thing is that they are not prone to jealousy, or they would probably kill each other after a couple of weeks. The key here is for Gemini to get back to planet Earth and respect the sensitive nature of their partner. They can build their security and be free to be who they are at the same time. Also, Virgo will have to open up, or they will never find a way to stay together. 1%

Gemini & Virgo Communication and Intellectuality: Even if they have a fight, they will still be good at talking. Their view on intellect could be a bit different though. Gemini's superficial nature could seem extremely stupid to their Virgo partner. This will set Gemini off to be even more irritating on purpose, because they are everything but stupid. In return, Gemini will see Virgo's obsession with details and think of them as crazy. They are ruled by the same planet, master of communication, Mercury. They may have their differences, but we wouldn't say their communication is bad. Virgo will probably not fight this, but sit in their room and cry, alone, in the dark. They need to find a way to recognize each other's "flaws" as good qualities and they will get there if each of them lets their partner help. When these minds come together, there is no problem to stay unsolved, for together, they are Mercury in all its glory. These signs have a mission to unite and transform heavenly ideas into real life, and lift human knowledge step by step, towards divinity. This is a relationship of the Air side of Mercury and its Earth side, grounded and well thought of. It shouldn't be forgotten that the sign of Virgo exalts Mercury so Virgo can exalt their Gemini partner, if only they wanted to land. 80%

Gemini & Virgo Emotions: Gemini is located between Taurus and Cancer, two strong, feminine, motherly signs. This means that the sign of Gemini is formed out of Taurus and is a base for Cancer. Virgo has a problem with Venus and is unemotional. Their biggest problems are speed and fear. Gemini is considered not that emotional. Virgo brings Venus to its fall and rationalizes everything. Still, there is an interesting emotional side to both of them. Gemini simply doesn't stay in one place for long enough to build an emotional bond or recognize how they feel, while Virgo tries hard to dismiss Venus and with it love and all earthly pleasures because they don't think they deserve any of it. Both signs have a small, fast planet in common, and if they touch each other's hearts, they will understand their partner's occasional need to run off, or run away. 55%

Gemini & Virgo Values: However, their approach to basic intelligence is different and they won't always recognize it in the same way. In fact their relationship can show us exactly how relative intelligence is. They are both fans of intelligence, resourcefulness and practicality. Their meeting point is actually in their emotional intelligence, not our typical one. However, they do value rationality, someone's ability to reason and one's practical use of their hands and their brain. 70%

Gemini & Virgo Shared Activities: Virgo will probably stay there for a bit longer, not even noticing their Gemini partner got lost and accidentally found an emergency exit that they

had to use to see how the big handle on huge red door works. The lack of self-value Virgo suffers from constantly will make them question every location and activity their Gemini partner would want to take them to. It is a good thing they both like to read, so they can visit a library every once in a while. Even if they try to hide it, their hesitation will be felt and Gemini will rather go alone than push someone to come along. Both of these signs are mutable and as such have a tendency to change locations, interests, relationships. This can allow them to stay on the move together, if only Virgo was open enough for exciting new experiences and Gemini finds a way to not get lost. 30%
Conclusion: Their mutual love for Mercury is what binds them and what tears them apart, because they both tend to over think things instead of following their hearts. The relationship of Gemini and Virgo can change as the wind, while both partners get lost and found on a daily basis. Both of them are mostly in their minds, each one in their own way, and need to respect each other to the point where no one's intelligence is judged on a superficial level. If they do fall in love, they will become a unification of Air and Earth Mercury – heaven on Earth. 40%

Gemini Compatibility with Libra

Gemini & Libra Sexual Intimacy and Compatibility: Fragile ego of Libra can be lifted by Gemini's charm and approach to sex. They seem to know how to make everything a little less serious and this will help their Libra partner open up and share their emotions through sex, too. When Gemini partner sees how mellow Libra can be, although they seem so tactful and cold, they won't have much choice but to share emotions as well. The main goal in their sexual life is, in fact, to balance these emotions. Gemini and Libra are both guided by the element of Air and this should give them a good start for their mental connection and verbal understanding. When it comes to sex, this is a plus, for they will both be free to communicate anything that bothers or satisfies them. Libra is a sign ruled by Venus, sexual, sensual and seductive, while Gemini is ruled by Mercury, having no sexual or emotional wisdom. The basis of a good sex life between them is their curiosity, for Libra is always curious about their partner, as much as Gemini is curious about everything else. Although Libra might be indecisive, Gemini won't have a problem thinking of a different approach and finding new techniques, words and adventures to spice up their sexual encounters. 80%

Gemini & Libra Trust: When they choose a partner, they choose them for their character and their straightforward nature. Libra has no reason to doubt their own judgment and will probably believe their Gemini in every case, except when their dishonesty is too obvious. When Libra decides to be with someone, even if it is after a long, hard inner battle, they will probably believe in their words and their actions. In return, Gemini will respect Libra's need to flirt in order to be accepted and loved by other people. Not only will they not find this threatening, but they will actually enjoy a consequence of this behaviour – their own freedom. 95%

Gemini & Libra Communication and Intellectuality: Although Gemini will, in most cases, just follow their rational nature and comment on things simply because they want to talk, it will be hard on Libra to overcome some of the things they might say. The sign of Libra is very sensitive to any sort of will imposing or criticism and will recognize it even when Gemini has no idea what their partner is imagining. Gemini partner is very opinionated and Libra has a tendency to take a lot of things their partner says as a personal insult. Since no two people can agree on everything every time each of them opens their mouth to speak, Gemini and Libra can have a very hurtful and tough communication due to the mutual lack of tolerance. Gemini partner would rather go blind than accept that they have a lack of tolerance, but the truth is they can be quite strict when it comes to someone's mental activity and their opinions. Libra is in most cases hurt enough by the pressure on their personality produced in their primal family, so they will have a very bad response to this behaviour even if Gemini meant nothing wrong. The main issue here is in the fact that Gemini lives to learn and teach what they have learned. They often present themselves as someone who knows things, and Libra can see this as their need to prove their intellectual dominance. Even though this isn't their intention in most cases, sometimes a single sentence Gemini said can be something Libra feels hurt by for years to come. Libra can learn from a teacher, from someone who has proven their worth, but hardly from their know-it-all partner. 60%

Gemini & Libra Emotions: They seem to be in sync while Libra partner searches for depth, and Gemini flies around looking for a new discovery. They won't even notice as love between them starts to show, one of them running around and the other thinking about reasons why they wouldn't be perfect together. We could say that neither of them is very emotional, but Libra is ruled by Venus, so there is a strong link to an emotional plain here. The way Libra's emotions develop is something really suitable for Gemini. The problem develops when they both talk too much about their emotions, while none of them stops to actually feel. They can remain detached and distant, unless Libra falls in love deeply enough to follow their Gemini partner wherever they go, and Gemini falls in love deeply enough for all words to lose meaning. 90%

Gemini & Libra Values: Gemini will value someone's creativity and intellectual strength and this is something Libra can't respond to if their Sun is in its lowest state. Their meeting point is in their value of intellect, however strange it may seem. They are both Air signs and give a lot of attention to their partner's mental personality and the way they think. While Libra will value consistency and someone being responsible and reliable, Gemini will be very different from this, with opposing values as well. They could find a way to tease each other with words, seduce each other and in the end find a way to communicate everything else – in case they both care enough. 55%

Gemini & Libra Shared Activities: However, this can end in an unfortunate way, leaving Libra with their energy drained and wishes not granted. Gemini just doesn't care about following their partner only because they want to be followed. Both of them need to keep their expectations low and let their partner surprise them by something new and exciting. Libra is your typical activity chameleon when they fall in love. They want to examine the world of their partner, beginning to end, and will gladly follow them around in all their activities expecting the same in return. They will enjoy various activities together, but in order to do so, they need to respect each other's limits and desires to begin with. 85% Conclusion: If Libra partner has trouble being alone and doing things by themselves, this isn't something Gemini will easily understand. Due to their lack of personal boundaries, Libra representatives will often let their Gemini partners lead the way until all of their energy is gone, they feel like they should only lie down and turn their brain off. If they want to work on their relationship and be happy, Libra needs to respect their Gemini partner enough to let them be their teacher, lover and a friend. Gemini and Libra partners are not exactly always a perfect couple, although their signs support each other. In return, Gemini will have to take care of their Libra partner, respecting their limits and their need for togetherness. 78%

Gemini Compatibility with Scorpio

Gemini & Scorpio Sexual Intimacy and Compatibility: Gemini is so far off from Scorpio's emotional world that good sex between them seems like something almost impossible to happen. This couple needs to be supported by other positions in their natal charts if they are to stand any chance of lasting in a loving, sexual relationship. Gemini can be superficial and there is no other sign who knows this better, than Scorpio. A sexual relationship between a Gemini and a Scorpio is like a connection of the deepest and the highest point on planet Earth. Their Air element combined with the rule of Mercury and its lack of emotion is close to Scorpio's worst nightmare. Scorpio is a sign of our deepest emotions and as such, linked to the most intimate side of sexuality. When they begin a relationship with Gemini, it probably never crosses their mind that such an asexual person can exist in the world. If they fall in love with each other, there is so much for both of them to learn. Scorpio gives a strong focus on their sex life and can be very creative when relaxed. Still, they have a tendency to make a dark, sadistic or masochistic atmosphere that Gemini can only laugh at. If their mutual respect is at a very high point, Gemini could teach Scorpio that not everything needs to be so fatalistic in their sex life. In return, Scorpio will give their Gemini partner depth and emotional vibe to sex that they have never encountered before. 1%

Gemini & Scorpio Trust: They like things clean and without a doubt, and are prepared to give unconditional honesty, expecting the same in return. Scorpio trusts everyone until they don't. They have this weird, possessive nature that can give ultimate trust to their partner until the first worm of suspicion is created, usually by flakiness and disrespect. It is ridiculous to expect such a definite honesty out of a Gemini partner when they are so prone to change and have no idea what they will feel or do tomorrow. Still, they have a knack for communication that can save the day, but it can only be used if Scorpio doesn't feel threatened because of their previous heartbreak they've never healed from. 5%

Gemini & Scorpio Communication and Intellectuality: This is something Gemini doesn't want to deal with if they don't have to. Unfortunately, there is not a lot that Scorpio thinks they can learn from Gemini. Although they might wish to be more superficial, when they get in contact with someone with less depth, their ego sparks and they feel dominant because they are just the way they are. They will hardly ever respect Gemini for this, and can try to feed on their personality in order to add quality to their own. It is a good thing that Gemini can communicate with anyone. They will be moved and intrigued by Scorpio's nature, and very curious as usual. This will be amplified by Scorpio's depth and interesting topics, up to the point where they get too dark and depressive. In case they share interests and have similar professional or educational directions, they could complement each other very well. Gemini would give ideas and discover new information, while Scorpio will dig in and give real essence to everything. Their communication can be inspirational if they get into this mode and start accepting each other's qualities. They have so much to give to each other and it would be a shame if they held their relationship in an ego conflict for too long. 20%

Gemini & Scorpio Emotions: We could say it's hard for them to synchronize, because the emotions Scorpio would give away freely are something Gemini would have to be swept off their feet for. They don't seem to have the same emotional scale and this can leave them both unsatisfied or pressured to a breaking point. Emotional lack of compatibility is what ruins their sex life and gives them both a headache. If one of them falls in love with the other, they will hardly have a good time if their feelings are not returned in the same proportion. The best option for their shared emotional world would be for both partners to give what they can and not expect anything in return. 1%

Gemini & Scorpio Values: The good thing in their relationship is that they both value strength of thought. Although Scorpio values many other things in someone's personality, they will be impressed by someone's intelligence and resourcefulness. Gemini will focus on the same thing, but have a slightly different assessment of someone's intellect. However, they can agree to have a shared point of same value, although other things they strive for will differ greatly. 20%

Gemini & Scorpio Shared Activities: Gemini will want to change the scenery and Scorpio will want to change their life, gladly taking Gemini's small steps toward this goal. Scorpio is a fixed sign, set in change of huge proportions, and Gemini's mutable quality will annoy them in most cases. A great thing with these two is their openness for change. Still, they can relate to their need for different experiences in life and the excitement they always search for. Even though they might not be excited by the same things, there will be enough excitement for both partners to choose along the path they decide to cross together. 40%

Conclusion: Gemini and Scorpio will usually annoy each other senseless. None of them will lightly understand their partner's personality. Gemini would get deep, emotional satisfaction they have never felt before and Scorpio would finally get the chance to rest their troubled soul, and realize that not everything needs to be taken seriously. This is a relationship of great lessons and an enormous capacity for personal growth of both partners. 15%

Gemini Compatibility with Sagittarius

Gemini & Sagittarius Sexual Intimacy and Compatibility: They will both enjoy their sexual relations, followed by laughter, creativity and joy. As two children in bodies of grown-ups, they could go through the feeling of shame together if they don't have much experience. When they meet a bit older, there is a slim chance that both of them didn't have enough sexual experiences and partners to understand their personal needs and desires. Gemini and Sagittarius have this strange approach to sex, childish and light as if they don't really care about it. When they get together, they usually get strangely involved in emotions none of them really understands. Their sex life is something to cherish, easy, open and with no pressure at any side. This can make them both a bit selfish, but if their communication keeps going, there is no reason why this would be a turn off for anyone. It is a strange thing, but sex is really not that important to these partners. They are looking for someone to complete their mental personalities, someone to talk to and give them a sense of purpose. This is why they could decide to stay friends after a breakup, for their starting premise was in building a strong relationship founded on their personalities, rather than their sexual or emotional natures. 90%

Gemini & Sagittarius Trust: Sagittarius is not someone who can tell a lie and keep a straight face, and they are usually really disturbed by the likes of other people. Gemini can tell a lie with such ease that they sometimes don't even know they're lying. If anyone can understand the need of their partner to not be faithful, it's these two. Strangely enough, this can lead to ultimate faithfulness, for there will be no more excitement in the secrecy and mystery of parallel relationships. When they get together, this all becomes something to have fun with and they could play a game of trust until they build it on strong foundations of mutual respect. 99%

Gemini & Sagittarius Communication and Intellectuality: As opposing signs they complement each other in general, but this is strongly sensed in this segment of their relationship. With Gemini's ideas and mind flow, there is nothing Sagittarius can't learn or share, being a student and a teacher at the same time. The curiosity goes both ways and they will spend days just learning about each other and absorbing shared experiences. This kind of understanding is truly something to cherish. A problem can surface when they are both preoccupied with chasing their personal values and don't see what they have with each other. The only thing that can interfere with the quality of their mental connection is the possible fear of intimacy that builds in the meantime. That strength of personal exchange stops being mental and starts being emotional at some point, and as two signs that aren't exactly emotional to begin with, they can be frightened by the intensity of emotions that are surfacing when they are together. In general, this is a couple you want to hang out with, every day. They will literally share happiness with one another and with those around them. They can inspire anyone to love and to smile, because when in love, they will laugh so sincerely and have so much fun together. The "here-and-there" nature of Gemini will get new meaning and purpose through the eyes of their Sagittarius, while the search for the ultimate truth can be so much easier for a Sagittarius with the mind of a Gemini. Their optimism and their eloquence will multiply, day after day, until one of them gets scared and decides to take off or death do them part. 99%

Gemini & Sagittarius Emotions: Both signs have a non-emotional feel to them, but their contact develops so much emotion that maybe neither one of them will be able to cope with it. It is kind of strange to think about the emotional side of the relationship between a Gemini and a Sagittarius. They are not used to feeling that much, and when they "click", Sagittarius could discover the new meaning of life and Gemini a synthesis that they've never had a chance to experience. This can truly be a fascinating love story, if only they don't run away from all that emotion. 95%

Gemini & Sagittarius Values: As opposing signs it might seem that Gemini is scattered and superficial, while Sagittarius is collected and deep, but in fact they have the same core in the fact that everything needs to make sense. There is this important thing they both value, things that make sense. Usually, we would connect this with the sign of Sagittarius, but Gemini has it in their approach to words and everyday actions. Their Mercury can't deal with senseless words, stories without meaning and purpose, whatever that purpose may be.

70%

Gemini & Sagittarius Shared Activities: This positive emotion and pure joy they can share, becomes something like a happy drug to both of them and they no longer want to be apart. Not only will they share every activity that any of them thinks of, but they will also laugh all the way, whatever they decide to do together. As two mutable signs, they understand each other's changeability and flexibility, perfectly capable to find all the right reasons why everything they do makes perfect sense. There is a point when they will get irritating to their surroundings, like two spoiled children without a care in the world, but while they are this happy – why would they care? 99%

Conclusion: Gemini and Sagittarius make an incredible couple, probably being the most innocent one of all oppositions in the zodiac. They don't often find each other right away, but at some point in life it is almost certain that a Gemini will find their Sagittarius and vice versa. Their relationship has a strong intellectual connection, in which they will gradually find deep emotions. There is no real prognosis how this will end though, because the emotions they feel could easily scare them away and their relationship could end only because of fear. If they decide to give in and find out what they could share, with Gemini's ideas and Sagittarius' beliefs, the sky is the limit. Or is it beyond? 92%

Gemini Compatibility with Capricorn

Gemini & Capricorn Sexual Intimacy and Compatibility: It is almost unbearable to watch these partners with their completely different philosophies while they try to manoeuvre their sex life. In order for Capricorn to experiment in sex, their partner needs to manage to really relax them and open their mind. With Gemini, they feel like taking care of a child heading for trouble, getting naked wherever they feel like it. In Gemini's and in Capricorn's humble opinion sex is one of them. Then Gemini comes along and starts explaining each position, the interesting overview on Kama Sutra and the beauty of outdoor sex. Although this is not actually the case, this is how it may seem to Capricorn, reliant on traditional values and always taking responsibility for their actions. In most cases, they will hardly even be attracted to one another. If they become sexual partners, there is a big chance that Gemini will find their Capricorn partner uncreative and stiff, while Capricorn would think of Gemini as too unconventional. The strangest thing in this combination of the signs is in the fact they will both probably consider each other boring. Yes, everyone would say Capricorn can be boring and Gemini is so interesting and fun, but actually, the lack of focus and deep feelings Gemini partner usually suffers from, is a huge turn off for Capricorn. All things considered, these two are not actually the best sexual partners among the zodiac signs, but could make a meeting point in a relationship with enough boundaries and enough creativity of both partners. 1%

Gemini & Capricorn Trust: A typical Capricorn representative will not be easily tricked. Gemini partner can have a tendency to flirt a lot and to consider "light adultery" normal. There are certain activities that don't require many words. This is thinking of a crazy person in the opinion of a Capricorn, and there are no "levels" of adultery in their world. What's clear is clear. Capricorn's trust equation that makes their life easy, they will tend to trust their Gemini. Actually, they will trust their own interpretation of what Gemini says. Capricorn always goes one step deeper than others, and Gemini rarely puts that much thought into their alibis. This is why Capricorn will read them with so much ease, probably knowing exactly where they've been and what they are dishonest about. Gemini, on the other hand, will find Capricorn so square and by the book, that they won't even doubt their faithfulness and honesty, even if they have something to hide. 50%

Gemini & Capricorn Communication and Intellectuality: Unfortunately, this doesn't mean a lot to a Capricorn when they recognize the lack of essence in the things their partner says. However, they will still have a lot to talk about because there is always that serious side to Gemini in one of their personalities that will have a thing or two to share with a strict and sometimes difficult Capricorn. It is a good thing that Gemini are interested in literally everything that exists in the world and outside of it. So if nothing else works, they can always talk about Space Stations, diamond stars and other galaxies. It is safe to say that Gemini can talk to anyone and settle any issue by communicating. Capricorn is interested in things that have deep, hidden meaning, looking at them as equations that should be solved and admiring problem solvers. They could spend their life in this strange analysis that is not so much focused on details as maybe Virgo's would be, but on the bridge between different worlds. Capricorn is fascinated by the before and after logic behind every little thing, and this is where Gemini can help them set a list to investigate. If they are at peace and don't look at each other as stupid, distant or boring, they could help one another build a better understanding for the world. Capricorn's steady, secure nature could teach Gemini how to make schedules and organize their thoughts and their actions, giving them a chance to move each thought a step further. In return, Gemini's childish approach to life can be something wonderful for Capricorn to incorporate in their life in order to be happier. 25%

Gemini & Capricorn Emotions: Although both of them can have relationships with other not-so-emotional signs in which they feel awakened with their partner, when they are with each other, they can be immune to each other's charm. Ruled by Mercury and Saturn, both signs are not that emotional, but the real problem is in the fact that they usually don't even spark some emotions in one another. There is just not that much to connect them and mostly their emotional relationship comes down to Gemini's dark thoughts and Capricorn's emotional distance. 1%

Gemini & Capricorn Values: Gemini truly values information in any form and shape, someone's ability to talk beautifully, to creatively use their hands and to implement ideas with a higher purpose. Capricorn values stability, punctuality and plain honesty. Although they will both be dazzled by the independence of their partner, the rest of their worlds rarely coincide that much. 5%

Gemini & Capricorn Shared Activities: If they do go for a walk, they will want to do this in order to get from point A to point B, or to have a healthier lifestyle. When Gemini walks, they never know where they'll end up. They might have started their route on the way to the supermarket, but one telephone call later, they are already in their car, heading to a different city. They would both walk a lot, that's true, but the divergence of their motives is almost unbelievable. Capricorn is a sign of useful things, and they will want to have useful activity, whatever it may be. It is a good thing that Gemini always wants to learn new things and Capricorn likes routine and dedication, so they have a strong base for constructive studying and problem solving. However, in most cases, their roads go in different directions. 10%

Conclusion: While Gemini needs someone to ground them and give them depth, when they look at Capricorn, they see someone old, unmovable and boring. Capricorn needs joy and relaxation in their life, but Gemini seems like a ball of uncontrollable, superficial opinions heading nowhere. Gemini and Capricorn partners are a very strange fit. Although they are both looking for things the other person has, they don't seem to recognize them in each other. In truth, they could have a valuable experience being together, sharing their different lives day after day. They might even find out that they actually work well together and have the ability to reach any goal that they think of. 15%

Gemini Compatibility with Aquarius

Gemini & Aquarius Sexual Intimacy and Compatibility: They will get lost on their way to somewhere and has sex there or somewhere else. But who cares when they are in search for kindred spirits and want to have a good time while at it. They will both be aroused by the intellectual side of their relationship and if they are to be satisfied, they have to consider each other intelligent. Neither Gemini nor Aquarius will ever be in a serious relationship with someone who is, in their opinion, stupid. Gemini and Aquarius could probably have sex by simple verbal stimulation. They don't need to get naked to have a sexual experience, although they will want to be naked all the time to set themselves free from all the human restrictions represented through clothes. Even something that they would call an "insignificant sexual encounter" has to be with someone with enough wit and something to say. They can have sex anywhere and none of them would care. Gemini is a bit childish and can be ashamed in certain situations, but when Aquarius takes over, Gemini will realize that there is no limit to their freedom of expression. These partners will try everything, communicate excessively and learn quickly about each other's body and the way to satisfy one another. Still, their relationship could lack emotion and true physical intimacy. This could lead to them pulling apart, often not aware that they both need something else in their partner. 85%

Gemini & Aquarius Trust: Aquarius finds lying ridiculous and Gemini will usually feel free enough not to lie. On the other hand, Aquarius understands one's need for privacy, for this is a sign where Neptune is exalted. They will both probably have this ultimate trust for their partner and are rarely deceived because of their premise to give and receive freedom as an absolute priority. Trust is a strange thing for this couple. We should emphasize that they will trust each other. None of them will have any satisfaction in storytelling or lying when there are so many interesting things to talk about with their weirdo partner, and so little to share that will be judged. 90%

Gemini & Aquarius Communication and Intellectuality: They stimulate each other's mind to such a point that they fire arguments they weren't aware existed in their thoughts. While Gemini will probably be fascinated by the belief system of Aquarius, always so rational and humane, Aquarius will have an opportunity to relieve some of their ego problem with their Gemini partner. The mutable quality of Gemini will allow them to adapt to some of those rigid Aquarian attitudes and opinions, even if they disagree. When Gemini and Aquarius engage in an intellectual debate, they are fun for everyone to watch. Gemini does have this mellow nature that understands the flow of the social touch with other people, and will rarely fight for their beliefs with someone they feel really close to. This is a good thing for their everyday life, but in general, this can present a problem because the authentic personality of Gemini could be shushed until they are not sure who they are anymore, once again. It is important for them to have enough flexibility for one another, however different their premises might be. Still, it is best if they share the same basic life philosophy, which they usually do, or they could get distant and lose interest in each other. As two Air representatives they find that communication is the solution to any problem, but aren't aware how far from Earth they might get with their ideas unrealized and their goals unreached because of too much talk, and too little action. 99%

Gemini & Aquarius Emotions: The unstable nature of Gemini can make them change their mind or their emotional state on a daily basis, and if they don't feel good in a relationship, they will set themselves free without over thinking the reasons why they had to do so. Aquarius is always in a rush to set themselves free from anyone or anything, so a breakup wouldn't really be something strange in their world. Gemini and Aquarius understand each other perfectly when it comes to their emotions. Usually this is true, but that doesn't mean this is what they both need. In most cases, their rational, mental natures will complement each other in an exciting way, but there is not much emotion to be built in the core of their relationship. It seems that both of these partners need to find someone a bit warmer in order to feel things more deeply and light their passionate hearts. They will much more often become friends than lovers, even if they were attracted to each other when they first met. 40%

Gemini & Aquarius Values: They both value intellect. The rest is just something that other signs worry about.

However, Aquarius can be very passionate about their humane beliefs and will often support them strongly. This is something Gemini can understand but rarely supports. Because of the fact that Aquarius partner values equality of the people as much as their own freedom, this can be their point of separation, even though Gemini partner does not really disagree. 95%

Gemini & Aquarius Shared Activities: They are so different than everyone else and represent a step that Gemini should climb if they want their life to be unbelievable. This is not always the case because some of the Aquarius' weirdness can be a show off, designed by their need to mend their turbulent Sun. It is exciting for them both to enter this relationship like Aquarius to surprise and Gemini to follow them wherever they go. Aquarius is the only sign capable to really surprise Gemini. Their main activity to be shared is movement. They could drive thousands of miles just to find a specific ice cream or for no reason at all. Mostly they can do anything together, from travelling and clubbing, to reading labels and instructions on the use of different kitchenware. 99%

Conclusion: Gemini needs a partner who doesn't bore them or make them feel inhibited. When you look at things this way, you could say that there is no better match for them than the fabulous Aquarius. Aquarius needs someone to understand their grandiose ideas and discuss each one with them, and also someone who doesn't make them feel inhibited. Who could do this better than Gemini? However, they could find themselves in a relationship that doesn't have enough emotion and compassion, and this is certain to surface as soon as the first disturbing thing happens in the life of one of these partners. They need to work on their emotional base and their non-verbal understanding if they want their relationship to last. 85%

Gemini Compatibility with Pisces

Gemini & Pisces Sexual Intimacy and Compatibility: They can be attracted to each other due to the fact that they are ruled by Mercury and Jupiter, the same planets that rule their opposing signs. Still, there is a big chance they won't even recognize each other as sexual beings or keep a distance from each other if they do. Gemini has a lot of creative potential, but isn't exactly in search for their one and only true love in order to have sex. It is a good thing there is so much creativity to Gemini's approach to sex, or they would really have difficulty making any sort of intimate connection with Pisces. Pisces, on the other hand, exalt Venus, and they only want to have sex with the love of their life, unless they've been disappointed too many times. If they meet after these numerous disappointments, Gemini will not find Pisces very attractive, for they will no longer have any childish energy or charm. If their sex life is supposed to be functional, both of them will have to find a way to be a bit more grounded than they normally are. Gemini will have to realize the truth behind their own emotional nature and give in to true intimacy, while Pisces will have to accept the differences of their partner instead of searching for a soul mate with predefined qualities. 15%

Gemini & Pisces Trust: They have completely different ways of dealing with their emotional relationships and their issues with self-image, and they will think of many different ways to bend the truth when they are together. Trust is already a weak spot in almost every relationship these two can have and when they get together, there is a chance they will have no idea how to create any trust between them. Unfortunately, when any of them lies, they won't have much success. Gemini is too smart to be lied to by their Pisces partner, and Pisces sense their Gemini partner's state too well to not realize when they're not telling the truth. Basically they both dip into each other's unconscious and see each other in the way that none of them looks at themselves. 1%

Gemini & Pisces Communication and Intellectuality: Gemini can decide to make a joke and Pisces will laugh without really thinking about it. Pisces will then say something to poke their Gemini and Gemini will laugh without thinking about it. It is as if they never really listen to each other and sink into a strange pool of superficial relationships and small talk. If they start discussing their deep thoughts and feelings, they might end up in a conflict that none of them anticipated. We could say that they idealize each other, but only to a point of recognition. There is always a fairytale or two to share between them and some fun to have when they go out. They will laugh together, but this is a strange connection with a lack of real communication. Neither Gemini nor Pisces will think of each other as their one true love unless they really are each other's one true love. So they will have this image of each other that is deviated from reality, due to the fact they don't really listen to each other. The only possible way for them to have a conversation with depth is in a situation in which they have absolute emotional intimacy, such as family members usually have. 20%

Gemini & Pisces Emotions: When they do fall in love, they are rarely on the same frequency and often only one of them has true emotions for the other. Gemini is one of the most rational signs in the zodiac and Pisces is the cureless romantic and one of the most emotional signs. They represent ideal candidates for an unreturned love scenario and can be a nuisance to everyone around them if they end up in a relationship with no emotional balance. 1%

Gemini & Pisces Values: In general, they will both hold on to what they know best and Gemini will value intellectual strength and won't get very disturbed by dishonesty for as long as their image of a relationship isn't disturbed. Pisces will value their partner's reliability and trust is very high on their list of priorities. They both value what they stand for and although Gemini values someone to listen to them and love them unconditionally, this is not the same as the passionate love Pisces partner wants to have. There is one thing they will share, hidden in the fact that they both value someone's ability to create. Even though this comes out of different views on creation, it can bond them in the act of creativity. Pisces would provide talent and inspiration and Gemini their resourcefulness and practicality. 5%

Gemini & Pisces Shared Activities: When we speak of Gemini and Pisces, we have to keep in mind that both signs are mutable. Although their interests might differ greatly, they could find activities to share due to their mutual need for movement of any kind. Pisces will normally dream about movement rather than actually move and this is exactly what their Gemini could teach them – how to make the first step. 15%

Conclusion: They are both usually positive enough to have a superficial enjoyable relationship and go well together at large social gatherings. They could both forget to call each other when they agreed to, and they can both change their opinions in two seconds, but they simply don't share the same goals. As a strongly mental and a strongly emotional sign, their lack of understanding can be hurtful for Pisces and sometimes for both of them. Gemini and Pisces are squaring signs that often don't have that much in common. If they do fall in love and start a romantic relationship, chances are they will not last very long. However, there is a beauty in the creative side of this relationship and if Gemini decides to truly listen to Pisces, they could help them use their talent in a constructive way. In most situations Pisces will just drain the energy out of their Gemini partner, especially if they end up in their fragile, needy mode that some other signs could understand much better than Gemini. If they are to succeed in their persistence to be together, they should work together and socialize a lot. The most important thing for both of them in this relationship is to reach for their emotional cores and give in to true intimacy, or they will never manage to communicate. 10%

PART-4
CANCER HOROSCOPE

(21st June – 22nd July):

Who cannot stick to any adhesion and changes like a season?

Who most perplexing character and let's go without any reason?

Element: Water

Quality: Cardinal

Ruler: Moon

Lucky Gem: Yellow sapphire, Pearl & Coral. If he suffers with sleepless nights, a pearl could be the perfect cure. Wearing the pearl ensures peace of mind, and brings all the good luck in the world.

Numbers: 2, 4, 6, 7, 11, 16, 20, 25

Lucky Number: Number 4 & 6 are his/her pick for good fortune.

Colour: Orange, White

Lucky Colour: white, Soak in the elegance of white.

Day: Monday, Wednesday, Thursday

Lucky Day: Wednesday; it is time to look ahead on Wednesday.

Lucy Flower: Larkspur.

General Insights: Deeply intuitive and sentimental, Cancer can be one of the most challenging Zodiac signs to get to know. Cancer is very emotional and sensitive, and they care about family and home. Cancer is sympathetic and is very attached to the people who surround him. People born under the Cancer sign are very loyal and empathetic people, able to empathize with your pain and suffering.

Because of the ruling planet the Moon, the many phases of its lunar cycle can deepen Cancers internal mysteries and create fleeting emotional patterns that the sensitive Cancer cannot control, especially when a child. This can show itself as mood swings, selfishness, manipulation and fits of rage.

Cancer is quick to help others and avoid conflicts. One of his greatest strengths is persistent determination. Cancer doesn't have great ambitions, because they are happy and content to have a loving family and tranquil and harmonious home. They often take good care of their co-workers and treat them as family.

Personal Quality: The Cancer is protective, jealous, sensitive, full of suspicion, not easy to understand for his/her moods, often fluctuate from sweet to cranky and lacks faith in others. He/She can be untidy, sulky, devious, and inclined to self-pity because of an inferiority complex but always ready to cooperate. He/She is very fond of food and is usually hearty eaters. He/She gets ancestral and sudden properties after the age of 35. He/She is the most family-centred persons and fiercely protective of loved ones. He/She possesses strong paternal and maternal instincts.

Positive Quality: He/She is lovely, tenacious, highly imaginative, loyal, emotional, sympathetic, persuasive, kind, faithful, loyal, honest, hard working and sensitive and leaves his unforgettable impression on others.

Negative Quality: He/She is sulky, moody, pessimistic, suspicious, manipulative, insecure, devious, moody, clinging, manipulative, overly emotional and insecure and inclined to self-pity and prone to a sense of personal inferiority and believes his/her views, opinions and behaviour to be impeccable, and beyond question or criticism.

Physical Appearance: He/She is small with round faces and possesses a tendency to gain weight. He/She has abundant shiny hair, expressive eyes and is economical with his/her gestures.

Relationships: He/She treasures emotional bonds and doesn't severe tie easily. He/She clings on to a failing relationship and finds it difficult to let go. He/She wears his/her hearts on his/her sleeve, and is prone to emotional excesses.

Career: He/She is best suited for counselling and charity work, good journalists, writers or politicians, archaeologist, caterer, dairy farmer, deep-sea diver, dietician, hotel worker, manufacturer, merchant small businesses. He/She works well with people and often adopts the role of a mediator.

Health: He/She has a weak digestive system and constantly suffers from heartburn and ulcers. He/She tends to become hypochondriacs.

Cancer likes: Art, home-based hobbies, relaxing near or in water, helping loved ones, a good meal with friends

Cancer dislikes: Strangers, any criticism of Mom, revealing of personal life

Ideal Partner: He/She seeks steady, stable and practical partners, and usually bonds best with Capricorn.

Compatible Zodiac Signs: Taurus, Virgo, Scorpio and Pisces

Incompatible Zodiac Signs: Aries, Gemini, Libra, Leo, Sagittarius, and Aquarius

Greatest Overall Compatibility: Scorpio, Pisces, Capricorn

Best for Marriage and Partnerships: Capricorn booster Gem is Pearl.

BUSINESS ASTROLOGY FOR CANCER

With penchant for trade or business these individuals know how to keep their clients happy. But with passing time they often turn careless & lose interest in their business. To take business to dizzy heights they should keep alive the rhythm of interest. Business booster Gem is Moon Stone.

Cancer Compatibility with Aries

Cancer & Aries Sexual Intimacy and Compatibility: This would be fine if the members of this universal Cancer family weren't convinced in their asexual nature as well. Their emotional characteristics allow only for sexual relationships with meaning and enough tenderness. Only when they meet the right person to set them free, they come to learn about the other aspects of their sexuality. We usually see the sign of Cancer as extremely asexual. Our family is presented by the sign of Cancer and the Moon, and it is a psychological challenge for all of us to understand that our parents are sexual beings. The problem with sexual relations with an Aries is that Aries partners are usually not that gentle to begin with. They need to learn to show emotion. For them intimacy is something built, not implied. If they manage to reconcile these huge differences at the beginning of their relationship and if none of them is forced to do anything they are not ready for, their attraction to each other should do the trick and their sexual relationship could become truly sensual and exciting for both of them. 70%

Cancer & Aries Trust: Usually the problem they encounter is a trust issue when it comes to intimacy. Aries has a different view on intimacy. In the eyes of their Cancer partner they can seem pushy and even aggressive with an attitude that doesn't lead to anything close to relaxed. As much as Cancer would like to understand the straightforward nature of Aries, it will be extremely difficult to see it as anything other than beastly. There is also a problem with the way they show and recognize emotions. The issue of trust is something different for this couple. They will rarely debate about their trust in each other's fidelity. It may be hard for an Aries partner to understand that they are loved if someone only asks annoying questions, tries to tie them down and doesn't want to have sex. On the other hand, Cancer will probably feel violated in every way, unless Aries partner slows down and has an atypical show of gentle emotions. Usually any sort of mistrust is a consequence of the lack of ability to believe in each other's feelings for one another, for they don't really recognize them well. 50%

Cancer & Aries Communication and Intellectuality: It is not their intention to react in this way, but they push each other's buttons and it is very hard for them to stay focused and solve the issue they talked about. Their interests differ too much, so even when they are trying to have a peaceful conversation about something impersonal, it is still a battle to keep the attention to the subject in question, whoever initiated the talk. Their only shared characteristic is the cardinal quality of both signs, which gives them a good understanding on each other's "ad hoc" personalities. Both of these partners have the same tendency to act on an impulse and cut the conversation short before they even got to the point of it. This will make it easier for the couple to recover from all of the possible conflicts and misunderstandings. Still, in the eyes of a Cancer partner, this type of relationship doesn't have a purpose and they might find themselves fighting in a way they don't feel comfortable with. As their signs are ruled by Mars and the Moon, it is an archetypal story of hurt and emotional pain, so their intentions have to be truly pure. They have to treat each other in a gentle, thought-out way, measuring every word they say. This can be exhausting for both of them, unless they fully accept the fact that they don't need to change their personality, only the way they express it and make a game out of it. 20%

Cancer & Aries Emotions: They are warm, passionate and have high expectations of their partner when it comes to scratching beneath the surface. Their boundaries may be too strict as they fear their own sensitivity and sometimes act like heartless soldiers. Cancer wears their emotions as a winter coat and hides them only when feeling ashamed to show them. They accept their emotional nature as a given and work toward realizing a personal world full of respect for their soft side. We often say that Cancer wants to have a family and raise children, but this is not due to their need to reproduce or stay in the house all day long, but because they need a safe haven for their emotional side and enough people to share their compassionate nature with. Aries and Cancer are both deeply emotional, although Aries is often described as if they had an emotional disability. Even though these approaches to their emotions seem different, they understand each other's depth and in most cases respect each other in this area of existence, in case Aries leaves their impatient nature out of their relationship. The problem appears when they are supposed to understand how they feel about each other, as feelings are not easily shown when dealing with partner's personality they don't fully understand. 70%

Cancer & Aries Values: While Aries gives a lot of significance to someone's state of energy, focus and consistency, Cancer values the ability to stay rational and stable, qualities they have a difficulty achieving, or being in a state of emotional balance. Their values aren't even connected, except for the fact they both have the idea that some sort of future balance, that can be quite hard to achieve, would make them better. 40%

Cancer & Aries Shared Activities: As much as Aries wants to devote to their physical body, sports and all the ways to keep their creative energy high, Cancer wants to sleep, dance and eat all they long. Mostly they share sexual activities and the time for rest, since Cancer probably has no intention of following that insane Aries pace.
30%
Conclusion: This relationship can be painful for both partners and needs a lot of work put into it in order to work. It requires both of the partners to adapt and make changes in their behaviour, while tip toeing around each other most of the time. It is not an easy road, but the rewards are such inner understanding of passion, full of emotion and the ability to create something truly unique. If they succeed, they will probably never be satisfied with a different partner.
47%

Cancer Compatibility with Taurus

Cancer & Taurus Sexual Intimacy and Compatibility: This is an instinctive assumption based on the fact that both signs don't care for Mars very much, meaning they don't care for instinctive sex. They would probably never have the urge to have sex just for the sake of it, but this doesn't in any case mean they are asexual. Taurus is a sign of physical pleasure. Ruled by Venus, the planet of feminine sexuality, Taurus needs to approach their sexual experiences with the same studiousness with which they would approach any other thing in their life. Someone might think that Taurus and Cancer are two of the most asexual signs in the entire zodiac. They need to see, touch, smell, feel everything on their partner's body and enjoy making them satisfied. With Cancer's need for closeness and the lack of ability to make their sexual life light and carefree, Taurus seems to have the perfect touch to relax them and build trust within their sexual relationship. With the lack of Mars comes the lack of initiative, and this could be their problem when it comes to sex. In case they both don't have a sexual drive stronger than their love for food, they could end up in an asexual relationship, in which they would lie around the house, cook, eat and gain weight all day long. 95%

Cancer & Taurus Trust: When Taurus and Cancer fall in love with each other, they base their entire relationship on the feel of their partner. There is not much that can be hidden from this sensitive, "sixth" sense these two can share when they connect. It would take a lot to break their trust and this would certainly mean the end of their relationship. In most situations, none of them has the need to betray their partner, for their goal is the same – love, family and home. 99%

Cancer & Taurus Communication and Intellectuality: Their most common topics would include love, home and children, except when they feel unready or when they are too young. It is important to understand that these are not the only issues on their minds, though. Whatever their current interests, they will communicate it in a slow, sensitive manner, leading a conversation to a point of deep mutual understanding. Still, Taurus can be truly stubborn at times. They share lots of different interests and will easily talk about their relationship. Neither of these signs talks much, but they are perfectly capable of understanding each other's silence and give each word a lot of significance. It doesn't really matter if they are right or not, for they simply close up for any further conversation as soon as one of their true convictions is touched. Cancer can't really do that much when this happens. They can try and be even more gentle and compassionate. In case they are close to enlightened, this will certainly work. If not, they will get really emotional and discover that their Taurus partner in fact drives them crazy. As a result, Taurus will see their partner as a real lunatic, waving their hands for no reason and showing no rational behaviour whatsoever. 80%

Cancer & Taurus Emotions: While Cancer feels, senses and takes care of their Taurus partner in the emotional realm, Taurus will give love back through physical tenderness, material security and the gentle touch of practical sense that Cancer needs. When this cycle happens several times, their love seems like a chain reaction that will never stop growing. If they meet in supporting circumstances, when they don't need to fight for each other or the possibility of their love, every emotion should simply build up on the previous one and things between them should run smoothly. Taurus and Cancer are the rulers of the entire warm, earthly emotional world. It is not just due to their sensitivity, but the combination of their emotional expression is something almost unimaginable. But if they stumble upon an obstacle of any kind, before their love for each other develops, they will probably be discouraged and never discover what they might have felt if only they fought for one another. If they do fall in love, they will not lack the energy to fight for their relationship, no matter the obstacles. 99%

Cancer & Taurus Values: Cancer is a Water sign, much more focused on the value of emotion, while Taurus will be turned to financial security. This usually reflects the fear Taurus feels when it comes to their material existence being in question. Most of all, they both value life and peace. Since the Moon is the ruler of the sign of Cancer and exalted in Taurus, they both deeply value all things the Moon represents, such as, family, compassion, understanding and bliss. However, there can be a certain divergence in their view of the material world. In the eyes of Cancer this may seem superficial for they have a tendency to think of material reality in an idealistic way. If they have a family together, these issues should settle, because the love of Cancer partner to their children combined with their desire to give them everything they need, easily shifts their perspective and teaches them about the true value of money. 80%

Cancer & Taurus Shared Activities: We could say that they might share every single activity any of them thinks of, but this is not that often called an "activity". Most of the time and especially if they both have demanding jobs, they will simply share the activity of sleeping, eating or doing nothing. This is not a question of laziness, but more of an exaggerated need for the pleasure of rest. When shared, it seems to multiply and grow beyond both of their rational minds. 90%

Conclusion: When they fall in love, they will rarely find the reason to separate, because of their shared emotional goals for love, understanding, family and the feeling of home. This is the relationship that seems like a perpetual mobile of love, in case both partners don't already have too much emotional baggage that makes them unable to give and receive this depth of emotion. Taurus and Cancer present the gentlest couple of the zodiac. Even if they do, with no obstacles on the way, they will likely learn to forgive and forget as the flow of their relationship takes them to what they always desired. 91%

Cancer Compatibility with Gemini

Cancer & Gemini Sexual Intimacy and Compatibility: This is not always the case, of course, but it is very rare for a Gemini partner to manage to relax their Cancer and make them join their sexual adventure. Still, there is a link between them, pretty strong for that matter. Cancer continues the sign of Gemini, and in a way their nature is a consequence of endless conversations and rational explanations. If Gemini partner has enough patience to talk to their partner about same things over and over, in time they could build enough intimacy to have a good sex life. In order for this scenario to develop, Cancer needs to talk, too. Gemini would go outside and have the weirdest sexual experiences, and Cancer would stay at home and wait for the night to be loved by their tender partner. It will not be enough to say a sentence or two and expect that their Gemini will understand how they feel. If they manage to communicate, Cancer could actually wake the depth of their Gemini, giving them enough support to express their more emotional sexual personality. If they want their relationship to work, their sex life needs to be somewhere in the middle like intimate enough and exciting enough – for both partners to be satisfied. 5%

Cancer & Gemini Trust: As a rational sign, they usually find no reason to lie or cheat on their partner, for as long as they don't feel threatened by intimacy they are not ready for, or tied down. The major task of Cancer in a relationship with Gemini is to let them be free. If they start living together, Gemini will have to stay out of their love nest and Cancer will spend a lot of time alone. Gemini is a sign ruled by Mercury, our little trickster. It is not easy to trust a Gemini partner, especially if one is trying to take away their freedom. This is a challenge, but both signs can have enough dedication to their personal values, and these partners should know better than to ruin their relationship with trivial lies. 25%

Cancer & Gemini Communication and Intellectuality: In their relation to a Cancer partner, they usually feel the need to open up a bit more and share things that they wouldn't with other people. There is a certain "motherly" glow around Cancer, male or female, that gives Gemini enough room to set their inner child free. This can actually be a wonderful relationship that lasts much longer than other Gemini relationships, because there is usually enough understanding in Cancer for their childlike partner. There is probably not a single sign in the zodiac that Gemini couldn't talk to, so their communication skills mark this category a bit higher in every case. That is, if other aspects of the relationship are satisfying enough. It is safe to say that Cancer and Gemini make good friends and this could give a push in the right direction toward their emotional and sexual understanding as well. 70%

Cancer & Gemini Emotions: It is often said that Gemini is a superficial sign, but in fact, they like to stay afloat and keep away from difficult conversations because of their need to move forward. Cancer can try to share their emotions too often for Gemini to have enough patience to understand them, and this can be devastating for their emotional relationship. The main negative possibility here would be in the fact that Gemini doesn't always listen. On the other hand, Gemini doesn't really show emotions in a way that Cancer will easily recognize. It is a good thing they are capable of finding emotion in anything and anyone, so they might have a chance to understand the emotional nature of their childish Gemini. 10%

Cancer & Gemini Values: When it comes to relationships, they are probably the most vulnerable to differences in this category, for differences here make their primary goals different. They have completely different systems in which they value things in life. While Gemini is a rational sign, giving value to all that comes out of their mind, Cancer is an emotional guru, giving value to things their heart beats for. Although they will both be motivated to find love, the way their partner acts in a relationship could be considered "wrong" due to its opposite nature from what each of them values. 1%

Cancer & Gemini Shared Activities: If they have enough of it to follow Gemini's chaotic schedule, they could find a lot of enjoyable things to do together. Cancer is a sign that exalts Jupiter, and its representatives like to travel, as much as they like to feel at home due to the sign itself. Although Gemini partner doesn't really have to go that far, they could find middle ground in travelling to places they both want to visit. Activities Gemini and Cancer could share mainly depend on the energy level of Cancer partner. However, the main characteristic of Gemini is their curiosity. It will lead them in all directions, they will feel the need to try everything and find more new, exciting experiences to share with someone. Unfortunately Cancer doesn't really enjoy this that much. While the relationship is new, it will all seem exciting and breathtaking. As time passes, Cancer will want to settle down, imagining them buying furniture together and raising children. It is not easy for them to accept that this nature Gemini has isn't something that will be different tomorrow. There is no settling down with a Gemini partner. This is just not their mission in life. 15%

Conclusion: When it comes to emotional or sexual relationships, there seems to be too many things that set them apart. In order for their relationship to last, they both need to make some adjustments. Gemini will hardly ever change their routine for someone, especially when they find someone's way of life boring, so the best thing to do here is to give them their freedom. Gemini and Cancer are next to each other in the zodiac, and they are likely to be next to each other in friendship. If Cancer falls in love deeply enough, they will understand what their Gemini partner needs, and won't hold them back even if they wished for them to be different. Gemini partner has to open their heart and listen to those few words that Cancer wants to share.

Even though they can speak about many things, when it comes to discussions of their relationship, their views on it are different. Gemini needs to keep it interesting and Cancer needs to be heard, as much as felt. If they give each other enough freedom and understanding, they could be like children in love for the first time. 21%

Cancer Compatibility with Cancer

Cancer & Cancer Sexual Intimacy and Compatibility: Cancer is a sign where Mars falls and it is not easy for them to have initiative. For a healthy sex life, Mars needs to be strong and these two don't necessarily show this strength. This could lead to their emotional relationship blossoming, but no sexual chemistry between them. It is a good thing they both don't need much experience or technical knowledge in their sexual encounters, because this gives them a chance to base their sex life exclusively on emotions they have for each other. When it comes to intimacy, Cancer is a master of achieving it with the right person. If two Cancer truly find each other, they will reach for each other's deepest emotional core within their sex life. It could be boring from the perspective of some other zodiac signs, but they really won't care if they get the confirmation they are loved through their physical contact. However, they do need someone to wake them up, motivate them and push their limits. No Cancer will be satisfied with a boring emotional life, even though they might seem that way. They will fight for their family, for intimacy and their safe haven, but they will not settle for something that doesn't excite them at all. In order for the relationship of two of them to succeed, it would be a good idea to experiment a little and show initiative at any time they feel the least bit sexual. 65%

Cancer & Cancer Trust: Cancer is a well-grounded sign, for its ruler is exalted in Taurus. They want an emotional stability that can be felt in the material world and understand that there is no such thing as perfection. Why not trust each other when they both don't care about their ego being lifted by adultery, a younger partner or search for some impossible love story. When they find a person they can see themselves with in years to come, they will accept their faults and make necessary compromises to build a loving family and a home for themselves. If two Cancers see each other in this way, they have no reason not to trust each other completely. 99%

Cancer & Cancer Communication and Intellectuality: When there are two of them, especially if they are in an intimate relationship, they can be quiet for days for as long as their inner feeling is good. Although they enjoy the routine of sharing some life details with their partner, their favourite time of day will probably be that cup of morning coffee when they don't have to talk at all. Some of the more rational signs could have a low opinion on the intellectual strength of these partners and they could seem as if they have nothing to say. If you are looking for a fan of non-verbal communication, look for Cancer. But anyone who is a bit more sensitive will feel that they look at each other with undivided attention and follow each other's movements and grins. They don't need words when they can smile to each other. Who cares about communication really? It can be so obsolete. 90%

Cancer & Cancer Emotions: When two Cancer representatives start a relationship, they will understand each other's emotional states perfectly. There is an almost inevitable issue in their primary families that needs to be resolved, and they will usually use each other to do so. As they are both ruled by the Moon, their mood changes will probably coincide, but the scope of emotions that the Moon represents is far bigger than most of us presume. Mostly Cancer is tagged as the most emotional sign of the zodiac. This is somewhat true, although the leading roles are divided by all of the Water signs. Cancer is the sign of family love and closeness, not so much the sensual, sexual love presented through Venus. They carry emotional state of their ancestors and it is not enough to tag them as just "emotional". The best relationship for two Cancer partners means to build a family, but only in case they have a good sex life. All of their emotions are best shared, understood and dealt with when they share a home and their life together. 99%

Cancer & Cancer Values: They will share the same values and understand each other perfectly in this segment of their relationship. Their similarity comes to focus right here, where they are free to value emotional clarity, peace and a calm, family life someone is able to create. This could keep them together in a loving relationship even if they maybe didn't seem perfect for each other in the first place. 99%

Cancer & Cancer Shared Activities: As two Cancer partners they don't have a problem to share any activity, but they could realize that they have a problem starting one. While their relationship is new and everything is exciting, they will both share their ideas on things they could do together, probably things from their individual routines. When the relationship settles, they might find themselves in stagnation, not moving at all, just because of their passive natures that have a tendency to spend a cosy evening at home. 55%

Conclusion: However, their mellow nature, ability to feel and have enough compassion for each other, makes them great candidates for marriage, children and the whole picket fence scenario. Their sex life and their shared activities could suffer a general lack of initiative, energy and movement. Because of this, they should both try not to end up in a boring everyday routine in which they only eat and sit in front of a TV as soon as they come home from work. Cancer is a sign of genetic inheritance and it is sometimes difficult to reconcile the genetic predispositions of two Cancer partners. It is important for them to have enough tender surprises and activities that build their physical relationship, or they might end up unsatisfied and not really understanding why. If they are troubled by this possible shortcoming of their relationship, as two tender individuals, they will manage to make each other feel wonderful, even if that means breaking up. 85%

Cancer Compatibility with Leo

Cancer & Leo Sexual Intimacy and Compatibility: Although they don't have much in common, in astrology they represent a husband and a wife and are the king and queen of the zodiac. Unfortunately we know how unsatisfying the sex between a king and a queen can be like. The sexual aspect of their relationship depends on the depth of their emotions. As highly emotional signs, each in their own way, they tend to show their love in different ways and this can be a bit hard to reconcile in their sex life. Cancer and Leo make a very interesting couple due to the fact they are the only signs in the zodiac ruled by the lights in the sky, both of which are not planets, but the Sun and the Moon. As a Fire sign, Leo is way more openly passionate and this could scare their Cancer away. Cancer is tender and sensitive enough to make their Leo partner feel guilty because of their nature, or Leo could simply have difficulty being tender in the way Cancer needs them to. It is as if a lion and a roe started a sexual relationship and although they don't want to hurt each other, their primal behaviour seems to pull them in that direction. Still, because of their rulers, they can get pretty close and share fine emotions in their sexual encounters. Although there won't be much excitement to them, they could be satisfying enough for both partners if they don't expect a wild sex life. To find middle ground they really need to be quiet and listen to each other's needs. 30%

Cancer & Leo Trust: This is somewhat true and Leo is a born performer, more or less supported to become one. However, their need to show off is something Cancer might find irritating, but not something to lose their trust over. It is said that Leo likes to be in the centre of everyone's attention. If Cancer feels loved, they have no reason to doubt their partner just because of their nature. Still, the differences between them might lead to a secret search for more compatible partners and this is easily sensed by both partners. 50%

Cancer & Leo Communication and Intellectuality: The Moon does reflect the light of the Sun, but it circles around the Earth. This would be fine if the Sun wasn't so used to the fact that everyone circles around it. This explains what happens in their relationship once Leo starts talking. Although they shine, Cancer pays much more attention to someone else or to the idea of earthly things they could have together, than to their Leo partner. Cancer and Leo, ruled by the Moon and the Sun, represent a subconscious and a conscious mind. Even if they share interests, they will often have a strangely different view on the same thing. They will often just drift apart as the conversation progresses. The base of their communication should be in things that have yet to be discovered. All the mysteries that are there to be solved could be a good starting point because they both need to understand things, each in their own area of dominance. The best possibility they have is in the things that need a light cast on them from two different angles. If they have enough respect for each other, they could learn a lot from each other's passive and active approach to life. 10%

Cancer & Leo Emotions: Cancer is a Water sign that represent motherly love and all emotions in one's family. Although this is not always a promise of happiness, the depth of all the love that hides within is magical. On the other hand, Leo is a Fire sign of joy, first loves, fun and sex. Their heart is warm and big, for Leo represents our inner child, and their loyalty is unchangeable when they decide to give it to someone. The main problem here is their connection, for Cancer needs to look into somebody's eyes and feel the love, while Leo wants to shout it out from the rooftops. We can say with certainty that their emotions are a truly beautiful thing. Both signs represent love and although it is not the same kind of love, it is an emotion, pure and simple. This could seem untrue to Cancer, as much as Leo could recognize in Cancer something that would rather tie them down than allow them to fly. Both of them have opposing signs ruled traditionally by Saturn and this is exactly why they could recognize true quality in their relationship, as each of them takes their opposing role to learn something. Usually this is not something that can last very long and they will probably both moves on to find someone who is more of an image from their seventh house. 45%

Cancer & Leo Values: They simply don't value the same things. Although this is the point where they separate their ways, it usually takes them long to realize this fact. Cancer values tenderness, emotions, family and a stable life with someone, while Leo values initiative, passion, energy and focus. There is rarely something that they will both value in the same way, or put in the same spot at their priority lists. 1%

Cancer & Leo Shared Activities: However, when they leave the house, they will want to visit different places. Cancer will want to visit their close friends, especially if they have babies, go for a walk by the lake or have a romantic evening at the movies. Whatever they do, they like it with a little intimacy to it. As if to oppose this, Leo will want to spend time at places where they could be seen. It helps that Leo would gladly sleep for 20 hours a day. This will make them available at home. They will also gladly be served by breakfast, lunch and dinner, and who better to prepare those than a caring Cancer. Their partner needs to shine to others and hold them by the hand while everyone else claps theirs. This is not often something Cancer will be ready to do, as they don't really like to be the centre of attention. On the other hand, Leo could compromise, but in the end they wouldn't be happy if their social needs aren't met. 35%

Conclusion: Although the Moon reflects the light from the Sun, the sign of Cancer doesn't really see Leo as the source of all their joy. They are a sign that should spread joy and love with an active approach to each one of their relationships. It will be probably because the Moon circles around the Earth, not the Sun. They are special, that's for sure. Both of them are strong individuals, each on their own plane. Their lack of understanding and emotional touch can be explained through the fact that both of them have a mission to spread love to the less fortunate signs of the zodiac. Not everyone is born with an emotional flow like Cancer and a huge, warm heart like Leo. If they kept all this love to themselves, some unfortunate souls would probably search for them aimlessly, and the world would be a much sadder place. 29%

Cancer Compatibility with Virgo

Cancer & Virgo Sexual Intimacy and Compatibility: The sign of Virgo brings Venus to its fall and suffers from a general lack of emotion. It is a rational sign with a lot to analyze, that rarely gives in to the first impulse or their fragile emotional state. They are to learn on how to feel safe enough to let their guard down and shut their mind off in order to feel and enjoy sex. When Cancer and Virgo get together, there is potential for a great, everlasting love. They have a possibility to become an inspirational contact of heart and mind, if only they give in to the opportunity to enter each other's worlds. Their sexual relationship seems to be a lecture on emotion. Cancer doesn't really understand how someone could have trouble getting in touch with their emotions and can have unrealistic expectations because of this. They will learn to understand their partner better and make a stronger sexual bond, realizing how different people can be. 95%

Cancer & Virgo Trust: Although Cancer is a cardinal sign, they are stable by nature, especially when it comes to emotional decisions they have made. If someone can help Virgo build their trust, it is their Cancer partner. If they have chosen Virgo to be their loving partner, they will have no reason to lie or cheat. This behaviour would only endanger their vision of a shared life and a loving family they want with the partner they chose. This is also a reason why Cancer won't have an initial problem with trusting Virgo. Their convictions are stronger than their doubt. 99%

Cancer & Virgo Communication and Intellectuality: The lack of words from Cancer certainly makes room for everything Virgo wants to say, but as signs ruled by the Moon and Mercury, they have a simple conflict of emotion versus logic. Although Virgo represents the grounded side of Mercury and that makes it much easier for them to communicate with someone like Cancer, they are still leaning a bit too much on their rationality rather than their heart. This is a tricky side of a relationship for a Cancer and a Virgo partner. However, both of these planets are ruling the human brain, like Mercury represents the core and Moon rules the rest of it. This is why when their topics and their intellectual strengths combine in a right way, and with their emotions to follow, they find an uncharted, new territory in which none of them has ever been. Their communication can become truly inspiring and magical if they accept each other's characters fully. 60%

Cancer & Virgo Emotions: If Cancer starts showing their emotions openly and with no restriction, Virgo might get scared and start analyzing every little thing to determine if any emotion is really there. There is probably no greater turn off for a Cancer than someone who rationalizes their own emotion. As if the fight to keep them afloat among the human race isn't enough, now they have to deal with a neurotic partner who dismisses them. The rational side of Virgo could keep their overall emotional status very low. The emotional side to Virgo is a deeply feminine side, usually ashamed to show her face, especially if Virgo partner is male. It becomes almost impossible for them to feel something if they are in any way pressured or feel mistrust with their partner. Virgo would rather be alone, with a right book, than with the wrong person, and it takes a lot of patience and rationality from their partner if they want to understand and wait for the ice to melt. If Cancer manages to do this, there is no reason for both of them not to resolve any other emotional issue that needs resolving, mutual or individual, while together. 65%

Cancer & Virgo Values: To Virgo, their Cancer partner will seem much more down to earth and rational than they are, and to Cancer, their Virgo partner will have a recognized soft spot in their heart. Other things won't matter much when they find each other, because there is no compromise they both wouldn't make to keep this love going. However, their different values might represent an obstacle to them getting to this point in a relationship. If they form a truly functional and beautiful relationship, they will value each other. As a deeply emotional and the most intelligent sign in the zodiac, one of them values family, love and understanding, and the other intellect, attention to detail and their health. It is not like they don't have a meeting point, but it will not be easily found if they both stubbornly keep to their habits and opinions. 50%

Cancer & Virgo Shared Activities: This is very often a relationship of two people who can manage without each other, so they will not be bound by shared activities as much as some other signs might be. If one of them wants to do something, unless truly insecure, they won't take it as a personal insult if their partner doesn't want to do the same thing. Although Virgo is a mutable sign and they can be pretty hard to follow, they do belong to the element of Earth and will be capable to wait for their Cancer partner to decide whether they will want to join them or not. This is where their rational compatibility will really come in handy, as they make arrangements for their time together. Usually there will be just enough movement and romance for both of them to feel special, and this is certainly not the main concern in their relationship. After all, they would both gladly go to a movie and eat some popcorn, so there is always something they can do together, even if their needs are very different. 90%

Conclusion: The main problem of their relationship is in the possible conflict between emotional Cancer and reasonable Virgo. If they manage to overcome this, accepting each other's shortcomings and learning to incorporate some rationality or some emotion into their lives, they could end up in an inspiring relationship that will last for a very long time. Cancer and Virgo can have a wonderful connection and are usually brought together by sexual understanding. In a way, they complement each other as much as the heart complements the mind. If they share a spark of love, it would be a shame to miss the opportunity for happiness just because of someone's irrational expectations or someone's closed heart. 77%

Cancer Compatibility with Libra

Cancer & Libra Sexual Intimacy and Compatibility: Both signs have trouble accepting Mars and this leads to a lack of passion and initiative in their sex life. However, the tactful and careful nature of Libra can really soothe Cancer. This is mostly because of the fact that Saturn, the ruler of Cancer's opposing sign, is exalted in Libra. Although some initiative might be lacking, there is a chance for Cancer and Libra to function very well in their sex life if they spend enough time together. Cancer and Libra might seem as if they are really far apart. The problem with their sexual connection is in their element, more than anything else. Cancer belongs to the element of Water and Libra is an Air sign. Although Libra partner might be extremely patient and nice, there is still a speed to the element of Air that Cancer might have trouble adjusting to. Things they would like to try out will differ greatly, for Cancer needs emotional connection and Libra needs contact, touch and experience before they get too emotional. Libra is not often inspired by the nature of Cancer and won't normally fall in love with them at first sight. Their sex life can be very good if they already share deep emotions, so it would be best if they started a relationship out of friendship, already knowing each other to some point and sharing some feelings besides possible attraction. 40%

Cancer & Libra Trust: While Cancer wants a quiet family life with no interference from other people, Libra can't seem to stay away from other people, seeking their affection and approval day after day. At some point their Cancer partner has to ask themselves if this is the kind of partner they want to have children with. Cancer can be one of the most trusting zodiac signs, but there is an irritating side to Libra's nature that they rarely stay immune to. On the other hand, Libra finds the whole approach to a romantic relationship Cancer has a bit unrealistic. This can easily lead to a lack of trust in Libra, especially if the Sun is in final degrees of the sign. 30%

Cancer & Libra Communication and Intellectuality: Although Libra is in charge for the upper, spiritual nature of Venus, it is still a sign of relationships and the Moon will only emphasize the need for closeness and harmony. Their communication will not be too difficult, but there is a chance they won't share many interests or respect each other enough for it to have true quality. Their main problem could surface if they start making unrealistic plans together. The expectations of Cancer with Capricorn as their opposing sign would be extremely practical and strict. As signs ruled by Venus and the Moon, it is safe to presume that their relationship is in a way very important. Libra does exalt Saturn, but it is not an Earth sign, and usually they remain in the field of ideas instead of a practical approach to materialization. Cancer doesn't understand how someone can be so far from reality and this could lead to a strange passive conflict that could endanger their entire relationship. Cancer needs to realize that Libra has its place among the Air signs, in the field of ideas - not necessarily their realization. It is important to remember that Libra has a troubled Sun and usually looks for a partner with more fire and passionate energy than Cancer normally has. However ugly that might sound, they need someone to feed on, and they normally have strong relationships with fire signs which produce enough energy even for their loved ones, or Air signs who don't really care about it. The sign of Cancer unconsciously calculates and distributes energy to their inner priority list, and they will rarely have an excess big enough to shower Libra in it. 50%

Cancer & Libra Emotions: However, their emotional context is very different, for Cancer is looking for a love on Earth and Libra is looking for someone to take to heaven. This is exactly where the spiritual side of Libra's Venus takes away their true chance to share a life together. It is a Venus that doesn't need to eat or sleep, for as long as it feels the unique, balanced love. Cancer will understand this idea to some point, but probably look for a more grounded partner when they realize that this is a pattern that isn't about to change. Both Moon and Venus represent emotions, and they are both highly emotional signs. As two cardinal signs, they could have a long unsatisfying relationship because they are both waiting for a groundbreaking moment to set them free. In some cases they should be advised to make a change if they are not satisfied, and search for someone who could make them happier. 15%

Cancer & Libra Values: As both of them are signs of relationships, i. e. family relationships or relationships with partners and they will both value a pleasant and joyful connection between two people. If they find it with each other, they will certainly have trouble letting it go, both of them understanding that it is not that easy to find. However, their entire system of values differs greatly beyond the point of relationships, and while Cancer will value tenderness and care, Libra will value responsibility and platonic love. That doesn't sound promising, does it? 20%

Cancer & Libra Shared Activities: Cancer is a sign that exalts Jupiter and although they can be perfectly happy with a grumpy partner, they would appreciate them not to have a need to impose their grumpiness on them. There is a secret desire to every Cancer to travel the world, while having the safe base to come home to, but Libra could destruct their beliefs and lead them to doubt they simply don't want to deal with. Cancer and Libra could do many things together, but it is questionable if they will want to. With Cancer's mild and mellow temperament, Libra will probably feel even more inadequate and grumpy, for they need someone to be their opposition and defy them in order to have respect and sexual charge. 10%

Conclusion: Probably the biggest restriction in the relationship between Cancer and Libra is in things they want from their partner. Cancer wants someone responsible, to take them by the hand if needed and complement their emotional nature with practicality. Libra wants someone who is full of life, energized, strong and full of initiative to follow their ideas. They can really disappoint each other if any expectations are set wrongly at the beginning of their relationship. The best way for them to build a love that is to last, is for both partners to hold on to their independence whatever happens. If they focus on love and worry about earthly things each on their own, Cancer could "compromise" on heavenly love, as much as Libra would like to have a family. 28%

Cancer Compatibility with Scorpio

Cancer & Scorpio Sexual Intimacy and Compatibility: Cancer can usually understand the need of their Scorpio partner to express their deepest, darkest emotions in their sex life. If Cancer partner doesn't get scared or too forced to do something they are not ready for, a sexual relationship between Cancer and Scorpio can be deeply satisfying for both partners. This is a relationship of two Water signs and because of this their sex life needs to reflect all of their emotional connection or a lack of it if there is any. The sign of Scorpio is associated with death and all kinds of bad things, but all of their maliciousness comes from their emotional and sexual repression. When they fall in love, they will both need to express their feelings and the intimacy they might share is incredible. However, Scorpio is a sign in which the Moon falls and this is the ruler of the sign of Cancer. If Scorpio's need to bury their emotions is too intense, there is a great chance they will be too rough or insensitive on their partner. This is something Cancer will have difficulty coping with and could lead to Cancer's need to separate because they could simply get tired from all the special or aggressive sexual requirements their Scorpio partner has. 90%

Cancer & Scorpio Trust: If they feel betrayed in any way, they can start showing all of those maleficent sides of their nature and become truly possessive and jealous. Cancer partner usually wants someone to share a life with and they will have no reason to cheat or lie to their partner. As all water signs, they could both fear telling the truth to a certain point, but this doesn't necessarily need to speak of their unfaithfulness or the beginning of the end of their relationship. When Scorpio falls in love, trust is one of the most important things they are looking for. Usually they will both be able to give each other enough security to feel safe and build the trust they both need not to feel hurt or betrayed. 95%

Cancer & Scorpio Communication and Intellectuality: This can influence their sex life and make it much better, or much worse, depending on how their need for mystery is expressed. Their communication is very good, for as long as emotions are not the main theme of a conversation. They can finish each other's sentences if they have any need to talk in the first place. Cancer and Scorpio usually understand each other without words. The depth they both have, although it might not be visible at first in Cancer partner, makes them able to talk about anything at all. In the case when Cancer wants to run from negative experiences and Scorpio from their emotions, they could have trouble forming a relationship at all. Still, this is a very rare scenario and even if they have these tendencies, they will probably help each other deal with them and give each other the exact mental stimulation they both need. 99%

Cancer & Scorpio Emotions: Cancer lives buried in their emotions, positive or negative, capable of using them in their everyday routine as an incorporated part of their life. Scorpio can have trouble understanding how this works exactly, because they have a tendency to dismiss emotions, thinking that this is the only way to reach a certain goal. The middle ground they need to find is a place where they are both free to follow these needs. This is a tricky territory for a couple like this one. Emotions have to be a way of living, as much as they can interfere with our goals. Both of these partners need to learn to lose control, as well as gain it again, in order to be able to let things flow and change in the way they are supposed to. 70%

Cancer & Scorpio Values: Scorpio represents change and values it most of all, even if they are not fully aware of this. It can be difficult for these partners to coordinate their personalities if they are both not flexible enough to understand their differences and the depth each of them has behind these superficial needs. Cancer values their inner peace and wants a stable life with a family they can rely on. Scorpio can fear emotion to the point of agony and if Cancer recognizes this, they will be able to approach them in the best way possible and discover their true need for security and emotional balance. 25%

Cancer & Scorpio Shared Activities: It doesn't really matter what Cancer and Scorpio will do if they both feel good with each other. They have to share emotions and protect their loved ones, like Cancer due to their motherly need to protect people they love, and Scorpio in order to set good boundaries on what they think is right. If they create their own little world, they can be found in any situation together, dealing with things as one being. Scorpio usually likes some dangerous activities and Cancer will have difficulty adjusting to those, but if their emotional core is good, they will have a quiet understanding of each other's needs, however destructive they might get. 95%

Conclusion: A relationship between a Cancer and a Scorpio can go from one extreme to another, and although Cancer

partner will try hard to stabilize it, it might be too difficult if Scorpio doesn't have enough respect for their own emotions. When they find an emotional link, they can go very deep in search of true love, and unite on a level that is unreachable for other zodiac signs. This can make them speak without words, understand each other's thoughts with only one shared glance and be synchronized in their approach to their future together.

If their emotions aren't shared on a deepest possible level, or Scorpio partner refuses to deal with them, it could be too hard for Cancer to handle the self-destructive nature of their partner. Their connection needs to be sincere and pure, in order for both of them to be ready to give in to this intense emotional contact. 79%

Cancer Compatibility with Sagittarius

Cancer & Sagittarius Sexual Intimacy and Compatibility: If they do, against odds, they could find an interesting shared sexual language that none of them anticipated will be found. The changeable nature of Sagittarius can be somewhat difficult for Cancer to understand because of their opposite need for emotional security. If trust between them is reached in any possible way and true emotions are shared, this characteristic of a Sagittarius partner will become a spice to their sex life, rather than its destructive force. If they have enough emotional security with one another, their sex life could be very fun. Cancer is a sign that exalts Jupiter, and will probably make their partner feel special. Cancer and Sagittarius will almost never get attracted to each other. On the other hand, Sagittarius will make things light, fun and although the lack of depth could bother Cancer, passion and warmth they bring into their sex life might just be enough to compensate. The only way their relationship can succeed is for Cancer to let go of their preconceptions and allows some change and fun enter their strict sex zone. Although they can seem mellow most of the time, they have a tendency to hold on to secure patterns when it comes to things that can make them feel shame or insecurity. In return, Sagittarius will have to lower their expectations on Cancer's own changeability and sexual creativity, and be satisfied with lovemaking instead of a sexual adventure. 40%

Cancer & Sagittarius Trust: Deities associated with this planet were considered great lovers, always on the chase for different women, goddesses, nymphs, and whoever seemed attractive enough. More often than not, Sagittarius representatives have the need to show their seductive skills to everyone around them and we could call this a "Zeus' complex". Although the sign of Cancer loves Jupiter very much, emotions make it impossible for them to understand this flirty need of their partner to win the hearts of everyone around them. Not only will this disturb Cancer's trust, but it will also affect the trust Sagittarius has in the understanding they will get from their partner. Sagittarius is a sign ruled by Jupiter. This could be the source of many conflicts and misunderstandings, and could finally lead to the point where their relationship has no purpose or future at all. 1%

Cancer & Sagittarius Communication and Intellectuality: Their mutual love for it will give them plenty of things to talk about and a deep understanding of each other's reasoning. Their minds need to find the synthesis of things that surround them and their belief systems can be quite similar. The wonderful thing of their mutual love of Jupiter is exactly in the similarities between their minds and their ways of thinking. Cancer can seem a bit slow from Sagittarius' perspective, as much as Sagittarius can seem superficial or too "philosophical" to Cancer. Both of these signs strive for knowledge, pure and simple. They can easily overcome these issues if they find a passion they share and usually if these two choose the same profession, they have many things to talk about. They understand each other's mind and the way their brain works, even when there is a large intellectual difference between them. If they happen to fall in love, communication is something they can always use as means to solve any other problem that appears in the course of their relationship. 60%

Cancer & Sagittarius Emotions: As elements of Water and Fire, they don't really spark each other's passion and the love between them will hardly ever be the same intensity, at the same time or the same pace. Sagittarius is a mutable Fire sign and they usually fall in love quickly and passionately. If their love is to last, their partner needs to surprise and impress them often, making the relationship exciting and unpredictable. Cancer, on the other hand, is a cardinal Water sign, and they will make sharp turns and huge changes, but much less frequently than their partner. This is not a combination of Sun signs that will fall in love very often. Cancers follow their feel of situations and people, and need time to build a relationship in which they feel secure enough to share emotions. When love happens between them, usually Sagittarius feels it first, wants to jump in and out and in and out enough times for Cancer to realize that they cannot build this sense of security to even consider their relationship a true love. If they are to build a love that will last, Sagittarius partner will have to slow down and wait until their partner decides how they feel. In return, Cancer will have to take a leap of faith and jump into a relationship that offers no security, to see if enough love can be found between them so they can stay together. 10%

Cancer & Sagittarius Values: Although they value different things and different characteristics in people around them, they have a strong link in the way they value knowledge. Cancer will value Sagittarius' honesty and their ability to act on emotional impulse, even if they don't understand the emotion behind the act. Sagittarius will value Cancers dedication to things they love and their incredible ability for compassion. 45%

Cancer & Sagittarius Shared Activities: Sagittarius will look at their partner as if the entire person just turned into a long, irritating pause. The way these signs use their energy is so different that it is not only difficult to find activities they will do together, but more importantly it is difficult for them to do things in a similar way. They might study together, but it would probably be a torture for Cancer to watch their partner move from one paragraph to another, only to find what they were really searching for at the end of some old book and then lose interest in the entire subject. This could make the possible impossible and they could be forever separated by a simple difference in their speed.5%

Conclusion: If attraction and love are born between them, they will rarely have a damaging relationship for any one of them, because their signs are ruled by the Moon and beneficent Jupiter. It is safe to assume that they will be good for each other, for as long as their relationship lasts, but it is rare for them to succeed in the long run if they don't have strong support from positions in their personal horoscopes. Cancer and Sagittarius are usually signs that aren't attracted to each other at all. As much as Cancer can reach the depth of their partner's faith, Sagittarius can widen their partner's horizons and make them much happier in their approach to the world. If they have feelings for each other, it would be a shame not to act on them and miss the opportunity to peacefully grow. 27%

Cancer Compatibility with Capricorn

Cancer & Capricorn Sexual Intimacy and Compatibility: The patience Capricorn has for their partner is something Cancer really needs to relax and start feeling sexual to begin with. Capricorn needs someone who acts on true emotion, but also someone who doesn't take sex lightly. There are Capricorn representatives who have changed many partners, but they will probably never stay with the one that isn't family oriented and emotional when it comes to physical relations. Intimacy Cancer can create is exactly compatible to what Capricorn lacks. Cancer and Capricorn are opposing signs and there is a strong attraction between them. When they get together, a passion awakens and they both become perfect lovers for one another. There is a lack of love, home and warmth in the sign of Capricorn, and Cancer partner can heal this with their highly compassionate approach. This could lead to thawing of Capricorn's emotional state and uplift the state of their sex life significantly. 99%

Cancer & Capricorn Trust: Not only is the sign of Pisces in their third house representing the way they think, but they are also often led by panic in their intimate relationships.

When a Capricorn representative falls in love, they will understand that their partner needs to see trust and this is what they'll show. However, they could have trouble believing anything their partner says until some consistency is proven or their stories checked out with other people. Luckily, in Cancer's world there is often nothing that ugly and secretive to find, for their moral values are as high as the exaltation of Jupiter in their sign. Although Capricorn might seem trustful, they are probably one of the least trusting signs of the zodiac. For as long as Cancer feels Capricorn's devotion, they will not question their actions. This is why Cancer could easily sense their partner's lack of trust, pretend that they didn't notice, and find it endearing rather than repulsive. 99%

Cancer & Capricorn Communication and Intellectuality: Not literally, of course, but they often share their image of a relationship their distant relatives had, maybe centuries ago. There is information in our emotional body, about each emotion our ancestors have felt and didn't know how to deal with or how to use. This is where Cancer and Capricorn connect, as signs of the family we come from, and the family we will create. These partners could feel like they have known each other before they've actually met. Their mutual affection will seem familiar and warm, as if they grew up in a same house, even though their circumstances might be completely different. This is a couple that has the strangest thing in common like genetics. This could make them able to talk about anything, for there is closeness to the relationship of these two signs that is unexplainable to all others. However, if this emotional bonding doesn't come at first impulse, Capricorn could be very difficult to talk to from Cancer's perspective. They need to connect on a very deep level, or they will have opposite goals and Capricorn could seem like a career obsessed lunatic with no emotion what so ever, while Cancer could seem like a clingy housewife (no matter if male or female). They should both remember that if they see each other in this negative light, they are probably only hiding from their own, inner opposite side, dismissing the chance to be complete. 70%

Cancer & Capricorn Emotions: Cancer and Capricorn are a love story their ancestors had, waiting to be resolved. Although this could sound like a dream come true and could in fact create very strong emotions in both partners, there is almost always a karmic debt to be repaid before they could say they are truly happy together. These signs represent the axis of Jupiter's exaltation and fall and it is important to understand that their emotional states are closely linked to their expectations from each other and their relationship. These two are considered one of the most and one of the least emotional signs of the zodiac. One of them should be family oriented and the other turned to their career. Still, their emotions often run wild as soon as they lay eyes on each other. In time, they will both fight for security and stability of their relationship, and although it might be hard for them to reconcile these primal emotional differences, they will in most cases simply – find a way. The emotional depth of Capricorn is really hard to reach, but Cancer partner can approach this as their life challenge. When they get tied to each other, it is almost impossible for them not to get married, have children and the entire earthly love package. Still, they could take so much of each other's energy if they tried to change each other. It is best for them to accept each other's personalities as inevitable and impossible to change. This could lead to a more sensible future than the one in which they are both simply tired of each other. 75%

Cancer & Capricorn Values: They both value stability and practical sense. As opposing signs, they can seem to have opposing values, but this is not really the case. They both need stability in their lives and will value people who give them the sense of security. This is probably something they will value most in each other, the ability of both not to quit or give up, however hard things might get. 70%

Cancer & Capricorn Shared Activities: Cancer is not very picky when it comes to activity choices their partner has, for as long as they are not imposed on them or too aggressive for their taste. Capricorn is careful and will plan their activities well in advance, so both of them will have time to adjust to the idea or change their minds if they realize that this is not what they want. It will be very easy for them to agree on what they want to do together and find such activities, only if they respect each other's personalities. Capricorn will not want to go shopping for house decorations, no more than Cancer will want to go three nights without sleep just because of a project at work. If they respect each other's boundaries, their time spent together should be truly satisfying for both. 90%

Conclusion: Cancer and Capricorn are usually bound to relive the love story of someone who lived before their time. This deeply seeded need to mend what is broken in our family tree is something we all carry within, but these Sun signs are predestined to handle karmic debts and residue emotions from their families. They will have to deal with problems first if they want to be free of the past, and only after they have repaid what needed to be repaid, will they be able to truly choose one another. In most cases this is a once in a lifetime love for both partners, and they will probably choose each other without a doubt. 84%

Cancer Compatibility with Aquarius

Cancer & Aquarius Sexual Intimacy and Compatibility: Although Cancer is considered the most sensitive sign of the zodiac, governed by the Moon, they can be quite rough and distant when they feel the need to set strong boundaries. Aquarius, on the other hand, is known as an innovator, someone to make the change, but in fact, they are a fixed sign, pretty set in their ways and as a paradox, like unchangeable. A sexual relationship between Cancer and Aquarius can be stressful for both partners. When they engage in sexual activity, Cancer could be so stressed that they will have to set those boundaries and Aquarius will not be able to make the needed change to be gentler to their Cancer partner. There is too much energy in Aquarius that needs to be grounded through their physical activity and this includes sex. Cancer doesn't really understand this and is convinced that in sexual relations with someone you love only emotions should be shared. If Aquarius finds a way to slow down and not force anything on their partner, and if Cancer allows their rational mind to take over for some of the time they spend together, they might share an exciting sexual experience. Cancer will bring emotions and tenderness to their sex life and Aquarius won't ever let boring routine take over. If they compromise on experimenting and emotional exchange, they could even start having fun. 1%

Cancer & Aquarius Trust: With Aquarius, they might feel stressed to share things and this could present in either way issue when it comes to trust. The liberal nature of Aquarius could seem crazy to a Cancer, and their partner's honesty about their craziness won't help the inner feeling of mistrust for their possible actions. Cancer is usually loyal and honest, except in situations when they are scared of the aggressive reaction of their loved one, or of hurting them badly. It is a complicated thing for them, because none of them wants to lie, but still they don't seem to trust the future they might share. 35%

Cancer & Aquarius Communication and Intellectuality: The mind of a Cancer is sensitive enough to pay attention to details and interpersonal relationships when Aquarius fails to do so. They could make grand ideas come true, especially those that need a lot of people involved on their way to become real. However, they might have trouble talking to one another in the same tone or understanding each other in the first place. Cancer and Aquarius are able to join forces in intellectual activity. Cancer is ruled by the Moon, the fastest heavenly body in the sky, but they are not fast to recognize what hides behind Aquarius' words. There is difficulty for Aquarius to express their inner state and this is something Cancer has trouble understanding. The best beginning of their relationship is guaranteed if Aquarius sees their Cancer partner as a weird human being that needs to be examined. This will allow them both enough space to get to know each other well, and this could influence all other areas of their relationship. If this happens, Aquarius will approach those strange activities Cancer needs as if they weren't ordinary at all. After all, not everyone can drink a morning coffee in total silence with their partner and enjoy this silence as much as these partners can. 55%

Cancer & Aquarius Emotions: The unconventional nature of Aquarius interferes with Cancer's need to stay in a peaceful environment, and this is something they will find hardest to reconcile. That homey, cosy feeling Cancer needs can be deeply disturbed by the rebellious Air sign of Aquarius. They will bring stress and too much information in their life, and speed that cannot be handled by a subtle state of deep empathy Cancer has to live with daily. The way they show love is very different, but it can be wonderfully focused on their kids and the family they build if they get to this point. There is no sign in the zodiac predestined for a family life such as Cancer. In a relationship with Aquarius, they would take over the most of everyday activities and responsibilities. In return, their children would get a childhood without boundaries and a life of free choices that no other couple can give. This is a consequence of the difference between them and the tolerance they have to build in order to stay together. When they do fall in love, they will not be so quick to end their relationship. Aquarius will approach it as a kind of challenge and understand the stability and love they get from this partner. Cancer will realize that they have never been this free to actually be themselves instead of living in a symbiotic relationship they are easily sucked into. Once they form a strong bond, it will be very hard for both of them to let it go. 50%

Cancer & Aquarius Values: This is a fine connection between their worlds and if it is nurtured it could be just enough for them not to be set apart by other values they hold on to. Cancer does value stability, intimacy and family, while Aquarius values their freedom, intellect and new technology. Cancer values knowledge almost as much as Aquarius values information. There is a difference between their worlds that might seem impossible to overcome, but if they hold on to their love of distances and travel or if they learn together, they could easily get over the fact that their values on other things differ so much. 10%

Cancer & Aquarius Shared Activities: While Cancer will want to stay at home, go for a picnic in the park or to a furniture store, Aquarius will look for the highest skyscraper, wish for a new laptop and read anything that falls into their hands.

The key activity that could truly connect them is travel. Although Cancer seems homey and unmovable, this is not exactly true. Due to the fact that it is a sign of Jupiter's exaltation, they will want to travel far. Aquarius will always want to board a plane and it would be perfect if they could parachute to a location where Cancer would safely land in a Boeing 747. 25%

Conclusion: Their relationship can be too stressful for Cancer partner and the lack of intimacy will most probably tear them apart. However, the link between them can actually be wonderful when found, and they could open up such interesting new perspectives for one another if this happens. They both want to learn new things and could travel far if a strong base is made at home, so Cancer can remain peaceful. For this couple to move in a positive direction, Aquarius needs to understand how unusual their partner is, and try to experiment on being homey while having fun. We could say that Cancer and Aquarius are not your usual happy couple in most cases. Cancer will have to take over the main set of responsibilities to hold on to the idea of their home as a base from which they can move wherever they want. In the end, Cancer might discover an unbelievable joy of freedom and Aquarius might develop closeness. If these partners can be silent together, sipping on their morning coffee, this is in most cases the first step to success. 31%

Cancer Compatibility with Pisces

Cancer & Pisces Sexual Intimacy and Compatibility: Their sexual connection is usually primarily emotional. Pisces partner might seem a bit weird and kinky to Cancer, but they should have a feel for each other, strong enough for both of them to enrich their sexual relationship with their own quality. Cancer will bring intimacy into their sex life and the meaning behind the act. Cancer and Pisces are almost always brought together by a romantic love. They will nurture their partner and care about their pleasure, giving them a stable and a safe approach to a healthy sex life. Pisces will bring in change, creativity, inspiration and probably a lot of sensuality due to the fact that this is the sign that exalts Venus. The beauty of this connection is in the emotion they share and the way they cherish each other and respect each other's sensitivity. Their main problem might arise because Cancer can be somewhat traditional when it comes to sex and Pisces partner doesn't really understand this. Pisces' need to connect and feel love is larger than any sort of rule humankind might have made for love. However, in most cases they will be tender enough to inspire their Cancer partner to let go of their rigid attitudes and shame, and give in to the beauty of sexual exchange of emotions. 85%

Cancer & Pisces Trust: Pisces don't really understand marriage except as a part of a fairytale ending or because of all that lace, and Cancer will usually want a wedding as a crowning of a loving relationship. This could be recognized as pressure to some point and this could lead to Pisces partner getting scared. When Pisces get scared, they somehow fail to tell the truth even on silly things in their life, because they feel the need to distance themselves from any pressure they might feel. It is a good thing that Cancer is usually not aggressive or pushy, or they could easily get dishonesty from Pisces as a response to their tendency to create intimacy and a happy home at any cost. It is a good thing that Cancer understands this and easily separates lies from intimacy. Whatever the situation, they will both probably be patient enough to have just enough trust in one another for their relationship to work out. 70%

Cancer & Pisces Communication and Intellectuality: Cancer is looking for someone with clarity on the use and the practicality of everything they mention. Pisces is everything but focused on practicality in most everyday situations. If Pisces partner learns to be more silent, relying on their feelings, and starts fighting for what they wish for, they could sweep their Cancer off their feet. As changeable as Pisces are, they always have something to talk about. This can be inspiring or irritating to Cancer who would maybe rather deal with "real information". Usually they communicate just fine, but there are situations in which they could float away on an idea made out of words. Unfortunately, their entire relationship will not last very long if only words are spoken but deeds don't follow. Cancer's opposing sign is after all Capricorn, and they need a partner able to constructively use things, situations and emotions. 85%

Cancer & Pisces Emotions: Everything that seems easygoing and positive might have a hidden negative note in the Pisces world, and Cancer feels rather than listens, which makes them a perfect companion for someone like Pisces. When they sense this deep understanding, Pisces partner will return the favour by absolute tenderness and finally open up to their Cancer partner. Cancer can understand the sensitive nature of their Pisces partner better than anyone else. When they find this shared point of intimacy where true emotions are shared, this will affect all other segments of their relationship and be a fuel for it to have a fairytale ending. 99%

Cancer & Pisces Values: This is where the difference in their character really comes to focus. As much as they will both value being loved and cared for, Cancer will value a stable emotional situation and a cosy home to come to, while Pisces will probably value any chance for an emotional rollercoaster more. It is often said that Pisces idealize partners and different things in life, but in fact they get depressed when there is no magic and perfect beauty surrounding them. If their day to day life with a Cancer partner becomes anything similar to a boring routine, they will find a way to run off, find a lover or create any sort of truly exciting circumstances. 25%

Cancer & Pisces Shared Activities: A relationship with a Pisces partner is always exciting and inspirational, and Cancer will give it strength, stability and roots. At the beginning, this may seem like a great arrangement, but in time, Pisces might want too much activity for what Cancer partner really needs. This wouldn't be much of a problem if they would say this to their partner without fear of any one of them getting hurt. When they meet and start their relationship, they will probably have a lot of things to do together. If they start bending the truth, Cancer will feel their trust beginning to fade and this could begin a series on problems between them, that could have been easily avoided. 70%

Conclusion: As two Water signs, Cancer and Pisces connect through emotions, usually as soon as they lay eyes on each other. This is one of the typical combinations of zodiac signs for love at first sight. Their main challenge is hidden in the changeable nature of the sign of Pisces, not because it is there, but because they might fear to show it. Their biggest problem lies in the fact that they give priority to different types of love in their life. If passion and sensual, sexual love isn't there, Pisces will rarely be satisfied with the love they get from their family, and Cancer would find a life without a family nest very depressing. A fine balance needs to be made between excitement and stability, and they could be one of the most wonderful couples of the zodiac – Cancer inspired and Pisces with a feel of home. 72%

PART-5
LEO HOROSCOPE

(July 23rd to August 22nd):

Who praises all his kindred and expects others to praise them too?

Who possess grace, dignity and generosity but cannot see their senseless view?

Element: Fire

Quality: Fixed

Ruler: Sun

Lucky Gem: Ruby and Topaz. The ruby is a miracle stone, and wearing it will bring him/her health and good fortune. He/She will also acquire the power to make instant decisions. The Ruby and Topaz can give him/her success and power in life.

Lucky Number: Number 1, 3, 4, 5, 6 & 9 will pull him/her out of trouble.

Colour: Gold, Orange, White, Red

Lucky Colour: Saffron. Let saffron lift his/he spirits.

Day: Sunday, Friday

Lucky Day: Friday. Preen, for he/she will rake in accolades on Friday.

Lucky stone: Peridot.

Flower: Gladiolus

General Insights: People born under the sign of Leo are natural born leaders. They are dramatic, creative self-confident, dominant and extremely difficult to resist. They can achieve anything they want, whether it's about work or time spent will family and friends. Leo is a fire sign, which means that he loves life and expects to have a good time. Like other fire signs, Sagittarius and Aries, Leo is also able to use his mind to solve even the most difficult problems and take the initiative in solving various complicated situations. Ruled by the Sun, Leo worships the Sun in all its forms which is also a metaphorical expression of the state of his ego. This can be good, because Leos can easily search for what they need. But, on the other hand it can be problematic when Leos ignore the problems and needs of others in order to fulfil their desires. Leo has a specific strength and "king of the jungle" status. Leo often has too many friends because he is very generous and loyal. Self-confident and attractive, Leo is able to unite many groups of people at every opportunity. Problems can arise, when Leo becomes too fond of his achievements. This zodiac sign can also be arrogant, lazy and inflexible, because he assumes that someone else will clean up after him. A healthy sense of humour can make the collaboration with other people, easier.

Personal Quality: He/She is creative, strong willed, self-confident, generous, warm hearted, loving, broad-minded and faithful. He/She has a powerful presence of mind and power of success in the conquests. He/She is called the true kings of the zodiac. He/She is manager because he dislikes subordination. He/She can be good as chairman or director in business because he is excellent organizers. He/She has self-confidence, alertness and hard struggling power. He/She never forgets the goal and tries hard to achieve it with patience, wisdom and hard labour. He/She is a dominant, always busy with some planning and work and cannot seat idle even for an hour. He/She is able to attain the top most position. He/She achieves full success after the middle age. He/She prefers position and honour to money. For position, he/she can forget any money. He/She always cares for others and other's interest or benefits except, where it is necessary to take care for him/he. He/She is born to lead. He/She is sometimes great saints. Many great Saints have born under the Leo sign. There is a reality in his/her love and he/she can do anything to please that he/she likes most. He/She will keep improving day by day after 30 years of age. He/She has a small family and their children are very brilliant and intelligent. He/She is straightforward and uncomplicated individuals. He/She is stubborn, and may suffer from short bouts of depression. He/She walks forward always, head held proudly and face turned towards the sun. He/She accumulates good amount of money, wealth and properties. His/Her fortune is good and he/she will never lack money in life.

Positive Quality: He/She is a good leader, creative, passionate, generous, warm-hearted, cheerful, humorous, and kind hearted, and confident, ambitious, loving, and honest and creative minded. He/She possesses a strong positive nature and doesn't shrink from any adverse circumstances. He/She can never bear dishonesty and injustice in life. He/She is witty and sets examples for others to follow. He/She is direct and to the point.

Negative Quality: He/She sometimes thinks himself/herself to be the only capable person in the total world. He/She believes in commanding only and does not care for others feelings. Some Leos think for money and profit and forget their other duties. It is very difficult to face his/her furies. He/She can be too sensitive to personal criticism, and when his dominance is threatened he can go into a sudden rage. He/She is conceited, arrogant, stubborn, self-cantered, lazy, and inflexible, and arrogant.

Physical Appearance: He/She has a distinguished stature and attracts attention easily. He/She normally possesses healthy skin and a well-sculpted body.

Relationships: He/She makes wonderful social companions and is passionate and faithful lovers, but very sensitive and gets hurt easily.

Health: He/She suffers from back and spinal problems, has tendency to be overweight by lack of exercise. He/She gets easily stressed and can suffer from heart ailments.

Ideal Partner: Hence, he/she gets best with Aquarius, but Taurus, Gemini and, Sagittarius women are ideal partner for him/her.

Compatible Zodiac Signs: Aries, Sagittarius
Incompatible Zodiac Signs: Taurus, Scorpio and Aquarius
Greatest Overall Compatibility: Aries, Sagittarius
Best for Marriage and Partnerships: Sagittarius
Leo likes: Theatre, taking holidays, being admired, expensive things, bright colours, fun with friends
Leo dislikes: Being ignored, facing difficult reality, not being treated like a king or queen

BUSINESS ASTROLOGY FOR LEO

These natives are good leaders, organizers and executers. Disturbed by uncertainty or losses they roar like lion and get themselves into trouble. For success in business they should learn to swim with tides, take cares of others ideas, suggestions and avoid being forceful. Business booster Gem is Topaz.

Leo Compatibility with Aries

Leo & Aries Sexual Intimacy and Compatibility: They have similar sexual preferences and they definitely take each other seriously, whatever the level of their relationship. As they both have extremely strong personalities, they could fight and make up all the time, but enjoy it in a way some Water signs might find crazy. They have a sexual connection that cannot be interrupted, changed or faded through time, since they are both individual sources of energy, waiting for someone to follow. This is such a warm and passionate connection, in which sparks fly around all the time. Still, if one of them has issues with their ego, they could slump into an energy drain system, where they insult each other and destroy each other's confidence and libido. This is a very rare possibility, but it is always there when two signs that present an extraordinary soil for the Sun come together. 90%

Leo & Aries Trust: They both think highly of themselves and see how their surroundings react to their partner. Leo is a strong willed, attractive masculine sign, and no matter if male or female, they will have this magnetic aura around them most of the time. This will make Aries strangely jealous and possessive, ready to fight for what belongs to them. There are probably many trust issues in every Aries and Leo couple, but their strong convictions and the need for loyalty mend these problems in most cases. Exactly why this problem culminates, for no one can ever possess Leo, the King of the zodiac. Still, their mutual understanding for the passionate nature they share and the determination of both partners to solve any problem that stands in their way, might just make them stay together for years, building security and trust every day. 60%

Leo & Aries Communication and Intellectuality: As they get closer, it can be expected for them to put in an emotional note, but not a very gentle one for their emotions burn like the Fire element they belong to. This is a certain promise of a lot of fights, loud statements and interruptions. Interestingly, as two very passionate people, they usually get over their fights quickly and don't care much about specific words spoken in the heat of the moment. The moment they cool down, their relationship will easily go back to normal and their sexual life will blossom every time they fight. At first, their conversations will probably be unbelievably energized, full of admiration and respect. They are interested in similar things. Aries is interested in Leo, while Leo is interested in everything great about Leo. This is a very nice distribution of interests, since Leo will cherish all the things Aries will say and do for them, and give it all back multiple times greater. This is, of course, the scenario of two healthy individuals in these roles. If one of them has greater psychological issues, their communications will turn to senseless talk about one of them, and their relationship will become a battlefield for the ego where they both constantly need to prove something to each other. 90%

Leo & Aries Emotions: Both of these signs have a strongly positioned Sun, growing on the soil made out of the Fire element. They understand each other's emotional state perfectly, even when one of them would like to flee from the passionate need of the other to give resistance to something small and seemingly irrelevant. Their emotional natures are very similar. Sun is connected to love, pure and simple, not the nurturing kind but the creative, warm, passionate and playful love. Emotional compatibility of this magnitude is a cure for all other imperfections they might encounter, while evolving in time and spicing their sexual relations with even more warmth and emotion then the beginning of their relationship might have promised. 99%

Leo & Aries Values: It is almost as if these were the words that describe them. Anything concise needs strong Mars energy, the ruler of Aries, while Leo brings clarity to all. Their only problem is the need they both have to be the leader and the brave one in the relationship. It is not as if they both can't be brave, but they have a tendency to compare to one another and search for their role as a lead. What these two values most is a person's ability to stay concise and clear. This can come between them and manifest as a typical battle of the sexes (in case they are of the opposite sex) or as a fight for dominance of any kind, consuming the quality of their relationship only because of their need to be the one with dominant values in general. 95%

Leo & Aries Shared Activities: Aries loves to be active, walk, exercise and always feels excited when about to do something for the first time. Leo is a sign of fixed quality, a drama queen of signs, and the one that wants to show off and go where they can be seen. Leo's energy is easily focused to coffee shops and places where they can rest enjoy and be the centre of everyone's attention. This is a waste of time in the eyes of an Aries, always ready for something different and exciting. They have enough energy to do everything together, but the problem is they don't actually need the same daily routine. For as long as Aries has the will to initiate their shared activities and compromises for a show off from time to time, they can find things to do together and enjoy them. But if Aries partner gets tired and irritated by Leo's laziness, there is a big chance they will simply separate their activities all together. 65%

Conclusion: The relationship of Aries and Leo is passionate and turbulent, but they don't seem to mind an occasional fight and a sharp word. When they fall in love deeply, they are almost impossible to separate as they stubbornly hold on to the idea of their future together. Although they are not two of the most romantic believers in love, they are passionate in their beliefs and when they find love, they will fight for it until there is literally nothing left of their relationship. It is meaningless to advise gentle behaviour or looking for peace, because the entire world of their relationship is based on the element of Fire they share. It is pointless to look for peace, when the opposite of peace is what attracts them in the first place. For as long as they love each other and stay faithful and true, they will be tied up in a relationship they need to fight for every day. Their main objective is to find a way to enjoy the fight and have fun. 83%

Leo Compatibility with Taurus

Leo & Taurus Sexual Intimacy and Compatibility: This is mostly due to the fact that they can both be lazy. While Taurus likes to lie down and enjoy being loved, Leo likes to lie down and be served and taken care of. It is in the nature of both signs to spend time in a horizontal position and it might be hard for them to agree on who is to be on top. When motivated, they can both be excellent lovers that put a lot of energy into their sexual activities, but with one another, their sex life will most likely become a battle for personal satisfaction and rest. Their best possibility of a healthy sex life would be the one where both partners have already built their sexual identity and know how to satisfy themselves. The sexual relationship between a Taurus and a Leo can be in a way exhausting for both of them. In this case, sensual Taurus would take care of their Leo partner, while passionate Leo would bring excitement into their relationship. In this scenario they would both take care of their own personal needs, aware that they need to commit to their partner's satisfaction in order for a relationship to work. In general, they are a feminine and a masculine sign, and share a similar need for personal satisfaction. If they don't end up in a clinch in which they both have expectations and won't move until they are met, they could have a very rewarding sex life. After all, they are just two different sides of love, joy and life in colour. 50%

Leo & Taurus Trust: They understand that honesty is the base of any relationship that might last, and if they fall in love, none of them will want to jeopardize their future together. If, however, one of them has a habit to lie or cheat that they have developed in their previous relationships, they will probably continue the same behaviour in this one. It is of outmost importance that both of them develop their personalities and moral boundaries independently in order for them to be functional together. As two fixed signs, they are most likely to stand on their own two feet when it comes to telling the truth. Their main problem could be the lack of will on any side to change behavioural patterns that might arise. In case one of them believes that the other one could change, time could consume them through a sense of distrust that builds with disappointment because change doesn't occur. 60%

Leo & Taurus Communication and Intellectuality: It is hard to say who will be more annoying to whom. While Taurus holds on to their practical perspective, Leo holds on to their ego, and a conflict with no solution is born. Their fixed natures will make them hold on to their "side" however stupid it might be, with no actual intent of reconciliation. They both need someone with a not so rigid approach if they want to find the middle ground. Taurus will find a mutual language with Leo through their usual, materializing role. They are in luck because they are ruled by Venus and the Sun, both warm and with a tendency to be close to one another, because when it comes to their interests and their intellectual understanding, they could drive each other out of their minds. Any creative impulse of Leo could be followed by the realization plan thought out by their Taurus partner, if only they shared enough emotions to have patience for each other. Their creative strength is the strength of a Venus in combination with Sun, so we could say with certainty that they would create something in image of universal love. 5%

Leo & Taurus Emotions: Leo is a passionate sign that represents love as a power of creation and all we feel gravity toward. They are both a personification of love, each one of them in their own way. When they get together, they will rarely feel this love for one another. Taurus is a deeply emotional sign, in case they don't close up and live in their own little, safe material world. Maybe we could view this as their mission to give love to the less fortunate zodiacal signs, or maybe their emotional nature has to give more and receive less. Whatever the reason in most cases they simply don't fall in love with each other. There is a great possibility that they will simply stay in their own worlds, with no prospects of merging even in something close to a friendship. This is not because they don't like each other or feel some sort of hostility, but because they are like two islands in two different oceans. Each of them has their own nature, their own world with all its beauties, and they need someone closer to this world they hold on to. None of them has the role of a floating island in search for someone to merge with. If emotions are shared, they could be huge, but there would still be the issue of showing and recognizing them before we imagine a fairytale. 20%

Leo & Taurus Values: While Taurus values financial security and material beauty, Leo values everything shiny, bravery and someone's inner fire. There is a peace to Taurus that Leo doesn't want to understand, for it seems like a boring place to visit. Leo values peace, too, but for them it is hidden in a different, much more joyful place or in public, such as peace between entire countries and continents. Taurus and Leo have different views on value itself. To Taurus, Leo may seem like someone to strike a pose and have no depth at all, and although depth is not one of their primary values, it is still a very important one. They find nothing interesting in people without essence and neither does Leo, thinking of Taurus and how they don't open up to share anything deep. 1%

Leo & Taurus Shared Activities: Either they will lie down, sleep, eat and cuddle, or they will separate and do things without each other. Taurus is inert and loves to spend time on their couch while the rain outside falls and they hear nothing but the sound of a fireplace. Leo is guided by their nature, and a lion does sleep for 20 hours a day and plays for the rest of it. Even the fiercest of lions wait for their lioness to serve their food. Shared activities are very easy for both of them to find. The problem will occur when Taurus sets up a romantic image and Leo falls asleep, but this is still a scenario that could work. Their outside activities can be fun for both of them if they go out to fancy restaurants. This is where Leo can be seen, and act as a gallant person that deserves the best, while Taurus could enjoy really good food. Other than these, they probably won't have that many activities to share, but if they are not stubborn, they could enjoy everything that is not too demanding and physical together. 40%

Conclusion: The relationship of Taurus and Leo could be aggressively challenging if not for their warm natures ruled by Venus and the Sun. Although they are both signs of fixed quality with entirely different natures, if they gather enough patience before they enter their relationship, they have a chance to become your archetypal couple of a girl and a boy. When their masculine and feminine principles are in balance, they can use them to mend their sexual, intellectual and financial circumstances and really enjoy each other. 29%

Leo Compatibility with Gemini

Leo & Gemini Sexual Intimacy and Compatibility: They are a very good fit when it comes to sex, for Gemini gives their relationship ideas and excitement, while Leo brings in energy, creativity and love. Their sex life can be stimulated by their intellect and communication, for they both rely on their conscious Self and their mind. Everything that Leo would like to show, Gemini would gladly examine. If Leo feels right in intimate relations with their Gemini, as a fixed sign they will give them stability and a chance to last together for a very long time. Gemini is childish when it comes to sex, and rarely connects deep emotion with sexuality. Leo could be the right partner to teach them how to make a real intimate connection if they are not preoccupied with themselves. They will both want to experiment, have sexual encounters outdoors and will enjoy being naked. This is a perfect relationship for both partners to overcome shame and any sort of fear regarding intimacy and sexuality. 90%

Leo & Gemini Trust: One of them is distracted by everything and the other focused only on their own needs. Because of this they could end up in a situation in which their relationship lacks trust and it might take some time for them to notice, because they will not question one another at the beginning. Because of their natures, they could get lost, each of them satisfying their own needs. Both Gemini and Leo can be bad listeners when it comes to other people's needs. This will lead to all sorts of situations they don't want to share with each other, even adultery. It is of outmost importance for them to share and listen to one another from the beginning, in order for both of them to be able to satisfy their partner's needs instead of letting them slip away. 45%

Leo & Gemini Communication and Intellectuality: Leo does have a deep emotional background, but pays a lot of attention to words and intellectual strength of their partner. This is where Gemini jumps in as a faithful follower, to admire their Leo partner and teach them a thing or two. If they find themselves in an emotional relationship with each other, they will probably talk about everything else but their feelings. Leo might have the need to, in general, but Gemini partner will easily seduce them to go in another direction. Gemini and Leo are both rational and focused on their mental activity. They both don't have much use for sweet talk and if Gemini tries to use the way they have with words, it could work only up to a point where they start sounding fake. There is not much room for over thinking, and they will both probably say the first thing that comes to their mind. This is a great way to build trust between them, if they don't judge each other in the process and share some emotions along the way. Still, there is an issue of Leo's "perfect" Sun. This position gives so much energy to Leo that they sometimes feel obligated to burn everyone around them by imposing their will. This is a strange need of Leo, since the sign supports Sun greatly, but it is there. They usually have this idea to change the world and make it a better place, and this can drive their Gemini partner crazy in case they don't have enough room for their own opinions. 95%

Leo & Gemini Emotions: Although the mutable quality of Gemini partner doesn't really go well with the fixed quality of any sign, with Leo it could be a perfect match because of their warm, supporting and respectful nature. If Leo has enough patience to wait for Gemini's emotions to surface, they could get more than they've bargained for. The beauty of their relationship is the consciousness of both of them, leading to verbal display of emotions, once they feel safe with one another. The warmth that Leo is prepared to give to their Gemini partner will hardly leave them indifferent, while the charm and childish nature of Gemini is going to spark wonderful emotions in Leo, day after day. If they both recognize love, this is a wonderful love story full of support, respect and always something new to discuss. 85%

Leo & Gemini Values: Most of all they both value intelligence and clarity. As two signs ruled by planets in charge for our mental, rational behaviour, they will meet each other's needs perfectly. Gemini values the independence of their partner and their own freedom, and this is exactly what Leo can give them. On the other hand, Leo always values their partner's inner child and this is exactly what they will get in their Gemini. 99%

Leo & Gemini Shared Activities: This way, Leo can lift their activities to a "higher level" going to all places that Gemini would suggest, but in a fancy, expensive version. Although Gemini wouldn't really want to spend that much money, Leo doesn't mind paying for everything, for as long as they are not feeling used. Still, Leo can be very lazy. Gemini is always on the move, and has the need to do at least three different activities every day. It is a good thing Gemini wants to go everywhere and do everything, or these two might have a big problem finding activities they want to share. When Leo has time to rest, they will probably want to watch TV and move from the left to the right side of the couch all day long. This is something that could build a gap between their worlds, but usually they have enough respect for each other's needs that they can separate activities and be very happy together. None of them is needy and they don't want to spend each moment of their lives with their partner, so it will give them both enough freedom – for Gemini to move and for Leo to rest. 80%

Conclusion: They both consider their day best spent in laughter, and if they share friends, they could seem like a perfect couple. Their main challenge is the difference in their approach to change and they both need to make room for small adjustments in their behaviour if they want their relationship to last. Gemini and Leo can have so much fun that it could make the rest of the zodiac envious. Leo will need to make room for more movement and understand what seems to be "flakiness" of their changeable Gemini partner, while Gemini will have to understand that Leo is in fact keeping them together for however long they are meant to last. Their mutual respect can usually overcome any boundaries, and they should keep having fun and building their relationship on a solid foundation of childish joy. 82%

Leo Compatibility with Cancer

Leo & Cancer Sexual Intimacy and Compatibility: Although they don't have much in common, in astrology they represent a husband and a wife and are the king and queen of the zodiac. Unfortunately we know how unsatisfying the sex between a king and a queen can be like. The sexual aspect of their relationship depends on the depth of their emotions. As highly emotional signs, each in their own way, they tend to show their love in different ways and this can be a bit hard to reconcile in their sex life. Cancer and Leo make a very interesting couple due to the fact they are the only signs in the zodiac ruled by the lights in the sky, both of which are not planets – the Sun and the Moon. As a Fire sign, Leo is way more openly passionate and this could scare their Cancer away. Cancer is tender and sensitive enough to make their Leo partner feel guilty because of their nature, or Leo could simply have difficulty being tender in the way Cancer needs them to. It is as if a lion and a roe started a sexual relationship and although they don't want to hurt each other, their primal behaviour seems to pull them in that direction. Still, because of their rulers, they can get pretty close and share fine emotions in their sexual encounters. Although there won't be much excitement to them, they could be satisfying enough for both partners if they don't expect a wild sex life. To find middle ground they really need to be quiet and listen to each other's needs. 30%

Leo & Cancer Trust: This is somewhat true and Leo is a born performer, more or less supported to become one. However, their need to show off is something Cancer might find irritating, but not something to lose their trust over. If Cancer feels loved, they have no reason to doubt their partner just because of their nature. It is said that Leo likes to be in the centre of everyone's attention. Still, the differences between them might lead to a secret search for more compatible partners and this is easily sensed by both partners. 50%

Leo & Cancer Communication and Intellectuality: Even if they share interests, they will often have a strangely different view on the same thing. They will often just drift apart as the conversation progresses. The Moon does reflect the light of the Sun, but it circles around the Earth. This would be fine if the Sun wasn't so used to the fact that everyone circles around it. This explains what happens in their relationship once Leo starts talking. Although they shine, Cancer pays much more attention to someone else or to the idea of earthly things they could have together, than to their Leo partner. Cancer and Leo, ruled by the Moon and the Sun, represent a subconscious and a conscious mind. The base of their communication should be in things that have yet to be discovered. All the mysteries that are there to be solved could be a good starting point because they both need to understand things, each in their own area of dominance. The best possibility they have is in the things that need a light cast on them from two different angles. If they have enough respect for each other, they could learn a lot from each other's passive and active approach to life. 10%

Leo & Cancer Emotions: Cancer is a Water sign that represent motherly love and all emotions in one's family. Although this is not always a promise of happiness, the depth of all the love that hides within is magical. On the other hand, Leo is a Fire sign of joy, first loves, fun and sex. Their heart is warm and big, for Leo represents our inner child, and their loyalty is unchangeable when they decide to give it to someone. The main problem here is their connection, for Cancer needs to look into somebody's eyes and feel the love, while Leo wants to shout it out from the rooftops. We can say with certainty that their emotions are a truly beautiful thing. Both signs represent love and although it is not the same kind of love, it is an emotion, pure and simple. This could seem untrue to Cancer, as much as Leo could recognize in Cancer something that would rather tie them down than allow them to fly. Both of them have opposing signs ruled traditionally by Saturn and this is exactly why they could recognize true quality in their relationship, as each of them takes their opposing role to learn something. Usually this is not something that can last very long and they will probably both moves on to find someone who is more of an image from their seventh house. 45%

Leo & Cancer Values: They simply don't value the same things. Although this is the point where they separate their ways, it usually takes them long to realize this fact. Cancer values tenderness, emotions, family and a stable life with someone, while Leo values initiative, passion, energy and focus. There is rarely something that they will both value in the same way, or put in the same spot at their priority lists. 1%

Leo & Cancer Shared Activities: They will also gladly be served by breakfast, lunch and dinner, and who better to prepare those than a caring Cancer. However, when they leave the house, they will want to visit different places. Cancer will want to visit their close friends, especially if they have babies, go for a walk by the lake or have a romantic evening at the movies. Whatever they do, they like it with a little intimacy to it. As if to oppose this, Leo will want to spend time at places where they could be seen. Their partner needs to shine to others and hold them by the hand while everyone else claps theirs. It helps that Leo would gladly sleep for 20 hours a day. This will make them available at home. This is not often something Cancer will be ready to do, as they don't really like to be the centre of attention. On the other hand, Leo could compromise, but in the end they wouldn't be happy if their social needs aren't met. 35%

Conclusion: Leo is a sign that should spread joy and love with an active approach to each one of their relationships. How is it possible that Cancer is immune? It will be probably because the Moon circles around the Earth, not the Sun. They are special, that's for sure. Both of them are strong individuals, each on their own plane. Their lack of understanding and emotional touch can be explained through the fact that both of them have a mission to spread love to the less fortunate signs of the zodiac. Although the Moon reflects the light from the Sun, the sign of Cancer doesn't really see Leo as the source of all their joy. Not everyone is born with an emotional flow like Cancer and a huge, warm heart like Leo. If they kept all this love to themselves, some unfortunate souls would probably search for them aimlessly, and the world would be a much sadder place. 29%

Leo Compatibility with Leo

Leo & Leo Sexual Intimacy and Compatibility: Leo's warm nature and passionate approach to all things in life, including sex, will keep them satisfied together for a long time. However, there is a face behind the act to both of these partners that they might be too proud to show. They usually search for partners that can help them show their core in order to really connect instead of simply having sex as an instinctive act. The main problem that two Leo partners can have are their boundaries and the possible lack of respect they have for each other. While they would both enjoy being with someone who is so confident, they might hold on to the image of confidence for way too long, until all sorts of insecurities surface. The combination of two Leos can be difficult when it comes to intimacy, but their sex life might be excellent even when they are not intimate at all. It is difficult to develop closeness with all that fire in one place and when you think about it, there is only one Sun in our Solar system and everything revolves around it. Then what do you think, can it be possible to have two of them in one bed, circling around each other? 50%

Leo & Leo Trust: After all, they are ruled by the Sun, so how could this not be the case? Although they are usually open with other people, when they get together it is like a constant struggle for supremacy. This could lead to all sorts of "inflated stories", those that cannot be repeated and are almost always a product of fiction. Leo is a bad liar, in general, and it would be a shame for these partners to get to the point where they need to prove anything to each other. These two have a deep conviction that everything is clear in their lives. They have to learn how to be together with absolute focus on the other person. As soon as they turn away and start explaining how great they are, they have lost a chance for trust in advance. 70%

Leo & Leo Communication and Intellectuality: When they connect on a deeply personal level, they can find a special language of learning and find out so much about their own situations and other relationships. A wonderful thing in this relationship is the ability of both partners to shed a light on one another's important issues. The problem will arise when they start their ego battle to prove to one another who is right and who is wrong. When two Leo partners are in this type of conflict, it is impossible to resolve it because they both hold on to their points that both can be correct. In these situations they should try hard to find the middle ground or they might end up in a serious, lasting fight over an irrelevant thing. 65%

Leo & Leo Emotions: The Fire of Leo creates warmth, passion and creative energy. This can sometimes fail to be recognized as true emotion, especially by Water signs, and it is a good thing that two Leos understand each other's emotional depth perfectly. The truth is that Leo is an extremely emotional sign. It relies on Cancer and moves to Virgo, so this is a sign that has an impossible task to connect pure emotion to pure intellect. Their starting point, however, is emotion. Their main challenge in this emotional field is the way to express how they feel and how not to get burnt. As all Fire signs, Leo has a passionate nature that moves fiercely and they can sometimes regret not following their heart. Leo is a Fire sign of fixed quality, and they are pretty hard to change. They easily substitute emotion with passion and often burst into flames before they realize how they feel. When there are two of them, their relationship can seem like a chain nuclear reaction that has no emotional foundation. However, there is so much emotion underneath the surface if they choose to stay together, for only a Leo knows how they feel after the outburst has been shown to the world. 90%

Leo & Leo Values: In general, Leo values bravery, clarity and someone's inner strength. It is safe to presume that they will value each other because of these primary values. As two representatives of the same sign with such strong characteristics, they value similar things. What they will both value the most when they are together is their time for rest and their time for play. As if they were real little lions, these two are capable of truly enjoying their leisure time and each weekend could seem like paradise. 99%

Leo & Leo Shared Activities: This is exactly what a relationship of two Leos could look like. Although they would spend most of their time in bed and playing, they could actually have so much fun doing so that they don't need anyone else to keep them company. They are social beings but they don't care much about spending time with people who don't know how to enjoy life, and will in most cases enjoy their troop. Imagine a family of lions in the African Forest, lying around all day, then playing for a while, then licking their paws and calmly purring, well fed and happy. The only problem they might have is in their need to stay away from things that don't fit their character. As fixed signs, they will both be turned to a certain routine. Although the base for it will be the same for they are two Leos, they might do it in a different way. For example, if one of them likes to show their admired personality in a club down the street, maybe the other will want to go to a fine restaurant and show good manners instead. It is important for them to stay open for each other's suggestions because this is the only way their worlds can actually merge. 85%

Conclusion: Their main goal is to find true intimacy and understand each other's inner emotional beings. Leo has a habit to exaggerate and make drama out of small, irrelevant things, but this could be a good thing for their relationship because of their social status and the ability to support each other's theatrical needs. If they begin a battle for supremacy, it might be a good idea to set the territory that each of them is in charge for. Two Leo partners can do the impossible and this fact could keep them in a perfectly satisfying relationship for a long time. If one of them is the best at an emotional department, the other one can be the best in the sexual one. If they split their rules this way, it will be much easier for both of them to function and think of each other as worthy of the relationship. What might make a loving relationship between them impossible is the lack of respect. If they catch this disease, it might be best for them to part ways and search for different partners. 78%

Leo Compatibility with Virgo

Leo & Virgo Sexual Intimacy and Compatibility: Still, the shy nature of Virgo and their caution when it comes to choosing a sexual partner might make it difficult for them to find a language they both understand. Leo wants to be with a partner that makes them feel special and even more confident than they already are, and this is hard for Virgo to give. Their relationship can be quite challenging because the passionate nature of Leo doesn't give much space to Virgo to feel protected and secure about their choices. As two rational signs governed by pure consciousness they could easily agree on the way their sex life is supposed to look like. Their rationality might turn into an intellectual battle for sexual dominance, that is, if they ever reach the point in which they both want to have sex with each other. It is a good thing that Leo is a fixed sign, so they have a conservative note to them that suits Virgo. Still, they will rarely settle for Virgo's approach to emotions in their sex life, and they will probably both be unable to make an emotional connection that will keep them satisfied. In rare cases when a Virgo partner doesn't feel ashamed or attacked by a Leo partner, they might share a physically satisfying sexual relationship, but they could still both be too rational together to find any intimacy whatsoever. 5%

Leo & Virgo Trust: The problem might arise as Leo partner starts showing off and posing as "the king of the jungle". There is no reason for two conscious individuals not to trust each other in most situations. Too much attraction that Leo will summon is not something Virgo is comfortable with, and Virgo's initial problem with trust will blossom if their communication doesn't make up for the lost confidence. 65%

Leo & Virgo Communication and Intellectuality: The sign of Virgo belongs to Earth, and Leo belongs to Fire. This is why Leo is very passionate and fiery about their convictions, choices and everything that is important to them. On the contrary, Virgo is very practical, down to earth and usually too proud of their intellect to give into passionate, emotional outbursts. These signs together represent the king (Leo) and his followers (Virgo). In the same way, it represents the boss and his employees or a husband and his cleaning lady. The most important thing here is for both of them to remain respectful and tolerant of each other. Both Leo and Virgo are ruled by rational, conscious planets, and as such they are usually easy to talk to. However, their personalities are very different due to the difference in the element they belong to. If Leo shows any disrespect and starts giving orders, Virgo will run off, for this is not the relationship they were looking for. On the other hand, if Virgo partner fails to understand that they chose the king of the jungle, their relationship will not last very long because Leo needs to be acknowledged. 50%

Leo & Virgo Emotions: While Virgo rationalizes everything that happens in life, Leo has a perfectly rational answer to everything. Although Leo can be painfully aware of their need for intimacy, this doesn't mean they will be able to create it with ease, especially with someone like Virgo. This can be a great challenge for their relationship, for even when they are strongly attracted to each other and communicate well they don't seem to awaken each other's emotions. The hardest thing for a Leo-Virgo couple to find is emotional closeness. Although they can both be very intimate with other signs of the zodiac, they will rarely find this with each other. Leo will show his affection through a passionate, warm approach, full of attention and vigour. Virgo will be shy and have a hard time understanding this, while giving love through care that might seem ridiculous to a confident Leo. 1%

Leo & Virgo Values: Leo can be smitten by someone's mind, and this is exactly what Virgo has to give, in case they are not to closed up to show it. If they work together, they might create the exact atmosphere in which anything can be created, but only if they share similar professional interests. They will both value intelligence and one's ability to use their mind, and they will value this to the point of indisputable respect. Differences between them are still often too big to be overcome by a simple rational mind, and while Leo will value everything that shines, grand and striking things, Virgo will value someone's ability to be humble and modest. 35%

Leo & Virgo Shared Activities: Their roles in the zodiac support their cooperation and Virgo has no better boss than a Leo, in case they both don't have serious ego related problems. With just enough mutual respect, they could do just about anything together, for as long as it can be kept behind closed doors if this is what Virgo wants, and without being exposed to too many people. Leo wants to have the attention of their surroundings if they haven't been bruised and felt shame too many times, and Virgo doesn't feel this need. When Leo and Virgo work together, they will have no problem sharing activities however different their natures might seem. They will rather stay in the shades, behind a leader, someone smart, with a great vision. They like taking care of all the details in order for them both to get to a point of desired success. Since Leo represents our stomach, and Virgo our intestines, they will metabolize different experiences together, guided by Leo's leadership and passion and followed by Virgo's practical sense and attention to detail. 55%

Conclusion: They both tend to be too rational and their mental strength will rarely be a good foundation for a fairytale love they secretly wish for. Both of these signs have opposing signs linked to Neptune. Leo's opposing sign is Aquarius, the sign of Neptune's exaltation, while Virgo's opposing sign is the sign of Pisces, ruled by Neptune. Leo and Virgo form a constructive relationship that rarely serves their emotional natures. Both of them need someone perfect, someone made just for them, and if they just think for a second that they don't belong together, their search of perfection will prevail. It is rare for these partners to form a strong emotional or sexual bond, however well they might get along when it comes to work and communication. 35%

Leo Compatibility with Libra

Leo & Libra Sexual Intimacy and Compatibility: With Leo's confidence, and Libra's sexuality, they tend to inspire each other to become great lovers when together. Their sex life is usually filled with respect, and they feel free to try out new things with one another. If they found their relationship on a strong mutual attraction, they could enjoy a satisfying sex life for a very long time. Leo doesn't mind being seen and Libra is a sign that represents the public eye. When a Leo and a Libra come together, they don't need much time to build up a healthy sex life. Although this says something about their sexual preferences too, they will usually be well behaved in public. As soon as any restrictions show up, they will have to play out their passionate scenarios at any time, and in any place in which they get a chance to be alone if only for a minute. Libra is a sign of Saturn's exaltation and it is easy for them to wait and be rational, but with passionate Leo they find it hard to stay in control. 90%

Leo & Libra Trust: The problem here arises from their understanding of the Sun, for it rules Leo and falls in Libra. To add to that, Leo is a sign of Neptune's fall and Libra can often sense the dishonesty behind Leo's confident act, if there is any. The problem lies in the fact that they both like to be seen, but in an entirely different way. Leo wants to show everything they've got and Libra wants to get approval from other people. It is not that often for Leo and Libra to share a relationship filled with mutual trust. None of them understands the other, and this can become a reason to get jealous and mistrustful. If they wish to remain in a trusting relationship, they need to find approval and a suitable audience in each other to begin with. Only then will they be able to move on and look for these things in other people without arising suspicion. 40%

Leo & Libra Communication and Intellectuality: The Sextile between their Suns usually makes it possible for them to respect each other, and help each other build stronger personalities, free of judgment of any kind. Their elements of Fire and Air fit perfectly and there is a passionate approach of Leo for every idea of Libra. Their communication is fast and inspiring, although sometimes hard to ground through constructive ideas if Libra doesn't rely on its cold and rational relationship with Saturn. When it comes to the rational side of their relationship, Leo and Libra have a very nice way to support each other's personalities and communicate. The problem arises if Libra feels any sort of jealousy at their Leo partner for their sometimes unfounded confidence and that inner sense of security. The only way for Libra to learn how to feel confident as well is to accept this ability of Leo as the best part of their beautiful character. If Libra starts judging Leo, making assumptions on how their partner should behave but doesn't, their mutual respect will fade and they will both miss the point of their relationship. 85%

Leo & Libra Emotions: They will never end up in a relationship with no future, and their belief in love will move them towards marriage, children and growing old together, if only they share enough trust and love. Ruled by the Sun and by Venus, these signs represent one of the basic planetary cycles of love that is often connected to periods of eight years. These two signs represent our loving relationships and marriage, and when you look at this couple, you will see that their love for one another is real, obvious, shown and leading them in a certain direction. If they stay together longer than that, they might as well walk down the aisle and have a bunch of kids. 99%

Leo & Libra Values: Libra, on the other hand, values justice and one's ability to be the hero, which something they often think they lack. They are finely compatible when it comes to matters of the Sun and they complement each other well in a way that helps them both learn about expressing themselves and their abilities and strengths. Nothing holds greater value for Leo than someone's strong personality and their own pride and heroism. The problem with this couple is in their relationship toward Saturn, and while Leo represents its detriment, Libra exalts it. Although this can be a lesson to be learned, the challenge of responsibility they take on unequally can tear them apart. Leo needs to get serious and realize what their responsibility is to fit into the thing Libra values most – reliability and tact. 75%

Leo & Libra Shared Activities: This doesn't sound that fast, does it? If they share the same interests, they could have an endless field of possibilities for shared activities. They will mostly enjoy "red carpet" events and the fancy gatherings where they can both show one another to the world. The biggest problem in their choice of activities lies in Libra's indecisive nature that Leo simply doesn't understand, and usually doesn't have patience for. This is where they might give in to the temptation to "help" them decide, taking over the wheel and deciding instead of them. There is a strange similarity in the speed of these signs. Leo is a Fire sign and as such it shouldn't be slow as a Water sign or and Earth sign. Libra belongs to the element of Air, and it should be faster than any other element. But when you look at these two signs, you will see that Leo would like to sleep 20 hours per day, and Libra needs to think about everything twice and carefully choose activities and words they want to say. This can lead to mutual lack of respect, even though it seems like a little thing that no one would even notice. They need to give each other time, and stay as independent as possible. 60%

Conclusion: They have a lot to learn from each other, and the main goal of their relationship is to reach the point of shared respect and responsibility in a perfect balance of power. If you want to sum up the relationship between a Leo and a Libra, you have to understand that their bond involves the beautiful and challenging dignities of Saturn and the Sun. It will sometimes be hard for them to overcome the need for competing, trying to determine who is a better, smarter or a more capable person. Even if they don't, their relationship will be something to enjoy and show off in public. 75%

Leo Compatibility with Scorpio

Leo & Scorpio Sexual Intimacy and Compatibility: Leo is a passionate lover, warm, always in search for action and they can be quite casual when it comes to their sexual encounters. Scorpio is sex itself, and the depth of emotion that goes with it in its purest form. When they get together, they could have real trouble finding middle ground between their personalities. These partners can seem as if they've crashed into each other with no plan or purpose. This is a complicated relationship between two strong personalities with an incredible sex drive. If they are attracted to each other, this could drive them mad, for none of them will be able to realize their desires in a wanted way. If they end up having sexual relations, they could have misunderstandings on everything, from their verbal communication to their physical needs. They simply don't operate in the same ways and while Leo wants to be respected, Scorpio understands that all respect dies in the act of sex. It is extremely difficult for a Leo and a Scorpio to reach intimacy, because they have a different view on emotions. What Leo sees as love, Scorpio finds superficial and irritating, and what Scorpio sees as love, Leo finds depressing and irritating. They both need to give up control entirely if they want to find sexual satisfaction with each other. 5%

Leo & Scorpio Trust: This is not such a good thing when it comes to their ability to adapt and be flexible for each other, but it is a perfect thing for mutual trust. If they set the clear foundation in the beginning of their relationship, Leo transparent as they are and Scorpio direct and honest, they might trust each other without exception for a very long time. The positive side to this relationship when it comes to trust is in the fixed quality of both signs. That is if they both want to be open for this kind of relationship in the first place. 65%

Leo & Scorpio Communication and Intellectuality: Although this is not always the case, Leo wants to show the right image to the world, and Scorpio understands karma better than many other signs. This is why they will probably have enough respect for each other to communicate in a civilized manner. They are both obsessive in a way. Leo will never give up on chasing their passion, with enough energy to spark everyone around them, and Scorpio will hold on to things they care about, and obsessively fight for their goals. It is a good thing that these two signs can be so well behaved. If they share the same passion or interests, they will have something to talk about, obsessively. The depth that is typical for Scorpio is something that Leo tries hard to reach in their search for Unity. Their conversations can be very tense and irritating for both, but along the way they might realize that they give each other exactly what they both need. 30%

Leo & Scorpio Emotions: In some cases, they can be identified with hate, but the important thing to remember here is that hate is also love, in its "negative" form, and both of these partners will think that any emotion is better than no emotion. Their a bit torturous relationship can hold them together for a long time, even though they might be unhappy and aware that they could be happier with someone else. This is probably the most challenging relationship in the entire zodiac when it comes to emotions that these partners have for each other. In a way, Scorpio likes to be tied through negative emotions, for love sometimes has to hurt, and Leo sticks with their decisions because they rarely accept that they might have been wrong. This relationship can become a very difficult circle of suffering for both partners, especially if any one of them doesn't have their independent life, friends and finances. When they do, they might find a fine balance, for as long as they both have freedom to think that there might be better options for them with other people. If they don't find any, they could realize that they are perfect for each other as self-sufficient individuals, through a healthy approach. 1%

Leo & Scorpio Values: Although they understand clarity in different directions and depth, the main characteristics in people they wish to date are very similar. Often, they won't even recognize their similarity out of an emotionally unstable or obsessively stable position, completely different from a passionate, creative one. Both of these partners will value honesty and clarity. They have to deal with the value of creation against the value of destruction and this is not easily reconciled. The bridge between them is found in unconditional honesty. 35%

Leo & Scorpio Shared Activities: While they are in most cases interested in different things, Scorpio is ruled by Mars, a planet that Leo understands well and cooperates well with. They will both have the energy to follow each other's desires and they could actually have a lot of fun. Still, they usually stumble upon a problem when it comes to their understanding of certain circumstances and people around them. There is rarely a compromise between the positive, constructive approach of Leo and the often negative, sensitive approach of Scorpio, especially when none of them is exactly true. It is very interesting how a Leo and a Scorpio might organize their time together. It would be best for them to remain true to themselves, and understand that there is a middle ground between their opinions and feelings, however strange it might seem. 40%

Conclusion: This is in no way an easy relationship, and both partners can be stubborn and stiff in their opinions, life choices and ways they handle reality. If they want to remain in a loving relationship, they need to understand each other's way of expressing emotions and respect each other's needs however different they might be from those they are used to. When Leo and Scorpio start dating, they might not know exactly what they are to expect. When they find a way to love each other without conditioning, they might realize that they are in search for the same thing like Unity. 29%

Leo Compatibility with Sagittarius

Leo & Sagittarius Sexual Intimacy and Compatibility: When they start dating, their sexual relationship might come as a surprise for both of them, for they will feel liberated to be exactly who they are with each other. The best thing they could do is use the trine between their Suns and build-up each other's self-esteem, especially if they have been in demanding or disrespectful relationships prior to theirs. The best thing about their sex life is the passion they share. Leo is there to bring inner fire for the act of sex, and Sagittarius to fire up the expansion, the places, positions and horizons. As two fire signs, one of them fixed and one of them mutable, Leo and Sagittarius share a warm love for each other. They will both enjoy each other in a fiery way and respect each other's bodies, minds and entire personalities. If they stumble upon one another and love is born, their sex life could represent a perfect connection for both of them. 99%

Leo & Sagittarius Trust: Leo does like to be the centre of attention and feel attractive and desirable, but this is something a Sagittarius partner can provide in abundance. There is usually no reason for them to lose trust over time, except when their emotions start to fade. Sagittarius is a mutable sign, and as such, they can fall in and out of love quickly and frequently. Since they spark each other's sense of security and confidence, they will rarely show jealousy or misunderstand each other's actions. In case Leo starts feeling left out and unloved, the suspicion will rise, and what better way to respond to suspicion than by becoming suspicious yourself, Sagittarius might think. Although they both might be unaware of the root of their issues when trust is lost, it is usually a simple lack of love. 80%

Leo & Sagittarius Communication and Intellectuality: This is something that will help them communicate about almost anything, even though their interests might differ and their backgrounds as well. Leo has the ability to help Sagittarius when they get lost, and this could happen often if their plans are grand. Sagittarius will give Leo vision and the ability to understand the future of their current creative efforts. Together, they make an important part of the process of creation. As two highly aware individuals with a strong sense of Selves and their personalities, they could build up an incredible understanding. They can both be loud, communicate a lot, and this could make their relationship truly remarkable, deepening their intimacy through openness they share to get into each other's worlds. Leo and Sagittarius are both very focused on their mental activity. Leo because they are ruled by the Sun and this gives them a certain rational awareness and Sagittarius because they always aim higher from the Earth, philosophical and wide opinionated. Since they both have a strong personality, they will not feel threatened by each other's character and each other's strength of opinions and convictions. The only thing they might lack is the sensibility to outer influences and their fiery relationship could make them a bit too rough on each other, and on themselves. Still, the power of creativity and their active approach to life should keep them interested in one another and very well connected for a long time. 85%

Leo & Sagittarius Emotions: They will want to show their love, share their love and act on their impulses as much as they can. This can sometimes be too much, for no passive, fragile emotions or energy will be respected. Some balance would come in handy, especially if they often fight. Conflicts between them could be quite aggressive, not because they are that aggressive themselves, but because two fires build an even larger fire. It is almost like they might explode if they both go too far. When they fall in love, this seems like the warmest, cuddly love on planet Earth. In most cases, this will be enough to overcome any difficulties in their way, but sometimes these partners both tend to forget their actual sensitivity. As two Fire signs, Leo and Sagittarius tend to be very passionate and open in showing how they feel. They have to understand when the time has come to slow down, stay at home, talk about nothing at all and just be quiet. If they don't, they will probably turn to someone who can give them this kind of peace from time to time. 80%

Leo & Sagittarius Values: It is not easy to explain to a Leo why it is so good to run away from the world, travel in Greenland alone, and eat bugs somewhere in Asia, except if one wants to show their courage. On the other hand, Sagittarius doesn't really understand why they would go to fancy places and confront all the people that it is easier to run from. This is not a consequence of a lack of courage, but the lack of meaning they feel when they need to spend their time on tiresome people. They will most certainly value each other's strength of character and incredible personalities, the ability to warm each other up in every possible way and the passion they carry within, each for their own purposes. So although they value the same thing – courage, they see it through different eyes. 65%

Leo & Sagittarius Shared Activities: They have the energy and the need to search for knowledge and widen their horizons, but they don't exactly like to move that much. This is due to their fixed nature, and although they would like to visit any possible part of the world, they wouldn't do it at the same pace as Sagittarius, nor would they choose the same destinations. We would think that Leo likes to travel just as much as Sagittarius. Sagittarius, on the other hand, doesn't really understand why Leo wants to perform in front of so many people when there are starving children in Africa. These are simplified examples, but they serve us pretty well to understand how well they might work together, travel together, perform together, but only if they are open enough to ad purpose and strength to their approaches. 40%

Conclusion: This love is warm, passionate and inspiring, and they will have a chance to create, perform and have fun together for as long as they feel this way. However, Sagittarius partner might lose interest in Leo because they tend to get pushed away by their static, fixed nature. Leo and Sagittarius are a very good fiery combination of signs, and when two people with these Sun signs come together, they inevitably fall in love. The only way they might get to keep their passion and emotions going, is if they manage to listen to their softer emotions and remain tender and sensitive for one another. 75%

Leo Compatibility with Capricorn

Leo & Capricorn Sexual Intimacy and Compatibility: It will be a rare occasion when Leo is attracted to a Capricorn, but the other way around attraction seems more probable. However, they won't often get to the sexual part of their relationship, for even though they both might enjoy the chase, they will not see their future together. Leo is a warm, passionate sign, and Capricorn likes to be coolheaded and practical. Leo and Capricorn have one thing in common and it is their awareness of their Selves. This doesn't mean that Leo isn't at all practical, or that Capricorn isn't passionate, but they won't see each other as similar in any way. The rulers of these signs represent one of the archetypal conflicts of the zodiac, and tell the story of the fallen ego. This need could easily pull them both in a direction which will endanger their self-esteems and affect the image they have on their beauty and attractiveness. This is usually ignited by Leo's freedom of sexual expression that Capricorn fears, leading to the insecurity in both partners because they are not able to fit into each other's set of expectations. Their sex life can easily become boring for both partners, and what they often don't realize is how similar they actually are. The only way for them to have a healthy sex life is to share warmth and always bring new experiences, spicing things up. If they find themselves in a rut, they might stay there for a very long time, leading to the loss of libido and confidence, up to a loss of any sexual desire. 5%

Leo & Capricorn Trust: The depth Capricorn is prepared to go to makes Leo partner question their own motives and their whole personality. Lies seem to be impossible in this relationship, for each lied told, comes right back. As much as Capricorn sees behind the shine of Leo, Leo shines a light on Capricorn's darkness. Nothing stays hidden for long and as soon as one of them tries to stay secretive, mistrust is awaken. Capricorn knows that Neptune falls in Leo. This is exactly why they also know what hides behind the act in their Leo partner. However, in many situations they tend to trust each other because it gets so obvious that there is no reason not to. 40%

Leo & Capricorn Communication and Intellectuality: This is something that will not be easily reconciled and these partners could spend too much time trying to prove to one another why each of them has a point on what comes first. The problem is in the lack of understanding that each one of them has their own mission and their own role. It is futile to insist on someone's priorities changed when they should be different to begin with. If they respect each other enough to accept some pretty big differences, their communication might be very satisfying and fulfilling for both of them. Their priorities differ greatly, and they both have a strong set of personal priorities in their lives. Leo will help Capricorn find a more positive and creative view on every situation and Capricorn will give Leo the depth and the serious intentions they need. When they combine their abilities to organize, any plan made could be perfect. 60%

Leo & Capricorn Emotions: Warm emotions of Leo are easily cooled down and buried, and without the ability to express love, Leo can become pretty depressed. In return, the time Capricorn needs to build the emotional story they need, will be roughly interrupted by their fiery Leo partner. A relationship between a Leo and a Capricorn partner can be truly emotionally challenging, not because they don't love each other, but because they do. This could hurt them, or lead them to the opinion that Leo is not the right person for them, however attractive, smart, capable or beautiful they might be. The problem with this couple is in the way they build up emotions, and their best chances are in time and patience, things that Leo rarely possesses, and Capricorn rules. There is no other way to reach the heart of a Capricorn partner and discover that they can be warm too. If there has been too much pain in their prior emotional relationships, both partners could be almost too stubborn to get to the point where they might actually fall in love. 1%

Leo & Capricorn Values: Leo is not much of a plan maker, they would rather go with the flow and look only a couple of days in advance, and they respect Capricorn's ability to focus on the final destination and weigh every step of the way. Still, the sensitive, calm, emotional centre that Capricorn values is never found in a Leo and unless they are truly inspired by their Leo partner, they could take away their worth just because of preferences. Leo and Capricorn both value well organized people, presentations and plans. Leo values direct, open hearted people with big smiles, and as soon as they judge Capricorn for not smiling all the time, they might as well end the relationship. 50%

Leo & Capricorn Shared Activities: If Leo wants to settle down, they might find it interesting to spend time in a usual, Capricorn way. In return, if Capricorn needs some additional energy and vigour, they will gladly follow Leo in their chosen activities. The most important thing in their relationship is good timing. Activities these partners might share depend greatly on their priorities, once again. If it doesn't exist, they will simply resist, stubbornly, doing anything the other person wants to do. 5%

Conclusion: The main problem in their relationship is the set of priorities they might not share, and the passion or determination that both of them have. It is not an easy job, reconciling Saturn with the Sun, but it brings great benefits when it is done. The structure Leo could get and the creativity they might build on together could lift them to exactly what they desired, however their relationship might end. If they meet in the right moment, Leo and Capricorn might get along very well. They differ as much as the Earth and the Fire, but when they share a common goal, they are unstoppable. 27%

Leo Compatibility with Aquarius

Leo & Aquarius Sexual Intimacy and Compatibility: Leo is the king of the entire zodiac, and Aquarius seems to be there to bring down the king and fight for independence. Imagine the attraction and the passion between two such strong individuals, lying on the axis of Sun's rule and detriment. Their sex life is a struggle, a fight and an incredible experience for both. Liberating and yet warm and passionate, sensual but still interesting. When they find true emotion, Aquarius might actually end up respecting the king. The attraction is always great in relationships of opposing signs, and it is probably the greatest in a relationship of Leo and Aquarius. The beauty of their sex life is in things they can learn about their bodies, their confidence and the way they look at the act of sex. Through the struggle of insecurities and forced liberation, these are two partners to form a strong connection by a simple act of gravitation that the Sun has over Uranus. 99%

Leo & Aquarius Trust: However, these signs represent the axis of Neptune's exaltation and fall, and they will almost always have the challenge of trust and the search of truth in their relationship. Everything seems clear in a relationship between Leo and Aquarius when we look at it from a distance. Although they could find incredible understanding and freedom for both partners, usually when they separate they realize how little they have actually known about each other and how little trust they shared in the first place. 75%

Leo & Aquarius Communication and Intellectuality: While Aquarius is reaching for heroism, looking for ways to set free from repression, Leo was born a hero and sometimes doesn't even know it. If they end up fighting for the same cause, they could turn down entire governments and use their incredible force to change anything in the world. To get there, these two would have to stop the battle they have with each other, because energy can be scattered on their unnecessary fight for dominance in a relationship. Both Leo and Aquarius are heroes in their own way. Leo is a sign ruled by the Sun and has the ability to give clarity to any situation. No matter how confused they might be or how lost they may sound, if you take a closer look to the time spent with them, you will see that they've brought clarity in your life. Aquarius, on the other hand, understands the necessity of change and they seem to carry around a spark to ignite and excite any possible situation that they find worthy. This can be irritating to many, especially Leo, but in fact it is a necessity of liberation we all carry within. 90%

Leo & Aquarius Emotions: While Leo is the Sun, Aquarius is a lightning and it usually comes out on a rainy day. This is exactly what they need to understand – there is a time for both of them to shine and they don't endanger each other's chance to do so. When Leo falls in love, the entire world can feel it. The warmth pours out from the centre of their being and one has to be blind or senseless not to pick up the signals. Leo is exactly what Aquarius needs to find love. It is a strange thing how they find each other, on the grounds of their former relationships, to liberate and shine as if they have been searching for one another for many lifetimes. Aquarius can hide their emotions much better and often has trouble expressing and acknowledging how they feel. It is a good thing that Leo's warm emotional nature will melt even the coldest of hearts and there will be no safer place for Aquarius to share their love than in these fiery arms. The only thing that can endanger their emotional relationship is their everlasting ego battle and they should both pay attention not to be too proud to let go to love. 99%

Leo & Aquarius Values: Someone with a strong character, who knows exactly what she or he wants, cannot stay unnoticed by Leo or Aquarius. The deepest value they share is the value of individuality. Although they will not agree on many other things, this is the one that could connect them strongly, because they are both such strong individuals in the eyes of each other. 80%

Leo & Aquarius Shared Activities: From a different perspective, this should help Leo feel more free and confident, although it might not seem so in the beginning. They will both like to show off, each in their own way, and it is only important for them to set the territory for both partners to be expressed. Just like the Sun and the lightning don't go together, Aquarius should take over on a rainy day, in a depressive crowd or in places where they both feel as if they would drown. While Leo shines, Aquarius likes everything shiny. The need Leo has to show off might be a bit disturbed by Aquarius' tendency to show what others don't want to see. This is where Leo needs to give in and let their partner rule the sky if they are planning to keep the relationship going. There will be enough regular, shiny chances and days with no clouds for Leo to rule. 90%

Conclusion: Imagine what these partners could do together if they let each other lead the way when the territory of their rule is in front of them. They both need to learn to let go of the image they have about themselves and about each other, or they won't get very far stuck in their unnecessary ego battle. Warm and cold, hearted and smart, nuclear gravitation and vacuum in space, it cannot be easy to mend their differences or form a stable, loving relationship. Signs of Leo and Aquarius combined represent the ultimate creativity, famous scientific discoveries, the first man in an airplane and the first man on the Moon. The best thing they could do is find a cause they will support together. This would give them a focus on the outer world and allow them to deepen the inner emotional world of their relationship while fighting outside of it. 89%

Leo Compatibility with Pisces

Leo & Pisces Sexual Intimacy and Compatibility: Leo will seem like a brute, caring selfishly about their own needs, incapable of forming an intimate relationship with anyone, let alone Pisces. Although this is not true, it might be the obvious reality to Pisces if they end up in a relationship with Leo Partner. In return, Leo will think of Pisces as weak and unrealistic, completely separated from their own desires and the strength of their body or emotions. The truth is, they can both be incredible lovers but they will rarely discover this together. Their roles and characters seem to be too different for them to find a way to coexist in a satisfying sexual relationship. It is incredible how two signs that represent love, can be so wrong for each other. The main problem of their relationship is in the fact that the sign of Leo is a sign of the fall of Pisces' ruler, Neptune. In a practical sense, this means that Leo will burst the bubble of Pisces and endanger their sensitivity, idealism and go against their beliefs. This will ruin the romance between them and make it impossible for them to find any magic while they are together. Leo's openness and directness will make Pisces feel ashamed and rushed, and their sex life could be delayed indefinitely until Pisces partner feels secure enough to get naked. Because of differences in their approach to sex, Leo will in most cases seem like an insensitive brute, unless Pisces start understanding their emotional depth even though it is so different from theirs. The best way for these partners to find a language that can sustain their sex life is by building emotional trust first and worrying about sexual satisfaction later. 1%

Leo & Pisces Trust: Pisces is the sign ruled by Neptune, and Leo brings it to its fall. This is a difficult combination of signs to reach a point of mutual trust. They will both seem dishonest to one another, not because they lie, but because their characters seem unreal. Neptune is a planet of all deceit and mistrust in the world, making things around us seem foggy, unclear and fake. Leo will think of Pisces as if they were always on drugs, and Pisces will feel sorry for Leo and their lack of faith. 1%

Leo & Pisces Communication and Intellectuality: They could share many interests due to the creative power of both signs. Pisces will easily give inspiration to Leo, but the problem is in the way Leo might use it. The best way for them to create a safe surrounding for both partners, is to stick to the subject they are individually interested in. Leo is a warm sign, very passionate about their doings and their desires. Pisces will rarely show the same initiative to realize any of their dreams and this is their greatest difference. Ideals Pisces have could be shattered by the approach of Leo if they get too close to one another. Since Leo always shines a light on our virtues and shortcomings, they will not miss a chance to show their Pisces partner how unrealistic they are. This could help Pisces build a more realistic approach, but it could also affect their confidence and hurt them through a difficult perception of the world. 35%

Leo & Pisces Emotions: The Fire element Leo belongs to, makes them passionate and gives them the need to fight for their loved one and their emotions. Pisces is a Water sign, and much more passive, showing their passion through the flow of emotion. They will rarely fight for anything, convinced that perfection doesn't need fighting for and that real treasures are spontaneous and free of conflict. The middle point for these partners is in their realization that not everything needs to be won, as much as not everything should be uninfluenced. Both of these individuals are extremely emotional, each in their own way. Although true emotions are supposed to develop without difficulty, sometimes life is testing us to see if we really care. Still, this is not something that happens all the time, and sometimes things need to be let go because they don't belong to us, and we don't belong to them. 15%

Leo & Pisces Values: Pisces partner understands the necessity of lies, but still lives for clarity of the mind and the realization of their true inner Self. A great link between their worlds of values is in Leo's heroic nature that seems to have roots in a fairytale of Pisces. It is interesting how much both of these partners will value clarity and honesty. As much as they will both value their individual set of beliefs, they will be able to find middle ground in the grandiose character of Leo and the idealizing nature of Pisces. 20%

Leo & Pisces Shared Activities: Although Leo is a Fire sign, always ready to start something new, they will like to stick to their routine and show themselves in all the usual places every day. Pisces want to be invisible and they will change the scenery often in order for people not to recognize them. As a fixed and a mutable sign, they will have trouble synchronizing their need for changes and new activities. Although they could share some interests and have activities they would like to share, they will rarely stick to the same place and same actions together for very long. 10%

Conclusion: The problem isn't in their element or their quality, as much as it is in their connection through the fall of Neptune, the ruler of Pisces. If they get attracted to each other, they will be subjected to the risk of great damage to their beliefs, their inner faith and usually succumb to mutual disrespect because of a simple lack of understanding. Leo and Pisces seem to be put on this Earth to spread entirely different kinds of love. The beauty of their relationship could be developed through the fairytale approach of Pisces, if they build the heroic image of their Leo partner to the point in which other differences between them fade. 14%

PART-6
VIRGO HOROSCOPE

(August 23rd to September 22nd)

Who criticizes all she sees and would even analyse a sneeze? Who is observant, shrewd but hugs and loves her own disease?

Element: Earth

Quality: Mutable

Ruler: Mercury

Lucky Gem: Pearl, Topaz and Emerald. He/She should wear Emerald to crack a tough problem, to help him. The stone will bless him/her with all the intelligence he/she needs.

Lucky Number: 2, 3, 5, 6 &, 7 are for all the good luck.

Colour: White, Browns, Yellow, Beige, Forest Green

Lucky Colour: Get close to Nature. Wear earthy browns.

Day: Wednesday, Friday

Lucky Day: His/Her quest for perfection pays off on Friday.

Lucky Flower: Lavender

General Insights: Virgos are always paying attention to smallest details and their deep sense of humanity makes them one of the most careful signs of the zodiac. Their methodical approach to life ensures that nothing is left to chance. Virgos are often tender but also very careful. Virgo is an Earth sign, which prefers conservative and organized things, and those dependent on them. People born under the Virgo sign have very organized life, and even if they are very messy, their goals and dreams are put on strictly defined points in their mind. Since Mercury is the ruling planet of Virgo, this sign has a well-developed sense of speech and writing, as well as all other forms of communication. Many Virgos may choose to pursue a career as a writer or journalist. Virgo is often misunderstood, because of the symbolism of the name of this sign. Virgo experiences everything for the first time. Virgos always want to serve and please others, so they often choose to work as caregivers.

Personal Quality: The Virgo is critical, precise, easygoing, reliable, steady, helpful, intellectual, studious, logical, methodical, communicative, sciences, languages, takes a romance to new heights, good followers and the best employee one can ever have. He/She often dislikes delegating. He/She knows how to please the persons in power and position to get his/her work done easily and so, gets promotion very fast. He/She has a pleasant nature, colourful personality and sharp mind with a great sense of responsibility. He/She is precise, refined, and a lover of cleanliness, hygiene and order. It is not so easy to measure the depth of his nature. He/She knows to mould others in his shape by his clever activity. If somebody offends him/her, he/she does not show his real feelings on his/her face but act secretly to take revenge and hit back all his/her might when he/she gets the opportunity. The early part of his/her life is spent in struggles. The luck favours him/her at the age of 24, 36 to 42 years of age. He/She gets properties after the age of 40.

Positive Quality: He/She is good planners, practical and hard working, loyal, analytical, kind, hardworking, practical, trustworthy and able to do perfect work. He/She is a hard worker, conscientious and perfectionists. He/She is plain spoken and is able to express well. He/She leads a moderate life and does not like excesses.

Negative Quality: He/She is sometimes very critical and thinks himself/herself all in all. He/She has a penchant for turning molehills into mountains, Shyness, difficulties into stress and cleanliness into obsessive behaviour and a capacity for endless worry. He/She is miser, worried, overly critical of self and others, all work and no play and selfish and wishes others to follow them.

Physical appearance: He/She has good bone structure and is often highly photogenic. He/She is attractive, with beautiful eyes that sparkle with intelligence.

Relationships: He/She is truthful and loyal. He/She is tense in close relationships, which could badly upset his/her sex lives and makes it hard for him/her to become truly intimate with those he/she loves.

Career: He/She is usually happy working in a job that calls for precision, a shrewd mind, and logic. Dogged, analytical and intellectual, he/she is makes skilled and inspired research scientists, analysts or even literary critics. He/She can be at the best as a manager, a secretary, a lawyer and a trader.

Health: He/She is often martyrs to stomachs, and may suffer from Irritable Bowel Syndrome, from food allergies. Virgo rules the abdominal region, intestines, and the lower lobes of the liver, the spleen, the duodenum, and the sympathetic nervous system.

Ideal Partner: He/She gets attracted to opposites and vibe well with dreamy Pisces.

Compatible Zodiac Signs: Taurus, Capricorn, Pisces
Incompatible Zodiac Signs: Gemini, Sagittarius
Greatest Overall Compatibility: Taurus, Capricorn
Best for Marriage and Partnerships: Pisces
Virgo likes: Animals, healthy food, books, nature, and cleanliness
Virgo dislikes: Rudeness, asking for help, taking centre stage

BUSINESS ASTROLOGY FOR VIRGO

Virgos are perfectionists with little tolerance for shoddy work. They expect others to match there perfection. When confronted with obstacles they get upset easily. For attaining success in business they should learn to control their emotions and temperament. Business booster Gem is Sapphire.

Virgo Compatibility with Aries

Virgo & Aries Sexual Intimacy and Compatibility: Aries may look at Virgo and think of Virgin Mary, her chastity and what we would call a total absence of sex. There is nothing more asexual for Aries than a person without an obvious sexual identity. To express their sexuality or feel sexual at all, Virgo needs patience, verbal stimulation and a lot of foreplay. This is where Aries comes in as a brute with no manners or tact what so ever, to sweep them of their feet with a passionate nature that looks superficial and completely unattractive. It is hard to say if Aries and Virgo would present the clumsiest or simply the worst couple when it comes to sex. The real question is – how did these two get attracted to each other in the first place? Their intimate life can be good only in case Aries accepts to wait and communicate about things they don't find important at all, or if Virgo was so disappointed in their previous relationships that they turned into a sexual predator, open for an interesting turn in their intimate life. 10%

Virgo & Aries Trust: Aries usually has the need to be honourable and straightforward, except in rare cases when they cannot contain their sexual appetites. In case they overcome their sexual difficulties and stay together against odds, their problems with trust shouldn't be significant. In most relationships, Virgo is obviously faithful and hates being lied to. This is why they have a need to be honest and ask for honesty in return. 70%

Virgo & Aries Communication and Intellectuality: When you think of a partner who brings out the best in you, Aries and Virgo are the worst possible match. The downside of an Aries partner is their impulsive nature, readiness to fight and the tendency to lose their mind over something that might not be that big of a deal. The possible downside of Virgo is hysteria and continuous, never ending talks, when they are not understood. It is a known fact that Virgo likes things clean. Well Aries is like an animal in their cage, especially if they are crazy enough to decide to live together. These two can be so annoying to each other that they might annoy everyone around them. This could lead to endless, pointless fights, because Aries will never change their nature, or their priorities, while Virgo will seem like a crazy person screaming, with gloves on and a huge bottle of antiseptic liquid in their hand. The good thing is that before they get to this stage of the relationship, they will probably find each other extremely repellent and break up instead. To good that could come out of this strange bond is their intellectual cooperation, in case they share the same interests or work. They will awake each other's intelligence, challenge each other's mind and probably think of entirely different, but constructive solutions for problems that might occur. 30%

Virgo & Aries Emotions: Their best chance for love would be the silent observation by Virgo partner for some time before they get together, because this would give a rational advantage in knowing the person they are starting a relationship with. It would also be good for Aries to think before acting, not a usual thing they would do. If they knew each other as friends, going through their emotional experiences with other people prior to them becoming a couple, they could know each other well enough to make their relationship work. When it comes to emotions, we could say that their emotional compatibility is better than their sexual one. Still, as Virgo is primarily an intellectual sign, a sign where Venus falls and the lack of emotion is evident, and Aries usually mixes up love and sexual attraction, it is hard to achieve a quality emotional connection between them. When Aries goes from friendship into a sexual relationship, they tend to be much more considerate and gentle. Without sexual involvement Aries is more tolerant and a better listener, so friendship will provide more substance to their romantic relationship. 20%

Virgo & Aries Values: This is what makes them great as colleagues, but this is not exactly the most important set of values a happy couple would share. Their relationship could be based on their joint business though. This would give more meaning to their conversations and everyday life. They both value hard work and ambition, as well as clear and sharply deduced information. Other things they value don't coincide that much. Aries is all for bravery and an attitude while Virgo thinks of these as stupid, unless they are a part of tradition or have historic significance. Virgo values intelligence while Aries thinks success has nothing to do with it and sees it as a possible reason for loneliness and sorrow. Still, these would rather be the reasons to tease each other and have a nice laugh, than they would have the capacity to tear their relationship apart. 50%

Virgo & Aries Shared Activities: There is always a chance that Virgo will use their health to get out of these activities and spend some time alone. This is a couple that could go for a run because it's healthy, spend time in the nature because it's healthy, think about their bodies together because it's healthy, have regular intimate relations, once a week, because it's healthy and to sum up, anything that's healthy would be easy to incorporate in their relationship. Also, it is a very good thing that Aries doesn't care much about "empty time", such as watching television or opening solitaire, because Virgo would rather read or clean than subject to these activities for the "permanent damage of the brain". It is a good thing that Virgo is a sign of mutable quality and always concerned about their health, or they would never think about following Aries to their activities. The problem in this area of their relationship is connected to activities that don't leave much sense of dignity or make careful Virgo feel scared, or the activities too boring for an Aries personality. 70%

Conclusion: Although in most cases they are not really meant to last, it can still be a fun experience if none of them takes their potential for a shared future too seriously. In case they take the best out of their relationship, giving it enough freedom and unpredictability, Virgo would incorporate some of Aries' energy, while Aries would allow Virgo to teach them how to organize their thoughts and communicate calmly. It's a good thing that the relationship between an Aries and a Virgo is never boring. This way they might come to the point where their relationship could actually last and the outcome depends on their ability to relax and have fun together. 42%

Virgo Compatibility with Taurus

Virgo & Taurus Sexual Intimacy and Compatibility: This is where Taurus gets in the picture as a hero setting their Virgo free. The gift of Taurus is their ability to relax their sexual partner by giving them enough attention and obsessing about their satisfaction. To their Virgo partner this seems almost unreal, for they would expect something rough and scary when it comes to sex. This is an ideal combination of partners for first sexual experiences, because Virgo can enter the world of sexuality in the gentlest way possible. The problem here can arise because of the nature of Virgo and their need to go into detail and analyze everything. Not only can they damage the spontaneity of their sex life, but they could also affect their Taurus' self-esteem by finding little flaws on their body and in their actions. The sexual relationship between a Taurus and a Virgo can be quite touching. Virgo partner is usually ashamed to show their sexuality, or their body for that matter. Virgo is a sign of virginity and is a place where Venus, representing all satisfaction, falls. The fear of being hurt is sometimes too big to handle and with Virgo's view of Venus they rarely understand the side of sexuality that is in relation to satisfaction and tenderness. Taurus is a sign ruled by Venus and their understanding of sex is quite different. They seem to have a mission to explain what tenderness is to those around them, and find someone like Virgo a perfect student for their teachings. They will gladly explain to their partner what the beautiful side of sex is, only if Virgo is ready to listen.

85%

Virgo & Taurus Trust: It is not easy to open up to such an enormous field of possibility when you feel so small. Taurus is much more relaxed and gives so much importance to the beauty of sex, so if Virgo doesn't feel adequate with their Taurus partner, it will not be easy for them to believe in their honesty or faithfulness. Virgo doesn't trust anyone with ease. This is due to the fact that Pisces are their opposing sign, and they see every partner in their life as a glimpse of the unknown. This mistrust will really hurt their Taurus partner, for they can't understand what they did to deserve it and they will probably blame it on Virgo's changeable nature, thinking that they are not that honest, either. 75%

Virgo & Taurus Communication and Intellectuality: It is often said that Taurus can be really stubborn and difficult to talk to, but it is almost certain that a Virgo will use their mutable quality to find different approaches in order to explain their point of view. Intellectual strength of Virgo is exactly what Taurus needs to build a better understanding of the world. As two Earth signs, they can both stick to their convictions and be too rigid not to accept another's point of view, but in most cases, the intellect of Virgo and the tenderness of Taurus can help them find a language they both understand whatever the situation. 90%

Virgo & Taurus Emotions: Since Virgo will not recognize their feelings right away, they will need time to set a strong emotional foundation. Because of the lack of trust and disappointment Virgo is almost always ready for, Taurus needs to stay put and never let them down in order to build the trust and let their feelings for each other evolve. If they are not both too stiff and too afraid to get hurt, they can build a strong and deep emotional relationship with mutual respect intact. The patience Taurus can have when they fall in love is what makes them such a good fit for Virgo. It is really important not to stay at a safe distance for too long, because they could easily build a relationship with no emotion and stay in it, unsatisfied, for years, even though they might have had potential to fall madly in love. 85%

Virgo & Taurus Values: With Taurus' ruler in fall at the sign of Virgo, their Earth to Earth understanding is a bit damaged. Since Venus represents all value, Virgo could show what Taurus would recognize as a lack of understanding in general. They don't exactly value the same things, but they will be okay for as long as the feminine side of Taurus isn't disrespected. However, they will value the nature of Earth element, stable, secure and slow, and this should give them enough time to mend the differences and find middle ground. 50%

Virgo & Taurus Shared Activities: The intimacy of their nature scares Virgo to the point of agony and they will quickly need a change of scenery not to feel like they are standing in one place for eternity. If Taurus approaches their usual activities in a way to respect Virgo's occasional obsession with their health, they could think of a number of things to do together, and complement each other very well. Taurus can truly seem lazy to their Virgo partner, especially when they are on a satisfaction spree and don't leave the house except if they are on their way to a nice restaurant. The preparation of healthy snacks would be just one of possible suggestions to satisfy the needs they both have. Virgo is a sign ruled by Mercury, and although it belongs to the element of Earth, they need to move. Taurus can be really static and it is important that they make a decision to follow their partner or they really won't have much of a future together. It is a good thing that Taurus is usually guided by inertia, so when they get used to movement, this will become a permanent state for both of them. Although you wouldn't connect these signs to travelling, when together, they could feel and follow the urge to travel the world. 55%

Conclusion: Virgo needs to be flexible enough to value their Taurus and give them the intellectual view on things they might idealize. In general, Taurus is there to teach Virgo about love, tenderness and sexuality. Their relationship could be a match made in heaven, only if they are not too scared of being hurt and too distrustful. If they do give in to each other and fall madly in love, they could be the combination of a clear heart, represented by Taurus, and a clear mind, represented by Virgo. What more would they need than each other? 73%

Virgo Compatibility with Gemini

Virgo & Gemini Sexual Intimacy and Compatibility: However, this affects them in different ways, for Gemini is a masculine sign, always ready to explore, while Virgo is a feminine sign, shy and sensitive. Their sexual relationship is hardly promising, but they both have the need to communicate. If they find a language they both understand, they might agree on the way their sexual life is to progress. However, there is a big chance that endless discussions will not lead to their mutual understanding, leaving them distant and not interested to share any sexual experiences. If they fall in love, they will use enough tenderness and respect to make their sex life work, but it will still rarely be satisfying for both partners. Gemini and Virgo are both ruled by Mercury, not a very sexual planet at first glance. They are both curious, but not in the same way, and have an extrovert vs. introvert conflict. While Gemini would often like to be free to get naked and run around the streets, Virgo would prefer if everyone kept their bodily fluids to themselves. To top it all, they both need verbal stimulation, but the words each of them would like to hear often aren't what the other person knows how to say. 5%

Virgo & Gemini Trust: The good thing is that they are not prone to jealousy, or they would probably kill each other after a couple of weeks. Gemini is not someone Virgo can trust. When Virgo starts doubting and analyzing everything, Gemini will simply fly away, in the best case scenario. If they have a touch through maleficent planets, such as Mars, this will not end there, but in endless fights and conflicts on trust and fidelity. This combination is very bad. With Virgo's trust issues and Gemini's trickster nature, both of them are about to go crazy. The key here is for Gemini to get back to planet Earth and respect the sensitive nature of their partner. They can build their security and be free to be who they are at the same time. Also, Virgo will have to open up, or they will never find a way to stay together. 1%

Virgo & Gemini Communication and Intellectuality: They may have their differences, but we wouldn't say their communication is bad. Even if they have a fight, they will still be good at talking. Their view on intellect could be a bit different though. Gemini's superficial nature could seem extremely stupid to their Virgo partner. This will set Gemini off to be even more irritating on purpose, because they are everything but stupid. In return, Gemini will see Virgo's obsession with details and think of them as crazy. Virgo will probably not fight this, but sit in their room and cry, alone, in the dark. They are ruled by the same planet, master of communication, Mercury. They need to find a way to recognize each other's "flaws" as good qualities and they will get there if each of them lets their partner help. When these minds come together, there is no problem to stay unsolved, for together, they are Mercury in all its glory. These signs have a mission to unite and transform heavenly ideas into real life, and lift human knowledge step by step, towards divinity. This is a relationship of the Air side of Mercury and its Earth side, grounded and well thought of. It shouldn't be forgotten that the sign of Virgo exalts Mercury so Virgo can exalt their Gemini partner, if only they wanted to land. 80%

Virgo & Gemini Emotions: There is an interesting emotional side to both of them. Gemini is located between Taurus and Cancer, two strong, feminine, motherly signs. This means that the sign of Gemini is formed out of Taurus and is a base for Cancer. How can it lack emotion than? Virgo has a problem with Venus, that is true, but does this make Virgo unemotional? Their biggest problems are speed and fear. Gemini simply doesn't stay in one place for long enough to build an emotional bond or recognize how they feel, while Virgo tries hard to dismiss Venus and with it love and all earthly pleasures because they don't think they deserve any of it. Gemini is considered not that emotional. Virgo brings Venus to its fall and rationalizes everything. Both signs have a small, fast planet in common, and if they touch each other's hearts, they will understand their partner's occasional need to run off, or run away. 55%

Virgo & Gemini Values: However, their approach to basic intelligence is different and they won't always recognize it in the same way. In fact their relationship can show us exactly how relative intelligence is. They are both fans of intelligence, resourcefulness and practicality. Their meeting point is actually in their emotional intelligence, not our typical one. However, they do value rationality, someone's ability to reason and one's practical use of their hands and their brain. 70%

Virgo & Gemini Shared Activities: Virgo will probably stay there for a bit longer, not even noticing their Gemini partner got lost and accidentally found an emergency exit that they had to use to see how the big handle on huge red door works. The lack of self-value Virgo suffers from constantly will make them question every location and activity their Gemini partner would want to take them to. It is a good thing they both like to read, so they can visit a library every once in a while. Even if they try to hide it, their hesitation will be felt and Gemini will rather go alone than push someone to come along. Both of these signs are mutable and as such have a tendency to change locations, interests, relationships. This can allow them to stay on the move together, if only Virgo was open enough for exciting new experiences and Gemini finds a way to not get lost. 30%

Conclusion: The relationship of Gemini and Virgo can change as the wind, while both partners get lost and found on a daily basis. Their mutual love for Mercury is what binds them and what tears them apart, because they both tend to over think things instead of following their hearts. Both of them are mostly in their minds, each one in their own way, and need to respect each other to the point where no one's intelligence is judged on a superficial level. If they do fall in love, they will become a unification of Air and Earth Mercury – heaven on Earth. 40%

Virgo Compatibility with Cancer

Virgo & Cancer Sexual Intimacy and Compatibility: They have a possibility to become an inspirational contact of heart and mind, if only they give in to the opportunity to enter each other's worlds. Their sexual relationship seems to be a lecture on emotion. The sign of Virgo brings Venus to its fall and suffers from a general lack of emotion. It is a rational sign with a lot to analyze, that rarely gives in to the first impulse or their fragile emotional state. They are to learn on how to feel safe enough to let their guard down and shut their mind off in order to feel and enjoy sex. When Cancer and Virgo get together, there is potential for a great, everlasting love. Cancer doesn't really understand how someone could have trouble getting in touch with their emotions and can have unrealistic expectations because of this. They will learn to understand their partner better and make a stronger sexual bond, realizing how different people can be. 95%

Virgo & Cancer Trust: Although Cancer is a cardinal sign, they are stable by nature, especially when it comes to emotional decisions they have made. If they have chosen Virgo to be their loving partner, they will have no reason to lie or cheat. If someone can help Virgo build their trust, it is their Cancer partner. This behaviour would only endanger their vision of a shared life and a loving family they want with the partner they chose. This is also a reason why Cancer won't have an initial problem with trusting Virgo. Their convictions are stronger than their doubt. 99%

Virgo & Cancer Communication and Intellectuality: The lack of words from Cancer certainly makes room for everything Virgo wants to say, but as signs ruled by the Moon and Mercury, they have a simple conflict of emotion versus logic. Although Virgo represents the grounded side of Mercury and that makes it much easier for them to communicate with someone like Cancer, they are still leaning a bit too much on their rationality rather than their heart. This is a tricky side of a relationship for a Cancer and a Virgo partner. However, both of these planets are ruling the human brain – Mercury represents the core and Moon rules the rest of it. This is why when their topics and their intellectual strengths combine in a right way, and with their emotions to follow, they find an uncharted, new territory in which none of them has ever been. Their communication can become truly inspiring and magical if they accept each other's characters fully. 60%

Virgo & Cancer Emotions: If Cancer starts showing their emotions openly and with no restriction, Virgo might get scared and start analyzing every little thing to determine if any emotion is really there. There is probably no greater turn off for a Cancer than someone who rationalizes their own emotion. As if the fight to keep them afloat among the human race isn't enough, now they have to deal with a neurotic partner who dismisses them. The emotional side to Virgo is a deeply feminine side, usually ashamed to show her face, especially if Virgo partner is male. The rational side of Virgo could keep their overall emotional status very low. It becomes almost impossible for them to feel something if they are in any way pressured or feel mistrust with their partner. Virgo would rather be alone, with a right book, than with the wrong person, and it takes a lot of patience and rationality from their partner if they want to understand and wait for the ice to melt. If Cancer manages to do this, there is no reason for both of them not to resolve any other emotional issue that needs resolving, mutual or individual, while together. 65%

Virgo & Cancer Values: To Virgo, their Cancer partner will seem much more down to earth and rational than they are, and to Cancer, their Virgo partner will have a recognized soft spot in their heart. Other things won't matter much when they find each other, because there is no compromise they both wouldn't make to keep this love going. However, their different values might represent an obstacle to them getting to this point in a relationship. As a deeply emotional and the most intelligent sign in the zodiac, one of them values family, love and understanding, and the other intellect, attention to detail and their health. If they form a truly functional and beautiful relationship, they will value each other. It is not like they don't have a meeting point, but it will not be easily found if they both stubbornly keep to their habits and opinions. 50%

Virgo & Cancer Shared Activities: This is very often a relationship of two people who can manage without each other, so they will not be bound by shared activities as much as some other signs might be. If one of them wants to do something, unless truly insecure, they won't take it as a personal insult if their partner doesn't want to do the same thing. Although Virgo is a mutable sign and they can be pretty hard to follow, they do belong to the element of Earth and will be capable to wait for their Cancer partner to decide whether they will want to join them or not. This is where their rational compatibility will really come in handy, as they make arrangements for their time together. Usually there will be just enough movement and romance for both of them to feel special, and this is certainly not the main concern in their relationship. After all, they would both gladly go to a movie and eat some popcorn, so there is always something they can do together, even if their needs are very different. 90%

Conclusion: The main problem of their relationship is in the possible conflict between emotional Cancer and reasonable Virgo. If they manage to overcome this, accepting each other's shortcomings and learning to incorporate some rationality or some emotion into their lives, they could end up in an inspiring relationship that will last for a very long time. Cancer and Virgo can have a wonderful connection and are usually brought together by sexual understanding. In a way, they complement each other as much as the heart complements the mind. If they share a spark of love, it would be a shame to miss the opportunity for happiness just because of someone's irrational expectations or someone's closed heart. 77%

Virgo Compatibility with Leo

Virgo & Leo Sexual Intimacy and Compatibility: The shy nature of Virgo and their caution when it comes to choosing a sexual partner might make it difficult for them to find a language they both understand. Leo wants to be with a partner that makes them feel special and even more confident than they already are, and this is hard for Virgo to give. Their relationship can be quite challenging because the passionate nature of Leo doesn't give much space to Virgo to feel protected and secure about their choices. As two rational signs governed by pure consciousness they could easily agree on the way their sex life is supposed to look like. Their rationality might turn into an intellectual battle for sexual dominance, that is, if they ever reach the point in which they both want to have sex with each other. It is a good thing that Leo is a fixed sign, so they have a conservative note to them that suits Virgo. Still, they will rarely settle for Virgo's approach to emotions in their sex life, and they will probably both be unable to make an emotional connection that will keep them satisfied. In rare cases when a Virgo partner doesn't feel ashamed or attacked by a Leo partner, they might share a physically satisfying sexual relationship, but they could still both be too rational together to find any intimacy whatsoever. 5%

Virgo & Leo Trust: The problem might arise as Leo partner starts showing off and posing as "the king of the jungle". There is no reason for two conscious individuals not to trust each other in most situations. Too much attraction that Leo will summon is not something Virgo is comfortable with, and Virgo's initial problem with trust will blossom if their communication doesn't make up for the lost confidence. 65%

Virgo & Leo Communication and Intellectuality: This is why Leo is very passionate and fiery about their convictions, choices and everything that is important to them. On the contrary, Virgo is very practical, down to earth and usually too proud of their intellect to give into passionate, emotional outbursts. These signs together represent the king (Leo) and his followers (Virgo). In the same way, it represents the boss and his employees or a husband and his cleaning lady. The most important thing here is for both of them to remain respectful and tolerant of each other. Both Leo and Virgo are ruled by rational, conscious planets, and as such they are usually easy to talk to. However, their personalities are very different due to the difference in the element they belong to. The sign of Virgo belongs to Earth, and Leo belongs to Fire. If Leo shows any disrespect and starts giving orders, Virgo will run off, for this is not the relationship they were looking for. On the other hand, if Virgo partner fails to understand that they chose the king of the jungle, their relationship will not last very long because Leo needs to be acknowledged. 50%

Virgo & Leo Emotions: While Virgo rationalizes everything that happens in life, Leo has a perfectly rational answer to everything. Although Leo can be painfully aware of their need for intimacy, this doesn't mean they will be able to create it with ease, especially with someone like Virgo. This can be a great challenge for their relationship, for even when they are strongly attracted to each other and communicate well they don't seem to awaken each other's emotions. Although they can both be very intimate with other signs of the zodiac, they will rarely find this with each other. The hardest thing for a Leo-Virgo couple to find is emotional closeness. Leo will show his affection through a passionate, warm approach, full of attention and vigour. Virgo will be shy and have a hard time understanding this, while giving love through care that might seem ridiculous to a confident Leo. 1%

Virgo & Leo Values: Leo can be smitten by someone's mind, and this is exactly what Virgo has to give, in case they are not to closed up to show it. If they work together, they might create the exact atmosphere in which anything can be created, but only if they share similar professional interests. They will both value intelligence and one's ability to use their mind, and they will value this to the point of indisputable respect. Differences between them are still often too big to be overcome by a simple rational mind, and while Leo will value everything that shines, grand and striking things, Virgo will value someone's ability to be humble and modest. 35%

Virgo & Leo Shared Activities: Their roles in the zodiac support their cooperation and Virgo has no better boss than a Leo, in case they both don't have serious ego related problems. With just enough mutual respect, they could do just about anything together, for as long as it can be kept behind closed doors if this is what Virgo wants, and without being exposed to too many people. Leo wants to have the attention of their surroundings if they haven't been bruised and felt shame too many times, and Virgo doesn't feel this need. They will rather stay in the shades, behind a leader, someone smart, with a great vision. When Leo and Virgo work together, they will have no problem sharing activities however different their natures might seem. They like taking care of all the details in order for them both to get to a point of desired success. Since Leo represents our stomach, and Virgo our intestines, they will metabolize different experiences together, guided by Leo's leadership and passion and followed by Virgo's practical sense and attention to detail. 55%

Conclusion: rarely be a good foundation for a fairytale love they secretly wish for. Both of these signs have opposing signs linked to Neptune. Leo's opposing sign is Aquarius, the sign of Neptune's exaltation, while Virgo's opposing sign is the sign of Pisces, ruled by Neptune. Both of them need someone perfect, someone made just for them, and if they just think for a second that they don't belong together, their search of perfection will prevail. Leo and Virgo form a constructive relationship that rarely serves their emotional natures. They both tend to be too rational and their mental strength will It is rare for these partners to form a strong emotional or sexual bond, however well they might get along when it comes to work and communication. 35%

Virgo Compatibility with Virgo

Virgo & Virgo Sexual Intimacy and Compatibility: It is a good thing they can communicate and make their sex life much better through the art of speech, but unless they are madly in love to begin with, they could both simply think of their sexual relations as inadequate. The main problem with the sign of Virgo is their search for faults, and this can be a real deal breaker when it comes to sex. When two Virgo partners enter a sexual relationship, this need for criticism multiplies, and what's worse, they seem to motivate each other to grow it and make it even stronger. None of these partners realizes that this takes away their emotional or sexual satisfaction, and makes them feel tense. In most cases they just realize that they don't feel relaxed with one another, often not even knowing why. The good thing in a sexual relationship of two Virgos is their respect for each other's sense of shame and the pace they want their relationship to move in. It would be funny to say that two Virgos don't understand each other when it comes to sex, but true attraction and the emotion behind the act, seem to be missing. Their mutable quality will help them make the necessary changes to make things work, and even if their sex life isn't satisfying in the beginning, they might be able to adapt until they reach the point of satisfaction. Creativity is the key here. And their training to live in the moment, without noticing every single detail that doesn't work between them. 30%

Virgo & Virgo Trust: Since they share the same set of convictions on trust that needs to be built, usually quite traditional, they often end up resolving these issues together. When these partners come together, they stand face to face to their own issues with trust awaken by the sign of Pisces in their seventh house. Still, they both need to remember that as soon as one partner questions the other, the favour will be returned, and the circle of mistrust can suddenly escalate to the point where they both start feeling the need to hide. 70%

Virgo & Virgo Communication and Intellectuality: Similar to the sign of Gemini, Virgo can be quite eloquent and smart; both of them ruled by Mercury, the master of communication. Since the sign of Virgo is related to the written word, much more than to the spoken one, these partners will probably want to text each other all the time, for as long as their chats don't become dull and boring. They are both highly intellectual, but also very quick to dismiss someone else's intellectual strengths if they differ from their own. When it comes to evaluating someone's words, Virgo can be extremely critical and focused on details that most other signs wouldn't see as important at all. If there is something Virgo is capable for, it is communication. The beauty of the relationship of two Virgos is in their shared understanding for the importance of details. Unfortunately, this can be the thing that will distance them from the bigger picture and make them preoccupied with things that really aren't important. 99%

Virgo & Virgo Emotions: The biggest challenge for them is to keep the love burning after their brains interfered with the process their hearts should have kept to themselves. As two representatives of a mutable sign, these partners change quickly and they often end up in a situation where love at first sight brings them together, but they stay together even when emotions between them are long gone. The love between two Virgo partners can be strangely rational. They are often a couple that meets at the perfect time – when they are both ready or old enough to start a family, or when they both ended relationships they were exhausted by for years. In order for them to keep the flame going, or break up, at least one of them has to have enough faith to believe they will make the right decision whatever they do. If both of them start questioning everything, they will both probably get nowhere at all. 25%

Virgo & Virgo Values: If there was a sign to show us how different similar people can be, it is the sign of Virgo. These partners have their own opinions and thoughts on everything. It will be very hard for them to find a partner, even if it is another Virgo, to coincide fully with their system of value. In most general issues of life, they will agree, and they will both value intelligence, capability and one's focus on details. We could say that their values match perfectly, but nothing with Virgo matches perfectly. They might have a hard time adapting to each other's emotional or professional values, especially if their choices of profession are too far off. 80%

Virgo & Virgo Shared Activities: When it comes to everyday things and the routine they both care about, they will find excitement and joy in most of their activities. Their choices make them feel proud and happy they have found someone who shares their appreciation for certain "smart" activities, but the challenge in their choices is hidden in the fact that none of them likes Venus very much. This fact could lead them to a point of apathy, where none of them lives their life in colour, to the fullest, threatening their creative energy. When it comes to everyday things and the routine they both care about, they will find excitement and joy in most of their activities. Their choices make them feel proud and happy they have found someone who shares their appreciation for certain "smart" activities, but the challenge in their choices is hidden in the fact that none of them likes Venus very much. This fact could lead them to a point of apathy, where none of them lives their life in colour, to the fullest, threatening their creative energy. Two Virgos can really do anything together. They need to remain in love, creative and romantic, or it will be very hard for them to truly enjoy the time they spend with one another. 85%

Conclusion: Because of their shared tendency for sacrifice, the lack of faith they have in themselves, and the tendency to rationalize everything with value, they might easily end up in a relationship where none of the partners is actually in love, or satisfied. When Virgo decides to be with another Virgo, we can assume that their relationship is a product of one of two possible things, the first one being the need for stability and their rational decision to be with one another, and the second one being the unexplainable force of love at first sight. Whatever the case, both partners are quite rational and belong to the sign of mutable quality, so their emotions can change very fast. It is imperative for them to act according to their hearts if they want their love to last. 65%

Virgo Compatibility with Libra

Virgo & Libra Sexual Intimacy and Compatibility: Virgo is an Earth sign ruled by Mercury, relying highly on their intellect, while bringing Venus to its fall. Libra is an Air sign ruled by Venus, and couldn't be more distant from Virgo's shy nature or practicality when it comes to sex. Most often, they are not even attracted to each other, and when they do begin a sexual relationship they have to face the challenge of speed. Virgo will want to move slowly, and even though their mutable quality makes them pretty adaptable, it is often not enough to reach Libra's speed of Air. They might find a middle though, when Libra relies on their exaltation of Saturn and slows down, while Virgo adapts and changes more quickly. Most often, they are not even attracted to each other, and when they do begin a sexual relationship they have to face the challenge of speed. Virgo will want to move slowly, and even though their mutable quality makes them pretty adaptable, it is often not enough to reach Libra's speed of Air. They might find a middle though, when Libra relies on their exaltation of Saturn and slows down, while Virgo adapts and changes more quickly. The sexual relationship between a Virgo and a Libra just doesn't seem like a good choice. They will probably be driven crazy by each other, one of them trying obsessively to keep things clean and looking as if they were scared of any emotional contact, and the other strict in their search for spiritual love and a partner they can really talk to. If they find their perfect timing, both partners might still end up unsatisfied. Virgo was expecting a fairytale connection seeing the image Libra shows and they will find so many faults in their partner's approach. Libra's self-esteem doesn't really tolerate that much criticism and they will probably see Virgo as boring, stiff and unaware of any emotional connection they wanted to make. For this sexual relationship to work, they need to coordinate the emotions they give and receive. Only then will they be able to satisfy each other. 1%

Virgo & Libra Trust: If this wasn't the case, Virgo would dismiss each word they say to begin with. There is a lot of tension between them, for they don't understand each other's primal natures. This is not so much an issue of trust, but it comes down to it when the dust settles. There is no way to explain to shy, introvert Virgo, why Libra has to have everything out in the open. It is even less clear why so much attention must be brought to their relationship, or why does their Libra partner flirt with everyone that looks decent. It is a good thing Libra exalts Saturn and their need to be fair is, in many cases, more than obvious. On the other hand, Libra doesn't really trust the moodiness of Virgo and think of their partner's fear of expression as if there is something they want to hide. 25%

Virgo & Libra Communication and Intellectuality: This can keep them in a pretty good place for quite some time, especially if they share the same professional interests or have a goal to support each other in learning or advancements of any kind. Even if they touch the subject of emotional contact, they will still see eye to eye for a while. Both of them value rational choices and smart moves. But then, true emotions will surface in one of these partners, and they might realize that communication between them is no longer possible. This can keep them in a pretty good place for quite some time, especially if they share the same professional interests or have a goal to support each other in learning or advancements of any kind. Even if they touch the subject of emotional contact, they will still see eye to eye for a while. Both of them value rational choices and smart moves. But then, true emotions will surface in one of these partners, and they might realize that communication between them is no longer possible. They will understand each other perfectly for as long as they speak of taking responsibility and serious matters in life, such as professional choices and their income. As soon as one of them has to deal with a personal issue, the other one seems to freeze, losing all ability for compassion of closeness. The problem here lies in the fact that their relationship is based on their mental connection and this interferes with actually loving one another, and providing a kind word at an important time. 60%

Virgo & Libra Emotions: There is no couple more prone to separate emotionally than these two, and this can lead to so much dissatisfaction in both of them, with both partners unable to end the relationship they don't feel good in. Virgo is a sign where Venus falls and this is the ruler of Libra and a planet that, shoulder to shoulder with the Moon, represents our emotions. The value and stability of emotions that needs to be found by Libra, seems to lose all meaning when rational Virgo comes along. Libra will instantly fit into the expectations of their partner, trying hard to be that voice of reason, more rational than they actually are. Their emotional contact leans on the point of their communication for they are in a strange coexistence. As time goes by, the emotions they have kicked away will cumulate, returning in a sudden wave that cannot be stopped. There is no way to control the emotional nature of Libra, and this is something they will both learn in this relationship. 1%

Virgo & Libra Values: This is a couple that will value all the same things until they discover that this simply isn't true. It is a strange deceit that comes between them, as if they were both able to practice something entirely out of their reach. The guilt they will subconsciously feel for dismissing each other's (and their own) emotions will take out the satisfaction between them and leave them in the state of wonder. These two people who value so many similar things, value something entirely different by the end of the relationship. 30%

Virgo & Libra Shared Activities: Virgo is a mutable sign and they will adapt to any desire their partner has with ease. Libra is tactful and thoughtful enough for Virgo not to feel bad about their choices. There is rarely anything extreme about their contact and they will mostly enjoy the usual, relationship activities with one another. Even their pace can be synchronized well enough, as Virgo is much faster than the rest of the Earth signs, while Libra's indecisiveness makes them quite slow. They need to pay special attention when Virgo helps with decision making, taking even more of Libra's already bruised ego. This can lead to serious problems, not only in deciding on future activities, but in their mutual respect. 65%

Conclusion: In general, this relationship can sometimes work, and these partners can synchronize their pace, choose appropriate activities and build a satisfying sex life with enough patience and care. They could have a deep problem with emotional understanding though, and the thing they will find most difficult to reconcile is their fragile egos. Virgo, willing to please, will easily take over the responsibilities and decisions that Libra needs to take on. Virgo and Libra can form a very satisfying intellectual bond, for as long as they respect each other's feelings. This will lead to a feeling of inferiority in Libra and the loss of respect toward their Virgo partner. If this issue is left unresolved, their relationship might end because of disrespect they were both unaware of in the beginning. 30%

Virgo Compatibility with Scorpio

Virgo & Scorpio Sexual Intimacy and Compatibility: Even though Scorpio can be too rough of Virgo, making them feel uncomfortable and even violated in a way, in most relationships between representatives of these two signs, there is enough rationality to the approach of Virgo to make this contact possible. What we often fail to understand is the fact that Scorpio is a Water sign and as such is deeply emotional. Virgo looks for someone emotional to share a life with, and if they share this emotion of Scorpio through their sexual relations, they will both find sex between them extremely satisfying. If there is something Scorpio would like to fight for, it is the chastity of Virgo. This is a very interesting couple in the domain of sexual activity like one of them hiding their sexuality, and the other acting as sex itself. The best time for Virgo and Scorpio to create enough safety and emotion in their sexual encounters, is in the situation where they are each other's first truly emotional experience. If they surprise each other with the power of emotions beneath the surface, that both of them seem to carry around, they will have a hard time ever separating from one another. The biggest problem of these partners is in their relation to Venus, and this can lead to loveless acts of sex that both partners are not truly satisfied with. They need to show love and be tender enough, enjoying themselves enough, or they might have to move on to someone they love more. 65%

Virgo & Scorpio Trust: There is a strong understanding here, for one of them fears betrayal more than anything, while the other hates it and gets vindictive as soon as any sign of dishonesty is in sight. Trust is a very challenging issue for both of these signs and this is something they can finally talk about with each other. The best thing about their connection is in their ability to understand each other in silence, not ever wanting to let each other down. 90%

Virgo & Scorpio Communication and Intellectuality: Scorpio represents a deep silence of the flow of a river, and they will both have a strong urge to jump into the depths of silence together. Their intellectual contact is stimulating, often strongly influencing their sex life and their truly deep emotions. It is almost as if they wouldn't be able to form a relationship without this ability for non-verbal communication that makes them perfect for each other. Both of these signs are prepared to go all the way - Virgo in their intellectual depth and Scorpio in everything in life. Virgo is a talkative sign, ruled by Mercury the planet of communication, but they hold on to a much more quiet and intellectual side of Mercury than we might anticipate. This will inspire both of them to search for all sorts of answers together, analyzing each other's psyche and determining the source of their problems with the world, or with each other. Through carefully chosen words, they can help each other heal or regenerate from difficult or even devastating experiences. It is a good thing for both of these signs to have each other in the time of need. 99%

Virgo & Scorpio Emotions: We could say that this ability hides in both of them and the dig up goes both ways. The problem here is in the fact that they remind each other of their imperfections. Scorpio is a sign in which the Moon falls and at the same time the sign of Venus' detriment. All emotions get lost here, as if Scorpio is a black hole that cannot get enough. If someone can reach the emotions hidden behind the extremely rational approach of Virgo, it is Scorpio. Virgo partner is already sensitive and when in love, does everything they can to satisfy their Scorpio partner. This can feel like investing into a black hole with no gratitude whatsoever. Still, there is no other sign that can sense the needs of Scorpio better than Virgo, and no other sign that can dig up the emotions in Virgo better than Scorpio. An emotional relationship between them can turn out to be truly dark and difficult, but also incredibly strong and intimate. The only thing that can bore their emotions to death is the criticism they are both prone to. 75%

Virgo & Scorpio Values: Most of the time they will agree on things they value most, although they might stumble upon a huge problem when they get to the point of throwing out the trash. Even though Scorpio doesn't normally accumulate things, and loves throwing them away, those they hold on to can be quite disgusting to a Virgo. Both of these partners will value depth, intellectual most of all. There is nothing in the world that is as exciting as conversations that are so intense and so challenging for their minds. Just imagine as their first child is born and Scorpio wants to frame that dried out residue of an umbilical cord. Do you think Virgo would want to wake up to this in their apartment every morning? 70%

Virgo & Scorpio Shared Activities: Virgo will clean, that's a fact, simply because a clean house creates a clear mind, and Scorpio won't have much trouble fitting in, unless their personal belongings are questioned. However, when they choose places they want to visit, or clubs they want to go to, their choices will differ greatly. It is not hard for them to compromise to keep the relationship going, but it can be quite dark and demanding for both partners. For the same reason their values might differ, their daily routine might differ too. If they don't start willingly hanging out in some tidy cemeteries, they might run out of options that would actually keep them both interested and happy. 55%

Conclusion: In general, there is a problem that these partners share when it comes to Venus, and their relationship is often a reflection of these troubles. This can lead to all sorts of emotional blackmail, their tendency to control each other's lives, and if not this, than constant criticism that makes them both feel guilty or simply sad. The best thing they can do is decide that they will value each other and be thankful for each other in this relationship. That changeable nature of Virgo will be settled down by the fixed quality of their Scorpio partner, who will keep their relationship exciting for a very long time. If they develop a strong sense of gratitude, their relationship might be extremely deep, exciting and truly appreciated by both partners. 76%

Virgo Compatibility with Sagittarius

Virgo & Sagittarius Sexual Intimacy and Compatibility: Even though Virgo can be quite demanding and critical, especially from the point of view of Sagittarius, their sex life can be satisfying for both. The good thing about their connection lies in a fact that these signs are ruled by planets that also rule their opposing signs. This means that they will feel attraction and a need to begin a sexual relationship in the first place. Just like all mutable sign combinations, these partners could have a lot of fun. The main problem here is in the difference in their elements. Virgo is an Earth sign, and as such, doesn't often take too many risks. Sagittarius is a Fire sign, and they will passionately force things until they reach their goal. This doesn't work well in their sexual contact, for Virgo might feel pushed into things they don't want to do, and Sagittarius might be turned off by the practical and static nature of Virgo. The most important thing these partners should remember is that they both need room to be who they are. With two such giving people, sex life comes down to who will satisfy whom best, as soon as they deal with unrealistic expectations. 30%

Virgo & Sagittarius Trust: As friends, they can be unshakeable about their convictions and hold on to some traditional values together, but as soon as they start a romantic relationship, both of them seem to start feeling trapped. Virgo doesn't look like a zodiac sign that will easily feel trapped, but their mutable quality makes them impatient and always in search for change. They cannot be held in one place for long, any more than a Sagittarius can. If you dig for the biggest problem in the relationship between a Virgo and a Sagittarius partner, you will realize that it is their lack of trust, not only in each other, but in their entire relationship. The main difference being the degree of sacrifice they are willing to make. Out of these emotions, both partners will start feeling the need to be with someone different, and this is a relationship with probably the biggest potential for adultery, unless incredible guilt stops them first. Communication followed by mutual respect is their only chance of building a trustful bond. 1%

313

Virgo & Sagittarius Communication and Intellectuality: Virgo will bring all the little pieces into their intellectual connection, while a Sagittarius will have vision and help create a bigger picture. Even though they don't complement each other anywhere close to their opposing signs, the intellectual excitement will be equally important to them from the start. The most relevant fact for these partners to remember is that their respect is the most important thing to hold on to. If they disrespect one another, Virgo will observe their Sagittarius partner as a weirdo, stupid enough to run away from anything that has depth, while Sagittarius will look at their Virgo partner as a weirdo, stupid enough to hold on to irrelevant things. There is so much to say when Virgo and Sagittarius come together, and even though these partners might spark each other's need to talk excessively, all the time, they will both feel quite good about it. If they hit the zone of real understanding, they will be excited about the use of their minds and the beautiful conclusions and philosophy they can create together. They need to remember that each of them has a different role, and that for each role, these "stupid" characteristics represent the best possible base. 65%

Virgo & Sagittarius Emotions: In some rare situations, their mutable natures allow them to move in the same pace with enough respect to stay in an emotional bond that satisfies them both. Both of these partners are considered unemotional, but this is mostly because of their need to rationalize, analyze and use their minds to explain everything that happens to them, rather than rely on their hearts or gut feelings. This will often be a problem, for Virgo needs someone truly emotional so they can show their own deep feelings. This is not exactly a couple that will often end up in a happily ever after, even though they both wish to find the right person for this more than anything. In most cases, their vision of a fairytale ending differs too much for them to have it with one together. Sagittarius seems to be uninterested in needs of Virgo or simply unaware of them because they act as if they are purely rational. The trick here is for both partners to see behind the act in order to find each other's hearts and understand what they can expect from one another. 10%

Virgo & Sagittarius Values: This is why they will both treasure someone able to adapt, change and move, which is definitely something they will find in each other as they start their relationship. As highly mental signs, they will also both value clarity of mind and intelligence, in general. Still, their approach to intellectual value is different, and as much as Virgo values depth and detailed analysis, Sagittarius will value the width of one's mind. Virgo and Sagittarius will strangely have similar values based on their mutable quality. Even though they differ in other things they value greatly, Virgo valuing practicality and Sagittarius vision and focus, there is enough common ground here for both of them to feel good when together. 50%

Virgo & Sagittarius Shared Activities: This is why they will both treasure someone able to adapt, change and move, which is definitely something they will find in each other as they start their relationship. As highly mental signs, they will also both value clarity of mind and intelligence, in general. Still, their approach to intellectual value is different, and as much as Virgo values depth and detailed analysis, Sagittarius will value the width of one's mind. Virgo and Sagittarius will strangely have similar values based on their mutable quality. Even though they differ in other things they value greatly, Virgo valuing practicality and Sagittarius vision and focus, there is enough common ground here for both of them to feel good when together. 35%

Conclusion: There are many challenges in their way, the biggest being their emotional lack of understanding and their possible lack of respect. Still, when they find a way to show emotions and share them in the same pace and in an understandable way, they could actually have a lot of fun together. Their communication is often exciting and they both have a lot to say to each other, but their rationality may distract them from an actual search for love. The relationship between a Virgo and a Sagittarius is not a usual happy ending emotional story. If they discover how well they complement each other, they might be able to stay together for a long time. 32%

Virgo Compatibility with Capricorn

Virgo & Capricorn Sexual Intimacy and Compatibility: Very often these partners don't get to have sex, because they will have more reason not to, than to give in. The beauty of their sex life, when they manage to synchronize, is in the depth both partners are capable of, that directly links to the depth of emotions they will show through the act of sex. Their main goal is to find someone who doesn't take sex lightly, someone who is not superficial toward them and cherishes them as they should be cherished. There is a certain shyness to both of them, and this is something that will make them go crazy for one another, if they only reach the point behind the rational distance they normally share. Virgo will bring enough change to their sex life, as a mutable sign, ready to experiment with someone who is so reliable and respectful. A sexual connection between a Virgo and a Capricorn might be great if they both weren't so stiff and strict when it comes to sex. Even though they don't lack the patience or the understanding for each other, there always seems to be just that one shred of pure emotion missing in their contact. This is a perfect relationship for both partners to relax and try out new things, if they find a way to open up in the beginning. 65%

Virgo & Capricorn Trust: There is nothing shady about them, nothing unreliable or quick to turn to deceit. Virgo usually has no reason to be unfaithful, except when they suffer from their own lack of faith and emotion that cannot be controlled. Even if this is the case, a Capricorn partner will inspire them to be the best they can be, and as faithful as possible. Capricorn is a sign that can be trusted, and Earth signs understand this best. They will need some time to get used to each other's habits and build the trust they both wish for. When they do, they will rarely break it for anyone or anything else. 99%

Virgo & Capricorn Communication and Intellectuality: The flow between Earth signs in its clearest form is sometimes unbearable for other zodiac signs and this is something Virgo and Capricorn truly enjoy. The depth of mind in both of these signs will fascinate them at first, excite them and make their communication incredibly interesting and informative. They both like a good, respectful debate, and in each other, they can find a perfect adversary. These are signs that make one cycle of communication complete, Virgo deciding what's next to discuss and Capricorn deciding when the subject is resolved. They are like a perfect mechanism, like gears fitting in together to solve any equation the world has to give. When someone from the Air or the Fire realm observes these two as they talk, this conversation might seem extremely boring. Their passion lies within these roles and when they find an understanding in other areas of their relationship, the intellectual one can be stimulating to the point of absolute bliss. This is a couple who knows that any problem is there to be solved and anything broken is there to be fixed. 90%

Virgo & Capricorn Emotions: They do have some emotional issues, but not the same ones, and this helps them find an approach to each other that they both understand. Their relationship needs time, most of all, and the emotions between them need to build, just like trust. With the calm, practical, physical passion rising between them, both partners start building their confidence. With confidence, they feel much more liberated to experiment in life and sex, and this gives true quality to all of the areas of their relationship. The most incredible thing this couple shares is their discovery of one another. In time, they will peel layer by layer from each other's hearts, and be more and more fascinated by what none of them noticed before. Both of these signs are considered unemotional. Virgo brings Venus to its fall and Capricorn is the sign of the Moon's detriment. Just like a mathematical equation, they represent a mystery box to each another, and they need to open it, bow by bow, side by side, until they unravel the treasure hidden inside it. 65%

Virgo & Capricorn Values: They value depth and this is something they will find incredibly soothing in one another, for they will both feel like they don't need to pretend to be shallow anymore. They will both value practicality, grounding, money and rational investments. The main difference they have to resolve here is in the value of

Capricorn's goals, for they might be ready to do too much from Virgo's point of view, in order to reach them. Both Virgo and Capricorn will value calm, rational behaviour and choices, and one's ability to remain smart however unbearable the situation might be. In return, Capricorn doesn't understand Virgo's lack of motivation and their lack of need to claim the leading position. 80%

Virgo & Capricorn Shared Activities: Even though they are both Earth signs and this will allow them to follow each other with the same energy, they don't connect that well on the choice of places they want to visit. They will both want to go to a history museum and learn a lot of information there, but Capricorn often doesn't want to deal with doctors, let alone the calorie counting and green tea. There is no other sign to understand Virgo's need to sacrifice better than Capricorn, but if they don't feel responsible for their partner's activities, they will rarely follow them to depression. Where Capricorn wants to go up, Virgo wants to go forward. The most important thing to do here is hold on to positive activities and to a routine that makes them both healthy and happy. 65%

Conclusion: Even if everything between them seems too slow for some other zodiac signs, they build respect, trust and love, on the foundation of mutual analysis and detailed examination. Virgo and Capricorn belong to the element of Earth and follow each other's pace perfectly. The search for perfection can be ended in this relationship, for they give each other enough time, and listen to each other well enough to meet the expectations that need to be met. Both of these partners can be stiff and lose sight of the importance of the emotional, mellow approach to life, and this relationship can make them rough and too strict. Still, in most cases, they will give each other enough time to grow out of this and grow old together. 77%

Virgo Compatibility with Aquarius

Virgo & Aquarius Sexual Intimacy and Compatibility: Their natures find it very hard to support each other, they are both intellectual but in a completely different way, and they will probably ruin any chance of a good sexual relationship by over thinking everything, each of them in their own direction. Aquarius really holds on to their spontaneity, but who's to say what does "spontaneity" really mean? They are rational when it comes to sex, and yet spontaneous? That sounds strange, doesn't it? The truth is, they choose when to be spontaneous and their intellectual strength often gives them the image of spontaneity because they've seen the result faster than other, "not-so-spontaneous" people. This is in no way an easy sexual relationship and unless some strong support is provided by their natal charts, Virgo and Aquarius will rarely be attracted to each other enough to start a sexual relationship at all. This could ruin their sex life, because Virgo is one of those people and will usually think long and hard about starting a sexual relationship. Over thinking is a true turn-off for Aquarius, although they do it too just in a quicker pace, and they will rarely find Virgo's analysis sexy in any way. Shy, thoughtful, sensitive Virgo will have trouble understanding their nude, weird and often too fast Aquarius partner. It is almost certain that none of them will have enough patience to build their sex life with someone so different from what they need. 1%

Virgo & Aquarius Trust: Although Virgo can be tough on trust, with Aquarius it is obvious even to Virgo that a lack of trust would lead nowhere. Still, there is a great chance they will drift apart, even when their first contact is passionate and strong, because they could simply start feeling that both of them need someone different. Their rational natures usually connect them in a trusting relationship, because they both find it stupid to lie or not trust their partner. If they want to hold on to their mutual trust, they need to keep their relationship fresh and accept each other for exactly who they are. 50%

Virgo & Aquarius Communication and Intellectuality: Aquarius has no trouble dealing with Mercury, in most situations, and will most certainly like this adaptable Virgo quality. As a fixed sign, Aquarius is usually set in their ways and this can be difficult to handle for someone so willing to sacrifice their happiness, such as Virgo. Their communication should mostly be good and their topics similar. They will share interest in many things and usually be excited about similar details. However, they belong to the most different elements of all – to Earth and Air. As an Earth sign, Virgo can be very slow, too thorough and rarely inspired enough for Aquarius to even feel the need to share their ideas with them. On the other hand, Aquarius can seem unrealistic or even crazy to their Virgo partner. The best way for Virgo and Aquarius to function and be satisfied with their relationship, is to take each other seriously enough. With Virgo's attention to detail, any plan for Aquarius' brilliant mind to come to life is possible. Virgo is ruled by Mercury and has a mutable quality to it, which gives it this changeable, moveable and adaptable nature. Their communication might be tricky and the lack of compassion Aquarius suffers from will sometimes hurt Virgo, but they still have a great opportunity to combine their minds and form a universal intelligence capable of creating anything at all. 40%

Virgo & Aquarius Emotions: In case they do fall in love, they will have to deal with a constant fight for freedom and routine. Virgo is a sign of health and our daily routine, and its representatives can often be obsessed by their every meal or every check up they've had with their doctor. Aquarius will in most cases avoid doctors at all costs and the fact that they exalt Neptune will mostly turn them to all sorts of alternative medicine, rather than anything typical that Virgo might hold on to. The entire emotional world of their relationship could come down to Virgo worrying for their irresponsible Aquarius partner, and the lack of gratitude they might get in return. The emotional rollercoaster Aquarius gladly offers is something Virgo will probably despise. In fact, Aquarius usually doesn't need to be taken care of in this way. This is a complicated emotional relationship because the worrying of Virgo degrades the personality of Aquarius and the best of intentions could have damaging consequences. The biggest problem in the relationship of Virgo and Aquarius is in the fact they both heavily rely on their rational mind. This leaves no room for the joy of seduction, love and satisfaction, and usually they both need a partner with more warmth, life or emotion to them so they could both be happier. 35%

Virgo & Aquarius Values: This doesn't mean they will find the same people, actions or thoughts intelligent and they could often have opposite opinions on someone in their surroundings. Still, they could motivate each other to develop their intellectual strengths and hold on to this asset if other things in their relationship aren't that good. Virgo and Aquarius will both value intelligence and a clear mind most of all. Other things they value aren't really similar and while Virgo would always choose practicality, Aquarius would choose the unknown and a not so understandable reality. 30%

Virgo & Aquarius Shared Activities: Their taste in many things can be almost the same, because the same attention to detail Virgo cherishes so much, makes some people great in their art and this is what fascinates Aquarius. The work of any artist with a great mind could connect them and they could easily be interested in similar shows, galleries and plays. However, Virgo is too cautious and predictable and most of the time they will have trouble fitting in that Aquarius' too exciting, unpredictable world. 25%

Conclusion: Their strongest meeting point is in their rationality and communication, and this can be used to overcome many problems that their differences result in. Unfortunately, in most cases they will not have enough chemistry to start a relationship, let alone stay in a sexually satisfying one for very long. Aquarius runs everything from like practical, worried about health and earthly things, down to Earth, cleaning obsessed maniac. Imagine how incredibly irresponsible, chaotic and unrealistic Aquarius looks to them. If they take each other seriously, they might create incredible things together, as their great minds merge. 30%

Virgo Compatibility with Pisces

Virgo & Pisces Sexual Intimacy and Compatibility: Since they also represent the axis of Venus' fall and exaltation, we can conclude that their relationship always has a lesson on Venus to teach. These partners have a task to find the place of physical intimacy in which they will both be relaxed to be exactly who they are. Virgo partner will usually be shy, trying to show their sexuality through rational behaviour, and Pisces will see right through this. On the other hand, Pisces will fear close physical connection with another person, and this will be practically dismissed by Virgo. As they both learn that they cannot hide who they are, they will have no choice but to set themselves free from any fear and shame, giving in to the wonderful sexual experience Venus has to offer. Virgo and Pisces are opposing signs and their attraction is very strong. This is a couple that will never have instinctive sex, however passionate they might get. Virgo's analytical mind wouldn't allow for them to act "like animals", and this is something that Pisces will find so humanlike and attractive. Virgo will mostly be attracted by the purity of sex with Pisces, who truly approach it as an act of love, free from prejudice and following their inner feeling, wherever it leads. 99%

Virgo & Pisces Trust: In order to have a healthy relationship in which they both trust each other, they will both need to be secure and confident enough to be honest. Both of these partners will easily give in to dishonesty, although their convictions are the opposite of their behaviour. It can be torture for both of them if any one of the partners tells a lie. Virgo can have some serious trust issues that Pisces won't actually help them overcome. Fortunately, they are both aware that some secrecy might even spark their relationship and give it more passion. This is why they will usually get over small intimacy outbreaks in order to trust each other on a higher level. 65%

Virgo & Pisces Communication and Intellectuality: When they start their relationship, they are bound to realize how similar they actually are, even though they seem so different. The mutable quality of their signs will allow them to jump from topic to topic, both of them staying interested in the flow and the outcome of their conversations. The best person to pull Virgo out of their obsessive analysis is Pisces, with their smile and their wider picture. Pisces will give their Virgo partner faith, teach them how belief can form reality and help them be free from too much caution and fear from failure. They complement each other best through communication and intellectual stimulation. Virgo often has this inner battle in which nothing they know, think or do is good or valuable enough. Pisces are able to inspire and find value in everything in life, and those insecurities and emotional problems of Virgo may seem like something needless that damages the self-esteem of everyone around them. They will do their best to help their partner reach the point of inner security in which they understand their worth. In return, Virgo will help Pisces reach the actual materialization of their incredible talents. They might do so through nagging and constant criticism, but in the end, Pisces will have many things to be thankful for. Not only does this relationship represent the axis of Venus' exaltation and fall, but it also represents the axis of the exaltation and fall of Mercury. As much as Virgo has trouble with Venus, Pisces have trouble with practical Mercury and their mind can send mixed signals making them lost and confused. Virgo will help them build an inner sense of intellectual security, in return for their emotional one. 85%

Virgo & Pisces Emotions: Understanding between them could reach points of perfection and this is something that both of them will probably be unable to find with anyone else. Still, it is important for them not to build too much expectation around this idea of perfection, for no one can meet these requirements in real life. If they go astray, losing touch with their partner's true personality, they will easily get disappointed and have the need to end their relationship. There is no other sign in the zodiac that can awaken the emotional depths of Virgo better than Pisces. With two mutable signs everything moves fast and changes are inevitable. If they want to remain in a stable relationship, they need to find a fine balance between rationality and emotion, reality and dreams and love each other as if they were perfect just the way they are. 95%

Virgo & Pisces Values: This is their meeting point and it can make them divine, or constantly dissatisfied with the need to change everything about their partner. Both of these partners will value flexibility, someone's ability to adapt and change, and they will most certainly value the love they get from their partner. They share a great love of perfection. As much as Virgo will value one's perfect mind, Pisces will value a perfect emotion. Their differences in approach to one's beliefs and convictions might be huge, and acceptance between them needs to be unconditional. 75%

Virgo & Pisces Shared Activities: They are after all ruled by Mercury, and have this need to see and feel everything that this Earth has to offer. Virgo can be grumpy about doing anything from a fairytale of Pisces, but they will follow out of curiosity. Once they get into the magical world of Pisces, they could discover the beauty of life they were fully unaware of. When these partners are together, they make each other feel like anything is possible, for Pisces understand endless possibilities as much as Virgo makes things come true, understanding reality better than many others. 99%

Conclusion: This makes them partners with greatest challenges and the greatest potential for love in the entire zodiac. They need to find a fine balance of rationality and emotions, each one individually and together through their relationship. In many cases this is not a couple that will last very long, as their mutable quality makes them changeable enough to disregard the entire relationship quickly if they aren't satisfied. Virgo and Pisces represent the axis of the exaltation and fall of both Venus and Mercury. They need to realize that perfection they seek might not be presented in the form they expect. If they stay together for long enough to understand the benefits of their contact, they might discover that the love between them is the only true love they could find in this lifetime. 86%

PART-7
LIBRA HOROSCOPE

(September 23rd to October 22nd)

Who is easygoing, sociable and keeps you waiting for half the day?
Who puts you off with promise gay and compromises all the way?
Element: Air
Quality: Cardinal
Ruler: Venus
Lucky Gem: Diamond and Blue Sapphire. He/She does love the real thing, and wearing a diamond can bring him/her wealth and can make him/her a better person.
Lucky Number: Number 1, 2, 4, 6, 7 & 21 will fill with joy.
Lucky
Colour: Blue, Green
Lucky Colour: Royal Blue. He/She is can rule over the world with royal blue.
Day: Tuesday, Friday
Lucky Day: Tuesday. Pack your bags for a holiday on Tuesday.
Lucky Flower: Aster.
General Insights: People born under the sign of Libra are peaceful and fair, and they hate being alone. Partnership is very important for Libra -born, and with their victorious mentality and cooperation, they cannot stand to be alone. The Libra is an Air sign, with expressed intellect and a keen mind. They can be inspired by good books, insurmountable discussions and interesting people. The ruling planet of Libra is Venus, who is a lover of beautiful things, so the quality is always more important than the quantity for people born under the Libra sign. They are often surrounded by art, music and beautiful places. They are cooperative by nature, so they often work in teams. Libra is fascinated by the balance and symmetry. Libra-born prefers justice and

equality, and they cannot tolerate injustice. They avoid indulging in all types of conflicts and prefer to keep the peace, where this is possible. They like to do everything in pairs and not alone. The biggest problem for Libra-born is when they are forced to choose sides, because they are very indecisive and sometimes they forget that they have their own opinion.

Personal Quality: The Libra is diplomat, impartial, sociable, cheerful, charming and sensitive to the needs of others. He/She is affectionate, polite in behaviour, cooperative, peace loving and sacrificing. He/She is natural arbitrators and diplomats, justice, honest and hard worker. He/She can win over his/her staunchest enemies with the help of his/her sweet voice. He/She has an idealistic and generally peace loving nature. He/She is the most civilized of the twelve signs. He/She has financial stability in the life. He/She is sure to have properties of his/her own but keep away from others' properties. He/She is more interested in making friends than enemies, and is willing to go along with others and do whatever it takes to maintain a relationship. He/She is a sensual lover and does not like any interference in the matters of love and marriage. Discord makes him/her totally insecure, and uncomfortable. He/She likes harmony in his/her life, and will do whatever it takes to have it.

Positive Quality: He/She always maintains a sweet relation with others. He/She is usually sympathetic, cooperative, diplomatic, gracious, fair-minded, social, kind, loving nature and artistic. He/She does not like injustice, quarrels and disagreements. He/She is fair minded and loyal and have reach taste or good sense of humour. He/She does not hurt other person's feelings and likes to help the person in need.

Negative Quality: He/She is insincere, indecisive, avoids confrontations, will carry a grudge, self-pity, and jealous and likes self admirations. He/She does not have argument power well even he is right. He/She tries to keep away from truth and painful experience. He/She can be frivolous, flirty and quite shallow. He/She is fickle minded, dependent, indecisive, sulking, and likes peace at any cost.

Physical Appearance: He/She is smart and attractive. He/She has sweet open faces with laughing eyes, and a devastating smile. He/She tends to be plump rather than angular or skinny.

Relationships: He/She highly understands his/her companions and he meets her with his own innate optimism. His/Her married life is delightful with the Gemini and Cancer girls. He/She gets special co-operation from Gemini, Aquarius, Sagittarius and Leo natives. He/She is not very good at handling relationships.

Career: He/She gets success in life as a businessman, engineer, lawyer, chartered accountant or a doctor. He/She is attracted to careers in the luxury trades, including fashion, beauty and design. He/She also makes good diplomats and counsellors. He/She is successful as writers, composers, fashion designers, interior decorators, critics, administrators, lawyers, and in civil services.

Health: This sign rules the kidneys, the lumbar region of the spine, the skin, the urethras, which are the tiny ducts running between the kidneys and the bladder, and the verso-motor system. He/She tends to suffer from nervous stomachs and ulcers. He/She needs to drink plenty of water in order to flush out the toxins from kidneys.

Ideal Partner: He/She gets along with the best ones and the gentle that captivate attention for life.

Ideal Partner: He/She gets along with the best ones and the gentle that captivate attention for life.

Compatible Zodiac Signs: Gemini, Aquarius
Incompatible Zodiac Signs: Aries, Cancer and Capricorn
Greatest Overall Compatibility: Gemini, Aquarius
Best for Marriage and Partnerships: Gemini,
Libra likes: Harmony, gentleness, sharing with others, the outdoors
Libra dislikes: Violence, injustice, loudmouths, and conformity.

BUSINESS ASTROLOGY FOR LIBRA

Good contacts and tactful dealings help Libra individuals to get investors and clients in no time. But their indecisiveness and impatience can shock others. A positive attitude with determined approach is their key to success in Business world. Business booster Gem is Coral.

Libra Compatibility with Aries

Libra & Aries Sexual Intimacy and Compatibility: Attraction they feel toward each other is great, but their signs combined present passive-aggressive behaviour in general and as a couple they could have a tendency to hurt each other in intimate relations. Ruled by Mars, Aries is a sign of Saturn's debilitation and Libra exalts it, so their main issue is the lack of emotion and poor boundaries when it comes to sex. Saturn can cool things a bit too much and be a challenge to overcome in their attempts to get sincerely close. Although Aries and Libra are both signs of masculine nature, they are a primal opposition of the zodiac and present a relationship between Mars and Venus, planets in charge of our sex life. When they engage in intimate relations, it is expected for all their libido and possible problems with sexual expression to surface. When they connect through real emotion and respect each other's boundaries, they have a potential for a very good sex life, as Aries gives initiative and energy to indecisive Libra, lifting their libido and Libra awakens the fineness of Aries, teaching them how to be selfless lovers and enjoy thinking about the satisfaction of their partner. 80%

Libra & Aries Trust: Libra partner has a problem with insecurity in general and needs to show their worth through relationships with different people. They love to be loved and seem to be hungry for the approval of those around them. Aries finds this stupid but easily gets jealous and threaten their mutual sense of stability and belief in other person's choices. Due to Libra's lack of confidence, it is also possible that they will doubt everything their partner does. Since Aries doesn't put much time or thought into their actions, the lack of conversation about every single detail from their personal life could easily arise suspicion in the mind of a Libra. Trust is not their forte and problems with it could torture them for years. The most important thing here is for Libra to work on their self-esteem and keep their focus on their own life instead of trying to blend into the life of their partner. 40%

Libra & Aries Communication and Intellectuality: Their opposition covers the points of debilitation and exaltation of Saturn and Sun, and this is mainly shown in their communication and everyday functioning. This means that their role in each other's life is quite simple like Aries needs to boost their Libra partner's spirits all the time, showing them how capable and brave they can be, while Libra takes on the responsibilities of their Aries partner and shows them how to reach a certain goal. All of this can be quite tiresome at times, especially if one of them has a problem with this unconditional role play, or doesn't recognize the effort of their partner. Aside from possible conflicts Libra tries to flee from most of the time, their communication usually serves to feed the hungry Sun of a Libra partner, or Aries' hungry Saturn. Mostly they will talk about their daily activities and events since they don't share many interests. While talking about different activities and people, they find a common language as Aries helps Libra not to obsess about others and Libra helps Aries to understand different views than their own. Their communication might be great if they were in the same profession or at least share a workplace, because that would cover the basic interests they share and give them more space to find the middle between their opinions. 55%

Libra & Aries Emotions: This is something every Libra needs, as they have trouble letting their guard down. Libra, on the other hand, has enough depth to look inside Aries personality instead of superficially examining their behaviour. Of all the zodiac signs, Libra is probably one of the few that have good understanding of the nature of Aries. They don't understand their actions and their way of display of emotion, but the core of emotion and sensitive personality is easy to reach from their perspective. As crazy as this may seem with lack of qualities their relationship might suffer, this is a couple that understands each other very well when it comes to emotions. Aries can awake Libra's ability to show them because of their own openness. It is safe to say that this is a couple that could solve any issue with love they have for each other and although their troubles could be great, this is possibly such a deep emotional connection that all problems fade next to it. 99%

Libra & Aries Values: Capricorn is ruled by Saturn that exalts in Libra. In the practical sense, this means that Libra helps Aries achieve their goals, while following necessary values. This is an interesting observation because the sign of Aries is the sign of Saturn's debilitation and doesn't seem to understand the set of values or exact steps that would lead them to their goal. It is almost as if Libra knew the way for Aries to reach their goals by discovering new values in relationship with them. Our values set the direction that leads us to our goal for personal development. Aries has a goal in the sign of Capricorn for this is the sign in their tenth house. In general, their individual values are different in so many ways, but it is exactly the purpose of their relationship to question them and set them straight. Aries values direct, energized approach and outspoken people. Libra values tact, fineness and prestige. While Aries gives their best to live in the now, Libra examines the past to set distant targets in the future. They have a lot to learn from each other, but if they do, they might just set their mutual values somewhere in the middle. 70%

Libra & Aries Shared Activities: They want to do opposite things most of the time, and the only activity they always agree on sharing is sexual activity. Although this is a pillar for a good relationship and everything else they cannot share might seem irrelevant for some time, they need to find a way to do something else they both enjoy. Even if they don't, their relationship might work, but only if Libra partner lets go of their idea that they need to include their partner in everything they do. This is the couple that finds it very difficult to coordinate their activities. Aries could help Libra by supporting their independence in any possible way, while accepting involvement in a part of activities Libra cares about. 30%

Conclusion: Aries and Libra are the couple of the zodiac, as much as any other opposing signs, for they are each other's seventh house, house of relationships. Even more so if we acknowledge the fact that Libra is the sign of relationships in general. Any problem they might have with each other is something to be worked on, because it shows what their personal problem with any relationship is. When they are madly attracted to each other and fall in love, there is almost nothing that could separate them, no matter the differences. However difficult it might be to reconcile these two natures, remember that this is a primal opposition that represents partners by signification. Wouldn't we all like to find the middle ground with our loved one? They need to work on their bond, that's a fact, but their relationship is a promise of a perfect fit of two souls meant to be together. 62%

Libra Compatibility with Taurus

Libra & Taurus Sexual Intimacy and Compatibility: While Taurus loves to be comfortable and relies on their sense of touch and taste, Libra will want everything pretty and rely on their eyesight and the sense of smell. They do connect in a way, but in most cases they have this different approach to Venus as a planet of sexual pleasure. The main difference between these signs is in their exalting planet. Taurus exalts the Moon and Libra exalts Saturn. It is like they adore opposite things and while Taurus will care for emotions and tenderness in a sexual relationship, Libra will rely on its depth and good timing. It will not be easy for them to understand what the other person wants and they could both end up seeming needy to one another like Taurus to Libra because of their emotional neediness and Libra to Taurus because of their physical one. The relationship between Taurus and Libra has a special kick to it, since both signs are ruled by Venus, but represent its completely different characteristics. However different they might be, they are still two signs ruled by Venus and can be fairly attracted to each other. As a feminine and a masculine sign, they could mend their differences and try to learn about "the other side of Venus" instead of expecting the impossible from one another. They are both gentle lovers who like their relationships without stress and drama, so with enough patience they could be a really good fit. 35%

Libra & Taurus Trust: Even if they are Libra's first choice, there is still a matter of flirting with so many other people. With really insecure Libra specimens, it is almost impossible to have a trusting relationship, for their need for acceptance can go a long way and even suck them into unfaithfulness. Taurus' trust can be damaged with Libra's need to be liked by everyone, especially if they are not sure if they want to be with their Taurus partner in the first place. If Libra can't make this basic decision, the uncertainty would be a punch for the ego of Taurus and it will be very hard to recover after they realize they are not wanted with certainty. Their quality however, is in their outlook on justice, and they will rarely act on their insecurities, but still, who could be sure when the vibe is so unstable, especially when someone as stable as Taurus tries to blend in. 30%

Libra & Taurus Communication and Intellectuality: On one hand you would have Taurus, never doubting their character, never moving and annoyingly unchangeable. On the other, you would have Libra, indecisive and never certain of what they want. It seems impossible for them not to jump on each other's nerves daily. The main challenge here is their primary opinion on each other. As two sides of Venus, these signs represent a peasant girl (Taurus) and a city lady (Libra). When you combine this with the fact that Libra has a fallen Sun, their ego problem would easily make them feed on their refined, city image and they could criticize the "peasant girl" for her lack of style and her rusticity. These two will drive each other crazy. They simply don't understand that the fact that Taurus likes things comfortable doesn't always reflect on their lack of style and can't accept the differences in their appearance. You could say that this is quite superficial, but it is actually a really deep problem with insecurity. Libra is not criticizing Taurus because they are such a fine person, but because they are afraid they are not. Taurus will easily get insecure because of this critical view and dip into their guilt trip, almost as if they always searched for someone to wake the guilt in them, but without the ability to change and accept criticism as constructive. Even if this is not something any of them will say out loud, it can be felt in their relationship, even by those around them. 5%

Libra & Taurus Emotions: Although Taurus and Libra both are looking for someone to sweep them off their feet, they will rarely have this with each other. Taurus will usually decide not to give enough space for Libra to discover them, while Libra will spend too much time looking for faults. If they are attracted to each other just enough they could fall crazy in love, but in most cases, they are both too careful to end up in a loving relationship. As signs ruled by Venus, they are both complemented by signs ruled by Mars and normally look for a partner with initiative, to have a fast, exciting start and not get enough time to think things over. 25%

Libra & Taurus Values: Of course they both want true, magical, mystical love with Venus exalted in Pisces, but Taurus values tenderness and touch on their way to get there, while Libra values responsibility and seriousness. However, their final goals are the same and they do value one's ability to love them, most of all. Venus, their ruler, represents value itself, so we could say that they value same things because of the similarities shown through what their ruler likes or not, but in different ways. This could be their real point of connection if they do fall in love with one another. 40%

Libra & Taurus Shared Activities: Although Taurus doesn't really have to go to matinees and art shows with no soul, they could have a nice time at a cosy art gallery where art work in warm colours is exhibited. Libra will rarely go for a walk in the mud, but they could make a tour downtown, where they could both be seen wearing their new outfits. It is not hard to find activities to share for this couple, for as long as they are not boring to one of them. With their mutual love for beautiful things and love in general, they will find a nice way to spend their time together if they are open enough to make some changes to their usual routine. 65%

Conclusion: Taurus should be careful, too, for their need to feel guilt could blossom with a Libra. This relationship is a lesson both of them will never forget, especially if they manage to build enough understanding and tenderness between them. Taurus is to wake your inner fears and bring them all to surface. The only way they could ever be happy would be to embrace what they don't want to deal with in their own inner worlds. If they do this, well you can imagine what a Venus complete would be like. 33%

Libra Compatibility with Gemini

Libra & Gemini Sexual Intimacy and Compatibility: When it comes to sex, this is a plus, for they will both be free to communicate anything that bothers or satisfies them. Fragile ego of Libra can be lifted by Gemini's charm and approach to sex. They seem to know how to make everything a little less serious and this will help their Libra partner open up and share their emotions through sex, too. When Gemini partner sees how mellow Libra can be, although they seem so tactful and cold, they won't have much choice but to share emotions as well. The main goal in their sexual life is, in fact, to balance these emotions. Libra is a sign ruled by Venus, sexual, sensual and seductive, while Gemini is ruled by Mercury, having no sexual or emotional wisdom. The basis of a good sex life between them is their curiosity, for Libra is always curious about their partner, as much as Gemini is curious about everything else. Gemini and Libra are both guided by the element of Air and this should give them a good start for their mental connection and verbal understanding. Although Libra might be indecisive, Gemini won't have a problem thinking of a different approach and finding new techniques, words and adventures to spice up their sexual encounters. 80%

Libra & Gemini Trust: When they choose a partner, they choose them for their character and their straightforward nature. Libra has no reason to doubt their own judgment and will probably believe their Gemini in every case, except when their dishonesty is too obvious. In return, Gemini will respect Libra's need to flirt in order to be accepted and loved by other people. When Libra decides to be with someone, even if it is after a long, hard inner battle, they will probably believe in their words and their actions. Not only will they not find this threatening, but they will actually enjoy a consequence of this behaviour – their own freedom. 95%

Libra & Gemini Communication and Intellectuality: Although Gemini will, in most cases, just follow their rational nature and comment on things simply because they want to talk, it will be hard on Libra to overcome some of the things they might say. The sign of Libra is very sensitive to any sort of will imposing or criticism and will recognize it even when Gemini has no idea what their partner is imagining. Who's to say which one of them is right? Since no two people can agree on everything every time each of them opens their mouth to speak, Gemini and Libra can have a very hurtful and tough communication due to the mutual lack of tolerance. Gemini partner would rather go blind than accept that they have a lack of tolerance, but the truth is they can be quite strict when it comes to someone's mental activity and their opinions. Libra is in most cases hurt enough by the pressure on their personality produced in their primal family, so they will have a very bad response to this behaviour even if Gemini meant nothing wrong. When Libra decides to be with someone, even if it is after a Gemini partner is very opinionated and Libra has a tendency to take a lot of things their partner says as a personal insult. The main issue here is in the fact that Gemini lives to learn and teach what they have learned. They often present themselves as someone who knows things, and Libra can see this as their need to prove their intellectual dominance. Even though this isn't their intention in most cases, sometimes a single sentence Gemini said can be something Libra feels hurt by for years to come. Libra can learn from a teacher, from someone who has proven their worth, but hardly from their know-it-all partner. 60%

Libra & Gemini Emotions: They seem to be in sync while Libra partner searches for depth, and Gemini flies around looking for a new discovery. They won't even notice as love between them starts to show, one of them running around and the other thinking about reasons why they wouldn't be perfect together. We could say that neither of them is very emotional, but Libra is ruled by Venus, so there is a strong link to an emotional plain here. The problem develops when they both talk too much about their emotions, while none of them stops to actually feel. The way Libra's emotions develop is something really suitable for Gemini. They can remain detached and distant, unless Libra falls in love deeply enough to follow their Gemini partner wherever they go, and Gemini falls in love deeply enough for all words to lose meaning. 90%

Libra & Gemini Values: Gemini will value someone's creativity and intellectual strength and this is something Libra can't respond to if their Sun is in its lowest state. Their meeting point is in their value of intellect, however strange it may seem. They are both Air signs and give a lot of attention to their partner's mental personality and the way they think. While Libra will value consistency and someone being responsible and reliable, Gemini will be very different from this, with opposing values as well. They could find a way to tease each other with words, seduce each other and in the end find a way to communicate everything else – in case they both care enough. 55%

Libra & Gemini Shared Activities: They want to examine the world of their partner, beginning to end, and will gladly follow them around in all their activities expecting the same in return. However, this can end in an unfortunate way, leaving Libra with their energy drained and wishes not granted. Gemini just doesn't care about following their partner only because they want to be followed. Both of them need to keep their expectations low and let their partner surprise them by something new and exciting. Libra is your typical activity chameleon when they fall in love. They will enjoy various activities together, but in order to do so, they need to respect each other's limits and desires to begin with. 85%

Conclusion: If Libra partner has trouble being alone and doing things by themselves, this isn't something Gemini will easily understand. Due to their lack of personal boundaries, Libra representatives will often let their Gemini partners lead the way until all of their energy is gone, they feel like they should only lie down and turn their brain off. Gemini and Libra partners are not exactly always a perfect couple, although their signs support each other. If they want to work on their relationship and be happy, Libra needs to respect their Gemini partner enough to let them be their teacher, lover and a friend. In return, Gemini will have to take care of their Libra partner, respecting their limits and their need for togetherness. 78%

Libra Compatibility with Cancer

Libra & Cancer Sexual Intimacy and Compatibility: Both signs have trouble accepting Mars and this leads to a lack of passion and initiative in their sex life. However, the tactful and careful nature of Libra can really soothe Cancer. This is mostly because of the fact that Saturn, the ruler of Cancer's opposing sign, is exalted in Libra. Although some initiative might be lacking, there is a chance for Cancer and Libra to function very well in their sex life if they spend enough time together. The problem with their sexual connection is in their element, more than anything else. Cancer belongs to the element of Water and Libra is an Air sign. Although Libra partner might be extremely patient and nice, there is still a speed to the element of Air that Cancer might have trouble adjusting to. At first glance Cancer and Libra might seem as if they are really far apart. Things they would like to try out will differ greatly, for Cancer needs emotional connection and Libra needs contact, touch and experience before they get too emotional. Libra is not often inspired by the nature of Cancer and won't normally fall in love with them at first sight. Their sex life can be very good if they already share deep emotions, so it would be best if they started a relationship out of friendship, already knowing each other to some point and sharing some feelings besides possible attraction.40%

Libra & Cancer Trust: While Cancer wants a quiet family life with no interference from other people, Libra can't seem to stay away from other people, seeking their affection and approval day after day. At some point their Cancer partner has to ask themselves if this is the kind of partner they want to have children with. Cancer can be one of the most trusting zodiac signs, but there is an irritating side to Libra's nature that they rarely stay immune to. On the other hand, Libra finds the whole approach to a romantic relationship Cancer has a bit unrealistic. This can easily lead to a lack of trust in Libra, especially if the Sun is in final degrees of the sign. 30%

Libra & Cancer Communication and Intellectuality: Their communication will not be too difficult, but there is a chance they won't share many interests or respect each other enough for it to have true quality. Their main problem could surface if they start making unrealistic plans together. The expectations of Cancer with Capricorn as their opposing sign would be extremely practical and strict. Libra does exalt Saturn, but it is not an Earth sign, and usually they remain in the field of ideas instead of a practical approach to materialization. Cancer doesn't understand how someone can be so far from reality and this could lead to a strange passive conflict that could endanger their entire relationship. As signs ruled by Venus and the Moon, it is safe to presume that their relationship is in a way very important. Although Libra is in charge for the upper, spiritual nature of Venus, it is still a sign of relationships and the Moon will only emphasize the need for closeness and harmony. Cancer needs to realize that Libra has its place among the Air signs, in the field of ideas and not necessarily their realization. It is important to remember that Libra has a troubled Sun and usually looks for a partner with more fire and passionate energy than Cancer normally has. However ugly that might sound, they need someone to feed on, and they normally have strong relationships with fire signs which produce enough energy even for their loved ones, or Air signs who don't really care about it. The sign of Cancer unconsciously calculates and distributes energy to their inner priority list, and they will rarely have an excess big enough to shower Libra in it. 50%

Libra & Cancer Emotions: However, their emotional context is very different, for Cancer is looking for a love on Earth and Libra is looking for someone to take to heaven. This is exactly where the spiritual side of Libra's Venus takes away their true chance to share a life together. It is a Venus that doesn't need to eat or sleep, for as long as it feels the unique, balanced love. Cancer will understand this idea to some point, but probably look for a more grounded partner when they realize that this is a pattern that isn't about to change. Both Moon and Venus represent emotions, and they are both highly emotional signs. As two cardinal signs, they could have a long unsatisfying relationship because they are both waiting for a groundbreaking moment to set them free.

In some cases they should be advised to make a change if they are not satisfied, and search for someone who could make them happier. 15%

Libra & Cancer Values: If they find it with each other, they will certainly have trouble letting it go, both of them understanding that it is not that easy to find. As both of them to the signs of relationships like family relationships or relationships with partners and so they will both value a pleasant and joyful connection between two people. However, their entire system of values differs greatly beyond the point of relationships, and while Cancer will value tenderness and care, Libra will value responsibility and platonic love. That doesn't sound promising, does it? 20%

Libra & Cancer Shared Activities: Cancer is a sign that exalts Jupiter and although they can be perfectly happy with a grumpy partner, they would appreciate them not to have a need to impose their grumpiness on them. There is a secret desire to every Cancer to travel the world, while having the safe base to come home to, but Libra could destruct their beliefs and lead them to doubt they simply don't want to deal with. Cancer and Libra could do many things together, but it is questionable if they will want to. With Cancer's mild and mellow temperament, Libra will probably feel even more inadequate and grumpy, for they need someone to be their opposition and defy them in order to have respect and sexual charge.

10%

Conclusion: Cancer wants someone responsible, to take them by the hand if needed and complement their emotional nature with practicality. Libra wants someone who is full of life, energized, strong and full of initiative to follow their ideas. They can really disappoint each other if any expectations are set wrongly at the beginning of their relationship. Probably the biggest restriction in the relationship between Cancer and Libra is in things they want from their partner. The best way for them to build a love that is to last, is for both partners to hold on to their independence whatever happens. If they focus on love and worry about earthly things each on their own, Cancer could "compromise" on heavenly love, as much as Libra would like to have a family. 28%

Libra Compatibility with Leo

Libra & Leo Sexual Intimacy and Compatibility: With Leo's confidence, and Libra's sexuality, they tend to inspire each other to become great lovers when together. Their sex life is usually filled with respect, and they feel free to try out new things with one another. If they found their relationship on a strong mutual attraction, they could enjoy a satisfying sex life for a very long time. Leo doesn't mind being seen and Libra is a sign that represents the public eye. When a Leo and a Libra come together, they don't need much time to build up a healthy sex life. Although this says something about their sexual preferences too, they will usually be well behaved in public. As soon as any restrictions show up, they will have to play out their passionate scenarios at any time, and in any place in which they get a chance to be alone if only for a minute. Libra is a sign of Saturn's exaltation and it is easy for them to wait and be rational, but with passionate Leo they find it hard to stay in control. 90%

Libra & Leo Trust: The problem here arises from their understanding of the Sun, for it rules Leo and falls in Libra. To add to that, Leo is a sign of Neptune's fall and Libra can often sense the dishonesty behind Leo's confident act, if there is any. The problem lies in the fact that they both like to be seen, but in an entirely different way. Leo wants to show everything they've got and Libra wants to get approval from other people. None of them understands the other, and this can become a reason to get jealous and mistrustful. It is not that often for Leo and Libra to share a relationship filled with mutual trust. If they wish to remain in a trusting relationship, they need to find approval and a suitable audience in each other to begin with. Only then will they be able to move on and look for these things in other people without arising suspicion. 40%

Libra & Leo Communication and Intellectuality: The Sextile between their Suns usually makes it possible for them to respect each other, and help each other build stronger personalities, free of judgment of any kind. Their elements of Fire and Air fit perfectly and there is a passionate approach of Leo for every idea of Libra. Their communication is fast and inspiring, although sometimes hard to ground through constructive ideas if Libra doesn't rely on its cold and rational relationship with Saturn. The problem arises if Libra feels any sort of jealousy at their Leo partner for their sometimes unfounded confidence and that inner sense of security. When it comes to the rational side of their relationship, Leo and Libra have a very nice way to support each other's personalities and communicate. The only way for Libra to learn how to feel confident as well is to accept this ability of Leo as the best part of their beautiful character. If Libra starts judging Leo, making assumptions on how their partner should behave but doesn't, their mutual respect will fade and they will both miss the point of their relationship. 85%

Libra & Leo Emotions: They will never end up in a relationship with no future, and their belief in love will move them towards marriage, children and growing old together, if only they share enough trust and love. These two signs represent our loving relationships and marriage, and when you look at this couple, you will see that their love for one another is real, obvious, shown and leading them in a certain direction. Ruled by the Sun and by Venus, these signs represent one of the basic planetary cycles of love that is often connected to periods of eight years. If they stay together longer than that, they might as well walk down the aisle and have a bunch of kids.
99%

Libra & Leo Values: They are finely compatible when it comes to matters of the Sun and they complement each other well in a way that helps them both learn about expressing themselves and their abilities and strengths. The problem with this couple is in their relationship toward Saturn, and while Leo represents its detriment, Libra exalts it. Nothing holds greater value for Leo than someone's strong personality and their own pride and heroism. Libra, on the other hand, values justice and one's ability to be the hero, but something they often think they lack. Although this can be a lesson to be learned, the challenge of responsibility they take on unequally can tear them apart. Leo needs to get serious and realize what their responsibility is to fit into the thing Libra values most like reliability and tact. 75%

Libra & Leo Shared Activities: Leo is a Fire sign and as such it shouldn't be slow as a Water sign or and Earth sign. Libra belongs to the element of Air, and it should be faster than any other element. But when you look at these two signs, you will see that Leo would like to sleep 20 hours per day, and Libra needs to think about everything twice and carefully choose activities and words they want to say. This doesn't sound that fast. If they share the same interests, they could have an endless field of possibilities for shared activities. They will mostly enjoy "red carpet" events and the fancy gatherings where they can both show one another to the world. There is a strange similarity in the speed of these signs. The biggest problem in their choice of activities lies in Libra's indecisive nature that Leo simply doesn't understand, and usually doesn't have patience for. This is where they might give in to the temptation to "help" them decide, taking over the wheel and deciding instead of them. This can lead to mutual lack of respect, even though it seems like a little thing that no one would even notice. They need to give each other time, and stay as independent as possible. 60%

Conclusion: They have a lot to learn from each other, and the main goal of their relationship is to reach the point of shared respect and responsibility in a perfect balance of power. It If you want to sum up the relationship between a Leo and a Libra, you have to understand that their bond involves the beautiful and challenging dignities of Saturn and the Sun. will sometimes be hard for them to overcome the need for competing, trying to determine who is a better, smarter or a more capable person. Even if they don't, their relationship will be something to enjoy and show off in public. 75%

Libra Compatibility with Virgo

Libra & Virgo Sexual Intimacy and Compatibility: Libra is an Air sign ruled by Venus, and couldn't be more distant from

Virgo's shy nature or practicality when it comes to sex. They will probably be driven crazy by each other, one of them trying obsessively to keep things clean and looking as if they were scared of any emotional contact, and the other strict in their search for spiritual love and a partner they can really talk to. Most often, they are not even attracted to each other, and when they do begin a sexual relationship they have to face the challenge of speed. Virgo will want to move slowly, and even though their mutable quality makes them pretty adaptable, it is often not enough to reach Libra's speed of Air. They might find a middle though, when Libra relies on their exaltation of Saturn and slows down, while Virgo adapts and changes more quickly. The sexual relationship between a Virgo and a Libra just doesn't seem like a good choice. Virgo is an Earth sign ruled by Mercury, relying highly on their intellect, while bringing Venus to its fall. If they find their perfect timing, both partners might still end up unsatisfied. Virgo was expecting a fairytale connection seeing the image Libra shows and they will find so many faults in their partner's approach. Libra's self-esteem doesn't really tolerate that much criticism and they will probably see Virgo as boring, stiff and unaware of any emotional connection they wanted to make. For this sexual relationship to work, they need to coordinate the emotions they give and receive. Only then will they be able to satisfy each other. 1%

Libra & Virgo Trust: If this wasn't the case, Virgo would dismiss each word they say to begin with. There is a lot of tension between them, for they don't understand each other's primal natures. This is not so much an issue of trust, but it comes down to it when the dust settles. There is no way to explain to shy, introvert Virgo, why Libra has to have everything out in the open. It is even less clear why so much attention must be brought to their relationship, or why does their Libra partner flirt with everyone that looks decent. It is a good thing Libra exalts Saturn and their need to be fair is, in many cases, more than obvious. On the other hand, Libra doesn't really trust the moodiness of Virgo and think of their partner's fear of expression as if there is something they want to hide. 25%

Libra & Virgo Communication and Intellectuality: This can keep them in a pretty good place for quite some time, especially if they share the same professional interests or have a goal to support each other in learning or advancements of any kind. Even if they touch the subject of emotional contact, they will still see eye to eye for a while. Both of them value rational choices and smart moves. But then, true emotions will surface in one of these partners, and they might realize that communication between them is no longer possible. They will understand each other perfectly for as long as they speak of taking responsibility and serious matters in life, such as professional choices and their income. As soon as one of them has to deal with a personal issue, the other one seems to freeze, losing all ability for compassion of closeness. The problem here lies in the fact that their relationship is based on their mental connection and this interferes with actually loving one another, and providing a kind word at an important time. 60%

Libra & Virgo Emotions: There is no couple more prone to separate emotionally than these two, and this can lead to so much dissatisfaction in both of them, with both partners unable to end the relationship they don't feel good in. Virgo is a sign where Venus falls and this is the ruler of Libra and a planet that, shoulder to shoulder with the Moon, represents our emotions. Their emotional contact leans on the point of their communication for they are in a strange coexistence. The value and stability of emotions that needs to be found by Libra, seems to lose all meaning when rational Virgo comes along. Libra will instantly fit into the expectations of their partner, trying hard to be that voice of reason, more rational than they actually are. As time goes by, the emotions they have kicked away will cumulate, returning in a sudden wave that cannot be stopped. There is no way to control the emotional nature of Libra, and this is something they will both learn in this relationship. 1%

Libra & Virgo Values: It is a strange deceit that comes between them, as if they were both able to practice something entirely out of their reach. This is a couple that will value all the same things until they discover that this simply isn't true. The guilt they will subconsciously feel for dismissing each other's (and their own) emotions will take out the satisfaction between them and leave them in the state of wonder like that the two people value so many similar things, value something entirely different by the end of the relationship? 30%

Libra & Virgo Shared Activities: There is rarely anything extreme about their contact and they will mostly enjoy the usual, relationship activities with one another. Even their pace can be synchronized well enough, as Virgo is much faster than the rest of the Earth signs, while Libra's indecisiveness makes them quite slow. Virgo is a mutable sign and they will adapt to any desire their partner has with ease. Libra is tactful and thoughtful enough for Virgo not to feel bad about their choices. They need to pay special attention when Virgo helps with decision making, taking even more of Libra's already bruised ego. This can lead to serious problems, not only in deciding on future activities, but in their mutual respect. 65%

Conclusion: In general, this relationship can sometimes work, and these partners can synchronize their pace, choose appropriate activities and build a satisfying sex life with enough patience and care. They could have a deep problem with emotional understanding though, and the thing they will find most difficult to reconcile is their fragile egos. Virgo, willing to please, will easily take over the responsibilities and decisions that Libra needs to take on. Virgo and Libra can form a very satisfying intellectual bond, for as long as they respect each other's feelings. This will lead to a feeling of inferiority in Libra and the loss of respect toward their Virgo partner. If this issue is left unresolved, their relationship might end because of disrespect they were both unaware of in the beginning. 30%

Libra Compatibility with Libra

Libra & Libra Sexual Intimacy and Compatibility: They seem to fit perfectly when it comes to not crossing the line and being as moderate in their sexual expression as possible. This will help them build a strong relationship in time, if they don't start judging each other as soon as one of them jumps out of their strict sexual routine. As two partners ruled by Venus, they could both easily have a problem with initiative and this might be an obstacle impossible to overcome. A great thing about the relationship of two Libras is in their understanding of tasteful behaviour. When one of them starts relying on the masculine nature of their sign, a relationship can begin and the main objective here becomes satisfying one another. Their mental compatibility will allow them to understand each other's needs and talk about anything regarding their sex life, but their shared lack of confidence might make them judge each other in a way they wouldn't judge a different partner. 65%

Libra & Libra Trust: The uncertainty of their decisions can be well understood when it comes to everyday things, but as soon as they show it while choosing each other, everything ever said will be questioned. The fall of the Sun gives less light to these individuals, meaning it can be a challenge for them to see things clearly. Two Libras don't even know how many trust issues they have until they get together. This doesn't make them sense things any less, and they will feel something is wrong from a thousand miles distance, not aware what it is. Trust is a very sensitive issue when they get to this point, especially if they don't speak their mind waiting for resolutions. This is a couple most prone to misunderstandings of all couples in the zodiac. 35%

Libra & Libra Communication and Intellectuality: When this happens, one of these partners will become fully unaware of their behaviour, acting like a vampire that drains willpower from their partner, day after day. In the worst case scenario, both of them will feel powerless all the time, because of the disrespect they feel for each other, but lack initiative and seem unable to resolve the situation that burdens them. For as long as they respect each other enough, as two Air signs, communication between them will seem endless. They will always have something to share, in most cases talking excessively about other people. We can say with certainty that their communication will develop to something they will both enjoy, for as long as they don't start feeding their bruised egos on one another. When they start pointing to each other's flaws, it might be best to pull back and realize that nothing truly needs to be changed. Their mutual acceptance is the only thing that can keep their mental relationship in a good spot and their conversations flowing without judgment and unrealistic expectations. 80%

Libra & Libra Emotions: On the other, Libra is a sign that exalts Saturn, and this gives them both a cold side, one that will not easily allow them to build an intimate relationship. The Air element they belong to won't help much, since they will do their best to rationalize things until they lose any emotional value. Two Libras sometimes make an impression of two people who refuse to be in love, out of social or intellectual expectations that cannot be met. On one hand, two signs ruled by Venus seem to be made for love, and this speaks of their physical and spiritual closeness that can be made through their emotional contact. There is that stubbornness in Libra that isn't obvious at first glance, and if something could tear this couple apart, it is their need to stick to their convictions. As if they are incapable to look at the bigger picture, they will sometimes shove their emotions under the rug, only to hold on to what they know. It is strange how sure of their knowledge and intelligence they can be when they get together, as if they finally found someone that seems weaker. In order to stay together, two Libra partners need to turn off their brains and let go to their feelings and their sexual relations. This is the only way for them to build intimacy, or they might both resist it out of simple fear and lack of confidence. 50%

Libra & Libra Values: Venus is a planet that speaks of value in general, and this is a strong connecting point to them, especially since both of them lack the awareness of real value of money. The combination of Venus' rule and Saturn's exaltation makes them share the same values, gathered through similar experiences and relationships. These partners value dedication, a spiritual approach to love, fineness and moderate choices, reasonable behaviour that will not be judged by those around them. When it comes to this side of their relationship, two Libra representatives seem to be a perfect fit. 99%

Libra & Libra Shared Activities: The problem arises before they get to this point, while they try to decide and make an agreement on where to go in the first place. Both of them will want to stick to a certain secure routine, rarely prepared to meet too many new people. But if their routines don't coincide, they will have to make compromises in order to move from a point of stagnation. At first glance, we might assume that they will like to show off, inspire others by the love they show to the world and be on the move to create a certain image doing everything side by side. When they find a shared routine and share friends, shared activities won't be a problem anymore and they will follow each other with much more ease. 80%

Conclusion: When two Libras start dating, it might be difficult for them to find a purpose of their contact, for they both seem to share a mission and a goal that is connected to other people. The sign of Libra is a sign of relationships and they often have a mission to teach others about relating to one another. If they find a meeting point, combining their activities and sticking to their shared values, they will have a tendency to become a perfectly balanced couple. The only thing missing in both of them, very hard to develop, is the sense of mutual respect with no passive judgment or expectations. Both of them are susceptible to this problem with their surroundings, and when together, these problems will easily multiply. If they let each other be who they are, they might become an inspiration for all of us, teaching us what a productive relationship really is. 68%

Libra Compatibility with Scorpio

Libra & Scorpio Sexual Intimacy and Compatibility: They are linked through their animalistic natures, continuing one another, ruled by planets that also rule their opposing signs. This is a complicated sexual contact, for they are ruled by Venus and Mars, as if they were made for each other. However, they seem to lack the touch of the element they both need, separated by the quick, superficial element of Air in Libra and emotional, slow element of Water in Scorpio. Even if they don't make each other happy in other parts of their relationship, when a Libra partner lets go to their instinctive, animalistic side, they easily become one with their Scorpio partner. This is an intense couple in every way and their sexual relationship is something truly exciting and challenging at the same time. Their sex life can be incredibly emotional and demanding, for the pull of their energies is extremely strong and makes them both obsessive and possessive to one another. They will both often get so tied that their relationship seems impossible to break. These ties are created through their physical contact, even if everything else in their relationship makes them deeply unsatisfied. 45%

Libra & Scorpio Trust: As if this wasn't enough, that possessiveness of Scorpio is easily absorbed by Libra, and they will start acting in a similar way, obsessing about each time their partner wants to do anything alone. Even though this is the worst case scenario, this atmosphere will probably be present in any Libra Scorpio couple on a smaller scale. They are not ready to live like this every day. If Scorpio cannot trust someone, it is Libra. How could they when there seems to be an obvious need in their Libra partner to show how attractive and sensual they are to the rest of the world. If they accept this as a way of sharing true, deep love, than their trust issues might not be much of a problem however weird they might seem. 1%

Libra & Scorpio Communication and Intellectuality: Their communication is the exact place where the connection of Libra to Saturn comes in handy. This will give Libra the ability to slow down, breathe in, and understand the impulsive acts or words of their Scorpio partner. There is just enough depth in both of these signs, even though Libra is an Air sign and quite distant from planet Earth. This will allow just enough understanding between them in a rational sense. The problem they have to deal with is in the fact that they both represent a detriment for each other's rulers. There is nothing Scorpio can say that Libra won't be able to rationalize. This will easily lead to the dismissal of each other's personalities, especially since Libra has to deal with personality issues through the fall of the Sun, and Scorpio doesn't accept anything that isn't whole. Libra will have trouble understanding the aggressively clear side of Scorpio, as much as Scorpio won't understand the dishonest act of their Libra partner. Their differences could be very difficult to reconcile if they get too close to each other and start meddling in their personal lives and decisions. 55%

Libra & Scorpio Emotions: Libra looks for an active, fiery partner that will awaken their life force and bring light into their life. Scorpio looks for someone physical, sensual and practical, and although Libra can meet these needs to some point due to Saturn's exaltation and the rule of Venus, in most cases this is not primarily who they want to be. When a spark exists between them, it will easily build up to a fire, burning entire cities, forests and everyone on its way including both of them. That nuclear energy of Scorpio is ignited by its preceding sign of Libra, because it carries information on all unresolved issues in relationships and all of the anger that hasn't been let out because of the forced need for appropriate behaviour. There is an incredible intensity to their feelings when they do fall in love. In many cases, love simply won't happen. In other words, Libra will accumulate the unexpressed sexual energy and it will burst in Scorpio, leading to a lifelong war if enough is pushed under the rug. This will be construed as ultimate love, making their emotional relationship ridiculously intense. They need to understand that love is about tenderness as much as everything else and if they don't have any, maybe it is not love but simple physical attraction and a need to vent from restrictions. 25%

Libra & Scorpio Values: However, the rest of their systems of value don't coincide that much and they will differ on behaviour and expectations of other people. With Libra frowning at indecent behaviour, Scorpio will have indecent friends, honest about their act and fully accepted and respected for that. Both of these partners will value consistency and commitment and this is something that will connect them in the first place. It seems to Libra that Scorpio does everything out of the ordinary just to seem special, while Scorpio sees Libra as doing everything ordinary just to fit in. The main problem here is in the value they give to opinions of other people and this will not be easy to overcome. 10%

Libra & Scorpio Shared Activities: Scorpio will awaken the animalistic side of their Libra partner, and their time together will most certainly be focused on both partners following their instincts. Scorpio is a sign of the exaltation of Uranus and they will lack patience for indecisiveness of Libra. This is why their needs won't fit very well in other areas of their lives, and they might not be able to meet each other's expectations. If there is an activity a Libra and a Scorpio can share all the time, it must be sex. An exception to this rule occurs when Libra turns to the dark side and gives in to the need of Scorpio to take them to the Underworld. No other partner will ignite this need in Scorpio as a Libra partner. 35%

Conclusion: Both of these partners will have to face their dark sides through this bond, and although this can lead to an incredible and intense sex life, and emotions that no one else can understand, it might lead them both to a depressive hole they won't easily get out of. The relationship of Libra and Scorpio is in no way easy and light. The only way for this couple to last in a satisfying and gentle relationship, is for both partners to build a strong individual, independent life, or they will get sucked into the whirlpool of karmic emotions and obsessive, negative expectations. 29%

Libra Compatibility with Sagittarius

Libra & Sagittarius Sexual Intimacy and Compatibility: They are a very good match when it comes to sexuality, for no partner here feels pressured and there is just enough room for both of them to grow, develop, build their self-esteem and feel secure in each other's arms.

Ruled by two benefic planets, Venus and Jupiter, their main objective is to form an enjoyable sexual relationship, with a primary goal to make each other happy. For this goal, they will experiment a lot, and try out new things, everything followed by a smile and a sense of lightness, as if sex wasn't really a big deal in the first place. The seriousness of Libra linked to its exaltation of Saturn will give their entire relationship endurance and stability, while their ruling Venus working together with Jupiter, gives enough romance, sexual desire, tenderness and might lead them to a fairytale ending. The intensity of emotional contact and intimacy between a Libra and a Sagittarius will mostly depend on other factors in personal charts, but they will most certainly enjoy their sexual relationship. This combination of planets forms Neptune in a way, and speaks of the growth of satisfaction leading to orgasmic pleasure, even though both signs might not seem at all sexual to some other members of the zodiac. 90%

Libra & Sagittarius Trust: They can both go to extremes, either having unrealistic faith in each other or mistrusting every word and every action that is made. The only way to keep the image of trust for these signs seems to be to always stay in a fairytale, unrealistic state, and this is something a Sagittarius will never want to do. Rulers of Libra and Sagittarius are closely linked to Neptune and the challenge of trust is one of the most important experiences that this relationship gives. If truth isn't lived, nothing in the world is beautiful for a Sagittarian Sun. As soon as they start their search for something different, Libra will sense the change and become frustrated by their inability to create oneness with a partner they love. 5%

Libra & Sagittarius Communication and Intellectuality: Even if this isn't something with a promising future, for no one can run from their true nature, it will bring them both joy and happiness at least for a little while. Libra partner will be able to relax next to someone who doesn't judge, and Sagittarius partner will feel like their energy is well focused on someone that needs some youth, warmth, light, optimism and creativity in their life. For as long as they don't brush on ego problems, their communication and intellectual compatibility are a given. It is wonderful to watch how soft Libra gets, forgetting about Saturn and their own responsibility, as Sagittarius' childish nature melts their heart. The main problem that will eventually surface and need to be dealt with, is in the forces of their Suns. Libra's Sun is weak, and they will easily give the wheel to someone else who will make positive decisions and moves for them. Sagittarius has too much fiery energy in their Sun, active, taking action and always prepared to give some of it even if nobody asked for it in the first place. This could lead to a subtle, hidden, will imposing and a character shift that will leave them both bruised for respect when a light is finally shed on the issue. 85%

Libra & Sagittarius Emotions: It is not easy for any one of them to find love and share it with someone. They are, after all, an Air and a Fire sign. Even though Libra is ruled by Venus, it is linked to the mental processes, social adaptation and communication through its element, while Sagittarius has passionate feelings, but uses their head, spreading their philosophy, more than actually feeling. When they get together, they seem to be able to find a balance in which both of them use their heads just enough, and give each other enough room for love to be born. This is one of the most compatible couples when it comes to the emotional side of their relationship. This is a bond that gives both partners the opportunity to understand how deep their emotions can go, as beneficent rulers make way for feelings to surface in a supporting atmosphere. Even though their relationship is not always meant to be the one they will stay in for life, it could prepare them for a love they seek, giving them a glimpse of what they are capable of. 99%

Libra & Sagittarius Values: Libra doesn't seem like a creative person to others, but a Sagittarius sees their intellect through communication and motivates them to show their warmth. This leads to shared value of their entire relationship and an intellectual understanding that gives them room to build their shared philosophy. These partners will value the strength of mind in a way that is understandable only to them. Even if they don't start their relationship in the same place, they will have the opportunity to build similar values in time, showing each other what's truly important. 75%

Libra & Sagittarius Shared Activities: Libra wants to stick to their usual routine, and make fieldtrips to things that interest them from time to time. Sagittarius wants to move from any routine and live a life travelling the world. There are exceptions to this rule, of course, and there are uplifted Libras that will want to travel the world, as much as there are Sagittarius representatives that want to follow a certain trail, while fantasizing about their reality. Even though we could easily assume that a Libra and a Sagittarius will have lots of things to do together, there is a great chance that their choices of activities won't be so similar. However, in most cases, their needs won't fit that well and they will probably face the challenge of their usual ego battle while choosing what to do together. 70%

Conclusion: However, there is an archetypal battle between them, for Saturn exalts in Libra and doesn't really care for his son, Jupiter, the ruler of Sagittarius. This could easily lead to a struggle for supremacy and a battle to reach the ruling position among them. This comes as a continuation of Libra's bruised Sun and a Sagittarius will fit in perfectly with the need to give away every sense of pride out of some childish convictions. The only way for them to be happy together, is to respect each other fully and let each other do what they are meant to do. The relationship of Libra and Sagittarius is in most cases a beneficent bond that allows these partners to develop their emotional, inner worlds and build their lives without negative influences. Libra should stick to their relationship and love, ruled by Venus, while Sagittarius should stick to their convictions and width, ruled by Jupiter, multiplying the love Libra provides. 71%

Libra Compatibility with Capricorn

Libra & Capricorn Sexual Intimacy and Compatibility: Ruled by Venus and Saturn, they tell the tale of a soldier that had to leave his wife and came back after years of waiting. When it comes to sex, this is a combination that could point to a lack of sexual activity, even though both of these signs find sex extremely important in their lives. To begin with, they might feel no attraction at all, and even start a relationship on a basis formed in friendship, only to realize that there is no chemistry between them. If lack of attraction doesn't stop them, usually something else will. It is a combination that gives in to outer circumstances and things that are out of their control. Both of them could feel pressured and their self-esteem could suffer greatly. When we are speaking of a sexual relationship between a Libra and a Capricorn, the first thing that comes to mind is waiting. There is an understanding between them ignited by Saturn's exaltation in Libra. This makes them both understand good timing and usually won't allow them to make a mistake expecting what should not be expected. In case they overcome all of the obstacles and form a strong bond through their personal natal positions, Libra and Capricorn can have sexual relations that are pretty conservative, routinely approached, and only satisfying if they both let go of their strict premises and conditions. 15%

Libra & Capricorn Trust: Even though Libra can sometimes have questionable motives, a Capricorn partner will make them turn to Saturn completely and feel guilty at the smallest glimpse of a possible lie. The only possible problem surfaces when Capricorn is too strict from the start, making their Libra partner feel inadequate, judged, or even scared of the consequences of their actions. A strange thing in a relationship of Libra and Capricorn is a really high level of trust between them. This could make their relationship dishonest, not because there is actually something to hide, but because Libra partner feels the need to protect themselves by holding on to their privacy. 80%

Libra & Capricorn Communication and Intellectuality: Even though Libra loves Capricorn because of Saturn's exaltation, this is shown in the most unusual way, for they seem to feel the need to speak out of spite. This can be a long battle, with no winners or losers, simply two people always building walls between each other, for reasons that aren't clear to anyone around them. The biggest obstacles to their understanding are the elements they belong to. Air and Earth are too far apart and it seems unclear to these partners how to reach each other on any issue in life. Still, there is a prudence to both of them that might give them just enough depth and understanding to have very interesting discussions and motivate each other to build a better foundation for every next debate. We wouldn't say that Libra is that stubborn, but when they find themselves in a relationship with a Capricorn, they suddenly become headstrong and sometimes even impossible to talk to. If they remain rational in their mental relations, they could have a lot of fun that other signs wouldn't be able to understand. The satisfaction they will both get from serious problem solving might lead them to a point where they find a solution together, Libra puts it in words and Capricorn puts it in action. There is probably nothing in the world that could raise their egos higher than situations in which they managed to resolve something by a simple shared effort. 35%

Libra & Capricorn Emotions: Libra is a sign ruled by Venus and their emotions come naturally, but usually restricted and held back due to the seriousness of their nature and the judgment of others they fear. Capricorn has a mission in life to accept all emotion, and in most cases, unless enlightened, they will be this judgmental force that holds Libra down. As if this wasn't enough, the entire situation will feed Capricorn's ego and make them think they are right about their approach, leading them further away from their focus point. This is a couple that has to work hard on finding a shared language to show how they feel and still respect each other. The hardest thing to reconcile in the relationship between a Libra and a Capricorn, are the ways they approach their feelings. The emotional nature of Capricorn makes them distant for many, but completely untouchable for Libra as soon as they start dismissing their feelings. The only thing that can be done here is find a point of absolute respect and acceptance of all emotions and their manifestations. If they allow each other to break things, get angry, cry, make scenes in public or give in to hysteria, they might find a way to express their love in a way that will be correctly understood. 1%

Libra & Capricorn Values: This can help them overcome any differences and opposing attitudes, values or convictions, for each of these partners will be ready to understand the set of responsibilities they have toward each other. As Air and Earth signs, both pretty set in their ways, Libra and Capricorn will differ greatly in the value of words and deeds. Libra will communicate and think that their mind is their biggest asset, while Capricorn won't really care if results aren't manifested through the material world. The most important values Libra and Capricorn share are the value of time and taking responsibility. This is a good training for a Libra partner to find grounding, but it usually won't be pleasant for any of them in a romantic relationship. 50%

Libra & Capricorn Shared Activities: There is a great chance they will be lulled by their relationship to the state of hard work and lazy rest, with no effort for anything creative or inspiring. The best thing these two can do together is be boring to the rest of the world. They need to keep their passions lit and create a weekly routine that will make them get out of the house and do something fun. 25%

Conclusion: This doesn't mean they won't enjoy the trouble of being together, or stay in a relationship for a very long time, but this is most certainly not a bond that many other signs would engage in. If we want to choose the best word to describe the relationship between a Libra and a Capricorn partner, we would have to say, it is hard. Their biggest challenge is the lack of respect for emotional value that is usually initiated by Capricorn, but easily continued by Libra. If they find a way to share, show and understand each other's emotions, everything else will seem like a piece of cake. 34%

Libra Compatibility with Aquarius

Libra & Aquarius Sexual Intimacy and Compatibility: The problems with Libra's Sun bring too much concern for opinions of other people, so representatives of this sing go to extremes when it comes to the way they show their sexuality. They will either be afraid to be judged and seem too asexual, or have a need to show it "in spite" of everyone's opinion and this can be quite repulsive for their partner. When it comes to sexuality, Aquarius has an entirely different approach – they simply don't care about anyone's opinion. They live their life in a constant search for freedom from any taboos or restrictions, and this will help Libra forget about other people, at least for some time. Although their sex life can be quite liberating for Libra, it can also be a bit challenging for Aquarius because they will be the one who has to fight against Libra's need to fit in. Aquarius can truly help Libra express their sexuality. However, as two Air signs, they will both tend to be free to express their sexual desires to each other. They will like to experiment, learn about each other and their own inner desires and communicate with ease. Their sexual relations should be a strong pillar of their entire relationship, although they will usually think of their verbal ways to get along as the most important for their bond. 90%

Libra & Aquarius Trust: Their insecurities coincide very well, and they will usually help one another move through them, but the trust between them needs to be built, it is not implied. Both of these signs like to be attractive to different people and they should find a way to communicate this need in the right way. The problem can arise when Libra starts to get attached and becomes emotionally dependent on their partner. Because of their righteous natures they could trust each other without exception, if they were only that sure of themselves. This is not something Aquarius will easily deal with and it could damage the trust of both partners, in each other and their entire relationship. 85%

Libra & Aquarius Communication and Intellectuality: They are both stiff in their convictions and won't easily change their mind once they are set on it. Aquarius won't have such a good time waiting for Libra to make any decision in their life, no more than Libra will enjoy the spontaneous, unpredictable nature of Aquarius. Since they both rely on communication, they will have a lot to talk about and will usually find a language to solve all of their issues. Still, it won't be easy for them to reconcile some of the differences in their approach to things. Libra is indecisive but once they make a decision, they will rather stick to it than salvage their relationship, even if it is a simple meal in question. Aquarius will have a tendency to walk away as soon as they feel pressured into anything, even if it is that same meal. Libra and Aquarius both have certain images to maintain. Libra wants to look nice and act nice for others, while Aquarius wants to go in the opposite direction of everyone, sometimes even if there is no reason to do so. So basically, they could have an enormous problem about lunch if we talk about extremes. Still, they usually understand each other well on usual, daily things and have similar convictions that help them handle big changes in life well. In time, as they get to know each other, their personalities will get along better and they could realize that they respect each other to a point that is unreachable with anybody else. 40%

Libra & Aquarius Emotions: This is something they will get in a relationship with Aquarius and it could help them both build a very strong emotional bond. They have strangely different goals in life, but if they harmonize them, their emotional bond should be very strong and develop much faster than we would anticipate. The biggest obstacle that could present in their way is marriage, at some point in their relationship. Libra is a sign that represents marriage and finds it very meaningful as the institution Saturn would support. Aquarius might think of it as obsolete, even run from it and they will probably enter it only for practical reasons. Libra is a sign ruled by Venus and this makes them emotional in a way, but we should remember that it is also a sign of Saturn's exaltation and detachment is something that makes them feel very good. It is important not to make pressure to any of the partners when this point in their relationship arises, or they might both feel repulsed and angry, leading to unnecessary conflicts and even the end of their relationship. 80%

Libra & Aquarius Values: This could represent a big problem in their relationship, and Libra partner could seem clingy and not at all independent, while Aquarius might seem like an uncontrollable lunatic who would do anything to destroy all relationships in the world. As much as Libra will value togetherness, Aquarius will value solitude. They both value communication and intellectual strengths enough to talk about their needs and desires, and this should help them overcome their differences. 50%

Libra & Aquarius Shared Activities: Libra will have trouble deciding what they want to do and this could drive their partner crazy. If something can launch Aquarius into the orbit it's the lack of spontaneity and Libra can sometimes be the opposite of spontaneous. These partners could end up in a relationship in which only Aquarius pulls the strings and Libra follows. Aquarius will want to do anything, really, for as long as their life doesn't fall into a boring routine. This wouldn't really be a good solution for any of the partners, and Aquarius will have to learn to show some patients in order for respect between them to remain intact. 65%

Conclusion: It can be quite difficult for their troubled Suns to get along and they will often have difficulty adjusting to each other's character and finding deep respect for one another. The best cure for any problem in their relationship is usually in time, but with Aquarius' need for spontaneity they often won't last long enough for time to mend what gets broken. There is a strong understanding between a Libra and an Aquarius partner due to their shared element of Air. Whatever their story, they will have a lot of exciting things to live through together and if they fall in love, it would be a shame for a couple such as this one, not to give their relationship a try however it might end. 68%

Libra Compatibility with Pisces

Libra & Pisces Sexual Intimacy and Compatibility: Since they touch each other through this sensual, loving planet, they might find true sexual satisfaction together. They can both be selfless lovers, caring more about the satisfaction of their partner than their own. Tenderness shouldn't be an issue here for none of these partners will like too much aggression and roughness anyway. They could discover many different sexual preferences that they weren't aware of before, through a connection of very different natures. If they weren't connected by Venus, it would be very difficult for them to form a relationship on any kind, for their approach to life and sexuality is completely different. Libra partner wants someone strong, passionate and confident, while Pisces partner wants someone gentle, compassionate and aware of their feelings. Air and Water, Libra and Pisces seem to have almost nothing in common. However, we shouldn't forget their connection through Venus, the ruler of Libra, exalted in Pisces. Libra will want their sexual experiences fast and exciting, and Pisces will want them slow and sensual. The main issue of speed is usually overcome by the quickly changeable nature of Pisces, except in cases when they are too shy to jump into a sexual relation with someone as openly sexual as Libra. 50%

Libra & Pisces Trust: The need Libra has to be accepted and liked by other people will be wrongly interpreted by Pisces, for they don't understand how someone's self-confidence can be that low. That sparkly, always in love, childish, flirty nature of Pisces will be a huge turn off for Libra, who will not be able to trust someone who openly shows their interest in other people. They will not understand each other well enough to share much trust. The only way for these partners to remain in a trusting relationship is to approach it casually and build their understanding and trust from zero, as if they have never had any relationships prior to this one. 1%

Libra & Pisces Communication and Intellectuality: Unfortunately, they will often want to help them grow and change who they are so they could be "happier". The problem here is in the fact that Libra doesn't know or try to understand what would make Pisces truly happy. This is some basic disrespect and it could ruin the foundation of their entire relationship. The main challenge here is to remain in a respectful bond, however crazy Pisces might seem to their Libra partner, or however stiff and boring Libra might seem to Pisces. Their communication can be inspiring for as long as they don't try to change each other, explain to each other how they should think or feel in certain situations, or even worse, teach each other how to behave. In many cases, Libra will cherish the optimism and the childish naïve nature of their Pisces partner. Pisces partner can be quite direct and spontaneous and this might endanger the image Libra is trying to maintain in the eyes of others. If they find each other's actions inspiring in any possible way, they might reach a point in which they will truly communicate without judgment. 5%

Libra & Pisces Emotions: This is an everlasting love waiting to happen, and the kind that could be born between these partners in case they both overcome their ego. It will be a rare occurrence and their rationalities will weigh them down, for Libra does exalt Saturn, and Pisces are ruled by Jupiter. Their minds will probably be filled with all sorts of irrelevant information until they decide that it is simply too hard for them to be together. If they pass the point of disrespect and set strong boundaries, nurturing each other's individuality and self-sufficiency, they might reach the point of emotional interaction. Their emotional connection is mostly presented through the adoration of Venus in the sign of Pisces. As if they didn't have enough challenges, this will usually happen at a different time. Libra will rush into an emotional show off, realizing that they have found love, but Pisces will not feel any love until all the dust settles and they have the time to close their eyes and feel. If their timing is off, they will probably end their relationship on good terms, not expecting much from each other in the future. 15%

Libra & Pisces Values: With Venus at focus, they will both need to be loved and cherish those who know how to show it, hot to enjoy life, food and sex, and how to make their loved ones satisfied. This could give them a strong foundation for a sexual relationship if they are attracted to each other in the first place. The rest of their characters will differ greatly and while Libra will value consistency and stability, Pisces will value spontaneity and one's ability to follow their heart. They will both value love and this is something that will connect them over everything else. If they both believe that they have a mission here on Earth, and it happens that their missions cross paths, they could inspire each other to fight for what they value most – their names in the stars. 60%

Libra & Pisces Shared Activities: The problem here is that by the time Libra decides if they want to do something or not, Pisces will change their minds five times, not sure if this is the best thing for them, for their relationship, for their life's purpose and mission, etc. The questioning of Libra will raise suspicions in Pisces with painful ease. They will manage to find activities they will both like, and if nothing else works they can always turn to any sort of art. They want to do anything while they are together. Indecisive Libra is exactly what their Pisces partner doesn't need in order not to feel completely lost in life and all of their chosen activities. They could drain so much strength and confidence out of each other just trying to make a plan for the night. 40%

Conclusion: They perceive it in two different ways and they will often not respect each other enough to find the beauty of Venus in one another. They could have real trouble adjusting to their partner's speed, and the mutable quality of Pisces often won't help them open up any faster to build a relationship in the pace that would fit their Libra partner. Libra and Pisces have a meeting point in the beauty of Venus. Both Libra and Pisces can selflessly be interested in the satisfaction of their partner, and this should help them stay on the good side of their relationship whatever happens between them. If they move past the disrespect and the unrealistic expectations from each other's personalities, they might find that they share real love. 29%

PART-8
SCORPIO HOROSCOPE

(October 23rd to November 21st):

Who has an intense and powerful nature and keeps ready an arrow in his bow?
Who is self centred, wilful, proud, and detective and if you prod him, lets it go?
Who is a fervent friend, a subtle foe?
Element: Water
Quality: Fixed
Ruler: Pluto
Lucky Gem: Coral, Opal. He gathers courage from the coral. The stone could also make him/her rich and also gets confident.
Lucky Number: Discover magical powers of Number 2, 3, 7, 8 & 9.
Colour: Midnight Blue, Maroon, Scarlet, Red, Rust
Lucky Colour: Midnight Blue, Maroon. Kiss those blues away with midnight blue.
Day: Sunday, Tuesday
Lucky Day: Sunday. There will be sudden windfall on Sunday.
Lucy Flower: Chrysanthemum.
General Insights: Scorpio-born is passionate and assertive people. They are determined and decisive, and will research until they find out the truth. Scorpio is a great leader, always aware of the situation and also features prominently in resourcefulness. Scorpio is a Water sign and lives to experience and express emotions. Although emotions are very important for Scorpio, they manifest them differently than other water signs. In any case, you can be sure that the Scorpio will keep your secrets, whatever they may be. Pluto is the planet of transformation and regeneration, and also the ruler of this zodiac sign. Scorpios are known by their calm and cool behaviour, and by their mysterious

appearance. People often say that Scorpio-born is fierce, probably because they understand very well the rules of the universe. Some Scorpio-born can look older than they actually are. They are excellent leaders because they are very dedicated to what they do. Scorpios hate dishonesty and they can be very jealous and suspicious, so they need to learn how to adapt more easily to different human behaviours. Scorpios are brave and therefore they have a lot of friends.

Personal Quality: Scorpion is ruthless, mysterious, magnetically, attractive and emotional intimacy and is the most intense and passionate. He/She also has immense degree of willpower and is highly tenacious. He/She is of a secretive, timid, retiring nature, one that does not talk of his affairs. He/She is honest and independent nature person with hard struggles in life. He/She does not believe in accumulating the illegal wealth. He/She has strong will power and a natural quality of leadership. He/She does not believe in empty promises. He/She can be vindictive, dangerous enemies and possesses a strong streak of venom. He/She starts good earning at the age of 24 and after 40 years of age acquire properties. He/She is self contained and self centred and seethes and doesn't give up the enmity. He/She may burst any moment. He/She is too demanding, too unforgiving of faults in others. He/She is very jealous. He/She is the symbol of sex and passionate lovers.

Positive Quality: He/She is brave, resourceful, passionate, stubborn, a true friend, courageous, sincere and loving, subtle, imaginative, powerful, generous, loyal, passionate, exciting, and magnetic. He/She is the person with the fixed mind and achieve goal directly by his deed. He/She is not afraid of obstacles because he has a strong will power.

Negative Personality: He/She is proudly, distrusting, jealous, secretive, violent, over sensitive and careless. He/She is jealous, resentful, obstinate, compulsive, obsessive, brooding, secretive, revengeful, possessive, and extremist and can appear cold and impassive.

Health Concerns: He/She is prone to ailments of the liver and kidneys, stones and gravel in the bladder or genitals, and other genital ills such as pianism, abscesses, boils, carbuncles, fistulas, piles, ruptures and ulcers.

Physical Appearance: He/She is always striking and has a magnetic face and dress elegantly. There is a strange mysticism and magnetism in his personality, which is enchanting to the beholder.

Relationships: He/She likes to stay away from his/her family and leads an independent life. His/Her marital life with Pisces, Taurus, and Virgo and Cancer girls will be pleasant.

Career: He/She is traditionally associated with jobs such as mining and detective work. He/She makes demanding bosses. He/She might consider job such as Analyst, Criminologist, Detective, Doctor, Enforcer, Hypnotist, Insurance agent, Investigator, Lab technician, Private investigator, Psychiatrist, Psychologist, Researcher or Scientist.

Scorpio likes: Truth, facts, being right, long time friends, teasing, and a grand passion

Scorpio dislikes: Dishonesty, revealing secrets, passive people

Ideal Partner: Scorpios fits best with Taurus. This sign gives him the material and emotional security he craves.

Compatible Signs: Cancer, Virgo, Capricorn, Pisces

Incompatible Zodiac Signs: Taurus, Leo and Aquarius

Greatest Overall Compatibility: Cancer, Pisces

Best for Marriage and Partnerships: Cancer

BUSINESS ASTROLOGY FOR SCORPIO

Scorpio can work long hours and withstand pressure to succeed in Business. They follow their passion to an extent that other often misunderstands them. For a successful business they should work upon to handle people and situations tactfully. Business booster Gem is Black Quartz Crystal.

Scorpio Compatibility with Aries

Scorpio & Aries Sexual Intimacy and Compatibility: Aries is our first breath, Scorpio is the last. They are two sides of the same coin, both ruled by Mars, a planet of instincts, necessities of the body and sexuality as one of these. When they are in a sexual relationship, it can be difficult to set all of the aggression aside. Not only are they both ruled by Mars, but Scorpio is ruled by Pluto, too. Pluto is known for its destructive qualities, usually related to sexual repression and it can intensify all things, sex primarily. So they are basically a combination of everything we don't want to deal with when it comes to sex, taboos and instinctive sexual behaviour. This is a contact that lacks pleasures and tenderness of Venus. Both signs are the opposite of one's ruled by Venus and represent positions where Venus is in detriment. We could say that this means "lack of love", but it is not quite that simple. Since Scorpio is a Water sign, it is connected to our deepest, darkest ability to love. Aries and Scorpio are signs with an unbreakable bond. Scorpios need to feel emotion in their sexual experiences, but due to suppressive nature of our society, can live out some weird sexual scenarios that may seem "sick and twisted" to more conservative zodiac signs. It is a good thing that Aries rarely belongs to this category, for it is a sign where all conservative and rigid opinions have fallen with Saturn. If Aries and Scorpio find an understanding inside their sexual relationship, they will probably become the atomic bomb of all sexual experiences you can think of. Still, it is hard for them to find their shared language. They are, in fact, completely different. Aries likes things "straight" and simple. Scorpio, on the other hand, has a slight need to manipulate, play a game of seduction and takes sexual relations very seriously. They always want to transcend all of their previous sexual experiences and find someone they can merge their Soul with, to possess and adore until they die. Aries is much more simple and masculine when it comes to sex. It is a physical need that needs to be met. They usually have to build emotion inside a sexual relationship as they get to know their partner. This relationship's real possibility exists only if they share the need to satisfy one another and treat each other with enough tenderness. 50%

Scorpio & Aries Trust: As opposed to sexual compatibility, this issue is easy for them. "If you lie, you die." Not literally, of course, but a small lie could easily end their relationship. They are both jealous and possessive by nature. Aries likes

to win and be the best lover and partner anyone has ever had. Scorpio wants to be the only one that was ever loved by their Aries partner. If they have doubts about each other's actions, it is very likely they won't last very long. 90%

Scorpio & Aries Communication and Intellectuality: From the perspective of Aries, this is something nobody should think about, let alone talk about all of the time. This wouldn't be expressed as boredom (although this is always an option with Aries), but more as a need to act and stop obsessing about everything. Scorpio is too dark and difficult for Aries, as much as Aries is too shallow from Scorpio's perspective. Aries will probably tap their foot impatiently while Scorpio goes on and on about all those deep and meaningful things. What they both enjoy though is their shared ability to give a lot of information in only a sentence or two, but this could interfere with their communication even more, since they might say everything they need to in a couple of minutes and have nothing to talk about afterwards. 20%

Scorpio & Aries Emotions: Aries will probably never know or understand what happened in Scorpio's emotional world because they simply didn't sense anything. They don't have a strong affection to emotion in general and they are both trying hard to be strong and unemotional. This is due to the archetypal "battle" of Mars with the Moon – the rejection of one's emotional Self and too much roughness in order to survive. It is really easy for Scorpio to get hurt here. It seems like they jumped into this relationship only for this reason, so they can repay some sort of a karmic debt. Since there is no one here to keep the emotional balance between them, it will be very easy for them to openly "cut" one another, possibly many times, before one of them decides to cut their bond entirely. 1%

Scorpio & Aries Values: They part ways in their further processing of these. While Aries considers something is done with as soon as it's cleared, Scorpio will dig for reasons why it would be unclear, or was unclear in the first place. So when together, they would both feel the need to clear things up, but Scorpio will obsess about them even when issues are solved and find new details that need to be cleared up, again and again. They need to be productive and fully independent, or they will drive their Aries partner crazy. When it comes to bravery, Aries thinks of bravery as a knight's tale, something to show when you are wearing your sword, while Scorpio thinks it is brave to sink into the darkness of the mind, go to the underground, the underworld or challenge the devil himself. It is a good thing they both value bravery and things that are concrete and clear. This is exactly where the difference in their deep levels of the nature of Mars comes to light. Although everything seems the same, nothing is even remotely close to being similar at all, as soon as you scratch beneath the surface. 40%

Scorpio & Aries Shared Activities: You could say that their main shared activity is sex. Everything else is secondary anyway. 99%

Conclusion: Fire evaporates Water, just like Aries shatters Scorpio's feelings. Water damps down Fire, just like Scorpio wears Aries out. Think of this combination of signs through the most aggressive image of Fire and Water element. Fire evaporates Water, just like Aries shatters Scorpio's feelings. Water damps down Fire, just like Scorpio wears Aries out. They seem to bring out the worst in each other and this is nobody's fault, it is just hard to reconcile so much focused energy that moves in two different directions. Their relationship is like the process of nuclear fusion and often just too much to handle. 48%

Scorpio Compatibility with Taurus

Scorpio & Taurus Sexual Intimacy and Compatibility: We wouldn't primarily link Taurus to sexuality, but it does represent sensuality and is a sign that governs physical pleasure. Their relationship is a connection of deepest emotions and sexuality that no other couple in the zodiac is privileged to have. Any sort of sexual frustration could lead to a pretty dark approach to their sex life. Scorpio has this depressive need to die naked and sweaty in the arms of a loved one, while Taurus has the need to be loved this much. It may even sound romantic, but carries with it all unsolved emotional issues as baggage into their sexual encounters. This doesn't mean their sex life will be bad. On the contrary, they will both find it fantastic, because they will fill it with all sorts of emotions, good or bad. In the end, emotion will be the only thing that is left and sex will be a way to connect rather than a means to personal satisfaction. As all opposing signs, Taurus and Scorpio can be madly attracted to each other, more so because of the sexual nature of their signs. This can become an obsession and even an addiction. As signs of fixed quality, when they click, it is impossible to separate them, and no one would want to when you consider the possible vengeance of Scorpio. They represent the basic contact between sexual planets Venus and Mars, while being from the physical and emotional realms as an Earth and a Water sign. They are the signification of a deeply intimate relationship and a very rich sex life, for as long as Scorpio is tender enough and Taurus ready to experiment. 95%

Scorpio & Taurus Trust: We wouldn't exactly say that Scorpio is insecure, but their deep emotional nature makes them question everyone's motives in caution not to get hurt. There is a fine line between two possibilities in a relationship of Taurus and Scorpio. The first possibility would be the one in which Taurus partner is really closed up, unreachable and too quiet. This could wake the suspicious nature of Scorpio and their obsessive interrogations will damage their mutual trust even more then they lacked it in the first place. The second option would be for Taurus partner to be open just enough to share what Scorpio needs to hear. As all opposing signs, Taurus and Scorpio can be madly Scorpio rarely trusts anyone but themselves unconditionally and in a relationship with Taurus they need to build the sense of security. If they manage to find this fine balance, they shouldn't have a problem. As they get more and more intimate, Taurus will feel secure enough to share everything Scorpio wants to know and Scorpio will realize that their stable and unchangeable Taurus won't disappoint them. 80%

Scorpio & Taurus Communication and Intellectuality: We should keep in mind that opposing signs complement each other perfectly and their communication should be exciting, challenging and something to enjoy if they are both confident enough. While Scorpio would go in depth about all those things Taurus doesn't seem to care about, they could be very surprised to find that behind the tender and alive nature of Taurus, there is a deep understanding of everything natural going on, however dark it might seem. In return, Scorpio will show Taurus the value of life from their perspective. As all opposing signs, they seem completely different and as if they have nothing in common. Taurus will find it incredible how Scorpio as a sign of death and destruction, can understand the depth of life and emotion better than any other sign in the zodiac. 75%

Scorpio & Taurus Emotions: Not only do these two represent the axis of Moon's special dignities, exalted in Taurus and fallen in Scorpio, but they also have Venus as a ruler on one hand and the intense element of Water on the other. When they fall in love, they become an image of eternal love. This emotional connection is really something to deserve. There is no better personification of Had, the god of the underworld in ancient Greece, and Persephone like an idea of immortal love that can never die. 99%

Scorpio & Taurus Values: The depth of their belief system goes as far as planet Earth's core and if they begin their relationship on the same page, this could be what binds them for years. They value life and love in a way that no other sign understands. Although their perspectives differ when it comes to material and emotional values, their core is the same and everything else can be adjusted. 99%

Scorpio & Taurus Shared Activities: As a fixed sign, they are, as much as Taurus, static and inert. There is a lot of energy to Scorpio, nuclear energy lies within their sign, but when it comes to everyday life, they tend to repeat patterns and blend in what mostly other people find "normal". They do however need new, exciting and breathtaking experiences from time to time, but they would be ok having them alone if their Taurus partner wasn't interested. Although Scorpio is a sign of change, this doesn't mean they are not very slow in their everyday routine. Of all possible activities, they will mostly share sexual ones and all experiences of physical pleasure. They will both enjoy discovering how far their sexual desire could lead and this will keep them busy most of the time. 85%

Conclusion: This has to be the focus of their relationship, for they can't seem to understand platonic and imaginative relationships when they get together. There is no such thing as a platonic experience of romance, when the whole point of romance is to get physical. It is very possible that they will build their sexual life to the point where no other partner could ever satisfy their needs. Taurus and Scorpio are both signs of deepest physical pleasure, each in their own way. This could lead to a possessive relationship with no way out, although they probably wouldn't want to get out even if they could. The entire experience can be too dark for the Taurus partner, especially if their practical sense is challenged by Scorpio's character. In case they are both independent and ready to blend with someone else, they could be the perfect connection between sexual and emotional, the one that we all wish for. 89%

Scorpio Compatibility with Gemini

Scorpio & Gemini Sexual Intimacy and Compatibility: Gemini is so far off from Scorpio's emotional world that good sex between them seems like something almost impossible to happen. This couple needs to be supported by other positions in their natal charts if they are to stand any chance of lasting in a loving, sexual relationship. Gemini can be superficial and there is no other sign who knows this better, than Scorpio. Their Air element combined with the rule of Mercury and its lack of emotion is close to Scorpio's worst nightmare. Scorpio is a sign of our deepest emotions and as such, linked to the most intimate side of sexuality. When they begin a relationship with Gemini, it probably never crosses their mind that such an asexual person can exist in the world. If they fall in love with each other, there is so much for both of them to learn. A sexual relationship between a Gemini and a Scorpio is like a connection of the deepest and the highest point on planet Earth. Scorpio gives a strong focus on their sex life and can be very creative when relaxed. Still, they have a tendency to make a dark, sadistic or masochistic atmosphere that Gemini can only laugh at. If their mutual respect is at a very high point, Gemini could teach Scorpio that not everything needs to be so fatalistic in their sex life. In return, Scorpio will give their Gemini partner depth and emotional vibe to sex that they have never encountered before. 1%

Scorpio & Gemini Trust: They have this weird, possessive nature that can give ultimate trust to their partner until the first worm of suspicion is created, usually by flakiness and disrespect. They like things clean and without a doubt, and are prepared to give unconditional honesty, expecting the same in return. This is where Gemini comes and asks a simple question – how can I be honest when I don't know what my truth is in the first place? It is ridiculous to expect such a definite honesty out of a Gemini partner when they are so prone to change and have no idea what they will feel or do tomorrow. Scorpio trusts everyone until they don't. They have a knack for communication that can save the day, but it can only be used if Scorpio doesn't feel threatened because of their previous heartbreak they've never healed from. 5%

Scorpio & Gemini Communication and Intellectuality: They will be moved and intrigued by Scorpio's nature, and very curious as usual. This will be amplified by Scorpio's depth and interesting topics, up to the point where they get too dark and depressive. This is something Gemini doesn't want to deal with if they don't have to. Unfortunately, there is not a lot that Scorpio thinks they can learn from Gemini. Although they might wish to be more superficial, when they get in contact with someone with less depth, their ego sparks and they feel dominant because they are just the way they are. They will hardly ever respect Gemini for this, and can try to feed on their personality in order to add quality to their own. In case they share interests and have similar professional or educational directions, they could complement each other very well. It is a good thing that Gemini can communicate with anyone. Gemini would give ideas and discover new information, while Scorpio will dig in and give real essence to everything. Their communication can be inspirational if they get into this mode and start accepting each other's qualities. They have so much to give to each other and it would be a shame if they held their relationship in an ego conflict for too long. 20%

Scorpio & Gemini Emotions: If one of them falls in love with the other, they will hardly have a good time if their feelings are not returned in the same proportion. We could say it's hard for them to synchronize, because the emotions Scorpio would give away freely are something Gemini would have to be swept off their feet for. They don't seem to have the same emotional scale and this can leave them both unsatisfied or pressured to a breaking point. Emotional lack of compatibility is what ruins their sex life and gives them both a headache. The best option for their shared emotional world would be for both partners to give what they can and not expect anything in return. 1%

Scorpio & Gemini Values: Although Scorpio values many other things in someone's personality, they will be impressed by someone's intelligence and resourcefulness. Gemini will focus on the same thing, but have a slightly different assessment of someone's intellect. The good thing in their relationship is that they both value strength of thought. However, they can agree to have a shared point of same value, although other things they strive for will differ greatly. 20%

Scorpio & Gemini Shared Activities: Gemini will want to change the scenery and Scorpio will want to change their life, gladly taking Gemini's small steps toward this goal. Scorpio is a fixed sign, set in change of huge proportions, and Gemini's mutable quality will annoy them in most cases. Still, they can relate to their need for different experiences in life and the excitement they always search for. A great thing with these two is their openness for change. Even though they might not be excited by the same things, there will be enough excitement for both partners to choose along the path they decide to cross together. 40%

Conclusion: To Gemini, their partner will seem too depressed and dark for no apparent reason, and for Scorpio, this could be an experience with no purpose or depth. If they do fall crazy in love, they could connect through their mutual love of change and give each other the exact things they lack.

Gemini would get deep, emotional satisfaction they have never felt before and Scorpio would finally get the chance to rest their troubled soul, and realize that not everything needs to be taken seriously. Gemini and Scorpio will usually annoy each other senseless. None of them will lightly understand their partner's personality. This is a relationship of great lessons and an enormous capacity for personal growth of both partners. 15%

Scorpio Compatibility with Cancer

Scorpio & Cancer Sexual Intimacy and Compatibility: Cancer can usually understand the need of their Scorpio partner to express their deepest, darkest emotions in their sex life. If Cancer partner doesn't get scared or too forced to do something they are not ready for, a sexual relationship between Cancer and Scorpio can be deeply satisfying for both partners. This is a relationship of two Water signs and because of this their sex life needs to reflect all of their emotional connection or a lack of it if there is any. When they fall in love, they will both need to express their feelings and the intimacy they might share is incredible. However, Scorpio is a sign in which the Moon falls and this is the ruler of the sign of Cancer. The sign of Scorpio is associated with death and all kinds of bad things, but all of their maliciousness comes from their emotional and sexual repression. If Scorpio's need to bury their emotions is too intense, there is a great chance they will be too rough or insensitive on their partner. This is something Cancer will have difficulty coping with and could lead to Cancer's need to separate because they could simply get tired from all the special or aggressive sexual requirements their Scorpio partner has. 90%

Scorpio & Cancer Trust: If they feel betrayed in any way, they can start showing all of those maleficent sides of their nature and become truly possessive and jealous. Cancer partner usually wants someone to share a life with and they will have no reason to cheat or lie to their partner. When Scorpio falls in love, trust is one of the most important things they are looking for. As all water signs, they could both fear telling the truth to a certain point, but this doesn't necessarily need to speak of their unfaithfulness or the beginning of the end of their relationship. Usually they will both be able to give each other enough security to feel safe and build the trust they both need not to feel hurt or betrayed. 95%

Scorpio & Cancer Communication and Intellectuality: This can influence their sex life and make it much better, or much worse, depending on how their need for mystery is expressed. Their communication is very good, for as long as emotions are not the main theme of a conversation. They can finish each other's sentences if they have any need to talk in the first place. The depth they both have, although it might not be visible at first in Cancer partner, makes them able to talk about anything at all. In the case when Cancer wants to run from negative experiences and Scorpio from their emotions, they could have trouble forming a relationship at all. When Scorpio falls in love, trust is one of the most Cancers and Scorpio usually understands each other without words. This is a very rare scenario and even if they have these tendencies, they will probably help each other deal with them and give each other the exact mental stimulation they both need. 99%

Scorpio & Cancer Emotions: Cancer lives buried in their emotions, positive or negative, capable of using them in their everyday routine as an incorporated part of their life. Scorpio can have trouble understanding how this works exactly, because they have a tendency to dismiss emotions, thinking that this is the only way to reach a certain goal. The middle ground they need to find is a place where they are both free to follow these needs. Emotions have to be a way of living, as much as they can interfere with our goals. This is a tricky territory for a couple like this one. Both of these partners need to learn to lose control, as well as gain it again, in order to be able to let things flow and change in the way they are supposed to. 70%

Scorpio & Cancer Values: Scorpio represents change and values it most of all, even if they are not fully aware of this. It can be difficult for these partners to coordinate their personalities if they are both not flexible enough to understand their differences and the depth each of them has behind these superficial needs. Cancer values their inner peace and wants a stable life with a family they can rely on. Scorpio can fear emotion to the point of agony and if Cancer recognizes this, they will be able to approach them in the best way possible and discover their true need for security and emotional balance. 25%

Scorpio & Cancer Shared Activities: They have to share emotions and protect their loved ones like Cancer due to their motherly need to protect people they love, and Scorpio in order to set good boundaries on what they think is right. If they create their own little world, they can be found in any situation together, dealing with things as one being. It doesn't really matter what Cancer and Scorpio will do if they both feel good with each other. Scorpio usually likes some dangerous activities and Cancer will have difficulty adjusting to those, but if their emotional core is good, they will have a quiet understanding of each other's needs, however destructive they might get. 95%

Conclusion: When they find an emotional link, they can go very deep in search of true love, and unite on a level that is unreachable for other zodiac signs. This can make them speak without words, understand each other's thoughts with only one shared glance and be synchronized in their approach to their future together. If their emotions aren't shared on a deepest possible level, or Scorpio partner refuses to deal with them, it could be too hard for Cancer to handle the self-destructive nature of their partner. A relationship between a Cancer and a Scorpio can go from one extreme to another, and although Cancer partner will try hard to stabilize it, it might be too difficult if Scorpio doesn't have enough respect for their own emotions. Their connection needs to be sincere and pure, in order for both of them to be ready to give in to this intense emotional contact. 79%

Scorpio Compatibility with Leo

Scorpio & Leo Sexual Intimacy and Compatibility: Leo is a passionate lover, warm, always in search for action and they can be quite casual when it comes to their sexual encounters. Scorpio is sex itself, and the depth of emotion that goes with it in its purest form. When they get together, they could have real trouble finding middle ground between their personalities. These partners can seem as if they've crashed into each other with no plan or purpose. If they are attracted to each other, this could drive them mad, for none of them will be able to realize their desires in a wanted way. If they end up having sexual relations, they could have misunderstandings on everything, from their verbal communication to their physical needs. They simply don't operate in the same ways and while Leo wants to be respected, Scorpio understands that all respect dies in the act of sex. This is a complicated relationship between two strong personalities with an incredible sex drive. It is extremely difficult for a Leo and a Scorpio to reach intimacy, because they have a different view on emotions. What Leo sees as love, Scorpio finds superficial and irritating, and what Scorpio sees as love, Leo finds depressing and irritating. They both need to give up control entirely if they want to find sexual satisfaction with each other. 5%

Scorpio & Leo Trust: This is not such a good thing when it comes to their ability to adapt and be flexible for each other, but it is a perfect thing for mutual trust. If they set the clear foundation in the beginning of their relationship, Leo transparent as they are and Scorpio direct and honest, they might trust each other without exception for a very long time. The positive side to this relationship when it comes to trust is in the fixed quality of both signs. That is if they both want to be open for this kind of relationship in the first place. 65%

Scorpio & Leo Communication and Intellectuality: Although this is not always the case, Leo wants to show the right image to the world, and Scorpio understands karma better than many other signs. This is why they will probably have enough respect for each other to communicate in a civilized manner. They are both obsessive in a way. Leo will never give up on chasing their passion, with enough energy to spark everyone around them, and Scorpio will hold on to things they care about, and obsessively fight for their goals.

If they share the same passion or interests, they will have something to talk about, obsessively. The depth that is typical for Scorpio is something that Leo tries hard to reach in their search for Unity. It is a good thing that these two signs can be so well behaved. Their conversations can be very tense and irritating for both, but along the way they might realize that they give each other exactly what they both need. 30%

Scorpio & Leo Emotions: In some cases, they can be identified with hate, but the important thing to remember here is that hate is also love, in its "negative" form, and both of these partners will think that any emotion is better than no emotion. Their a bit torturous relationship can hold them together for a long time, even though they might be unhappy and aware that they could be happier with someone else. In a way, Scorpio likes to be tied through negative emotions, for love sometimes has to hurt, and Leo sticks with their decisions because they rarely accept that they might have been wrong. This relationship can become a very difficult circle of suffering for both partners, especially if any one of them doesn't have their independent life, friends and finances. This is probably the most challenging relationship in the entire zodiac when it comes to emotions that these partners have for each other. When they do, they might find a fine balance, for as long as they both have freedom to think that there might be better options for them with other people. If they don't find any, they could realize that they are perfect for each other as self-sufficient individuals, through a healthy approach. 1%

Scorpio & Leo Values: Often, they won't even recognize their similarity out of an emotionally unstable or obsessively stable position, completely different from a passionate, creative one. Both of these partners will value honesty and clarity. Although they understand clarity in different directions and depth, the main characteristics in people they wish to date are very similar. They have to deal with the value of creation against the value of destruction and this is not easily reconciled. The bridge between them is found in unconditional honesty. 35%

Scorpio & Leo Shared Activities: While they are in most cases interested in different things, Scorpio is ruled by Mars, a planet that Leo understands well and cooperates well with. They will both have the energy to follow each other's desires and they could actually have a lot of fun. They usually stumble upon a problem when it comes to their understanding of certain circumstances and people around them. There is rarely a compromise between the positive, constructive approach of Leo and the often negative, sensitive approach of Scorpio, especially when none of them is exactly true. It is very interesting how a Leo and a Scorpio might organize their time together. It would be best for them to remain true to themselves, and understand that there is a middle ground between their opinions and feelings, however strange it might seem. 40%

Conclusion: This is in no way an easy relationship, and both partners can be stubborn and stiff in their opinions, life choices and ways they handle reality. When Leo and Scorpio start dating, they might not know exactly what they are to expect. If they want to remain in a loving relationship, they need to understand each other's way of expressing emotions and respect each other's needs however different they might be from those they are used to. When they find a way to love each other without conditioning, they might realize that they are in search for the same thing like Unity. 29%

Scorpio Compatibility with Virgo

Scorpio & Virgo Sexual Intimacy and Compatibility: Even though Scorpio can be too rough of Virgo, making them feel uncomfortable and even violated in a way, in most relationships between representatives of these two signs, there is enough rationality to the approach of Virgo to make this contact possible. What we often fail to understand is the fact that Scorpio is a Water sign and as such like deeply emotional. Virgo looks for someone emotional to share a life with, and if they share this emotion of Scorpio through their sexual relations, they will both find sex between them extremely satisfying. The best time for Virgo and Scorpio to create enough safety and emotion in their sexual encounters, is in the situation where they are each other's first truly emotional experience. If they surprise each other with the power of emotions beneath the surface, that both of them seem to carry around, they will have a hard time ever separating from one another. If there is something Scorpio would like to fight for, it is the chastity of Virgo. This is a very interesting couple in the domain of sexual activity like one of them hiding their sexuality, and the other acting as sex itself. The biggest problem of these partners is in their relation to Venus, and this can lead to loveless acts of sex that both partners are not truly satisfied with. They need to show love and be tender enough, enjoying themselves enough, or they might have to move on to someone they love more. 65%

Scorpio & Virgo Trust: There is a strong understanding here, for one of them fears betrayal more than anything, while the other hates it and gets vindictive as soon as any sign of dishonesty is in sight. Trust is a very challenging issue for both of these signs and this is something they can finally talk about with each other. The best thing about their connection is in their ability to understand each other in silence, not ever wanting to let each other down. 90%

Scorpio & Virgo Communication and Intellectuality: Scorpio represents a deep silence of the flow of a river, and they will both have a strong urge to jump into the depths of silence together. Their intellectual contact is stimulating, often strongly influencing their sex life and their truly deep emotions. It is almost as if they wouldn't be able to form a relationship without this ability for non-verbal communication that makes them perfect for each other. Both of these signs are prepared to go all the way - Virgo in their intellectual depth and Scorpio in everything in life. This will inspire both of them to search for all sorts of answers together, analyzing each other's psyche and determining the source of their problems with the world, or with each other. Virgo is a talkative sign, ruled by Mercury the planet of communication, but they hold on to a much more quiet and intellectual side of Mercury than we might anticipate. Through carefully chosen words, they can help each other heal or regenerate from difficult or even devastating experiences. It is a good thing for both of these signs to have each other in the time of need. 99%

Scorpio & Virgo Emotions: We could say that this ability hides in both of them and the dig up goes both ways. The problem here is in the fact that they remind each other of their imperfections. Scorpio is a sign in which the Moon falls and at the same time the sign of Venus' detriment. All emotions get lost here, as if Scorpio is a black hole that cannot get enough. Virgo partner is already sensitive and when in love, does everything they can to satisfy their Scorpio partner. This can feel like investing into a black hole with no gratitude whatsoever. There is no other sign that can sense the needs of Scorpio better than Virgo, and no other sign that can dig up the emotions in Virgo better than Scorpio. If someone can reach the emotions hidden behind the extremely rational approach of Virgo, it is Scorpio. An emotional relationship between them can turn out to be truly dark and difficult, but also incredibly strong and intimate. The only thing that can bore their emotions to death is the criticism they are both prone to. 75%

Scorpio & Virgo Values: There is nothing in the world that is as exciting as conversations that are so intense and so challenging for their minds. Most of the time they will agree on things they value most, although they might stumble upon a huge problem when they get to the point of throwing out the trash. Even though Scorpio doesn't normally accumulate things, and loves throwing them away, those they hold on to can be quite disgusting to a Virgo. Both of these partners will value depth, intellectual most of all. Just imagine as their first child is born and Scorpio wants to frame that dried out residue of an umbilical cord. Do you think Virgo would want to wake up to this in their apartment every morning? 70%

Scorpio & Virgo Shared Activities: Virgo will clean, that's a fact, simply because a clean house creates a clear mind, and Scorpio won't have much trouble fitting in, unless their personal belongings are questioned. However, when they choose places they want to visit, or clubs they want to go to, their choices will differ greatly. It is not hard for them to compromise to keep the relationship going, but it can be quite dark and demanding for both partners. The same reason their values might differ, their daily routine might differ too. If they don't start willingly hanging out in some tidy cemeteries, they might run out of options that would actually keep them both interested and happy. 55%

Conclusion: In general, there is a problem that these partners share when it comes to Venus, and their relationship is often a reflection of these troubles. This can lead to all sorts of emotional blackmail, their tendency to control each other's lives, and if not this, than constant criticism that makes them both feel guilty or simply sad. The best thing they can do is decide that they will value each other and be thankful for each other in this relationship. That changeable nature of Virgo will be settled down by the fixed quality of their Scorpio partner, who will keep their relationship exciting for a very long time. If they develop a strong sense of gratitude, their relationship might be extremely deep, exciting and truly appreciated by both partners. 76%

Scorpio Compatibility with Libra

Scorpio & Libra Sexual Intimacy and Compatibility: They are linked through their animalistic natures, continuing one another, ruled by planets that also rule their opposing signs. This is a complicated sexual contact, for they are ruled by Venus and Mars, as if they were made for each other. However, they seem to lack the touch of the element they both need, separated by the quick, superficial element of Air in Libra and emotional, slow element of Water in Scorpio. Even if they don't make each other happy in other parts of their relationship, when a Libra partner lets go to their instinctive, animalistic side, they easily become one with their Scorpio partner. Their sex life can be incredibly emotional and demanding, for the pull of their energies is extremely strong and makes them both obsessive and possessive to one another. This is an intense couple in every way and their sexual relationship is something truly exciting and challenging at the same time. They will both often get so tied that their relationship seems impossible to break. These ties are created through their physical contact, even if everything else in their relationship makes them deeply unsatisfied. 45%

Scorpio & Libra Trust: As if this wasn't enough, that possessiveness of Scorpio is easily absorbed by Libra, and they will start acting in a similar way, obsessing about each time their partner wants to do anything alone. Even though this is the worst case scenario, this atmosphere will probably be present in any Libra Scorpio couple on a smaller scale. They are never ready to live like this every day. If Scorpio cannot trust someone, it is Libra. How could they when there seems to be an obvious need in their Libra partner to show how attractive and sensual they are to the rest of the world. If they accept this as a way of sharing true, deep love, than their trust issues might not be much of a problem however weird they might seem. 1%

Scorpio & Libra Communication and Intellectuality: Their communication is the exact place where the connection of Libra to Saturn comes in handy. This will give Libra the ability to slow down, breathe in, and understand the impulsive acts or words of their Scorpio partner. There is just enough depth in both of these signs, even though Libra is an Air sign and quite distant from planet Earth. This will allow just enough understanding between them in a rational sense. The problem they have to deal with is in the fact that they both represent a detriment for each other's rulers. This will easily lead to the dismissal of each other's personalities, especially since Libra has to deal with personality issues through the fall of the Sun, and Scorpio doesn't accept anything that isn't whole. There is nothing Scorpio can say that Libra won't be able to rationalize. Libra will have trouble understanding the aggressively clear side of Scorpio, as much as Scorpio won't understand the dishonest act of their Libra partner. Their differences could be very difficult to reconcile if they get too close to each other and start meddling in their personal lives and decisions. 55%

Scorpio & Libra Emotions: Libra looks for an active, fiery partner that will awaken their life force and bring light into their life. Scorpio looks for someone physical, sensual and practical, and although Libra can meet these needs to some point due to Saturn's exaltation and the rule of Venus, in most cases this is not primarily who they want to be. When a spark exists between them, it will easily build up to a fire, burning entire cities, forests and everyone on its way including both of them. That nuclear energy of Scorpio is ignited by its preceding sign of Libra, because it carries information on all unresolved issues in relationships and all of the anger that hasn't been let out because of the forced need for appropriate behaviour. In other words, Libra will accumulate the unexpressed sexual energy and it will burst in Scorpio, leading to a lifelong war if enough is pushed under the rug. This will be construed as ultimate love, making their emotional relationship ridiculously intense. There is an incredible intensity to their feelings when they do fall in love. In many cases, love simply won't happen. They need to understand that love is about tenderness as much as everything else and if they don't have any, maybe it is not love but simple physical attraction and a need to vent from restrictions. 25%

Scorpio & Libra Values: Both of these partners will value consistency and commitment and this is something that will connect them in the first place. However, the rest of their systems of value don't coincide that much and they will differ on behaviour and expectations of other people. With Libra frowning at indecent behaviour, Scorpio will have indecent friends, honest about their act and fully accepted and respected for that. It seems to Libra that Scorpio does everything out of the ordinary just to seem special, while Scorpio sees Libra as doing everything ordinary just to fit in. The main problem here is in the value they give to opinions of other people and this will not be easy to overcome. 10%

Scorpio & Libra Shared Activities: Scorpio will awaken the animalistic side of their Libra partner, and their time together will most certainly be focused on both partners following their instincts. Scorpio is a sign of the exaltation of Uranus and they will lack patience for indecisiveness of Libra. This is why their needs won't fit very well in other areas of their lives, and they might not be able to meet each other's expectations. If there is an activity a Libra and a Scorpio can share all the time, it must be sex. An exception to this rule occurs when Libra turns to the dark side and gives in to the need of Scorpio to take them to the Underworld. No other partner will ignite this need in Scorpio as a Libra partner. 35%

Conclusion: Both of these partners will have to face their dark sides through this bond, and although this can lead to an incredible and intense sex life, and emotions that no one else can understand, it might lead them both to a depressive hole they won't easily get out of. Both of these partners will have to face their dark sides through this bond, and although this can lead to an incredible and intense sex life, and emotions that no one else can understand, it might lead them both to a depressive hole they won't easily get out of. The relationship of Libra and Scorpio is in no way easy and light. The only way for this couple to last in a satisfying and gentle relationship, is for both partners to build a strong individual, independent life, or they will get sucked into the whirlpool of karmic emotions and obsessive, negative expectations. 29%

Scorpio Compatibility with Scorpio

Scorpio & Scorpio Sexual Intimacy and Compatibility: Their sexual energy and inner tension is something often hard to handle individually, and when they get together, this either multiplies to infinity, or they find an absolute understanding. Usually, we can predict the first option. In general, every Scorpio needs someone to balance them, for they go to extremes of all kinds, and when together, they will rarely have the patience or the tenderness to balance anything, let alone each other. Their sex life is intense, often amazing, although everything around it seems to be falling apart. They will fight to make up, manipulate obviously to make each other angry, and do things out of spite. When we talk about spicing one's sex life, we have to understand that these two are spices themselves, and there is rarely anything boring about them. Two Scorpio partners can be a dream comes true when it comes to sex, as much as they can be each other's worst nightmare. In order for their sexual relationship to work out, both partners need to stick to that inner tenderness and emotional closeness when together, or they might have to end things as if torn apart by nuclear energy. The key here is to slow down, breathe in, and spend time in each other's cosy arms. 65%

Scorpio & Scorpio Trust: As much as each Scorpio wants to be involved in their partner's life, no Scorpio wants to be controlled or let anyone else be involved in their own life. None of them will understand the lack of trust coming from their partner, lifting their ego high, knowing that there is no reason why they would continue sharing anything with the other Scorpio anymore. The problem here is in the fact that neither of them seems to be aware of what they want from their partner and this can turn into a real battle for supremacy. What a strange relationship of two possessive, want-to-know-all, striving-for-freedom partners. Often enough, they will see each other as true and honest, leaving their insecurities aside, trusting each other without too many words. 40%

Scorpio & Scorpio Communication and Intellectuality: No one can understand a Scorpio intellectually as much as another Scorpio. Their topics can easily become dark, not because they both want to talk about depressing things, but because they understand each other in areas other people don't want to deal with. This is a good way for both of them to discover that they are not alone, and it can be healing for each partner for as long as emotional expectations are not involved. If they work together and compete to advance to the same position or feel like they endanger each other's status in any way, their contact can become truly unpleasant and turn to arguing, dismissal of anything said, or in one word like disrespect. When they start a battle against each other, it cannot seem to stop, until one of them is beaten "dead" and there is nothing more that can be done for them to get back in the game. If they see each other as adversaries, they will sting each other for victory, meaning they will do whatever it takes to win. When emotional baggage is not a part of their everyday communication, the depth both of their minds have will be an incredible stimulus for each conversation they have. If they distance themselves from a situation, they might realize that none of them actually endangers the other. To find mutual understanding, they need to lead separate lives and give each other enough room to do so. 70%

Scorpio & Scorpio Emotions: This makes representatives of this sign turned to the dark emotional issues and this is something they will both understand in each other. The problem with Scorpio is in its battle with the Moon and the fact that emotions aren't approved here. This can make them both intolerant for weakness, rough in their approach to one another and too judgmental toward each other's emotional needs, even though they both actually share the same needs. They will both be faced with emotions they don't want to deal with, simply because that is their individual role in other people's lives. This could lead to numerous conflicts, but it could also be the base for incredible personal evolution of each partner, and an opportunity to be with someone who truly understands the depth of their hearts. Scorpio is a Water sign and represents all dismissed emotions we don't want to deal with. The best way for them to approach this relationship is through deepest emotional acceptance and the tolerance for incredible difference in character, even though they are, strangely, the same. 55%

Scorpio & Scorpio Values: They will value rationality and emotional maturity, but those are things none of them can actually deliver all the time. They belong to the element of Water and are deeply emotional, while valuing each other's lack of emotion, shown at the beginning of their relationship. The problem with things they value is in the illusion of value they both share. They do value similar things, but they are not entirely rational or realistic in their choices. This makes them susceptible to judging their own reflection in one another. 90%

Scorpio & Scorpio Shared Activities: This is about the balance they both seek, needing a partner of opposite character and choices. Their contact could lead them to all the dark places, as they play out situations they couldn't play out with anyone else. Even though they will feel sad and dried out most of the time, as two members of fixed quality, they could hold on to each other for a long time. They won't have a hard time discussing or agreeing on where to go and what to do together, but their choices might not make them happy. The more time they spend together, the less energy they might have, for there is a silent, inner battle in both of them that drains it out of their systems. 75%

Conclusion: Scorpio and Scorpio have this tendency to bring out the worst in each other. Even though they can share the deepest understanding known to the entire zodiac, they can also get too dark and depressed together, sinking into their pool of unresolved emotions. Their emotional understanding is something worth cherishing, if they are both open for their own feelings and accept their own inner needs. 66%

Scorpio Compatibility with Sagittarius

Scorpio & Sagittarius Sexual Intimacy and Compatibility: The strength of character they share is something that will give them just the right amount of confidence when it comes to sex, and the creativity and openness of Sagittarius will be refreshing for the fixed nature of Scorpio. However, this often doesn't last very long and in time, in most cases, Scorpio starts thinking of their Sagittarius partner as unreliable and not to be trusted, while Sagittarius sees Scorpio as dark, pushy and too controlling. In order to remain in a healthy sexual relationship, both of these partners have to compromise, Scorpio finding a way to give freedom, while Sagittarius finding a way not to run away from the seriousness of their partner. In the best possible contact, Scorpio will give their sex life emotion and true physical intimacy, while Sagittarius will be there to give meaning and shake things up, representing the light at the end of a tunnel. There is a strange understanding between these two signs, as if they were one and the same, at least for a little while. Together, they can build an incredible sexual relationship, for both tend to be uninhibited about locations, positions and situations in which they wish to make love. 25%

Scorpio & Sagittarius Trust: Scorpio has the need to tie their partner down, even if they have that seemingly liberating view on love. There is nothing a Sagittarius will dread more than someone trying to control their life. If someone can spark the need of Sagittarius to be unfaithful, it is most definitely Scorpio, launching them further and further away by trying to come closer. An additional problem with this couple is in the quality of their signs, Scorpio being fixed and Sagittarius mutable. This makes it almost impossible for them to share a pace, and in order not to disappoint each other, they both might choose to lie. The strangest thing in this relationship is in the fact that these two signs are, without a doubt, the most honest signs of the zodiac. The biggest problem for a Scorpio and a Sagittarius in a relationship is trust. Their relationship seems to face them with the other side of their personality; one none of them wants to see. 1%

Scorpio & Sagittarius Communication and Intellectuality: Not only does Scorpio feel lighter, more optimistic about life and everything in it when communicating with a Sagittarius, but the depth they give to Sagittarius' mind and ways to reach conclusions is impossible for any other sign. Scorpio will face Sagittarius with any superficial or outdated views, while being compassionate enough to know how to do this without hurting them, and being fixed enough not to change their intent somewhere along the way when Sagittarius thinks of running away. They will enjoy each other's company for as long as expectations and emotional disagreements are not in focus, for they have an incredible thing to share like their search for truth. Two such strong individuals give each other exactly what each of them needs when it comes to their mental compatibility. The meaning they seek in all things in life, Scorpio going into depth and Sagittarius travelling wide will connect them through a strong bond that no other combination of signs can form. If they are on a shared mission, they can accomplish incredible things and have real epiphanies together. 80%

Scorpio & Sagittarius Emotions: The meaning they seek in all things in life, Scorpio going into depth and Sagittarius travelling wide will connect them through a strong bond that no other combination of signs can form. They will enjoy each other's company for as long as expectations and emotional disagreements are not in focus, for they have an incredible thing to share like their search for truth. If they are on a shared mission, they can accomplish incredible things and have real epiphanies together. 10%

Scorpio & Sagittarius Values: They can both feel like outcasts and value each other's decisions to differ from others out of self-respect. Still, in most cases they will easily consider each other invaluable in a way, for they cannot meet the expectations each of them has for his partner's personality. Since Scorpio is the sign of the exaltation of Uranus, and Sagittarius understands this through its third house, they both value freedom and one's ability to fight for their beliefs. The only way for them to value each other, is to focus on the positive characteristics and sides of one another and their entire contact. 35%

Scorpio & Sagittarius Shared Activities: Sagittarius will want to try anything new and Scorpio's approach to life is always new from their perspective, while Scorpio will enjoy the first impulse of optimism and fun. It is a good thing that Scorpio exalts Uranus, for this gives them enough love for change and exciting, new things that Sagittarius can bring into their lives. However, the fixed quality of Scorpio will make this exhilaration fade as soon as their relationship becomes routine, in any way. For as long as their communication is inspiring for both partners, they won't need much to be satisfied by their activities. As time passes, there is a great chance for Sagittarius to start feeling pressured, or simply be bored, and this doesn't give great promise of future. 30%

Conclusion: While they don't know each other well and everything seems new and incredible, Scorpio will see their Sagittarius partner as a ray of light that suddenly makes their life brighter and better, while Sagittarius will see that there is so much to learn and enjoy the depth of their Scorpio partner, followed by emotional attachment. In time, there is a strong chance they will slowly lose interest in one another, especially the mutable sign of Sagittarius for their fixed Scorpio partner. Scorpio and Sagittarius make a pretty great couple, for as long as they feel the first excitement at the start of their relationship. Even though their relationship might end on bad terms, it would be a shame not to give in to it and let it fascinate and exalt both of them for however long. 30%

Scorpio Compatibility with Capricorn

Scorpio & Capricorn Sexual Intimacy and Compatibility: The physical nature of Capricorn will help Scorpio ground their sexual needs with ease. The main problem of this couple is their relationship to the Moon, for they are signs of its fall and detriment. This "agreement" not to be too sensitive and emotional, can take out any real intimacy from their sex life, and make them too cold and distant, even though physically enjoying their relationship. They might even think that this is all they need, but their hearts won't agree, and other people will show up in their lives that show them how much they actually depend on intimacy. Both of these signs feel a gravitational pull toward their opposing signs, Taurus and Cancer, two of the most emotional signs of the zodiac. This explains their need to build real intimacy. For both of them, physical pleasure has to be achieved through tenderness and emotion, or they won't truly be satisfied. In general, Scorpio exalts Uranus, and they might be a bit frustrated by the conservative approach of Capricorn. Scorpio and Capricorn share a special sexual bond as signs in Sextile with each other and due to the fact that Capricorn exalts one of Scorpio's rulers, Mars. It is a good thing they can wait and slowly build up an atmosphere in which their Capricorn partner will be relaxed enough to try new things and experiment. The excitement of this sexual contact is something Capricorn will have trouble letting go. Scorpio, on the other hand, will enjoy the sense of security and patience they get from their partner, even if they openly express their sexuality. 60%

Scorpio & Capricorn Trust: Even though Capricorn representatives don't have to be that honest at all, their relationship with this direct and honest partner will make them feel like they should be as honest as possible too. Any lack of trust in their relationship is a consequence of the lack of intimacy, for they seem to lack the ability to sense each other deep enough to understand if they do trust one another or not. If there is a sign Scorpio can trust, it is the sign of Capricorn. This can be solved if each partner deals with their own insecurities individually and with an emotional effort to build intimacy. 90%

Scorpio & Sagittarius Communication and Intellectuality: Their similar pace and the patience Capricorn has, followed by the feel of Scorpio can help their understanding very much, but when they disagree on something, they could end up in a silent fight for years to come. There is nothing light or easy in this interaction, and even though they understand each other's depth of mind and a certain "what goes around comes around" view on life, they will rarely laugh, dance, and have fun together. They might think that this is not even something they need, but everyone needs to have some fun and smile, or life loses a lot of its meaning. Dark humour might savour their situation, day after day, and if they have the same friends it will be much easier for them to enjoy life together. All friendships Scorpio makes, become long-term with the help of Capricorn when respected enough, and this could help them build a wonderful surrounding full of understanding people who love them. In general, they get each other's need for silence and patiently approach each other until each of them opens up. Fixed, unmovable Scorpio, in the state of constant metamorphosis and evolution in the same direction, can be a bite "too big to chew" for their stubborn, earthly, long lasting in everything Capricorn. This can open the door for a respectful communication and intellectual understanding that lasts for a very long time, if they learn to control all the negative convictions that surface when they are together. 70%

Scorpio & Capricorn Emotions: The biggest problem in a relationship between a Scorpio and a Capricorn is their emotional contact, simply because they both tend to have emotional problems, dismissing how they feel by dismissing the Moon. When they start their relationship, they will both give the impression of people who stand with their feet firmly on the ground, strong and rough when needed. They will rarely notice that this brings out the expectation to always be the strong person who they were in the beginning, and making them force things on themselves they are not ready for only to avoid showing any weakness. Fixed and unmovable Scorpio is in the state of constant. A lot of deep, emotional understanding is needed for them not to be forced to move even further from their life goal to find emotional balance. 35%

Scorpio & Capricorn Values: Scorpio is the sign of Venus' detriment and Capricorn brings a lot of guilt into it, so their combination of values is basically founded on feelings of guilt and the sense that nothing is ever good enough. It is truly interesting to watch this couple share values with such a difficult relation to Venus and the term of value itself. Even though this will be a good motivation for them to get better, every day, it is quite difficult to deal with in a healthy, loving relationship in which they should both discover they are good enough. 30%

Scorpio & Capricorn Shared Activities: They will focus their energy on constructive things in order to build the world they wish for themselves. This is not always the most happy, joyful place with rainbows and unicorns, but it is realistic, practical and most of all good for personal growth. If they start digging into the past, they might find shared therapy revealing, even if they don't need it for any obvious reasons. Together, Scorpio and Capricorn will strive for greatness. There is a need in them to dig out the truth, whatever it might be, and this will make it easy for them to spend quality time together without much doubt on what they want to do. 99%

Conclusion: They are both deep and don't take things lightly, and this will help them build a strong foundation for a relationship that can last for a long time. The relationship of Scorpio and Capricorn can be inspiring for both partners to search for the truth, dig up under their family tree and deal with any unresolved karma and debt. However, this exact thing can easily make their relationship too dark and unemotional, pull them both in a state of sadness and depression, or simply awaken their need to search for the light with someone else. 64%

Scorpio Compatibility with Aquarius

Scorpio & Aquarius Sexual Intimacy and Compatibility: As squaring signs, they should have a very troublesome contact, but the sign of Scorpio exalts the ruler of Aquarius, Uranus. These signs combined represent the ultimate sexual freedom, a place with no restrictions or taboos. They are a combination of Water and Air, of emotion and information, all combined in a strong scent of attraction. If they get tied to each other and break up, they could end up hating each other and despising everything they've shared in their sex life. It is very difficult for these partners to find a balance of passion, emotion and rational thinking. While Scorpio's sexuality is hungry, deeply emotional and pervasive, Aquarius wants to be free of any boundaries and emotion, and will have real trouble being with a possessive partner. Their sex life can be like a battle arena, or like a wonderland, depending on the flexibility of both of them and the depth of emotions they share. Contact between a Scorpio and an Aquarius can be truly intense. As two fixed signs, they will most certainly have trouble changing their natures and adjusting to a partner that is too different from them. 40%

Scorpio & Aquarius Trust: It impossible for two honest and straightforward individuals such as Scorpio and Aquarius to have such a problem to trust each other. The problem here shows its face when they get too close. As soon as Scorpio starts to assume that Aquarius should be tamer and belong to them in a loving relationship, it will result in a forceful rebellion and the counterattack of their partner. Things could really get out of control if any sort of manipulation takes place, and unspoken tendencies might tear them apart in a matter of minutes. 1%

Scorpio & Aquarius Communication and Intellectuality: None of them will want to have small talk or discuss their day at work. It is futile from their perspective, and although Scorpio likes to be in control of everything their partner does, it will be refreshing to talk to someone who says unusual things. The biggest quality of their relationship is an incredible connection of depth and width in only one couple. They will both have trouble understanding our society as it is, and have certain similar perspectives on anything out of the ordinary. Scorpio exalts Aquarius' ruler and this is why their relationship is the possibility for both of them to grow. Not only will Scorpio adore the intellectual strength of their partner, but they will also help them understand the way their ideas might be realized through a feeling of ultimate possibility. For as long as they don't give in to their stubborn, unmovable modes, these partners could have great conversations about all strange topics they can imagine. The weakest link in their relationship is their respect for each other, combined with their static natures. We would think that both of these signs are in connection to change and they couldn't possibly be static, but in fact, they are static in their way of change, and their biggest challenge is to stop for a minute and treasure what they've found in each other. 50%

Scorpio & Aquarius Emotions: It takes a lot of work and commitment to reach the emotional core of Aquarius, and it is impossible to get there without spontaneity and trust. Scorpio can be spontaneous in situations that are free of emotions, but will rarely let their love for someone be a part of a maybe-yes-maybe-no swing controlled by their partner. Aquarius will rarely tolerate or be with someone who tries to make them be more stable and down to earth, or anyone who quenches their desire to be free. As soon as they feel obligated to do anything, they will start pulling away and any emotion that might have been developing will suddenly be covered by the fear of commitment and the rut of everyday life. If they want to reach emotional balance, Scorpio has to be untied, realize that their partner will never belong to them and that they are free to leave anytime. If love happens between them, the most typical scenario is for Scorpio to fall into an obsessive mess of feelings towards their uninterested Aquarius partner. They will have to understand that this relationship might end tomorrow and there is nothing they can do about it but accept it. On the other hand, Aquarius will have to confront their emotional depth and be ready to make certain changes in their approach to romantic relationships, so they can steadily feel understand Scorpio's emotional nature. 1%

Scorpio & Aquarius Values: They will both value excitement and change and this will be a strong meeting point for their characters. Unfortunately, most of the other things they would value in their partner are completely different. While Aquarius values free spirit, communication and independence, Scorpio values commitment, sex and deep emotional connection. 30%

Scorpio & Aquarius Shared Activities: They will both like to take risks of any kind and their best date could be anything from parachute jumping to a night out in a casino. For as long as they stay out of their ego battle, they could find many things to do together. The best way for them to spend some quality time together is in some sort of intellectual activities and competitions, because this would allow them to manifest their possible hostility in a healthy way. 60%

Conclusion: The truth is, Scorpio is the sign of Uranus' exaltation and as such, it adores Aquarius in a way. In most cases, Scorpio partner will show their affection obsessively, but this might actually feel good for Aquarius. When we look at the sign of Aquarius, we will see that it exalts Neptune, the ruler of a Water sign of Pisces, and all of our assumptions on their lack of emotionality will drown in their ultimate love. The fact is they are both in a way outcasts and rebels. While Scorpio represents all of our emotions we don't want to deal with, Aquarius represents the way of thinking most of us are not ready for. Someone might say that this is a karmic relationship, that these partners were enemies in one of their previous lives and that they could fight until one of them falls dead. It is best to look at them as announcers of change, for this is exactly what they will bring into each other's lives. 30%

Scorpio Compatibility with Pisces

Scorpio & Pisces Sexual Intimacy and Compatibility: Scorpio is a sign that represents sex, as well as sexual repression, and depending on the upbringing and previous sexual experiences, they can be a bit rough on their sensitive Pisces partner. On the other hand, Pisces is a sign of orgasms, strange sexual experiences and all of the sexual weirdness.

If they understand the emotional depth of Scorpio, they might be much more resilient than we would assume. The biggest challenge for these partners is their relation to Venus, the planet of sensual physical satisfaction. Scorpio doesn't care for Venus very much, leading it to its detriment, while Pisces adore it through exaltation. This can be very unfortunate if Scorpio dismisses this emotional need of Pisces to be satisfied and loved at the same time. As two Water signs, both Scorpio and Pisces find it very important for emotions to be the most intense part of their sexual experiences. If Scorpio partner is aware of their animal nature and instinctive sexual desires and in any touch with their feminine side ready to show it, Pisces will easily find a way to blend in their sexual world. 70%

Scorpio & Pisces Trust: However, they will both be in search of their one, perfect love and this should bind them with certain honesty. As soon as one of them is cheated on or disappointed, their relationship should end, because none of these partners can handle the tainted image of love. Trust between them will be maintained for as long as Pisces have an idealistic approach, doing everything for their one true love. Suspicious Scorpio can easily become a clingy, control freak in a relationship with Pisces. When their image clears and they realize who they are with and what their relationship looks like, it might become very difficult for them to stay in loop with Scorpio's expectation of honesty. 65%

Scorpio & Pisces Communication and Intellectuality: With these two combined, it will be almost impossible to have a healthy conversation in which there will be no hurt, distance or anger. They will rarely fight, for Pisces partner usually has no reason to fight with anyone, but they could have a lot of misunderstandings that lead to their separation pretty quick. If Scorpio partner is tender enough and Pisces partner possesses the needed boundaries, their communication can be pretty exciting and magical. Both of these signs are linked to different types of magic, and they will both be interested in the "behind the scenes" view on everything that surrounds them. As they start to communicate and get out of their silent zones, they could easily get carried away in topics most signs wouldn't understand. The emotional approach to everything in their lives will help them understand each other when it comes to rational choices, too. The possible problems in communication between Scorpio and Pisces are either the roughness of Scorpio or excessive sensitivity of Pisces. The most superficial experiences will become something incredible to talk about, and the truth behind everything in life will be mesmerizing. They should hold on to the fascination with each other, instead of giving in to their weaknesses. 90%

Scorpio & Pisces Emotions: The sign of Pisces represents our oceans and seas, while Scorpio represents rivers. Each river flows into the ocean or the sea, and this reflects the emotional connection between these signs in the best possible way. Pisces partner will have the ability to disperse the intensity of emotion from their Scorpio partner. This will allow them both to breathe more easily, for as long as they don't cross the line and endanger the part of this depth that is loved by Scorpio. Scorpio, it is Pisces. There is an emotional depth to Scorpio that not everyone is ready to face and Pisces are ready to face anything in the field of emotions. This is a special connection in which Scorpio partner needs to focus their emotions and Pisces partner needs to give them a purpose. However difficult and dark they both might get, they will share a deep emotional understanding that should be followed to see where their relationship will lead. 99%

Scorpio & Pisces Values: Pisces represent all the fairytales in which a prince became a hero and married a beautiful girl. If any sign other than Scorpio is capable of understanding As much as Scorpio values someone's strength of character and depth, Pisces will value sensitivity and width. Their mutual love for a connection with emotions and the depth of their emotional connection, will give them just enough shared values to hold on to. Scorpio is a sign ruled by Mars and there is always a certain admiration for chivalry. Pisces partner has a mission to teach Scorpio how to reach their fairytale through chivalry, and they should both stay focused on creating their shared dreamland, royalty or not. 75%

Scorpio & Pisces Shared Activities: If Scorpio gets tied to their Pisces partner, this might become tiresome for both of them, for too much scattered activity of Pisces can be irritating for focused Scorpio and the obsessive nature of Scorpio might weigh Pisces down. Still, they will have enough energy to follow each other and it should be easy for them to find shared interests. The main problem with the time they spend together could be the unconscious negativity of Scorpio partner. When it comes to activities they could share, they will probably be inseparable whatever they do. It might endanger the positive, happy image of the world Pisces want to carry around and this could push Pisces partner away if their emotional connection is not strong enough to keep them together. 85%

Conclusion: They will both easily get carried away into an image of a fairytale love, and this image could keep them together for a very long time, even if they are both not that happy. As two Water signs, they will rely on their emotional judgments and understand this about each other, creating true intimacy. When Scorpio and Pisces come together, this relationship will probably give them both new insights on emotional possibilities. The challenge here is for the nature of Scorpio not to obsess and suffocate their changeable partner, and for Pisces to stop running away from negative emotions. 81%

PART-9
SAGITTARIUS
HOROSCOPE

(November 22nd to December 21st):

Who loves the dim, religious light and always keeps a star in sight?
Who is versatile enterprising and an optimist, both gay and bright?
Element: Fire
Quality: Mutable
Ruler: Jupiter
Lucky Gem: Topaz. If he/she is feeling invincible, thank the Hessonite for it. This is a stone of power, and the world is for you to conquer.
Lucky Number: He gets success with Number 2, 3, 5, 6 & 8.
Colour: Violet, Purple, Red, Pink
Lucky Colour: Red. Wear Reds for warmth and energy.
Day: Thursday
Lucky Day: Thursday. An old friend will brighten up an otherwise dreary Thursday.
Flower: Holly
General Insights: Curious and energetic, Sagittarius is one of the biggest travellers among all zodiac signs. Their open mind and philosophical view motivates them to wander around the world in search of the meaning of life. Sagittarius is extrovert, optimistic and enthusiastic, and likes changes. Sagittarius-born is able to transform their thoughts into concrete actions and they will do anything to achieve their goals. Like the other fire signs, Sagittarius needs to be constantly in touch with the world to experience as much as possible. The ruling planet of Sagittarius is Jupiter, the largest planet of the zodiac. Their enthusiasm has no

bounds, and therefore people born under the Sagittarius sign possess a great sense of humour and an intense curiosity. Freedom is their greatest treasure, because only then they can freely travel and explore different cultures and philosophies. Because of their honesty, Sagittarius-born is often impatient and tactless when they need to say or do something, so it's important to learn to express themselves in a tolerant and socially acceptable way.

Personal Quality: The Sagittarius is moralistic, impulsive, full of versatility and eagerness, and has a positive outlook towards life. He/She enjoys travelling and exploration. He/She is ambitious and optimistic, honourable, honest, trustworthy, truthful, generous and sincere with a passion for justice and truth. He/She has very charming voice and benevolent personality. He/She is 'Yes' man and never says 'No' to anyone. He/She doesn't get demoralized in his failure. He/She does not like to harm any person and does social work too. He/She is noted for longevity, intuitiveness and original thinkers. He/She cannot gain or be successful in Gambling, Races and Stock-exchange businesses. He/She can be successful in business of white and artistic gift items or textiles and metal. He/She earns wealth after 36 year of age and also gets parental property.

Positive Quality: He/She is honest, generous, idealistic, great sense of humour, tolerant, and friendly and trusts and respects people. He/She is kind and forgives the people easily is never proud.

Negative Quality: He/She is indiscipline, promises more than can deliver, very impatient, will say anything no matter how undiplomatic and never learns even from his/her mistake in the past. He/She likes gambling and lose money. He/She does not keep his promises and does not have foresightedness and hence, he/she is unsuccessful. He/She has a quick temper and a biting tongue. Physical Appearance: He/She has darting and piercing eyes that are always likely to flash with laughter. While not particularly fashion-conscious, but he/she looks trendy.

Relationships: He/She has a happy family life and Gemini or Arian girls will be suitable as a partner and also for success in his life. Leo and Libra can be helpful for him. He/She is tactless and can hurt with his/her brutal remarks.

Career: He/She is successful in social administration, in public relations, as scientists and as musicians inquisitive. He/She works best with a tactful, organized business partner. Here are some occupations that he might consider, such as, Academic, Adventure travel guide, Advisor, Astronaut, Consultant, Entrepreneur, Inventor, Humanitarian work, Market researcher, Senator.

Health: He/She often fails to look before where he leaps, and as a result suffer quite a few bruises, pulled muscles and broken bones.

Ideal Partner: He/She needs to spend his life with organized and tolerant people, so his vibe best with Aquarius or Libra.

Compatible Zodiac Signs: Aries, Leo, Aquarius and Libra

Incompatible Zodiac Signs: Gemini, Virgo and Pisces

Greatest Overall Compatibility: Aries, Leo

Best for Marriage and Partnerships: Aquarius

Sagittarius likes: Freedom, travel, philosophy, being outdoors

Sagittarius dislikes: Clingy people, being constrained, off-the-wall theories, details.

BUSINESS ASTROLOGY FOR SAGITTARIUS

Sagittarius starts with project with great enthusiasm but lacking in immediate results, they move on to the next project if it takes too long. Their behaviour can be inconsiderate at times. For success in business they should aspire to finish job in hand before moving to the next. Business booster Gem is Amethyst.

Sagittarius Compatibility with Aries

Sagittarius & Aries Sexual Intimacy and Compatibility: Sagittarius partner has this innate ability to make a joke out of almost anything. The seriousness of an Aries when sex is in question is something that gives Sagittarius a strong impulse to make a joke. These are two Fire signs, both very passionate, each one in their own way. Aries is passionate when it comes to action, new things and of course – naked people and specific sexual positions. Sagittarius is passionate about their cheerful personality. You have to understand that Sagittarius really only cares about their opinions, convictions and moral value. They can spend their entire life analyzing these to see if they are wrong or right and search for the universal truth. When it comes to their optimism and good mood, they passionately protect them from anything too serious or hard. When Aries and Sagittarius engage in sexual relations it can be quite funny. If they let someone taint them, it would shake their conviction that they should always smile and find a reason to be happy. Although Aries can be a bit vain about their sexual abilities and performance, in most cases Sagittarius is able to break this wall of strict, sexual tension and lead them to a more relaxed zone where they can relax and experiment. 95%

Sagittarius & Aries Trust: Usually they don't have to talk much to understand each other and can easily spot when the other one is lying. This makes it extremely hard to create a situation of mistrust, especially because of the feeling of security Sagittarius partner gives to Aries, by taking everything in with dignity and serenity. In most situations Aries feels they can share anything with their Sagittarius partner. The problem could appear if they have different views on the seriousness and depth of their relationship. If this is the case, usually a Sagittarius partner sees Aries as a short term, not that important partner. This is why they could easily cheat on them and probably wouldn't even call it cheating. Aries and Sagittarius are both aware of the excessive need for honesty in their life. In return, Aries partner that values their relationship more, would jump into their possessive nature with even more ease and never trust their Sagittarius partner again. 70%

Sagittarius & Aries Communication and Intellectuality: Their mutual understanding can be so deep, that even if they lack physical attraction, they would gladly substitute it with a life spent in this kind of intellectual relationship. They motivate and push each other wherever they might like to go. When they are together, they make each other feel as if nothing is impossible. While Aries gives initiative and focus, Sagittarius gives vision and faith. These signs are ruled by Mars and Jupiter, which means that they could have some disagreements on their convictions. In case these are not convictions they think of as their personality's foundation, this shouldn't be a huge problem. Still, it is possible for their set of beliefs to differ too much for them to even understand each other. This is a wonderful bond that is often seen in friendships that last for years. When this happens, they fight whenever and wherever they can, since none of them has the ability to let their convictions go. Aries because they want to win, and Sagittarius because convictions are their forte and something they have surely thought about a lot. 90%

Sagittarius & Aries Emotions: These are extremely warm signs, due to their corresponding element of Fire, open for any kind of activity just to share time together and feel that wonderful emotion in their stomach. This is a love that could last for a very long time, for as long as their respect their personal needs, individuality and the distance they possibly need from each other every once in a while. Although they are not considered very emotional, it is a mistake to assign emotionality only to the element of Water. This is an element which works from the heart and you can feel it in your chest. In search for an explanation of emotional nature of Fire signs, you should just imagine that warm feeling in your belly and that would be the best possible description. When they fall in love with each other, deeply and sincerely, it is almost possible for their passers-by to warm up in the middle of winter. Their emotions are active, warm and on the move. Always changeable but creative and there to move them anywhere they want to go. 90%

Sagittarius & Aries Values: In time they will both understand that Aries grows through this relationship and widens their entire system of values. When they started dating, Aries probably had this idea of honour and heroic "sweep off feet" logic. In time, they both must have realized that Sagittarius gives this idea a new step up and brings it into a world of royalty. Not only does Sagittarius value honourable and heroic people, too, but they value honourable people with blue blood that give money and food to the poor, every day. Aries partner values things that are brought up to a higher level by their Sagittarius. Their main difference is in the fact that Aries values things concise and clear, while Sagittarius will easily disperse and go around the point for days. This can be met through their mutual value of truth, so honesty can be their cure for anything. 70%

Sagittarius & Aries Shared Activities: This means that they easily get tired or just bored and they always need new and exciting stimulations. Sagittarius is a sign of mutable quality, ready to change whatever needs changing in order to feel good. When they get together, their activities can be shared and fun whatever they are. This has nothing to do with their needs and tendencies, but with the potential of their entire relationship. They can go for coffee and they would have fun, but they could also go bungee jumping together and have even more fun. It is all the same to them. Aries is a sign in which Saturn falls. They are fully capable of respecting each other's personality, so even if their wishes for certain activities differ, this would be easily dealt with. 99%

Conclusion: They might have to stand up to their environment and defend their feelings from others, but this won't shake them too much, for neither of them thinks that much about the opinion of others anyway. If they manage to mend their philosophical differences and respect each other's different opinions, they could become one of the warmest relationships in the zodiac. This is definitely a couple with lots of potential. Their main relationship advice would be to always tell the truth to each other and not go crazy about their healthy differences. Their differences are exactly the thing that could make their sexual life more exciting. 87%

Sagittarius Compatibility with Taurus

Sagittarius & Taurus Sexual Intimacy and Compatibility: Although this is a delusion, these signs are too far apart in their basic character to understand each other's sexuality. Sagittarius would probably think of Taurus as a person who eats and sleeps all day long. There is nothing sexual about it. It is interesting though, how two people ruled by two beneficent planets such as Venus and Jupiter can't seem to find sexual satisfaction. The fact is – they can. Although this is a rare scenario, they could actually use their attributes to enhance sexual pleasure Venus would offer. If they understood each other as two individuals who deserve respect, they could find the missing link for a very interesting and fulfilling sex life. Taurus would take care of their Sagittarius partner and keep them satisfied. In return they would get a cheerful soul who knows how to make their relationship exciting. When Taurus thinks of sexuality, Sagittarius is probably the last person on their mind. With their childish attitude that changes with the weather, there seems to be no room for any sexual activity in their life. There is so much to be learned about the "light side" of sexuality here, and this could be a fun experience if Taurus loosened up a little and Sagittarius slowed down. 25%

Sagittarius & Taurus Trust: It is true that representatives of this sign have no idea how to make up a lie, let alone tell it. Still, when it comes to romantic relationships, they often suffer from a Don Juan syndrome and can't get enough attention from one partner. Taurus finds this repulsive at best, and if a relationship with a Sagittarius partner begins and they start acting this way, there is a big chance they will get dumped. Trust between these partners isn't something to be questioned and analyzed. If they do trust each other they can have a wonderful, trusting relationship for a while, but there is still no guarantee on how long this could last. The sign of Sagittarius is considered something like a synonym for honesty. If they don't, it won't be mended whatever they try to do. Usually loss of trust here simply breaks up the relationship and they both go their separate ways with no regret. 5%

Sagittarius & Taurus Communication and Intellectuality: Taurus and Sagittarius both have this joy about them that can be awaken by their relationship. With the Moon exalted in Taurus and Jupiter exalted in Cancer (ruled by the Moon), there is a certain feel, a soul, a tenderness to share between them. Their approaches to life are different, their characters incomparable, but the joy they can feel toward some things is completely the same. If they don't find this shared feeling of joy, they could both learn what bad communication really is. There are so many beautiful things in the world, and so much to talk about when you think about them. In most cases they can talk about the weather and be fine, but when they have a problem, this turns into a ridiculous conversation that isn't really a conversation at all. Sagittarius will want to jump out of their skin while waiting for Taurus to finish the sentence, as much as Taurus will look at their Sagittarius partner as a source of all stupidity. Taurus is your countryside and Sagittarius is the world, so their problems could easily include disrespect because of their origin or their goals. Although they will rarely end up in a fight or use ugly words, it can sometimes be too obvious how much they don't care for each other's worlds and how far apart they really are. 50%

Sagittarius & Taurus Emotions: However, Sagittarius doesn't normally react with much emotion to static, from their perspective boring Taurus nature. If Taurus was a bit more prone to temporary infatuations or platonic relationships, they could fall in love with a Sagittarius enough to overcome the differences between them. Sagittarius on the other hand, is often infatuated and temporarily in love. There is a great chance they will fall in love with a Taurus if they like their physical appearances, but they won't last in those feelings for long enough to gently lead Taurus to mutual love. Their pace is off and they rarely get in sync with their emotions. They are both connected with the Moon in a way, so there are some feelings to be shared. Most relationships between a Taurus and a Sagittarius partner that manage to last, are those that started as a friendship and had a chance to develop emotionally for years without them actually being in a romantic relationship.25%

Sagittarius & Taurus Values: They could support each other's utopian worlds a bit too passionately, and this could lead to one of them, or both, being in a delusion about what reality is about. There is too much love and happiness in the world if they start sharing opinions and this can become like a drug to both of them. The combination of signs of Taurus and Sagittarius is a "flower child" full of love, understanding for the world and ultimately humane. The practicality of Taurus will usually break this pattern and hit a counter like attack with their reality checks and material issues so they can both remember where their values part ways like to security and utter lack of it. 60%

Sagittarius & Taurus Shared Activities: Taurus might not actually share food with joy, but they will certainly like to share the activity of eating. It could be quite easy for them to find other things to do together, too. The problem will surface the moment Taurus wants to go home and spend an evening in their warm bed, while Sagittarius' fun has just begun. Food is mostly what they both would share. They don't share the same passion toward the same things, and although they might have fun being together, their priorities are not the same. The Fire energy of Sagittarius will be put off by Taurus' Earth personality and this will be tedious for both of them. 20%

Conclusion: However, every positive needs a negative to complete it, and when we really observe, we can notice that often a Taurus and a Sagittarius don't even get attracted to each other. Taurus needs earthly pleasures in their relationships and as a fixed, Earth sign it is the slowest of all signs. This is not exactly someone who can easily understand the fast, changeable and fiery Sagittarius. The best possible scenario for their relationship would be for them to get to know each other very well and build a friendship without expectations, for years. In the end, this could result in deep understanding that would provide them both with enough patience to actually start a relationship that has a future. With their inner beauty and the understanding they share in search of the truth to life, these two might seem as a perfect couple. If not, they can always hold on to beauty in the world. Imagine how wonderful their world of creation could be if they joined their forces of good. 31%

Sagittarius Compatibility with Gemini

Sagittarius & Gemini Sexual Intimacy and Compatibility: When they get together, they usually get strangely involved in emotions none of them really understands. Their sex life is something to cherish, easy, open and with no pressure at any side. They will both enjoy their sexual relations, followed by laughter, creativity and joy. As two children in bodies of grown-ups, they could go through the feeling of shame together if they don't have much experience. When they meet a bit older, there is a slim chance that both of them didn't have enough sexual experiences and partners to understand their personal needs and desires. This can make them both a bit selfish, but if their communication keeps going, there is no reason why this would be a turn off for anyone. It is a strange thing, but sex is really not that important to these partners. They are looking for someone to complete their mental personalities, someone to talk to and give them a sense of purpose. Gemini and Sagittarius have this strange approach to sex, childish and light as if they don't really care about it. This is why they could decide to stay friends after a breakup, for their starting premise was in building a strong relationship founded on their personalities, rather than their sexual or emotional natures. 90%

Sagittarius & Gemini Trust: Strangely enough, this can lead to ultimate faithfulness, for there will be no more excitement in the secrecy and mystery of parallel relationships. Sagittarius is not someone who can tell a lie and keep a straight face, and they are usually really disturbed by the lie of other people. Gemini can tell a lie with such ease that they sometimes don't even know they're lying. Strangely enough, this can lead to ultimate faithfulness, for there will be no more excitement in the secrecy and mystery of parallel relationships. Sagittarius is not someone who can tell a lie and keep a straight face, and they are usually really disturbed by the lie of other people. If anyone can understand the need of their partner to not be faithful, it's these two. Gemini can tell a lie with such ease that they sometimes don't even know they're lying. When they get together, this all becomes something to have fun with and they could play a game of trust until they build it on strong foundations of mutual respect. 99%

Sagittarius & Gemini Communication and Intellectuality: This kind of understanding is truly something to cherish. A problem can surface when they are both preoccupied with chasing their personal values and don't see what they have with each other. As opposing signs they complement each other in general, but this is strongly sensed in this segment of their relationship. With Gemini's ideas and mind flow, there is nothing Sagittarius can't learn or share, being a student and a teacher at the same time. The curiosity goes both ways and they will spend days just learning about each other and absorbing shared experiences. The only thing that can interfere with the quality of their mental connection is the possible fear of intimacy that builds in the meantime. That strength of personal exchange stops being mental and starts being emotional at some point, and as two signs that aren't exactly emotional to begin with, they can be frightened by the intensity of emotions that are surfacing when they are together. He can understand the need of their partner. In general, this is a couple you want to hang out with, every day. They will literally share happiness with one another and with those around them. They can inspire anyone to love and to smile, because when in love, they will laugh so sincerely and have so much fun together. The "here-and-there" nature of Gemini will get new meaning and purpose through the eyes of their Sagittarius, while the search for the ultimate truth can be so much easier for a Sagittarius with the mind of a Gemini. Their optimism and their eloquence will multiply, day after day, until one of them gets scared and decides to take off or death do them part. 99%

Sagittarius & Gemini Emotions: Both signs have a non-emotional feel to them, but their contact develops so much emotion that maybe neither one of them will be able to cope with it. They are not used to feeling that much, and when they "click", Sagittarius could discover the new meaning of life and Gemini a synthesis that they've never had a chance to experience. It is kind of strange to think about the emotional side of the relationship between a Gemini and a Sagittarius. This can truly be a fascinating love story, if only they don't run away from all that emotion. 95%

Sagittarius & Gemini Values: They both value the things that make sense. As opposing signs it might seem that Gemini is scattered and superficial, while Sagittarius is collected and deep, but in fact they have the same core in the fact that everything needs to make sense. Usually, we would connect this with the sign of Sagittarius, but Gemini has it in their approach to words and everyday actions. Their Mercury can't deal with senseless words, stories without meaning and purpose, whatever that purpose may be.
70%

Sagittarius & Gemini Shared Activities: This positive emotion and pure joy they can share, becomes something like a happy drug to both of them and they no longer want to be apart. As two mutable signs, they understand each other's changeability and flexibility, perfectly capable to find all the right reasons why everything they do makes perfect sense. Not only will they share every activity that any of them thinks of, but they will also laugh all the way, whatever they decide to do together. There is a point when they will get irritating to their surroundings, like two spoiled children without a care in the world, but while they are this happy – why would they care? 99%

Conclusion: They don't often find each other right away, but at some point in life it is almost certain that a Gemini will find their Sagittarius and vice versa. Their relationship has a strong intellectual connection, in which they will gradually find deep emotions. Gemini and Sagittarius make an incredible couple, probably being the most innocent one of all oppositions in the zodiac. There is no real prognosis how this will end though, because the emotions they feel could easily scare them away and their relationship could end only because of fear. If they decide to give in and find out what they could share, with Gemini's ideas and Sagittarius' beliefs, the sky is the limit. Or is it beyond? 92%

Sagittarius Compatibility with Cancer

Sagittarius & Cancer Sexual Intimacy and Compatibility: If they do, against odds, they could find an interesting shared sexual language that none of them anticipated will be found. The changeable nature of Sagittarius can be somewhat difficult for Cancer to understand because of their opposite need for emotional security. If trust between them is reached in any possible way and true emotions are shared, this characteristic of a Sagittarius partner will become a spice to their sex life, rather than its destructive force. If they have enough emotional security with one another, their sex life could be very fun. Cancer is a sign that exalts Jupiter, and will probably make their partner feel special. On the other hand, Sagittarius will make things light, fun and although the lack of depth could bother Cancer, passion and warmth they bring into their sex life might just be enough to compensate. The only way their relationship can succeed is for Cancer to let go of their preconceptions and allows some change and fun enter their strict sex zone. Cancer and Sagittarius will almost never get attracted to each other. Although they can seem mellow most of the time, they have a tendency to hold on to secure patterns when it comes to things that can make them feel shame or insecurity. In return, Sagittarius will have to lower their expectations on Cancer's own changeability and sexual creativity, and be satisfied with lovemaking instead of a sexual adventure. 40%

Sagittarius & Cancer Trust: More often than not, Sagittarius representatives have the need to show their seductive skills to everyone around them and we could call this a "Zeus' complex". Although the sign of Cancer loves Jupiter very much, emotions make it impossible for them to understand this flirty need of their partner to win the hearts of everyone around them. Not only will this disturb Cancer's trust, but it will also affect the trust Sagittarius has in the understanding they will get from their partner. Sagittarius is a sign ruled by Jupiter. Deities associated with this planet were considered great lovers, always on the chase for different women, goddesses, nymphs, and whoever seemed attractive enough. This could be the source of many conflicts and misunderstandings, and could finally lead to the point where their relationship has no purpose or future at all. 1%

Sagittarius & Cancer Communication and Intellectuality: Their minds need to find the synthesis of things that surround them and their belief systems can be quite similar. The wonderful thing of their mutual love of Jupiter is exactly in the similarities between their minds and their ways of thinking. Cancer can seem a bit slow from Sagittarius' perspective, as much as Sagittarius can seem superficial or too "philosophical" to Cancer. They can easily overcome these issues if they find a passion they share and usually if these two choose the same profession, they have many things to talk about. Both of these signs strive for knowledge, pure and simple. Their mutual love for it will give them plenty of things to talk about and a deep understanding of each other's reasoning. They understand each other's mind and the way their brain works, even when there is a large intellectual difference between them. If they happen to fall in love, communication is something they can always use as means to solve any other problem that appears in the course of their relationship. 60%

Sagittarius & Cancer Emotions: As elements of Water and Fire, they don't really spark each other's passion and the love between them will hardly ever be the same intensity, at the same time or the same pace. Sagittarius is a mutable Fire sign and they usually fall in love quickly and passionately. If their love is to last, their partner needs to surprise and impress them often, making the relationship exciting and unpredictable. Cancer, on the other hand, is a cardinal Water sign, and they will make sharp turns and huge changes, but much less frequently than their partner. Cancers follow their feel of situations and people, and need time to build a relationship in which they feel secure enough to share emotions. When love happens between them, usually Sagittarius feels it first, wants to jump in and out and in and out enough times for Cancer to realize that they cannot build this sense of security to even consider their relationship a true love. This is not a combination of Sun signs that will fall in love very often. If they are to build a love that will last, Sagittarius partner will have to slow down and wait until their partner decides how they feel. In return, Cancer will have to take a leap of faith and jump into a relationship that offers no security, to see if enough love can be found between them so they can stay together. 10%

Sagittarius & Cancer Values: Although they value different things and different characteristics in people around them, they have a strong link in the way they value knowledge. Cancer will value Sagittarius' honesty and their ability to act on emotional impulse, even if they don't understand the emotion behind the act. Sagittarius will value Cancers dedication to things they love and their incredible ability for compassion. 45%

Sagittarius & Cancer Shared Activities: Sagittarius will look at their partner as if the entire person just turned into a long, irritating pause. The way these signs use their energy is so different that it is not only difficult to find activities they will do together, but more importantly it is difficult for them to do things in a similar way. They might study together, but it would probably be a torture for Cancer to watch their partner move from one paragraph to another, only to find what they were really searching for at the end of some old book and then lose interest in the entire subject. This could make the possible impossible and they could be forever separated by a simple difference in their speed. 5%

Conclusion: If attraction and love are born between them, they will rarely have a damaging relationship for any one of them, because their signs are ruled by the Moon and beneficent Jupiter. It is safe to assume that they will be good for each other, for as long as their relationship lasts, but it is rare for them to succeed in the long run if they don't have strong support from positions in their personal horoscopes. Cancer and Sagittarius are usually signs that aren't attracted to each other at all. As much as Cancer can reach the depth of their partner's faith, Sagittarius can widen their partner's horizons and make them much happier in their approach to the world. If they have feelings for each other, it would be a shame not to act on them and miss the opportunity to peacefully grow. 27%

Sagittarius Compatibility with Leo

Sagittarius & Leo Sexual Intimacy and Compatibility: When they start dating, their sexual relationship might come as a surprise for both of them, for they will feel liberated to be exactly who they are with each other. The best thing they could do is use the trine between their Suns and build-up each other's self-esteem, especially if they have been in demanding or disrespectful relationships prior to theirs. The best thing about their sex life is the passion they share. Leo is there to bring inner fire for the act of sex, and Sagittarius to fire up the expansion, the places, positions and horizons. As two fire signs, one of them fixed and one of them mutable, Leo and Sagittarius share a warm love for each other. They will both enjoy each other in a fiery way and respect each other's bodies, minds and entire personalities. If they stumble upon one another and love is born, their sex life could represent a perfect connection for both of them. 99%

Sagittarius & Leo Trust: Leo does like to be the centre of attention and feel attractive and desirable, but this is something a Sagittarius partner can provide in abundance. There is usually no reason for them to lose trust over time, except when their emotions start to fade. Sagittarius is a mutable sign, and as such, they can fall in and out of love quickly and frequently. Since they spark each other's sense of security and confidence, they will rarely show jealousy or misunderstand each other's actions. In case Leo starts feeling left out and unloved, the suspicion will rise, and what better way to respond to suspicion than by becoming suspicious yourself, Sagittarius might think. Although they both might be unaware of the root of their issues when trust is lost, it is usually a simple lack of love. 80%

Sagittarius & Leo Communication and Intellectuality: Leo because they are ruled by the Sun and this gives them a certain rational awareness and Sagittarius because they always aim higher from the Earth, philosophical and wide opinionated. This is something that will help them communicate about almost anything, even though their interests might differ and their backgrounds as well. Leo has the ability to help Sagittarius when they get lost, and this could happen often if their plans are grand. Sagittarius will give Leo vision and the ability to understand the future of their current creative efforts. Together, they make an important part of the process of creation. As two highly aware individuals with a strong sense of Selves and their personalities, they could build up an incredible understanding. They can both be loud, communicate a lot, and this could make their relationship truly remarkable, deepening their intimacy through openness they share to get into each other's worlds. Since they both have a strong personality, they will not feel threatened by each other's character and each other's strength of opinions and convictions. Leo and Sagittarius are both very focused on their mental activity. The only thing they might lack is the sensibility to outer influences and their fiery relationship could make them a bit too rough on each other, and on themselves. Still, the power of creativity and their active approach to life should keep them interested in one another and very well connected for a long time. 85%

Sagittarius & Leo Emotions: They will want to show their love, share their love and act on their impulses as much as they can. This can sometimes be too much, for no passive, fragile emotions or energy will be respected. Some balance would come in handy, especially if they often fight. Conflicts between them could be quite aggressive, not because they are that aggressive themselves, but because two fires build an even larger fire. It is almost like they might explode if they both go too far. When they fall in love, this seems like the warmest, cuddly love on planet Earth. In most cases, this will be enough to overcome any difficulties in their way, but sometimes these partners both tend to forget their actual sensitivity. As two Fire signs, Leo and Sagittarius tend to be very passionate and open in showing how they feel. They have to understand when the time has come to slow down, stay at home, talk about nothing at all and just be quiet. If they don't, they will probably turn to someone who can give them this kind of peace from time to time. 80%

Sagittarius & Leo Values: It is not easy to explain to a Leo why it is so good to run away from the world, travel in Greenland alone, and eat bugs somewhere in Asia, except if one wants to show their courage. On the other hand, Sagittarius doesn't really understand why they would go to fancy places and confront all the people that it is easier to run from. This is not a consequence of a lack of courage, but the lack of meaning they feel when they need to spend their time on tiresome people. They will most certainly value each other's strength of character and incredible personalities, the ability to warm each other up in every possible way and the passion they carry within, each for their own purposes. So although they value the same thing – courage, they see it through different eyes.65%

Sagittarius & Leo Shared Activities: They have the energy and the need to search for knowledge and widen their horizons, but they don't exactly like to move that much. This is due to their fixed nature, and although they would like to visit any possible part of the world, they wouldn't do it at the same pace as Sagittarius, nor would they choose the same destinations. Sagittarius, on the other hand, doesn't really understand why Leo wants to perform in front of so many people when there are starving children in Africa. We would think that Leo likes to travel just as much as Sagittarius. These are simplified examples, but they serve us pretty well to understand how well they might work together, travel together, perform together, but only if they are open enough to ad purpose and strength to their approaches. 40%

Conclusion: This love is warm, passionate and inspiring, and they will have a chance to create, perform and have fun together for as long as they feel this way. However, Sagittarius partner might lose interest in Leo because they tend to get pushed away by their static, fixed nature. Leo and Sagittarius are a very good fiery combination of signs, and when two people with these Sun signs come together, they inevitably fall in love. The only way they might get to keep their passion and emotions going, is if they manage to listen to their softer emotions and remain tender and sensitive for one another. 75%

Sagittarius Compatibility with Virgo

Sagittarius & Virgo Sexual Intimacy and Compatibility: Even though Virgo can be quite demanding and critical, especially from the point of view of Sagittarius, their sex life can be satisfying for both. The good thing about their connection lies in a fact that these signs are ruled by planets that also rule their opposing signs. This means that they will feel attraction and a need to begin a sexual relationship in the first place. The main problem here is in the difference in their elements. Virgo is an Earth sign, and as such, doesn't often take too many risks. Sagittarius is a Fire sign, and they will passionately force things until they reach their goal. This doesn't work well in their sexual contact, for Virgo might feel pushed into things they don't want to do, and Sagittarius might be turned off by the practical and static nature of Virgo. Just like all mutable sign combinations, these partners could have a lot of fun. The most important thing these partners should remember is that they both need room to be who they are. With two such giving people, sex life comes down to who will satisfy whom best, as soon as they deal with unrealistic expectations. 30%

Sagittarius & Virgo Trust: As friends, they can be unshakeable about their convictions and hold on to some traditional values together, but as soon as they start a romantic relationship, both of them seem to start feeling trapped. Virgo doesn't look like a zodiac sign that will easily feel trapped, but their mutable quality makes them impatient and always in search for change. They cannot be held in one place for long, any more than a Sagittarius can. The main difference being the degree of sacrifice they are willing to make. If you dig for the biggest problem in the relationship between a Virgo and a Sagittarius partner, you will realize that it is their lack of trust, not only in each other, but in their entire relationship. Out of these emotions, both partners will start feeling the need to be with someone different, and this is a relationship with probably the biggest potential for adultery, unless incredible guilt stops them first. Communication followed by mutual respect is their only chance of building a trustful bond. 1%

Sagittarius & Virgo Communication and Intellectuality: Virgo will bring all the little pieces into their intellectual connection, while a Sagittarius will have vision and help create a bigger picture. Even though they don't complement each other anywhere close to their opposing signs, the intellectual excitement will be equally important to them from the start. The most relevant fact for these partners to remember is that their respect is the most important thing to hold on to. If they disrespect one another, Virgo will observe their Sagittarius partner as a weirdo, stupid enough to run away from anything that has depth, while Sagittarius will look at their Virgo partner as a weirdo, stupid enough to hold on to irrelevant things. There is so much to say when Virgo and Sagittarius come together, and even though these partners might spark each other's need to talk excessively, all the time, they will both feel quite good about it. If they hit the zone of real understanding, they will be excited about the use of their minds and the beautiful conclusions and philosophy they can create together. They need to remember that each of them has a different role, and that for each role, these "stupid" characteristics represent the best possible base. 65%

Sagittarius & Virgo Emotions: In most cases, their vision of a fairytale ending differs too much for them to have it with one together. Still, in some rare situations, their mutable natures allow them to move in the same pace with enough respect to stay in an emotional bond that satisfies them both. Both of these partners are considered unemotional, but this is mostly because of their need to rationalize, analyze and use their minds to explain everything that happens to them, rather than rely on their hearts or gut feelings. This will often be a problem, for Virgo needs someone truly emotional so they can show their own deep feelings. There is so much to say when Virgo and Sagittarius come This is not exactly a couple that will often end up in a happily ever after, even though they both wish to find the right person for this more than anything. Sagittarius seems to be uninterested in needs of Virgo or simply unaware of them because they act as if they are purely rational. The trick here is for both partners to see behind the act in order to find each other's hearts and understand what they can expect from one another. 10%

Sagittarius & Virgo Values: This is why they will both treasure someone able to adapt, change and move, which is definitely something they will find in each other as they start their relationship. As highly mental signs, they will also both value clarity of mind and intelligence, in general. Still, their approach to intellectual value is different, and as much as Virgo values depth and detailed analysis, Sagittarius will value the width of one's mind. Virgo and Sagittarius will strangely have similar values based on their mutable quality. Even though they differ in other things they value greatly, Virgo valuing practicality and Sagittarius vision and focus, there is enough common ground here for both of them to feel good when together. 50%

Sagittarius & Virgo Shared Activities: This is why they will both treasure someone able to adapt, change and move, which is definitely something they will find in each other as they start their relationship. As highly mental signs, they will also both value clarity of mind and intelligence, in general. Virgo and Sagittarius will strangely have similar values based on their mutable quality. Their approach to intellectual value is different, and as much as Virgo values depth and detailed analysis, Sagittarius will value the width of one's mind. Even though they differ in other things they value greatly, Virgo valuing practicality and Sagittarius vision and focus, there is enough common ground here for both of them to feel good when together. 35%

Conclusion: There are many challenges in their way, the biggest being their emotional lack of understanding and their possible lack of respect. Still, when they find a way to show emotions and share them in the same pace and in an understandable way, they could actually have a lot of fun together. The relationship between a Virgo and a Sagittarius is not a usual happy ending emotional story. Their communication is often exciting and they both have a lot to say to each other, but their rationality may distract them from an actual search for love. If they discover how well they complement each other, they might be able to stay together for a long time. 32%

Sagittarius Compatibility with Libra

Sagittarius & Libra Sexual Intimacy and Compatibility: They are a very good match when it comes to sexuality, for no partner here feels pressured and there is just enough room for both of them to grow, develop, build their self-esteem and feel secure in each other's arms.

Ruled by two benefic planets, Venus and Jupiter, their main objective is to form an enjoyable sexual relationship, with a primary goal to make each other happy. For this goal, they will experiment a lot, and try out new things, everything followed by a smile and a sense of lightness, as if sex wasn't really a big deal in the first place. The seriousness of Libra linked to its exaltation of Saturn will give their entire relationship endurance and stability, while their ruling Venus working together with Jupiter, gives enough romance, sexual desire, tenderness and might lead them to a fairytale ending. The intensity of emotional contact and intimacy between a Libra and a Sagittarius will mostly depend on other factors in personal charts, but they will most certainly enjoy their sexual relationship. This combination of planets forms Neptune in a way, and speaks of the growth of satisfaction leading to orgasmic pleasure, even though both signs might not seem at all sexual to some other members of the zodiac. 90%

Sagittarius & Libra Trust: They can both go to extremes, either having unrealistic faith in each other or mistrusting every word and every action that is made. The only way to keep the image of trust for these signs seems to be to always stay in a fairytale, unrealistic state, and this is something a Sagittarius will never want to do. If truth isn't lived, nothing in the world is beautiful for a Sagittarian Sun. As stated above, rulers of Libra and Sagittarius are closely linked to Neptune and the challenge of trust is one of the most important experiences that this relationship gives. As soon as they start their search for something different, Libra will sense the change and become frustrated by their inability to create oneness with a partner they love. 5%

Sagittarius & Libra Communication and Intellectuality: Even if this isn't something with a promising future, for no one can run from their true nature, it will bring them both joy and happiness at least for a little while. Libra partner will be able to relax next to someone who doesn't judge, and Sagittarius partner will feel like their energy is well focused on someone that needs some youth, warmth, light, optimism and creativity in their life. For as long as they don't brush on ego problems, their communication and intellectual compatibility are a given. The main problem that will eventually surface and need to be dealt with, is in the forces of their Suns. Libra's Sun is weak, and they will easily give the wheel to someone else who will make positive decisions and moves for them. Sagittarius has too much fiery energy in their Sun, active, taking action and always prepared to give some of it even if nobody asked for it in the first place. It is wonderful to watch how soft Libra gets, forgetting about Saturn and their own responsibility, as Sagittarius' childish nature melts their heart. This could lead to a subtle, hidden, will imposing and a character shift that will leave them both bruised for respect when a light is finally shed on the issue. 85%

Sagittarius & Libra Emotions: It is not easy for any one of them to find love and share it with someone. They are, after all, an Air and a Fire sign. Even though Libra is ruled by Venus, it is linked to the mental processes, social adaptation and communication through its element, while Sagittarius has passionate feelings, but uses their head, spreading their philosophy, more than actually feeling. When they get together, they seem to be able to find a balance in which both of them use their heads just enough, and give each other enough room for love to be born. This is a bond that gives both partners the opportunity to understand how deep their emotions can go, as beneficent rulers make way for feelings to surface in a supporting atmosphere. This is one of the most compatible couples when it comes to the emotional side of their relationship. Even though their relationship is not always meant to be the one they will stay in for life, it could prepare them for a love they seek, giving them a glimpse of what they are capable of. 99%

Sagittarius & Libra Values: Libra doesn't seem like a creative person to others, but a Sagittarius sees their intellect through communication and motivates them to show their warmth. This leads to shared value of their entire relationship and an intellectual understanding that gives them room to build their shared philosophy. These partners will value the strength of mind in a way that is understandable only to them. Even if they don't start their relationship in the same place, they will have the opportunity to build similar values in time, showing each other what's truly important. 75%

Sagittarius & Libra Shared Activities: Libra wants to stick to their usual routine, and make fieldtrips to things that interest them from time to time. Sagittarius wants to move from any routine and live a life travelling the world. There are exceptions to this rule, of course, and there are uplifted Libras that will want to travel the world, as much as there are Sagittarius representatives that want to follow a certain trail, while fantasizing about their reality. Even though we could easily assume that a Libra and a Sagittarius will have lots of things to do together, there is a great chance that their choices of activities won't be so similar. However, in most cases, their needs won't fit that well and they will probably face the challenge of their usual ego battle while choosing what to do together. 70%

Conclusion: This could easily lead to a struggle for supremacy and a battle to reach the ruling position among them. This comes as a continuation of Libra's bruised Sun and a Sagittarius will fit in perfectly with the need to give away every sense of pride out of some childish convictions. The only way for them to be happy together, is to respect each other fully and let each other do what they are meant to do. The relationship of Libra and Sagittarius is in most cases a beneficent bond that allows these partners to develop their emotional, inner worlds and build their lives without negative influences. However, there is an archetypal battle between them, for Saturn exalts in Libra and doesn't really care for his son, Jupiter, the ruler of Sagittarius. Libra should stick to their relationship and love, ruled by Venus, while Sagittarius should stick to their convictions and width, ruled by Jupiter, multiplying the love Libra provides. 71%

Sagittarius Compatibility with Scorpio

Sagittarius & Scorpio Sexual Intimacy and Compatibility: The strength of character they share is something that will give them just the right amount of confidence when it comes to sex, and the creativity and openness of Sagittarius will be refreshing for the fixed nature of Scorpio. However, this often doesn't last very long and in time, in most cases, Scorpio starts thinking of their Sagittarius partner as unreliable and not to be trusted, while Sagittarius sees Scorpio as dark, pushy and too controlling. In order to remain in a healthy sexual relationship, both of these partners have to compromise, Scorpio finding a way to give freedom, while Sagittarius finding a way not to run away from the seriousness of their partner. In the best possible contact, Scorpio will give their sex life emotion and true physical intimacy, while Sagittarius will be there to give meaning and shake things up, representing the light at the end of a tunnel. There is a strange understanding between these two signs, as if they were one and the same, at least for a little while. Together, they can build an incredible sexual relationship, for both tend to be uninhibited about locations, positions and situations in which they wish to make love. 25%

Sagittarius & Scorpio Trust: There is nothing a Sagittarius will dread more than someone trying to control their life. If someone can spark the need of Sagittarius to be unfaithful, it is most definitely Scorpio, launching them further and further away by trying to come closer. An additional problem with this couple is in the quality of their signs, Scorpio being fixed and Sagittarius mutable. This makes it almost impossible for them to share a pace, and in order not to disappoint each other, they both might choose to lie. The biggest problem for a Scorpio and a Sagittarius in a relationship is trust. Scorpio has the need to tie their partner down, even if they have that seemingly liberating view on love. The strangest thing in this relationship is in the fact that these two signs are, without a doubt, the most honest signs of the zodiac. Still, their relationship seems to face them with the other side of their personality, one none of them wants to see. 1%

Sagittarius & Scorpio Communication and Intellectuality: Not only does Scorpio feel lighter, more optimistic about life and everything in it when communicating with a Sagittarius, but the depth they give to Sagittarius' mind and ways to reach conclusions is impossible for any other sign. Scorpio will face Sagittarius with any superficial or outdated views, while being compassionate enough to know how to do this without hurting them, and being fixed enough not to change their intent somewhere along the way when Sagittarius thinks of running away. Two such strong individuals give each other exactly what each of them needs when it comes to their mental compatibility. They will enjoy each other's company for as long as expectations and emotional disagreements are not in focus, for they have an incredible thing to share like their search for truth. The meaning they seek in all things in life, Scorpio going into depth and Sagittarius travelling wide will connect them through a strong bond that no other combination of signs can form. If they are on a shared mission, they can accomplish incredible things and have real epiphanies together.
80%

Sagittarius & Scorpio Emotions: They will enjoy each other's company for as long as expectations and emotional disagreements are not in focus, for they have an incredible thing to share like their search for truth. The meaning they seek in all things in life, Scorpio going into depth and Sagittarius travelling wide will connect them through a strong bond that no other combination of signs can form. If they are on a shared mission, they can accomplish incredible things and have real epiphanies together. 10%

Sagittarius & Scorpio Values: They can both feel like outcasts and value each other's decisions to differ from others out of self-respect. Still, in most cases they will easily consider each other invaluable in a way, for they cannot meet the expectations each of them has for his partner's personality. Since Scorpio is the sign of the exaltation of Uranus, and Sagittarius understands this through its third house, they both value freedom and one's ability to fight for their beliefs. The only way for them to value each other, is to focus on the positive characteristics and sides of one another and their entire contact. 35%

Sagittarius & Scorpio Shared Activities: Sagittarius will want to try anything new and Scorpio's approach to life is always new from their perspective, while Scorpio will enjoy the first impulse of optimism and fun. It is a good thing that Scorpio exalts Uranus, for this gives them enough love for change and exciting, new things that Sagittarius can bring into their lives. However, the fixed quality of Scorpio will make this exhilaration fade as soon as their relationship becomes routine, in any way. For as long as their communication is inspiring for both partners, they won't need much to be satisfied by their activities. As time passes, there is a great chance for Sagittarius to start feeling pressured, or simply be bored, and this doesn't give great promise of future. 30%

Conclusion: While they don't know each other well and everything seems new and incredible, Scorpio will see their Sagittarius partner as a ray of light that suddenly makes their life brighter and better, while Sagittarius will see that there is so much to learn and enjoy the depth of their Scorpio partner, followed by emotional attachment. In time, there is a strong chance they will slowly lose interest in one another, especially the mutable sign of Sagittarius for their fixed Scorpio partner. Scorpio and Sagittarius make a pretty great couple, for as long as they feel the first excitement at the start of their relationship. Even though their relationship might end on bad terms, it would be a shame not to give in to it and let it fascinate and exalt both of them for however long. 30%

Sagittarius Compatibility with Sagittarius

Sagittarius & Sagittarius Sexual Intimacy and Compatibility: This is a beneficent relationship, but it can be superficial and very short-lasting. Both of these partners think a lot, have their own liberal philosophy, and are easily pulled together in a series of one-night stands or casual sexual activity that no other sign would jump into. Sagittarius is a mutable sign, making both of them change with the slightest glimpse of a problem, and while they are satisfied with the unrestrained, liberating experience, they will remain as intimate as their minds allow them to. The problem they might have to deal with is the lack of emotion and depth in their contact. Sagittarius is not often superficial, but when with another Sagittarius, their minds throw them in all sorts of different directions and there is not much room for genuine emotion. They can be madly in love, but still lack consistency in their emotional contact. When you think of two Sagittarius partners in a sexual relationship, you might as well think of two teenagers that find everything funny. This will influence their sex life and make it as changeable as they both are. Even though this won't affect their overall satisfaction, it might cut their relationship short and make them understand they need to be with someone different. 65%

Sagittarius & Sagittarius Trust: Two Sagittarius partners don't trust each other? They consider themselves the most honest in the entire zodiac, and they will have the ability to understand each other in this honesty. Even though the sign of Sagittarius has a tendency to show affection to several people at the same time, enjoying everything in abundance, when two of them come together, they simply won't mind. The only possible reason for jealousy and mistrust in this contact rises from other personal pointers, rather than their Suns in the sign of Sagittarius. 99%

Sagittarius & Sagittarius Communication and Intellectuality: When they find shared interests and discover their similar convictions, there is nothing stopping them from exciting, passionate discussions in which it is finally easy for them to be who they are. One Sagittarius truly loves talking to another, and unless there is a hidden ego battle between them, they will rarely get angry or frustrated by anything their partner has to say. Each Sagittarius often has friends born in the same Sun sign, because no one else can understand their nature, cherish it, and awaken their inner child as their own reflection. When these Suns come together, their passion for things they do multiplies. The joy of this contact is something rarely anyone has a chance to feel except a Sagittarius with another Sagittarius. For as long as there's no judgment or a need to impose their will on one another, their time spent together will be incredibly valuable for both. Those smiles they share with everyone on this planet will be returned in just the right amount only by another Sagittarius. 99%

Sagittarius & Sagittarius Emotions: Each Sagittarius wants to be satisfied and happy, with no hidden intent, manipulation, dishonesty or any impurities. Their intentions are always good and they might help each other reach for their utopian goals when together. However, when these two begin their relationship, they understand the positivity and the optimism they share with clarity, but this lifts them even higher off the ground. If they feed each other's inconsistencies and needs to run from any seriousness and reality, they could truly get lost on their search for love. Sagittarius is a sign not often described as emotional, but in fact their ruler, Jupiter, finds the place of its exaltation in Cancer, the ruler of all emotion. We might say that the real goal of every Sagittarius is to find this inner emotional peace, to find home, without running from difficulty, sadness and any emotion that needs to be dealt with. The problem these partners have is not in the lack of love, but in the support of various directions that can move them away from their hearts. They often seem to need a bit more consistency to find the love they seek. 10%

Sagittarius & Sagittarius Values: Even when they stumble upon a disagreement, there is a great chance they will laugh it off and forget about it in a couple of hours. The positivity of this clash of fiery Suns is something that can overcome any value previously set, and they will easily adapt to one another and find a perfect compromise, even when they disagree. It is wonderful to watch two Jupiter's minions play with their convictions and share values as if it was the most common thing on Earth. The most important value they share is the one they both give to freedom of spirit and the goodness of humankind. When they find this point of shared utopia, there is nothing else that will truly matter. 99%

Sagittarius & Sagittarius Shared Activities: They are easily distracted by various things in their lives and aren't very fixed on their agreements, seeking freedom and understanding from one another. When they are faced with their own weaknesses, one of them being that lack of responsibility and reliability, they can really get annoyed and angry. This is not typical, but it is very possible in case one of them doesn't see themselves in a true light which can be expected from any Sagittarius. The best thing these partners could do is travel the world together, with a basis of a plan that is to be respected. They can basically do anything together, if they manage to find each other. No other sign can understand their need for travel, knowledge, width and distances, and this is something they should share and multiply when together. 70%

Conclusion: As two representatives of a mutable sign, they will adapt easily, but change their opinions and feelings toward each other with a similar ease. This doesn't always bring promise of a long-term relationship, for there is no partner to be the glue that holds them together. One Sagittarius will easily fall in love with the other and their passionate relationship can change very fast. This doesn't mean they won't enjoy each other's company, find many things to share while they are together, and laugh as children while being on the same path. If they discover the true happiness of two Jupiter affected people combined, they might lose interest in everyone else and find that point of needed balance to keep them together in their travels for as long as they live. 74%

Sagittarius Compatibility with Capricorn

Sagittarius & Capricorn Sexual Intimacy and Compatibility: Even when they are attracted to each other and form a sexual bond, after their time has passed they will probably feel like they shouldn't have been together. There is no logical explanation to this feeling, but it is present more often than not. Differences in their character can be strangely easy for them to handle, simply because a Sagittarius takes everything in with ease, and Capricorn feels responsible enough to understand their partner's immaturity as their own fault, in some strange way. Each Capricorn wants meaning and depth to their physical encounters, for they are slow, thorough and value their physical reality. Sagittarius often doesn't understand the pace at which a Capricorn wants to move in, nor do they see the importance of the physical world that Capricorn has the responsibility to. In the beginning of their relationship, if they share the same desires, they might not see how incompatible they actually are. Unfortunately, as time passes, it becomes pretty obvious that their archetypal battle reflects on their characters in a way that taints their sex life. There is something unbearable about the sexual contact of these partners. The only way they can ever remain in a healthy sexual relationship, is if Sagittarius respects the physical, as much as Capricorn loosens up and respects change that comes with their partner's Jupiter governed Soul. Their meeting point is in the sing opposing Capricorn, where the ruler of Sagittarius is exalted. In other words, their meeting point is in pure emotion. 5%

Sagittarius & Capricorn Trust: Capricorn feels this and recognizes the lack of inner honesty that doesn't seem to change. The problem here is in the fact that Capricorn is the sign of Jupiter's fall and this is the ruler of Sagittarius, as well as the traditional ruler of Pisces. The magic of life and the beliefs that lead in a certain direction seem to be lost on Capricorn. They know with certainty that the only things that give results are their rational mind and hard work. It is true that Sagittarius is one of the most honest members of the zodiac when it comes to their relationship with others, but they are rarely entirely honest with themselves. How can someone like Sagittarius explain to them that beliefs create their reality and that it is enough to believe in a good outcome, to affect the entire web of circumstances in a positive way? This issue comes down to a problem with trust, but in fact it goes much deeper than that. 10%

Sagittarius & Capricorn Communication and Intellectuality: Sagittarius has that optimistic smile that will definitely put a smile on Capricorn's face too, and it will get much easier for those fiery, creative ideas of Sagittarius to find their grounding through Capricorn's practical approach. With enough respect, this is a couple that links a visionary with a builder, and there is really nothing they cannot make when together. If they don't expect change from one another, they are very likely to be extremely intellectually compatible. The most beautiful thing in this contact is in their complementing protective roles. Both of these signs represent protection, Sagittarius ruled by the greatest benefic and Capricorn as our fence, our shell to the outer world. A Sagittarius and a Capricorn can be full of understanding for each other, if they don't jump into a battle over their belief systems. When they manage to build a functional core, these are partners that will never let anyone else affect their relationship. If they are in search of someone who will not allow interfering, meddling and any type of disrespect from other people, this relationship might be their best choice. 75%

Sagittarius & Capricorn Emotions: This is where their hearts meet and if there is enough faith in a Sagittarius, without any unrealistic expectations, they might fall in love deeply. Sagittarius and Capricorn can find a shared emotional language due to the fact that Capricorn needs someone like their opposing sign to complete them, and Sagittarius tends to become that sign as a place of Jupiter's exaltation. There is a small chance that a Sagittarius will be that mellow, tender person a Capricorn needs, but this is something that can be overcome with enough closeness and understanding of their differences. 55%

Sagittarius & Capricorn Values: Sagittarius is a mental sign, focused on philosophy and learning, always in search for unity, synthesis and that universal truth. Capricorn is the logical continuance of Sagittarius, as a practical tool that uses knowledge. If they don't find each other stupid, they will click in the same wavelength without much trouble, and discover that they share a certain depth and curiosity that isn't obvious at first glance. There is one important thing these partners agree on, and that is the value of intelligence. Most of their values differ greatly and their needs are often too far off. While one of them values freedom, width and creativity, the other values practicality, responsibility and focus. 20%

Sagittarius & Capricorn Shared Activities: The lack of relationship between their Suns helps their bond with a certain lack of disrespect. This leads them into a situation in which a Sagittarius finds their Capricorn partner interesting, as an extraterrestrial they have always wanted to meet. They are different enough to be interested in each other with genuine curiosity and a Sagittarius is always ready to try out something new. Capricorn will probably refuse many of the childish activities Sagittarius suggests, but it becomes fun to talk them into it and there is a note of laughter and joy in these attempts. Even though we would think that a Sagittarius will get bored and want to run off from their Capricorn partner, in most cases this doesn't happen. They are both smart enough and aware that their differences exist, which makes their entire story so exciting and refreshing for both. 60%

Conclusion: This is not your ideal relationship, and it will rarely be the one they both choose to stay in for the rest of their lives. Their understanding and acceptance of their differences is refreshing and fun for both partners, and they might have a good time while together, for however long. We cannot predict too much stability unless a Capricorn decides to make it, but the smile on Sagittarius' face and the ability they have to make their partner laugh, can be the pillar of their bond for as long as they both need it. 38%

Sagittarius Compatibility with Aquarius

Sagittarius & Aquarius Sexual Intimacy and Compatibility: Their attraction can be strong, especially when a Sagittarius partner is at a crossroads in their life and need confirmation of their freedom and sexuality. Their sexual relationship will be very fun, because they both like to experiment and learn new things. Their communication will usually give them both so much satisfaction that sometimes they both almost won't even need the act of sex in order to get satisfied. Although their sexual connection can be very satisfying for both partners, they could have trouble creating intimacy. Sagittarius partner will bring just enough warmth in their relationship, but the mutable quality of their sign will make them easily turn their focus to something else, while Aquarius partner still holds on to the same things. Aquarius acts in a way Sagittarius thinks and this is quite an asset in their sex life. They will both understand the necessity of change and incorporate it in their sex life. Still, the emotional bond and consequentially the intimacy between them could get weak and strong, on and off, too often for both of them to see each other as perfect partners. 80%

Sagittarius & Aquarius Trust: Sagittarius can be a sign prone to infidelity and Aquarius likes to be free to be available. With them both knowing these things about each other, they could easily start questioning if they should trust one another or not. Although they both find their relationships very dependent on the level of freedom they have, this is probably something they won't be able to give to each other when they decide to commit to their romantic relationship. They will sometimes know each other's minds too well for them to create trust from the sense of absolute freedom. The best remedy for the lack of trust in any of the partners is for both of them to realize they their relationship is just something out of the ordinary, casual and free from any restraints. 60%

Sagittarius & Aquarius Communication and Intellectuality: Aquarius partner can remain distant for a long time and

Sagittarius might feel like a little child, talking excessively about uninteresting topics and trying to make a connection. When they finally point in the right direction and choose to speak of something that awakens Aquarian interest, their conversations will become incredible. Both of these partners are rational and give a lot of attention to their chain of thoughts. Both of them are fast enough in coming to different conclusions. The contact between them will spark their need for intellectual sparring and they could end up in some great debates. When they share a love for something, they will talk about it passionately, excessively and find new ideas and solutions to incorporate in their approach to this subject. When Sagittarius and Aquarius find a mutual interest, it becomes the infinite source of new topics, information and could even change their life philosophies. The speed of Aquarius mixed with the passionate state of constant belief of Sagittarius, could make their relationship one of the most productive in the entire zodiac. 99%

Sagittarius & Aquarius Emotions: It is a good thing that Sagittarius is so changeable, or they would have trouble keeping up with their Aquarius partner. Another good thing in this contact is the rational nature of both signs and their focus on mental processes. This will allow them to communicate about their emotions, whatever they are, without any sense of guilt or emotional pressure. When their emotions start to build, it will take a long time before they are stable and both partners certain of their feelings for each other. Sagittarius will change their mind many times, probably going from one extreme to another because that is what they're inspired to do by their Aquarius partner. It is hard to set the scale for emotions in this relationship. None of these partners is that emotional on the surface, although Sagittarius can fall in and out of love quite often. On the other hand, Aquarius will need to form attachment first and then wait for the certainty of their partner's love. 85%

Sagittarius & Aquarius Values: They will both value wideness of one's mind, the optimism and the faith behind the brains, intelligence and vision. There are so many things they would agree on, starting from the usual like value of freedom, and moving on to their own qualities and expectations. As a sign of Neptune's exaltation, Aquarius has a special approach to honesty, and for a Sagittarius honesty is one of the things they value most. 90%

Sagittarius & Aquarius Shared Activities: They could have trouble reconciling their approach to religion and any religious activities could be the source of problems in their relationship, because their whole individual belief systems could be at stake. Still, in most situations they could have a lot of fun wherever they go. Sagittarius will easily put on a smile and follow any idea Aquarius has, for as long as it makes sense or they have something to learn from it. There will be activities that Sagittarius will want to commit to, while Aquarius will find them silly or even stupid. There will probably be no place strange enough for these two to discover their shared interests once again. 85%

Conclusion: They will get together when it is time for both of them to go through a change in their lives or leave a partner they feel restricted with. Their relationship is often a shiny beacon to everyone around them because it gives priority to the future and brings hope of a better time. The main challenge of Sagittarius and Aquarius lies in their rational natures. Although their minds will have a wonderful relationship, they could have trouble reaching real intimacy and closeness. A relationship between a Sagittarius and an Aquarius partner might seem like a same sex friendship to other people and whatever they might think of this, this is the type of relationship both of these partners might need. They both need to slow down and ask themselves how they feel before they end up in a heartless bond they find solace in as they run away from the world. 83%

Sagittarius Compatibility with Pisces

Sagittarius & Pisces Sexual Intimacy and Compatibility: As two mutable signs, there will be no end to their creativity and changes in positions, scenery and levels of commitment and intimacy. Their sex life will have ups and downs, excitements and disappointments, too many expectations and a lot of surprises. The best thing about their relationship is the positivity both partners share, and a lot of laughter and fun they will share in their sex life. Unfortunately, the level of intimacy will rarely be satisfying for any of these partners. Since they are both ruled by Jupiter, they will be faced with their rational natures and their convictions. If they ever manage to end up in a physical relationship, they will have a lot of fun. The main reason why their sexual relationship rarely comes true is over thinking of both partners. Sagittarius will wait for a grand emotion, grand gesture or any sort of passionate initiative from Pisces, while Pisces will wait for all of the pieces of the puzzle to fit in their perfect position. In most cases, neither of these things will happen and they won't move further from a platonic relationship. 30%

Sagittarius & Pisces Trust: This is the beauty of Jupiter's rule like everything makes sense in their relationship. Sagittarius partner is too passionate and loves to have a lot of options when it comes to relationships. They will rarely settle down with anyone who lacks a strong decision to win them over. Pisces, on the other hand, will be too sensitive while trying to show their imaginary strength. They could end up following their Sagittarius partner in adventures they are not ready for, sad because of the lack of emotional understanding from their partner and ready to open up to someone else. It will be very difficult for Sagittarius and Pisces to trust each other, but they will probably accept it as a perfectly normal thing. Regardless of the purpose of each little thing they do, they will often have twisted expectations from each other and this will lead to unintentional dishonesty. 1%

Sagittarius & Pisces Communication and Intellectuality: When we rule out emotional and physical sides of their relationship, a Sagittarius and a Pisces partner will be best friends, almost inseparable, for a while. There is no way to determine how long their relationship will last, and unless supported by fixed signs in their personal charts, they will rarely stay in it for long. Sagittarius is ruled by Jupiter, and traditionally so is Pisces. This is the biggest planet in the Solar system and as such, it has a great influence on the personality of these signs. They will share the same optimism, the same vision and pretty much the same delusions. These partners will be linked through extremely beneficent influences and they will most certainly share the same sense of humour, operate at the same speed and learn a lot from each other for however long they are together. Jupiter is a planet of knowledge, and they will be fascinated by the unknown they can share with each other. It will be very difficult for Sagittarius and Pisces to trust each. In time, they will realize what their differences are in the most unusual way. Sagittarius is a sign of convictions and will be more rational and reliable than their Pisces partner. At some point Sagittarius will start to form a distance because of expectations that haven't been met and the irresponsible, detached behaviour of their Pisces partner. In return, Pisces will have a simple feeling that this is no longer where they want to be. Both of them might never understand why, but they will simply separate with no ill intentions, and probably not much anger or hurt. The beginning of their separation lies within disrespect of each other's convictions and personalities. 85%

Sagittarius & Pisces Emotions: The relationship they will build through communication and understanding of each other's worlds will awaken emotions through excitement and the unexpected. They will laugh with each other with open hearts and share wonderful emotions for as long as they are in the beginning of their relationship. As soon as any problems start to arise, they will both feel their emotions fade, as if the entire relationship was superficial. Whatever the circumstances, it is important for both of them to remember that there is nothing superficial about this contact. The learning process and the beauty of their entire relationship shouldn't be forgotten, but kept as a base for all of their future relationships. They love each other in a strange way, idealizing each other, getting disappointed, choosing to stay apart even when they wish to be together. Their relationship will be an emotional rollercoaster for both partners, but it will rarely last very long. This is a complicated emotional contact because both partners easily fall in love, and the deepening of their relationship can make them both be swept off their feet. If the relationship ends in a disrespectful way, they could both lose a bit of their faith in love. 30%

Sagittarius & Pisces Values: If they connect through deep love, they will overcome this with ease and emotions they share will make Sagittarius understand their partner. In general, they will agree on many things. They will value each other's utopias, people with good hearts, knowledge, wide perspective and travels. The biggest difference here is the value Pisces partner gives to emotions, for Sagittarius often doesn't really understand that approach. They will understand each other's sense of not belonging and share the sense of a higher power. 65%

Sagittarius & Pisces Shared Activities: It is like they are both hungry for happiness in a dark world they have stumbled upon, and when they meet someone this fascinating it would be a shame to miss the opportunity to enjoy. For as long as they are in a fascination phase of their relationship, they will be inseparable. Both of them have the need to grab everything that is offered and leave nothing joyful unused, unsaid and left for tomorrow. In time they will realize that although they share the same need for movement and changes of scenery, they might not need the same contents in their lives and Sagittarius will turn to physical activity, philosophy and travel, while Pisces will usually go back to creative work and the pursuit of love. 90%

Conclusion: At first, it will be challenging for them to leave the platonic zone and start building a physical relationship. Once they get close to each other, their process of learning will begin and both partners will be fascinated by each other, thinking that their relationship could never end. They will easily idealize each other, think of their relationship as the perfect love, but this infatuation won't last very long because of their changeable natures. This is a relationship of two kindred spirits that often doesn't last very long. The fact is their relationship represents a moment in time when they have both deserved to smile. For as long as it lasts and they are happy, it will be cherished by both of them. 50%

PART-10 CAPRICORN HOROSCOPE

(December 22nd to January 19th):

Who takes what's due and climbs and schemes for wealth and place?

Who is confident, a resourceful and morns his brothers fall from grace?

Element: Earth

Quality: Cardinal

Ruler: Saturn

Lucky Gem: Blue Sapphire, Garnet and Diamond. This blue sapphire promises him/her all the luck he/she could wish for. He/She could soon be in for a lot of money, so better brush up on his/her financial skills.

Lucky Number: Number 1, 4, 6, 8, & 9 are his/her secret weapon for the success.

Colour: Blue, Brown, Grey, Black

Lucky Colour: Peacock Blue. Wear peacock blue for luck.

Day: Saturday, Friday

Lucky Day: Friday. Watch things falling into place perfectly on Friday.

General Insights: When it comes to professionalism and traditional values, Capricorn is the first. Capricorn is practical and is considered to be the most serious sign of the zodiac, which possess an independence that enables significant progress both on the personal level and in business. As an Earth sign, for a Capricorn there is nothing more important in life than family. Capricorn is a master of self-control and has the potential to be a great leader or manager as long as it is in the sphere of business. Saturn is the ruling planet of Capricorn, and this planet represents restrictions of all kinds. The influence of Saturn makes Capricorn-born practical and responsible, so they know how to save money for the future.

They are masters when they need to prove that they are right. People born under the Capricorn sign sometimes can be really stubborn. They strive to get to the top only with their experience. Problems may occur when the Capricorn is forced to be very close with his associates. Capricorn-born has a hard time accepting others' differences, and in these situations, there is a need to control people or to impose their traditionalist values. They think that they are the only ones who know how to solve a problem, but they must learn how to forgive others, to allow them to be who they are and to stop condemning them.

Personal Quality: The Capricorn is miserly, very ambitious, self-confident, loyal, snobbish, true workaholic and accepts hard work and reaches to greater heights. He/She makes superb administrators, and often raises to very high positions in his/her careers. He/She wants to get everything with money power. He/She can be temperamental and moody. He/She acquires wealth, dominant position and what else due to his/her smooth talking and hard working. He/She never gives up the thing what he/she had decided to get and takes rest only after completion of the work. He/She knows the value of money only and is always ready to face the consequences to grab the money. He/She saves money and does not spend without need. He/She has very wide circle of friends. He/She is resourceful, practical manager, works well in a disciplined environment. He/She can be frugal, possessing the ability to achieve results with minimum effort and expense. He/She manages several projects simultaneously. His/Her bright carrier begins after the age of 24 and between the ages of 35 to 48 years. He/She is among the wealthy persons. He/She moulds himself/herself as per the situation and hence gets success in life.

Positive Quality: He/She is honest, responsible, disciplined, self-control, good managers, very practical person and can be relied on. He/She does duty sincerely and finishes the work with great responsibility. He/She is goal oriented. He/She has great control and authority. He/She is loyal to intimates. Never impetuous, he/she considers business and personal relationships carefully before becoming involved.

He/She is family person, and family usually comes first, except where business is his/her primary concern.

Negative Quality: He/She is know-it-all, unforgiving, condescending, expecting the worst, and sometimes too bossy and very narrow-minded and thinks highly of him. He/She doesn't function well in subordinate positions. He/She can spread gloom and tension in a minute and is quite capable of depressing everyone else around him. Never really up, but often down, he/she needs a positive environment to enliven his/her spirits. He/She thinks as the wisest man in the world and likes to show the others. Sometimes, he/she is very proudly. He/She is stubborn, overbearing, unforgiving, inhibited, fatalistic, condescending.

Physical Appearance: He/She tends to be tall and have sharp, angular features. He/She generally has serene expressions and an air of tranquillity. He/She takes great care of appearance.

Relationships: He/She can be surprisingly passionate behind closed doors. However, he/she takes a while to warm up and is very cautious about relating to others. He/She is very unhappy with emotional scenes and upheaval, gets hurt easily, but thaws just as quickly if he/she finds that his/her partner is genuinely repentant. Cancer and Libra girls will be ideal for his/her life partnership. Taurus, Virgo and Libra people will be helpful to him.

Career: He/She is usually determined and ambitious, with a strong sense of discipline and a good head for business. He/She sees duty and law enforcement as of paramount importance. Here are some occupations that he/she might consider such as Administrator, C E O, Coach, Commissioner, Economist, Governor, Industrialist, Leader, and Manager, Mountain climber, Office manager, Official, Operations manager, Organizer, President, Professional, Programmer, Proprietor or overbearing.

Health: Traditionally, Capricorn problem areas are their joints, bones and teeth. He/She needs to ingest extra calcium, cod-liver oil and evening primrose oil supplements to help his/her flexibility.

Ideal Partner: Always seeking a loving and strong partner, he/she gets along very well with Taurus. Capricorn needs a strong, loving partner and bond best with Taurus.

Compatible Zodiac Signs: Taurus, Virgo

Incompatible Zodiac Signs: Aries, Cancer and Libra

Greatest Overall Compatibility: Taurus, Virgo

Best for Marriage and Partnerships: Taurus

Capricorn likes: Family, tradition, music, understated status, quality craftsmanship

Capricorn dislikes: Almost everything at some point

BUSINESS ASTROLOGY FOR CAPRICORN

In business, Capricorn moves forward with quiet, deliberate persistence. They seek positions where they can have great control and authority which is often misunderstood by peer group. Good control over emotions and behaviour is their key to success in Business world. Business booster Gem is Moss Agate.

Capricorn Compatibility with Aries

Capricorn & Aries Sexual Intimacy and Compatibility: Rulers of Aries and Capricorn are Mars and Saturn. These planets are considered archetypal or karmic enemies. When it comes to sexuality, it is mostly signified by Mars and its contact with Saturn may result in all sorts of physical and objective obstacles on the way to a healthy sex life. Saturn puts too much pressure on Mars and takes a lot of its energy. Their relationship will result in lack of sexual desire, the mutual feeling of incompetence or even impotence of one or both included parties. When this sort of relationship happens, it is in most cases triggered by some deep unconscious need to be held back and restricted when it comes to sexuality. As with everything that comes through the sign of Capricorn, with time Aries partner could achieve some sort of balanced state in which they are sexually satisfied and their instinctive needs are met. Unfortunately, Capricorn partner will lose their energy and the need to participate in this sort of sexual behaviour by that time. This will ultimately lead to their separation, for there is nothing light or easy with these two, especially when it comes to intimate matters. Because of the unconscious type of their relationship, they could be insanely attracted to each other, but in most cases their differences will keep them at a safe distance. This is a very difficult combination of signs when it comes to sexual compatibility. At their best, Capricorn will support Aries' libido and control their passion to burn as slowly as possible. At the same time, Aries would consider their partner a teacher and learn about their body and the way to satisfy them. Still, this is a balance that is extremely hard to achieve when the clash of these two hard personalities happens. 5%

Capricorn & Aries Trust: Even if they have deep misunderstandings in other areas of their relationship, they will rarely betray each other's trust. This is something they will easily take for granted though, for when they are together, they seem to lose awareness of the things between them that should be treasured. Since they are both in extremes and "all or nothing" types of people, it will be easy for them to trust each other. One of them should have the sense to remind the other from time to time about the qualities that their bond includes. 99%

Capricorn & Aries Communication and Intellectuality: Other than that, they don't have much to talk about. Rarely will a Capricorn partner allow their impulsive and from their perspective even stupid Aries partner to have their own opinions and value them as useful or practical. Although they will certainly respect their initiative and energy level, the rest of Aries behaviour is simply unacceptable in most matters. Since Capricorn has their feet on the ground and is capable to measure the situation rationally, they will hold on to their calm opinion about their partner's lack of tact and endurance or even of their idiotism. You can imagine how annoying this can be to an Aries, especially if you take into consideration their passionate need to set strong boundaries and be respected in every way possible. On the other hand, Aries will simply have no patience for their Capricorn partner. They will seem boring and as if they only want people in their life to be useful. This will be attributed to their selfishness and lack of emotion and heart. Both of them will be wrong in a way, for they would need to understand what they both could become if the right person or motivator came along. They should keep their conversation in touch with carrier goals, achievements at work and physical activity, for Capricorn exalts Mars, the ruler of Aries. The problem is they could remain stuck in this pointless ego battle, until they both get so tired that they will hardly think of another relationship ever again. 20%

Capricorn & Aries Emotions: Their problem is in the way they understand each other to begin with. At the beginning of their relationship they will both probably have an image of the other person as someone they could become after some effort is put into their growth. The problem is that no one here wants to change. Their emotions could easily be connected to this first image they've had, however unrealistic it might be. So this is exactly what we would call the lack of understanding, because these two could be crazy in love with each other's false image and crazy persistent to change one another. Their problem really isn't the lack of emotion to each other (although you could easily contribute their issues to emotional disability considering their Mars/Saturn nature), but the lack of understanding and acceptance of each other. In this special case it would be best not to call this paragraph "emotions", but "understanding". They are often too stubborn and narrow minded to see that there is something different than what fits into their boundaries of possible personal characteristics. 1%

Capricorn & Aries Values: Mostly they are in sync when it comes to serious view of people they are surrounded with. The problem they have is in their unrealistic expectations founded on the fact that they share some values. They both value independence, clarity and honesty, and in general their system of values is not what brings problems to their relationship. If Aries values someone's ability to endure and push themselves over all possible personal barriers, that doesn't mean they are ready to become this person, or have the control to be one. Similarly, Capricorn might value speed and focus of some people, but this doesn't mean they will jeopardize their depth or attention to detail just to be faster, or leave deep psychological needs unattended only to stay focused at some goal. 60%

Capricorn & Aries Shared Activities: Aries doesn't understand why Capricorn would spend all night doing something extremely boring, just to be thorough when it has no value in this particular matter. What Aries doesn't understand is that studiousness always has its value, as much as Capricorn doesn't understand the physical need of their body to rest and act on impulse from time to time. They can easily find activities they both like when they include physical movement. They both need to pay attention to their bodies and everyday habits, so it would be good for them to have the same physical activity every day at the same time together. This way they would motivate each other and Aries will lift up Capricorn's energy and Capricorn would give endurance to their shared efforts. Capricorn sometimes just doesn't understand why Aries wants to run at 6 in the morning. This could do wonders for their entire relationship, for they could understand each other's natures much better through this sort of basic activity. 40%

Conclusion: This is why their relationship might seem like a competition to ruin the relationship in the best possible way. It is hard to say who will get out of it a winner, for they will both feel lousy most of the time and be relieved that they finally separated. If they stubbornly decide they love each other too much to let each other go, both of them would probably bang their heads against a wall for years to come. Their only chance of success is unconditional respect and the wideness of their views and expectations. They could truly complement each other, but only in a scenario where they would look for good in one another and highlight each other's qualities. This is not an easy relationship. None of the partners has any trace of lightness and blissful ignorance. Unfortunately, the malefic nature of their rulers rarely allows for them to be this positive and acceptance oriented. If they got together, and whatever their story is, they should think about the things they could learn from each other instead of looking for each other's shortcomings, and always stay out of each other's business. 38%

Capricorn Compatibility with Taurus

Capricorn & Taurus Sexual Intimacy and Compatibility: This is exactly what could make them a perfect couple. In combination with other signs of the zodiac it can be hard for them to open up and feel the need to experiment, even though Capricorn will do their best to show how ingenious they are when it comes to sex. When they get together and get to know each other intimately, they will learn what it means to relax. Capricorn won't feel the need to show off and Taurus will let go of their fear of getting hurt. The problem in their relationship can be hidden in their understanding of the Moon, for Taurus exalts it, and Capricorn doesn't like it very much. They could have trouble connecting on an emotional level if Capricorn doesn't fall in love deeply enough or has trust issues. This will be multiplied by Taurus' need to be loved unconditionally that they show in an endless loop, scaring their Capricorn away. Their different approaches to the combination of sexual instincts and love are what could make a gap between them. Taurus and Capricorn can both be quite rigid when it comes to sex. Taurus has a problem with initiative and aggression, not understanding Mars that well, while Capricorn needs initiative, physical strength and supports Mars. In their sex life, this could lead to a lack of emotion from Capricorn partner, leading to the frustration of Taurus, scared away by their libido with no emotional foundation. This could go as far as impotence and a general lack of sexual desire in both partners, unless they hold on to intimate nature of their sexuality and approach each other as different individuals with certain needs. 85%

Capricorn & Taurus Trust: Even when they do tell a lie, in most cases it is an experiment with other human beings to see if they can guess where the truth lies. When they are intimately involved, they like things between them and their partner clean and true. Taurus can easily sense this and will feel secure enough to not give in to their occasional need to hide things from their partner. Taurus is ruled by Venus, a planet exalted in the sign of Pisces, so they have this understanding of importance of secrecy when they are in love. Capricorn just isn't into lying. They don't even judge it but find it unnecessary and stupid. With Capricorn, they can find a way to hide their intimacy from the rest of the world and stay true to their loved one for a very, very long time. 99%

Capricorn & Taurus Communication and Intellectuality: Their differences are exactly there to make them a perfect couple, because they complement each other in a more subtle way then their opposing signs. The deep understanding of the Moon is something Taurus is blessed with and Capricorn lacks in their core. The fear of emotion can easily become a daily routine of neglect toward their personal emotional needs. Taurus has a mission to teach Capricorn about the significance of tenderness one should always have for one self. In return, Capricorn will help Taurus deal with responsibility and teach them how to reach their goals with no distracting emotions. Although they have different natures, they understand each other very well and motivate each other to grow and each of them in their needed direction. It is not always easy for them to understand each other, but with enough compassion and openness to feel for the other person, they can support each other in a way no other pair of signs can. After all, they do belong to the element of Earth, and can make magic in our material reality when they reconcile their differences. 85%

Capricorn & Taurus Emotions: This is usually something like a pattern to be broken when they do begin a relationship, for they have enough time and patience for one another. From the perspective of Taurus, this might not be the best emotional contact they've ever had, but from Capricorn's perspective, things can't get much better than being loved by a Taurus partner. However, there is a dose of almost unbearable satisfaction Taurus will feel when their long-term digging reaches the emotional core of their Capricorn partner. When this contact is reached, they will rarely feel the need to separate from them again. It is difficult to say with certainty they will find emotional fulfilment with each other because they are both very careful when it comes to love. To Capricorn this may seem as if someone literally touched their heart and they will probably never want to let their Taurus partner go. 90%

Capricorn & Taurus Values: While Taurus would create and motivate, Capricorn would lead the way to success and financial security. Whatever their goals, they could easily reach them together due to the fact they share the same material values to begin with. Still, they don't have such a peachy situation when it comes to their approach to emotions and family. With shared sense for value of the material world, these two can get really far together. They should observe different sides of their personalities as complementing instead of destructive and find a way to coexist giving value to each other's shadow. 90%

Capricorn & Taurus Shared Activities: If any sign in the zodiac needs rest, it would be Capricorn. Their high ambition can lead them to a state of low energy and Taurus is there to mend their tired Soul with fine food and time for joy. On the other hand, if the creative, motivating side of Taurus is awaken by the striving nature of their Capricorn partner, they will become everything but lazy and make room for both of them to be satisfied and happy with what they've accomplished. You could say that Taurus is lazy and Capricorn never stops working, but this is not exactly the case. In short, they can do anything together, for as long as they hold on to a fine balance of activity and rest. 85%

Conclusion: With the ability to complement each other in a gentle, slow way, they are the most boring couple on the outside, with most exciting inner activity that stays hidden from the rest of the world. Taurus and Capricorn can form a relationship so deep that their creative power in the material realm could seem unreachable for other signs of the zodiac. If Taurus motivates their Capricorn partner, and Capricorn shows the way of accomplishment to their Taurus partner, they could work together, raise children and share a life with more fun than they are both used to, or simply form an unbreakable bond. When their deep emotions intertwine, they are bound to each other for eternity. 89%

Capricorn Compatibility with Gemini

Capricorn & Gemini Sexual Intimacy and Compatibility: Then Gemini comes along and starts explaining each position, the interesting overview on Kama Sutra and the beauty of outdoor sex. It is almost unbearable to watch these partners with their completely different philosophies while they try to manoeuvre their sex life. In order for Capricorn to experiment in sex, their partner needs to manage to really relax them and open their mind. With Gemini, they feel like taking care of a child heading for trouble, getting naked wherever they feel like it. Although this is not actually the case, this is how it may seem to Capricorn, reliant on traditional values and always taking responsibility for their actions. In most cases, they will hardly even be attracted to one another. If they become sexual partners, there is a big chance that Gemini will find their Capricorn partner uncreative and stiff, while Capricorn would think of Gemini as too unconventional. The strangest thing in this combination of the signs is in the fact they will both probably consider each other boring. There are certain activities that don't require many words, and in Capricorn's humble opinion sex is one of them. Yes, everyone would say Capricorn can be boring and Gemini is so interesting and fun, but actually, the lack of focus and deep feelings Gemini partner usually suffers from, is a huge turn off for Capricorn. All things considered, these two are not actually the best sexual partners among the zodiac signs, but could make a meeting point in a relationship with enough boundaries and enough creativity of both partners. 1%

Capricorn & Gemini Trust: This is thinking of a crazy person in the opinion of a Capricorn, and there are no "levels" of adultery in their world. What's clear is clear. Still, because of Capricorn's trust equation that makes their life easy, they will tend to trust their Gemini. Actually, they will trust their own interpretation of what Gemini says. Capricorn always goes one step deeper than others, and Gemini rarely puts that much thought into their alibis. This is why Capricorn will read them with so much ease, probably knowing exactly where they've been and what they are dishonest about. A typical Capricorn representative will not be easily tricked. Gemini partner can have a tendency to flirt a lot and to consider "light adultery" normal. Gemini, on the other hand, will find Capricorn so square and by the book, that they won't even doubt their faithfulness and honesty, even if they have something to hide. 50%

Capricorn & Gemini Communication and Intellectuality: Unfortunately, this doesn't mean a lot to a Capricorn when they recognize the lack of essence in the things their partner says. However, they will still have a lot to talk about because there is always that serious side to Gemini in one of their personalities that will have a thing or two to share with a strict and sometimes difficult Capricorn. It is a good thing that Gemini are interested in literally everything that exists in the world and outside of it. So if nothing else works, they can always talk about Space Stations, diamond stars and other galaxies. Capricorn is interested in things that have deep, hidden meaning, looking at them as equations that should be solved and admiring problem solvers. They could spend their life in this strange analysis that is not so much focused on details as maybe Virgo's would be, but on the bridge between different worlds. Capricorn is fascinated by the before and after logic behind every little thing, and this is where Gemini can help them set a list to investigate.

If they are at peace and don't look at each other as stupid, distant or boring, they could help one another build a better understanding for the world. It is safe to say that Gemini can talk to anyone and settle any issue by communicating. Capricorn's steady, secure nature could teach Gemini how to make schedules and organize their thoughts and their actions, giving them a chance to move each thought a step further. In return, Gemini's childish approach to life can be something wonderful for Capricorn to incorporate in their life in order to be happier. 25%

Capricorn & Gemini Emotions: Although both of them can have relationships with other not-so-emotional signs in which they feel awakened with their partner, when they are with each other, they can be immune to each other's charm. Ruled by Mercury and Saturn, both signs are not that emotional, but the real problem is in the fact that they usually don't even spark some emotions in one another. There is just not that much to connect them and mostly their emotional relationship comes down to Gemini's dark thoughts and Capricorn's emotional distance. 1%

Capricorn & Gemini Values: Gemini truly values information in any form and shape, someone's ability to talk beautifully, to creatively use their hands and to implement ideas with a higher purpose. Capricorn values stability, punctuality and plain honesty. Although they will both be dazzled by the independence of their partner, the rest of their worlds rarely coincide that much. 5%

Capricorn & Gemini Shared Activities: If they do go for a walk, they will want to do this in order to get from point A to point B, or to have a healthier lifestyle. When Gemini walks, they never know where they'll end up. They might have started their route on the way to the supermarket, but one telephone call later, they are already in their car, heading to a different city. It is a good thing that Gemini always wants to learn new things and Capricorn likes routine and dedication, so they have a strong base for constructive studying and problem solving. If they do go for a walk, they will want to do this in order to get from point A to point B, or to have a healthier lifestyle. When Gemini walks, they never know where they'll end up. They might have started their route on the way to the supermarket, but one telephone call later, they are already in their car, heading to a different city. It is a good thing that Gemini always wants to learn new things and Capricorn likes routine and dedication, so they have a strong base for constructive studying and problem solving. They would both walk a lot, that's true, but the divergence of their motives is almost unbelievable. Capricorn is a sign of useful things, and they will want to have useful activity, whatever it may be. However, in most cases, their roads go in different directions. 10%

Conclusion: Although they are both looking for things the other person has, they don't seem to recognize them in each other. While Gemini needs someone to ground them and give them depth, when they look at Capricorn, they see someone old, unmovable and boring. Capricorn needs joy and relaxation in their life, but Gemini seems like a ball of uncontrollable, superficial opinions heading nowhere. Although they are both looking for things the other person has, they don't seem to recognize them in each other. While Gemini needs someone to ground them and give them depth, when they look at Capricorn, they see someone old, unmovable and boring. Capricorn needs joy and relaxation in their life, but Gemini seems like a ball of uncontrollable, superficial opinions heading nowhere. Gemini and Capricorn partners are a very strange fit. In truth, they could have a valuable experience being together, sharing their different lives day after day. They might even find out that they actually work well together and have the ability to reach any goal that they think of. 15%

Capricorn Compatibility with Cancer

Capricorn & Cancer Sexual Intimacy and Compatibility: When they get together, a passion awakens and they both become perfect lovers for one another. The patience Capricorn has for their partner is something Cancer really needs to relax and start feeling sexual to begin with. Capricorn needs someone who acts on true emotion, but also someone who doesn't take sex lightly. There are Capricorn representatives who have changed many partners, but they will probably never stay with the one that isn't family oriented and emotional when it comes to physical relations. Intimacy Cancer can create is exactly compatible to what Capricorn lacks. Cancer and Capricorn are opposing signs and there is a strong attraction between them. There is a lack of love, home and warmth in the sign of Capricorn, and Cancer partner can heal this with their highly compassionate approach. This could lead to thawing of Capricorn's emotional state and uplift the state of their sex life significantly. 99%

Capricorn & Cancer Trust: Not only is the sign of Pisces in their third house representing the way they think, but they are also often led by panic in their intimate relationships. When a Capricorn representative falls in love, they will understand that their partner needs to see trust and this is what they'll show. However, they could have trouble believing anything their partner says until some consistency is proven or their stories checked out with other people. Luckily, in Cancer's world there is often nothing that ugly and secretive to find, for their moral values are as high as the exaltation of Jupiter in their sign. For as long as Cancer feels Capricorn's devotion, they will not question their actions. Although Capricorn might seem trustful, they are probably one of the least trusting signs of the zodiac. This is why Cancer could easily sense their partner's lack of trust, pretend that they didn't notice, and find it endearing rather than repulsive. 99%

Capricorn & Cancer Communication and Intellectuality: Not literally, of course, but they often share their image of a relationship their distant relatives had, maybe centuries ago. There is information in our emotional body, about each emotion our ancestors have felt and didn't know how to deal with or how to use. This is where Cancer and Capricorn connect, as signs of the family we come from, and the family we will create. These partners could feel like they have known each other before they've actually met. Their mutual affection will seem familiar and warm, as if they grew up in a same house, even though their circumstances might be completely different. This could make them able to talk about anything, for there is closeness to the relationship of these two signs that is unexplainable to all others. However, if this emotional bonding doesn't come at first impulse, Capricorn could be very difficult to talk to from Cancer's perspective. They need to connect on a very deep level, or they will have opposite goals and Capricorn could seem like a career obsessed lunatic with no emotion what so ever, while Cancer could seem like a clingy housewife (no matter if male or female). This is a couple that has the strangest thing in common like genetics. They should both remember that if they see each other in this negative light, they are probably only hiding from their own, inner opposite side, dismissing the chance to be complete. 70%

Capricorn & Cancer Emotions: Although this could sound like a dream come true and could in fact create very strong emotions in both partners, there is almost always a karmic debt to be repaid before they could say they are truly happy together. These signs represent the axis of Jupiter's exaltation and fall and it is important to understand that their emotional states are closely linked to their expectations from each other and their relationship. These two are considered one of the most and one of the least emotional signs of the zodiac. One of them should be family oriented and the other turned to their career. Still, their emotions often run wild as soon as they lay eyes on each other. In time, they will both fight for security and stability of their relationship, and although it might be hard for them to reconcile these primal emotional differences, they will in most cases simply and find a way. The emotional depth of Capricorn is really hard to reach, but Cancer partner can approach this as their life challenge. Cancer and Capricorn are a love story their ancestors had, waiting to be resolved. When they get tied to each other, it is almost impossible for them not to get married, have children and the entire earthly love package. Still, they could take so much of each other's energy if they tried to change each other. It is best for them to accept each other's personalities as inevitable and impossible to change. This could lead to a more sensible future than the one in which they are both simply tired of each other. 75%

Capricorn & Cancer Values: As opposing signs, they can seem to have opposing values, but this is not really the case. They both need stability in their lives and will value people who give them the sense of security. They both value stability and practical sense. This is probably something they will value most in each other, the ability of both not to quit or give up, however hard things might get. 70%

Capricorn & Cancer Shared Activities: Capricorn is careful and will plan their activities well in advance, so both of them will have time to adjust to the idea or change their minds if they realize that this is not what they want. It will be very easy for them to agree on what they want to do together and find such activities, only if they respect each other's personalities. Capricorn will not want to go shopping for house decorations, no more than Cancer will want to go three nights without sleep just because of a project at work.

Cancer is not very picky when it comes to activity choices their partner has, for as long as they are not imposed on them or too aggressive for their taste. If they respect each other's boundaries, their time spent together should be truly satisfying for both. 90%

Conclusion: This deeply seeded need to mend what is broken in our family tree is something we all carry within, but these Sun signs are predestined to handle karmic debts and residue emotions from their families. They will have to deal with problems first if they want to be free of the past, and only after they have repaid what needed to be repaid, will they be able to truly choose one another. Cancer and Capricorn are usually bound to relive the love story of someone who lived before their time. In most cases this is a once in a lifetime love for both partners, and they will probably choose each other without a doubt. 84%

Capricorn Compatibility with Leo

Capricorn & Leo Sexual Intimacy and Compatibility: It will be a rare occasion when Leo is attracted to a Capricorn, but the other way around attraction seems more probable. However, they won't often get to the sexual part of their relationship, for even though they both might enjoy the chase, they will not see their future together. Leo is a warm, passionate sign, and Capricorn likes to be coolheaded and practical. This doesn't mean that Leo isn't at all practical, or that Capricorn isn't passionate, but they won't see each other as similar in any way. The rulers of these signs represent one of the archetypal conflicts of the zodiac, and tell the story of the fallen ego. This need could easily pull them both in a direction which will endanger their self-esteems and affect the image they have on their beauty and attractiveness. This is usually ignited by Leo's freedom of sexual expression that Capricorn fears, leading to the insecurity in both partners because they are not able to fit into each other's set of expectations. Their sex life can easily become boring for both partners, and what they often don't realize is how similar they actually are. Leo and Capricorn have one thing in common and it is their awareness of their Selves. The only way for them to have a healthy sex life is to share warmth and always bring new experiences, spicing things up. If they find themselves in a rut, they might stay there for a very long time, leading to the loss of libido and confidence, up to a loss of any sexual desire. 5%

Capricorn & Leo Trust: The depth Capricorn is prepared to go to makes Leo partner question their own motives and their whole personality. Lies seem to be impossible in this relationship, for each lied told, comes right back. As much as Capricorn sees behind the shine of Leo, Leo shines a light on Capricorn's darkness. Nothing stays hidden for long and as soon as one of them tries to stay secretive, mistrust is awaken. Capricorn knows that Neptune falls in Leo. This is exactly why they also know what hides behind the act in their Leo partner. However, in many situations they tend to trust each other because it gets so obvious that there is no reason not to. 40%

Capricorn & Leo Communication and Intellectuality: This is something that will not be easily reconciled and these partners could spend too much time trying to prove to one another why each of them has a point on what comes first. The problem is in the lack of understanding that each one of them has their own mission and their own role. It is futile to insist on someone's priorities changed when they should be different to begin with. If they respect each other enough to accept some pretty big differences, their communication might be very satisfying and fulfilling for both of them. Leo will help Capricorn find a more positive and creative view on every situation and Capricorn will give Leo the depth and the serious intentions they need. Their priorities differ greatly, and they both have a strong set of personal priorities in their lives. When they combine their abilities to organize, any plan made could be perfect.60%

Capricorn & Leo Emotions: Warm emotions of Leo are easily cooled down and buried, and without the ability to express love, Leo can become pretty depressed. In return, the time Capricorn needs to build the emotional story they need, will be roughly interrupted by their fiery Leo partner. This could hurt them, or lead them to the opinion that Leo is not the right person for them, however attractive, smart, capable or beautiful they might be. The problem with this couple is in the way they build up emotions, and their best chances are in time and patience, things that Leo rarely possesses, and Capricorn rules. There is no other way to reach the heart of a Capricorn partner and discover that they can be warm too. A relationship between a Leo and a Capricorn partner can be truly emotionally challenging, not because they don't love each other, but because they do. If there has been too much pain in their prior emotional relationships, both partners could be almost too stubborn to get to the point where they might actually fall in love. 1%

Capricorn & Leo Values: Leo is not much of a plan maker, they would rather go with the flow and look only a couple of days in advance, and they respect Capricorn's ability to focus on the final destination and weigh every step of the way. Still, the sensitive, calm, emotional centre that Capricorn values is never found in a Leo and unless they are truly inspired by their Leo partner, they could take away their worth just because of preferences. Leo and Capricorn both value well organized people, presentations and plans. Leo values direct, open hearted people with big smiles, and as soon as they judge Capricorn for not smiling all the time, they might as well end the relationship. 50%

Capricorn & Leo Shared Activities: If Leo wants to settle down, they might find it interesting to spend time in a usual, Capricorn way. In return, if Capricorn needs some additional energy and vigour, they will gladly follow Leo in their chosen activities. The most important thing in their relationship is good timing. Activities these partners might share depend greatly on their priorities, once again. If it doesn't exist, they will simply resist, stubbornly, doing anything the other person wants to do. 5%

Conclusion: If they meet in the right moment, Leo and Capricorn might get along very well. The main problem in their relationship is the set of priorities they might not share, and the passion or determination that both of them have. It is not an easy job, reconciling Saturn with the Sun, but it brings great benefits when it is done. The structure Leo could get and the creativity they might build on together could lift them to exactly what they desired, however their relationship might end. They differ as much as the Earth and the Fire, but when they share a common goal, they are unstoppable. 27%

Capricorn Compatibility with Virgo

Capricorn & Virgo Sexual Intimacy and Compatibility: Even though they don't lack the patience or the understanding for each other, there always seems to be just that one shred of pure emotion missing in their contact. Very often these partners don't get to have sex, because they will have more reason not to, than to give in. The beauty of their sex life, when they manage to synchronize, is in the depth both partners are capable of, that directly links to the depth of emotions they will show through the act of sex. Their main goal is to find someone who doesn't take sex lightly, someone who is not superficial toward them and cherishes them as they should be cherished. There is a certain shyness to both of them, and this is something that will make them go crazy for one another, if they only reach the point behind the rational distance they normally share. Virgo will bring enough change to their sex life, as a mutable sign, ready to experiment with someone who is so reliable and respectful. A sexual connection between a Virgo and a Capricorn might be great if they both weren't so stiff and strict when it comes to sex. This is a perfect relationship for both partners to relax and try out new things, if they find a way to open up in the beginning. 65%

Capricorn & Virgo Trust: There is nothing shady about them, nothing unreliable or quick to turn to deceit. Virgo usually has no reason to be unfaithful, except when they suffer from their own lack of faith and emotion that cannot be controlled. Even if this is the case, a Capricorn partner will inspire them to be the best they can be, and as faithful as possible. They will need some time to get used to each other's habits and build the trust they both wish for. Capricorn is a sign that can be trusted, and Earth signs understand this best. When they do, they will rarely break it for anyone or anything else. 99%

Capricorn & Virgo Communication and Intellectuality: The flow between Earth signs in its clearest form is sometimes unbearable for other zodiac signs and this is something Virgo and Capricorn truly enjoy. The depth of mind in both of these signs will fascinate them at first, excite them and make their communication incredibly interesting and informative. They both like a good, respectful debate, and in each other, they can find a perfect adversary. These are signs that make one cycle of communication complete, Virgo deciding what's next to discuss and Capricorn deciding when the subject is resolved. They are like a perfect mechanism, like gears fitting in together to solve any equation the world has to give. Their passion lies within these roles and when they find an understanding in other areas of their relationship, the intellectual one can be stimulating to the point of absolute bliss. When someone from the Air or the Fire realm observes these two as they talk, this conversation might seem extremely boring. This is a couple who knows that any problem is there to be solved and anything broken is there to be fixed. 90%

Capricorn & Virgo Emotions: Virgo brings Venus to its fall and Capricorn is the sign of the Moon's detriment. They do have some emotional issues, but not the same ones, and this helps them find an approach to each other that they both understand. Their relationship needs time, most of all, and the emotions between them need to build, just like trust. With the calm, practical, physical passion rising between them, both partners start building their confidence. With confidence, they feel much more liberated to experiment in life and sex, and this gives true quality to all of the areas of their relationship. The most incredible thing this couple shares is their discovery of one another. In time, they will peel layer by layer from each other's hearts, and be more and more fascinated by what none of them noticed before. Both of these signs are considered unemotional. Just like a mathematical equation, they represent a mystery box to each another, and they need to open it, bow by bow, side by side, until they unravel the treasure hidden inside it. 65%

Capricorn & Virgo Values: They value depth and this is something they will find incredibly soothing in one another, for they will both feel like they don't need to pretend to be shallow anymore. They will both value practicality, grounding, money and rational investments. The main difference they have to resolve here is in the value of Capricorn's goals, for they might be ready to do too much from Virgo's point of view, in order to reach them. Both Virgo and Capricorn will value calm, rational behaviour and choices, and one's ability to remain smart however unbearable the situation might be. In return, Capricorn doesn't understand Virgo's lack of motivation and their lack of need to claim the leading position. 80%

Capricorn & Virgo Shared Activities: Even though they are both Earth signs and this will allow them to follow each other with the same energy, they don't connect that well on the choice of places they want to visit. They will both want to go to a history museum and learn a lot of information there, but Capricorn often doesn't want to deal with doctors, let alone the calorie counting and green tea. There is no other sign to understand Virgo's need to sacrifice better than Capricorn, but if they don't feel responsible for their partner's activities, they will rarely follow them to depression. Where Capricorn wants to go up, Virgo wants to go forward. The most important thing to do here is hold on to positive activities and to a routine that makes them both healthy and happy. 65%

Conclusion: Even if everything between them seems too slow for some other zodiac signs, they build respect, trust and love, on the foundation of mutual analysis and detailed examination. Virgo and Capricorn belong to the element of Earth and follow each other's pace perfectly. The search for perfection can be ended in this relationship, for they give each other enough time, and listen to each other well enough to meet the expectations that need to be met. Both of these partners can be stiff and lose sight of the importance of the emotional, mellow approach to life, and this relationship can make them rough and too strict. Still, in most cases, they will give each other enough time to grow out of this and grow old together. 77%

Capricorn Compatibility with Libra

Capricorn & Libra Sexual Intimacy and Compatibility: Ruled by Venus and Saturn, they tell the tale of a soldier that had to leave his wife and came back after years of waiting. When it comes to sex, this is a combination that could point to a lack of sexual activity, even though both of these signs find sex extremely important in their lives. To begin with, they might feel no attraction at all, and even start a relationship on a basis formed in friendship, only to realize that there is no chemistry between them. If lack of attraction doesn't stop them, usually something else will. It is a combination that gives in to outer circumstances and things that are out of their control. Both of them could feel pressured and their self-esteem could suffer greatly. Still, there is an understanding between them ignited by Saturn's exaltation in Libra. This makes them both understand good timing and usually won't allow them to make a mistake expecting what should not be expected. When we are speaking of a sexual relationship between a Libra and a Capricorn, the first thing that comes to mind is waiting. In case they overcome all of the obstacles and form a strong bond through their personal natal positions, Libra and Capricorn can have sexual relations that are pretty conservative, routinely approached, and only satisfying if they both let go of their strict premises and conditions. 15%

Capricorn & Libra Trust: Even though Libra can sometimes have questionable motives, a Capricorn partner will make them turn to Saturn completely and feel guilty at the smallest glimpse of a possible lie. The only possible problem surfaces when Capricorn is too strict from the start, making their Libra partner feel inadequate, judged, or even scared of the consequences of their actions. A strange thing in a relationship of Libra and Capricorn is a really high level of trust between them. This could make their relationship dishonest, not because there is actually something to hide, but because Libra partner feels the need to protect themselves by holding on to their privacy. 80%

Capricorn & Libra Communication and Intellectuality: This can be a long battle, with no winners or losers, simply two people always building walls between each other, for reasons that aren't clear to anyone around them. The biggest obstacles to their understanding are the elements they belong to. Air and Earth are too far apart and it seems unclear to these partners how to reach each other on any issue in life. Still, there is a prudence to both of them that might give them just enough depth and understanding to have very interesting discussions and motivate each other to build a better foundation for every next debate. If they remain rational in their mental relations, they could have a lot of fun that other signs wouldn't be able to understand. The satisfaction they will both get from serious problem solving might lead them to a point where they find a solution together, Libra puts it in words and Capricorn puts it in action. We wouldn't say that Libra is that stubborn, but when they find themselves in a relationship with a Capricorn, they suddenly become headstrong and sometimes even impossible to talk to. Even though Libra loves Capricorn because of Saturn's exaltation, this is shown in the most unusual way, for they seem to feel the need to speak out of spite. There is probably nothing in the world that could raise their egos higher than situations in which they managed to resolve something by a simple shared effort. 35%

Capricorn & Libra Emotions: Libra is a sign ruled by Venus and their emotions come naturally, but usually restricted and held back due to the seriousness of their nature and the judgment of others they fear. Capricorn has a mission in life to accept all emotion, and in most cases, unless enlightened, they will be this judgmental force that holds Libra down. As if this wasn't enough, the entire situation will feed Capricorn's ego and make them think they are right about their approach, leading them further away from their focus point. This is a couple that has to work hard on finding a shared language to show how they feel and still respect each other. The emotional nature of Capricorn makes them distant for many, but completely untouchable for Libra as soon as they start dismissing their feelings. The only thing that can be done here is find a point of absolute respect and acceptance of all emotions and their manifestations. The hardest thing to reconcile in the relationship between a Libra and a Capricorn, are the ways they approach their feelings. If they allow each other to break things, get angry, cry, make scenes in public or give in to hysteria, they might find a way to express their love in a way that will be correctly understood. 1%

Capricorn & Libra Values: This can help them overcome any differences and opposing attitudes, values or convictions, for each of these partners will be ready to understand the set of responsibilities they have toward each other. As Air and Earth signs, both pretty set in their ways, Libra and Capricorn will differ greatly in the value of words and deeds. Libra will communicate and think that their mind is their biggest asset, while Capricorn won't really care if results aren't manifested through the material world. The most important values Libra and Capricorn share are the value of time and taking responsibility. This is a good training for a Libra partner to find grounding, but it usually won't be pleasant for any of them in a romantic relationship. 50%

Capricorn & Libra Shared Activities: There is a great chance they will be lulled by their relationship to the state of hard work and lazy rest, with no effort for anything creative or inspiring. The best thing these two can do together is be boring to the rest of the world. They need to keep their passions lit and create a weekly routine that will make them get out of the house and do something fun. 25%

Conclusion: If we want to choose the best word to describe the relationship between a Libra and a Capricorn partner, we would have to say, it's hard. This doesn't mean they won't enjoy the trouble of being together, or stay in a relationship for a very long time, but this is most certainly not a bond that many other signs would engage in. Their biggest challenge is the lack of respect for emotional value that is usually initiated by Capricorn, but easily continued by Libra. If they find a way to share, show and understand each other's emotions, everything else will seem like a piece of cake. 34%

Capricorn Compatibility with Scorpio

Capricorn & Scorpio Sexual Intimacy and Compatibility: The physical nature of Capricorn will help Scorpio ground their sexual needs with ease. The main problem of this couple is their relationship to the Moon, for they are signs of its fall and detriment. This "agreement" not to be too sensitive and emotional, can take out any real intimacy from their sex life, and make them too cold and distant, even though physically enjoying their relationship. They might even think that this is all they need, but their hearts won't agree, and other people will show up in their lives that show them how much they actually depend on intimacy. Both of these signs feel a gravitational pull toward their opposing signs, Taurus and Cancer, two of the most emotional signs of the zodiac. This explains their need to build real intimacy. For both of them, physical pleasure has to be achieved through tenderness and emotion, or they won't truly be satisfied. In general, Scorpio exalts Uranus, and they might be a bit frustrated by the conservative approach of Capricorn. It is a good thing they can wait and slowly build up an atmosphere in which their Capricorn partner will be relaxed enough to try new things and experiment. The excitement of this sexual contact is something Capricorn will have trouble letting go. Scorpio and Capricorn share a special sexual bond as signs in Sextile with each other and due to the fact that Capricorn exalts one of Scorpio's rulers, Mars. Scorpio, on the other hand, will enjoy the sense of security and patience they get from their partner, even if they openly express their sexuality. 60%

Capricorn & Scorpio Trust: Even though Capricorn representatives don't have to be that honest at all, their relationship with this direct and honest partner will make them feel like they should be as honest as possible too. Any lack of trust in their relationship is a consequence of the lack of intimacy, for they seem to lack the ability to sense each other deep enough to understand if they do trust one another or not. If there is a sign Scorpio can trust, it is the sign of Capricorn. This can be solved if each partner deals with their own insecurities individually and with an emotional effort to build intimacy. 90%

Capricorn & Scorpio Communication and Intellectuality: Their similar pace and the patience Capricorn has, followed by the feel of Scorpio can help their understanding very much, but when they disagree on something, they could end up in a silent fight for years to come. There is nothing light or easy in this interaction, and even though they understand each other's depth of mind and a certain "what goes around comes around" view on life, they will rarely laugh, dance, and have fun together. They might think that this is not even something they need, but everyone needs to have some fun and smile, or life loses a lot of its meaning. Dark humour might savour their situation, day after day, and if they have the same friends it will be much easier for them to enjoy life together. All friendships Scorpio makes, become long-term with the help of Capricorn when respected enough, and this could help them build a wonderful surrounding full of understanding people who love them. In general, they get each other's need for silence and patiently approach each other until each of them opens up. Fixed, unmovable Scorpio, in the state of constant metamorphosis and evolution in the same direction, can be a bite "too big to chew" for their stubborn, earthly, long lasting in everything Capricorn. This can open the door for a respectful communication and intellectual understanding that lasts for a very long time, if they learn to control all the negative convictions that surface when they are together. 70%

Capricorn & Scorpio Emotions: When they start their relationship, they will both give the impression of people who stand with their feet firmly on the ground, strong and rough when needed. They will rarely notice that this brings out the expectation to always be the strong person who they were in the beginning, and making them force things on themselves they are not ready for only to avoid showing any weakness. The biggest problem in a relationship between a Scorpio and a Capricorn is their emotional contact, simply because they both tend to have emotional problems, dismissing how they feel by dismissing the Moon. A lot of deep, emotional understanding is needed for them not to be forced to move even further from their life goal to find emotional balance. 35%

Capricorn & Scorpio Values: Scorpio is the sign of Venus' detriment and Capricorn brings a lot of guilt into it, so their combination of values is basically founded on feelings of guilt and the sense that nothing is ever good enough. It is truly interesting to watch this couple share values with such a difficult relation to Venus and the term of value itself. Even though this will be a good motivation for them to get better, every day, it is quite difficult to deal with in a healthy, loving relationship in which they should both discover they are good enough. 30%

Capricorn & Scorpio Shared Activities: They will focus their energy on constructive things in order to build the world they wish for themselves. This is not always the most happy, joyful place with rainbows and unicorns, but it is realistic, practical and most of all good for personal growth. If they start digging into the past, they might find shared therapy revealing, even if they don't need it for any obvious reasons. It is truly interesting to watch this couple share values with Together, Scorpio and Capricorn will strive for greatness. There is a need in them to dig out the truth, whatever it might be, and this will make it easy for them to spend quality time together without much doubt on what they want to do. 99%

Conclusion: They are both deep and don't take things lightly, and this will help them build a strong foundation for a relationship that can last for a long time. The relationship of Scorpio and Capricorn can be inspiring for both partners to search for the truth, dig up under their family tree and deal with any unresolved karma and debt. However, this exact thing can easily make their relationship too dark and unemotional, pull them both in a state of sadness and depression, or simply awaken their need to search for the light with someone else. 64%

Capricorn Compatibility with Sagittarius

Capricorn & Sagittarius Sexual Intimacy and Compatibility: Even when they are attracted to each other and form a sexual bond, after their time has passed they will probably feel like they shouldn't have been together. There is no logical explanation to this feeling, but it is present more often than not. Differences in their character can be strangely easy for them to handle, simply because a Sagittarius takes everything in with ease, and Capricorn feels responsible enough to understand their partner's immaturity as their own fault, in some strange way. Each Capricorn wants meaning and depth to their physical encounters, for they are slow, thorough and value their physical reality. Sagittarius often doesn't understand the pace at which a Capricorn wants to move in, nor do they see the importance of the physical world that Capricorn has the responsibility to. In the beginning of their relationship, if they share the same desires, they might not see how incompatible they actually are. There is something unbearable about the sexual contact of these partners. Unfortunately, as time passes, it becomes pretty obvious that their archetypal battle reflects on their characters in a way that taints their sex life. The only way they can ever remain in a healthy sexual relationship, is if Sagittarius respects the physical, as much as Capricorn loosens up and respects change that comes with their partner's Jupiter governed Soul. Their meeting point is in the sing opposing Capricorn, where the ruler of Sagittarius is exalted. In other words, their meeting point is in pure emotion. 5%

Capricorn & Sagittarius Trust: Capricorn feels this and recognizes the lack of inner honesty that doesn't seem to change. The problem here is in the fact that Capricorn is the sign of Jupiter's fall and this is the ruler of Sagittarius, as well as the traditional ruler of Pisces. The magic of life and the beliefs that lead in a certain direction seem to be lost on Capricorn. They know with certainty that the only things that give results are their rational mind and hard work. Sagittarius can explain to them that beliefs create their reality and that it is enough to believe in a good outcome, to affect the entire web of circumstances in a positive way. It is true that Sagittarius is one of the most honest members of the zodiac when it comes to their relationship with others, but they are rarely entirely honest with themselves. This issue comes down to a problem with trust, but in fact it goes much deeper than that. 10%

Capricorn & Sagittarius Communication and Intellectuality: Sagittarius has that optimistic smile that will definitely put a smile on Capricorn's face too, and it will get much easier for those fiery, creative ideas of Sagittarius to find their grounding through Capricorn's practical approach. With enough respect, this is a couple that links a visionary with a builder, and there is really nothing they cannot make when together. If they don't expect change from one another, they are very likely to be extremely intellectually compatible.

The most beautiful thing in this contact is in their complementing protective roles. Both of these signs represent protection, Sagittarius ruled by the greatest benefic and Capricorn as our fence, our shell to the outer world. A Sagittarius and a Capricorn can be full of understanding for each other, if they don't jump into a battle over their belief systems. When they manage to build a functional core, these are partners that will never let anyone else affect their relationship. If they are in search of someone who will not allow interfering, meddling and any type of disrespect from other people, this relationship might be their best choice. 75%

Capricorn & Sagittarius Emotions: This is where their hearts meet and if there is enough faith in a Sagittarius, without any unrealistic expectations, they might fall in love deeply. Sagittarius and Capricorn can find a shared emotional language due to the fact that Capricorn needs someone like their opposing sign to complete them, and Sagittarius tends to become that sign as a place of Jupiter's exaltation. There is a small chance that a Sagittarius will be that mellow, tender person a Capricorn needs, but this is something that can be overcome with enough closeness and understanding of their differences. 55%

Capricorn & Sagittarius Values: Sagittarius is a mental sign, focused on philosophy and learning, always in search for unity, synthesis and that universal truth. Capricorn is the logical continuance of Sagittarius, as a practical tool that uses knowledge. If they don't find each other stupid, they will click in the same wavelength without much trouble, and discover that they share a certain depth and curiosity that isn't obvious at first glance. There is one important thing these partners agree on, and that is the value of intelligence. Most of their values differ greatly and their needs are often too far off. While one of them values freedom, width and creativity, the other values practicality, responsibility and focus. 20%

Capricorn & Sagittarius Shared Activities: The lack of relationship between their Suns helps their bond with a certain lack of disrespect. This leads them into a situation in which a Sagittarius finds their Capricorn partner interesting, as an extraterrestrial they have always wanted to meet. They are different enough to be interested in each other with genuine curiosity and a Sagittarius is always ready to try out something new. Capricorn will probably refuse many of the childish activities Sagittarius suggests, but it becomes fun to talk them into it and there is a note of laughter and joy in these attempts. Even though we would think that a Sagittarius will get bored and want to run off from their Capricorn partner, in most cases this doesn't happen. They are both smart enough and aware that their differences exist, which makes their entire story so exciting and refreshing for both. 60%

Conclusion: Their understanding and acceptance of their differences is refreshing and fun for both partners, and they might have a good time while together, for however long. This is not your ideal relationship, and it will rarely be the one they both choose to stay in for the rest of their lives. We cannot predict too much stability unless a Capricorn decides to make it, but the smile on Sagittarius' face and the ability they have to make their partner laugh, can be the pillar of their bond for as long as they both need it. 38%

Capricorn Compatibility with Capricorn

Capricorn & Capricorn Sexual Intimacy and Compatibility: We can see two partners that exalt Mars, meaning their libidos are strong and they have the need to follow their instincts. But we can also see two people who hold on to their restrictions, who deny themselves the right for satisfaction and choose to make rational decisions every single day. Practicality doesn't go well with sexuality, and while other partners can awake their sexual creativity and form an intimate bond with them, two Capricorn partners will rarely satisfy each other in a sexual and an emotional sense combined. On top of this, there is a fact that their sign is a ruler of time itself, and the most probable outcome of their contact is an endless wait for anything to happen. None of them lacks initiative as they do both exalt Mars, but when it comes to taboos and matters of the flesh, two Capricorn partners can't seem to get to the point in which they actually take their clothes off. When they start a sexual relationship, they could go to one of two extremes. It is hard to say anything about this couple, let alone imagine their sex life. Either they will fully understand each other's needs with very few words, or they will just hit the wall every time any partner wishes to create a more intimate atmosphere. 40%

Capricorn & Capricorn Trust: There is a strange need for a competition between these partners, and this won't exactly help them believe each other or trust their future all together. The problem here doesn't come down to lies, but to silences they decide to create. You don't trust a Capricorn even though you are a Capricorn. Probably because you know you are a better, more honest version of a Capricorn. If they form that quiet but tense air around them when they try to communicate, both of them will question each other to the point in which they lose sight of who their partner actually is. 80%

Capricorn & Capricorn Communication and Intellectuality: If they don't speak their mind, analyzing each other like guinea pigs, they won't get very far, and their respect will be shattered as soon as this becomes clear to both of them. In many cases, two Capricorns won't even feel the need to talk to each other that much. They would have a lot to say, both of them interested in each other's lives and stories, but with a fence in front of their faces that doesn't give them the opportunity to truly share. If they work on the same project and find themselves in a situation in which they have to talk, they will discover many things they agree on. There is so much two Capricorns have to discuss. For five minutes. Yes, their intellectual relationship can become a debate arena, but chances are greater for a silent tournament in which none of them can ever win. In order to build a relationship with the right amount of meaningful conversations, it is best for them to work together and solve the same situations with a shared mind. As they resolve equations of life with one another, they might start enjoying their communication a lot. 65%

Capricorn & Capricorn Emotions: These are partners that rely on their lack of emotion most of the time, always trying to be cold, controlling and rational. When they get together, they will most likely recognize these characteristics, but that won't make them any less annoyed by who their partner is. It is a good thing they share the same approach to all relationships, being unable to open up until someone lets the pressure drop and makes them feel safe and secure to express how they feel. If they truly fall in love with each other, they will need a lot of time to say this out loud, for both of them dread any public displays of emotion and don't feel confident enough for private displays either. They can be saved by a simple word of confidence, and by their understanding of each other's emotional depth that isn't easily reached. Respect, however, won't be enough for them to share the feelings that need to be shared, and they tend to leave each other be, in a certain silence, because of that understanding they have for similarities of their natures. Emotional contact between two Capricorns can be a very interesting thing. Unfortunately, this drives them even further apart as they fight to remain in an emotional connection by using only their brains. 30%

Capricorn & Capricorn Values: It is not easy for one Capricorn to find another Capricorn to share them. Even though they are members of the same Sun sign, every Capricorn is a specific individual with their own set of values that have to be set in stone. The rigid nature of Capricorn doesn't allow much understanding when it comes to behaviour that "isn't approved" and being a Capricorn doesn't exclude anyone from the primal equations each of them sets. They should stick to the values they share instead of questioning those they don't. 80%

Capricorn & Capricorn Shared Activities: It is hard to determine why when they have the time and the energy, and the only logical explanation seems to be – out of spite. Who would say that loyal, responsible Capricorn has this need inside their mind and their heart? When they loosen up and realize that their partner is not threatening them, the two of them will find many things to do together. They might not understand where they need to show up and what is truly important to their partner. Two Capricorn partners could share any possible activity that comes into their minds, but one of them, or both, simply won't. If they lose their closeness enough to lose the understanding of each other's needs and desires, there is a big chance they will have to end their relationship and find someone a bit more compassionate and mellow. 75%

Conclusion: One minus might give a plus with the other minus, but these two turn to whatever is the opposite of functional as soon as another dominant partner (Capricorn) comes into their life. The game of superiority they will have trouble containing can become the main stream of their relationship, leading them toward an inevitable end. The relationship of two Capricorn partners isn't really ideal. In order to stay together, they need to point their horns into someone or something else, and make room for emotion they both need in order to find balance. 62%

Capricorn Compatibility with Aquarius

Capricorn & Aquarius Sexual Intimacy and Compatibility: The main problem in their sex life will be their different pace, and this is mostly caused by the difference in their elements. As an Earth sign, Capricorn is slow and thorough. A representative of this sign will rarely jump into a sexual relationship without attraction and respect for another person, and will want to give their best performance when sex finally happens. Aquarius is an Air sign and this does make them kind of flaky and unreliable, although they are ruled by Saturn, the master of reliability. They will want things spontaneous and fast, without much thinking and as relaxed as possible. It is very rare for an Aquarius to have patience to wait for Capricorn to make a detailed plan, and this will be a great turn off for Capricorn because they don't like anything done in haste and the heat of the moment, especially when it comes to sex. They can both be very passionate when with the right partner, but the starting point in their approaches is usually simply too different to work out. It would be easy to make a simple assumption that Capricorn is traditional and restricting, while Aquarius is the opposite, but they are both traditionally ruled by the same planet and it would be silly not to understand their similarities. Fortunately, the respect they will have for each other could make them become very good friends and if they manage to find the right way to communicate, they might even build a quality sexual relationship on a foundation of friendship. 5%

Capricorn & Aquarius Trust: Their ideas of trust are different though, and it will be hard for both of them to accept each other's natures the way they were meant to be. They don't trust one another. One of them has stone cold convictions and the need to never be wrong or make a mistake of any kind, while the other has no fear of confrontation and values the image of truth among the human race too much to have the desire to lie. The lack of trust between them is not really the lack of trust they have in each other, but in the possibility this kind of relationship will work out. 80%

Capricorn & Aquarius Communication and Intellectuality: They have this distant, cold, silent agreement that they are both worth each other's respect and this can seem terrifying at times as they distance from each other in order to remain in this agreement. They would rather be in no relationship than look at each other in a different way, and because of this they could form a beautiful, lasting friendship. Still, it is important to remember how different they are. Although there is no real freedom for Aquarius if the rules of their preceding sign aren't followed, it is not easy for them to understand each other's way of living. An intellectual relationship between a Capricorn and an Aquarius can be kind of painful to watch when you are, for example, a Taurus or a Cancer. For as long as they hold on to shared interests and their mutual love for the seriousness of their bond, they will be able to maintain the image of a strong intellectual bond that makes them both satisfied. 65%

Capricorn & Aquarius Emotions: In general interpretations they are both supposed to be unemotional, detached and closed up for emotional interaction with other people. However, this is not exactly the root of their lack of emotional connection. As an Earth sign, Capricorn needs emotions to be shown in a physical, practical way. This is exactly why they are often described as selfish, taking care only of their own needs. It is not easily accepted by the more spiritual signs that someone gives so much attention to earthly things such as money, or any material or career oriented value. Aquarius belongs to the element of Air and it is a sign of ultimate overall faith, different than any religion or rule that a human might have created. They need heavenly love, someone to share all of their ideas with, so they can float together on a cloud to never land. The emotional side of a relationship between a Capricorn and an Aquarius is something really strange. They will not care much about money, food or even sex, if their elevated spirit gets wings and they get yet another chance to dream. In order to form an emotional bond, both Capricorn and Aquarius will have to accept the other side of reality. 5%

Capricorn & Aquarius Capricorn Values: It is sometimes easy to forget that the sign of Capricorn precedes Aquarius and that there would be no sense of liberation without enough pressure. Their roles are intertwined in a strange way and they could find themselves valuing the exact same things if they dig into their personalities a bit deeper. Capricorn values boundaries and Aquarius values freedom. They cannot ever possibly be in a loving relationship. To start with, they both value consistency and loyalty, and they will both consider all of the standard humane evaluations of people necessary. They will choose their relationships on the same grounds when it comes to long term commitments and rarely allow others to control them. 40%

Capricorn & Aquarius Shared Activities: Capricorn exalts Mars and knows exactly where to drain it and Aquarius doesn't really know what to do with it anyway. Even though they won't often choose the same activities, this should help them follow each other for long enough to determine what they both might like. None of these signs lacks energy. It is only necessary for Aquarius partner no to force, insist and speed up, and Capricorn needs not to inhibit, restrict and deny. 25%

Conclusion: Both of these sings are traditionally ruled by Saturn, but their roles in the zodiac are entirely different. Their most challenging point in a relationship is their emotional contact. If they are to stay together, Capricorn partner will have to separate from the ground, just a little, and Aquarius will have to come a bit closer to Earth. Capricorn and Aquarius might not find each other that interesting to begin with. They need to meet in the middle for Capricorn will be able to help Aquarius materialize their ideas, and Aquarius to be able to help Capricorn make the needed change in their life and turn to something new. 37%

Capricorn Compatibility with Pisces

Capricorn & Pisces Sexual Intimacy and Compatibility: Their sexual relationship is a contact of two powerful individuals, one of them extremely strict and rational, and another flexible and emotional but confident about their beliefs. Differences between them will create a strong attraction, almost as if they were opposing signs. The sex life these partners can share is unexplainable when their characters are superficially observed. But in a different way than a Cancer, Pisces can reach emotional depth of Capricorn by a simple feel. This is not a matter of compassion, but a matter of their deep inner truths. Capricorn isn't unemotional, however obvious their coldness might be, and Pisces can be quite rational, even though they seem lost in emotion. They can awaken the best qualities in each other and share strong intimacy through deep emotionally-rational understanding. Their sex life will move in a strangely spontaneous way. Capricorn will be inspired to let go and open up to their partner, while Pisces will easily ground their affections and find a way to show them through physical contact. There is probably no better way for a Capricorn to relax, than to enter a relationship with a Pisces partner. Capricorn will feel more casual and Pisces will start to gather themselves up. If they stay together for long enough, they could make a perfect blend of stability, trust and emotional excitement. 99%

Capricorn & Pisces Trust: This is not always the case, and the possible roughness of Capricorn can sometime induce Pisces to tell a lie or two. On the other hand, if Capricorn is closed up and unreachable, trust will be disturbed by a simple fact that they don't know each other that well at all. The beauty of their relationship is in their approach to trust when they realize who they are dealing with. Since both of the partners don't trust the world to open up easily, they will have to earn it, day by day, from each other. They will often understand each other well enough to respect their relationship and keep it clear of dishonesty. This might seem like a game or a competition to win over the trust of their partner. 85%

Capricorn & Pisces Communication and Intellectuality: They can both care a lot about communication in their relationship, but their bond will allow them to stop talking and start listening. They are both shy in a way, and in order to get to know each other, they will have to be very careful to pay attention to one another. It is a good thing they will both want to do so, intrigued by their partner's nature, excited about getting to know each other in depth. The problem in their communication can arise when Capricorn gives in to their rigid opinions and beliefs. The sign of Capricorn brings Jupiter to its fall, and Jupiter is traditionally the ruler of Pisces. This could truly endanger their entire relationship, for the strict and rational nature of Capricorn can damage the faith and the convictions of Pisces through simple disbelief. Pisces partner will rarely give up on their belief system, since this is what they live for, but they could question it and feel lonely because of the lack of understanding from their partner. If anyone can inspire a Capricorn, it is their Pisces partner. "What will be, will be" might become "I decide what will be" and this can suffocate the spontaneity and the inspiration that Pisces partner carries within. 65%

Capricorn & Pisces Emotions: The most wonderful side of their relationship is in the expectance of emotion, the constant growth and their ability to bring out the best in each other. Still, this doesn't mean any of them will actually change their ways, even though they will be the best they can be. Capricorn might remain grumpy, while Pisces might remain flaky and unreliable. These partners share a deep emotional bond that can be built for years until it truly blossoms. When they are too set in their ways, they could end up truly annoying each other as Capricorn quenches the beliefs and the magic of Pisces and Pisces disappoints reliable, earthly Capricorn.
90%

Capricorn & Pisces Capricorn Values: It is strange how someone like Pisces can value stable emotions that much, but they help them to finally rest their mind and their heart when they are in a stable emotional relationship. Capricorn will surprisingly value one's ability to be in touch with their emotions and have a clear vision of a positive outcome. Unfortunately, they will have a real problem in approaching the use of beliefs and emotions in everyday life, and while Capricorn will value coolheaded thinking and one's ability to be rational, Pisces will value the opposite. They will sometimes be too different, Pisces dreaming of a perfect love, Capricorn knowing it is impossible, Pisces imagining a God with a golden beard, Capricorn believing in this moment, or visiting a church because of tradition rather than belief. There is a certain consistency in their way of approaching their values. It won't be easy but they might value each other enough to overcome their differences. 45%

Capricorn & Pisces Shared Activities: Even though their interests are usually very different, Capricorn will be inspired to get inside the world of Pisces, as much as Pisces will want to solve the equation inside their Capricorn partner. In time, their activities will separate, as they realize that things Capricorn wants to do are boring to Pisces, while things Pisces like to do, drive Capricorn crazy – with no plan, no intent or any kind of seeming usefulness. When they begin their relationship, they will want to do everything together. They will both have a need to hold on to tradition, Capricorn for respect of tradition itself, and Pisces for romantic reasons, and this should help them build enough shared activities as time passes. 70%

Conclusion: If someone like Capricorn can be pulled into a crazy love story, exciting and unpredictable, this must be done by Pisces. In return, Capricorn will offer their Pisces partner stability, peace and some rest from their usual emotional tornadoes. There is a fine way in which Capricorn can help Pisces be more realistic and practical, while feeling more cheerful and optimistic themselves. Still, there are challenges in their contact, mainly represented through their love of Jupiter. It might be hard for them to reconcile their different approaches to religion, faith and their different belief systems. A relationship between Capricorn and Pisces tells a story about possibilities of inspiration. To overcome this, it is best if they both ask themselves and does their belief system work. And does the one of their partner also work? If they understand answers to these questions, they might find enough respect to leave each other's Jupiter intact. 76%

PART-11 AQUARIUS HOROSCOPE

(Jan. 20th to Feb 18th):

Who gives to all a helping hand but bows his head to no command?

Who are inventor, genius and superman and higher laws doth understand?

Element: Air

Quality: Fixed

Ruler: Saturn, Uranus

Lucky Gem: Amethyst, Blue Sapphire and Hessonite. If he/she is feeling invincible, think of hessonite for him. This is a stone of power, and the world is for him/her to conquer.

Lucky Number: He/She can be sure of success with Number 2, 3, 4, 7, 8 & 9.

Colour: Red, Blue, Blue-green, Grey, and Black

Lucky Colour: Red. Wear reds for warmth and energy.

Day: Thursday, Saturday, Sunday

Lucky Day: Thursday. An old friend will brighten up an otherwise dreary Thursday.

Lucky Flower: Violet

General Insights: Aquarius-born is shy and quiet, but on the other hand they can be eccentric and energetic. However, in both cases, they are deep thinkers and highly intellectual people who love helping others. They are able to see without prejudice, on both sides, which makes them people who can easily solve problems. Although they can easily adapt to the energy that surrounds them, Aquarius-born have a deep need to be some time alone and away from everything, in order to restore power. People born under the Aquarius sign, look at the world as if, it is a place full of possibilities. Aquarius is an air sign, and as such, uses his mind at every opportunity. If there is no mental stimulation, they are bored and lack a motivation to achieve the best result. The ruling planet of Aquarius, Uranus has a timid, abrupt and sometimes aggressive nature, but it also gives Aquarius visionary quality. They are capable of perceiving the future and they know exactly what they want to be doing five or ten years from now. Uranus also gave them the power of quick and easy transformation, so they are known as thinkers, progressives and humanists. They feel good in a group or a community, so they constantly strive to be surrounded by other people. The biggest problem for Aquarius-born is the feeling that they are limited or constrained. Because of the desire for freedom and equality for all, they will always strive to ensure freedom of speech and movement. Aquarius-born have a reputation for being cold and insensitive persons, but this is just their defence mechanism against premature intimacy. They need to learn to trust others and express their emotions in a healthy way.

Personal Quality: The Aquarian is eccentric, inventor, technical wizard, and computer and has strong convictions. He/She possesses ill health and likes to make the world a better place. He/She has spiritual bent of mind. He/She is friendly, humanitarian and original thinkers, but can be rather eccentric. He/She is far sighted and innovative. He/She will go out of his/her way to help when needed, but never gets involved emotionally. He/She gets respect and the dignity in the society due to his/her kind nature and service to the people. He/She is broad-minded and expects others the same. He/She is known by his own name and is not follower but makes his own path. He/She often adopts a life style that goes against the trends, because the odd and unique fascinate him/her. He/She is an active man who is always busy with some kind of mental and physical work. He/She can achieve masterly in artistry, writing, medical, management, police or intelligence work. He/She enjoys his/her own company and is recharged by this quiet time.

Positive Quality: He/She is kind hearted, progressive, original, independent, humanitarian, honest, kind, tolerant, cool, clear, logical people. He/She always thinks for welfare of everyone. He/She is a dedicated person and never bears injustice. He/She does not like to hurt other's feelings and are very helpful to the people in need.

Negative Quality: He/She is, sometimes, not efficient planners and the work undertaken is seldom completed. Sometimes, he/she thinks himself very clever and intelligent than others. He/She is an enigma, runs from emotional expression, temperamental, uncompromising, and aloof. He/She is quite aloof people and, sometimes, does not accept his/her fault.

Physical Appearance: He/She has clean-cut good looks and a ready smile that shows off his/her excellent teeth to advantage. He/She can eat any amount of junk food and still remain slim. He/She likes to wear unusual outfits.

Relationships: Physical relationship is not so important for him/her but he/she is very emotional lover and has in depth love. Gemini, Libra, Leo and Aries are the persons who will be helpful to him/her.

Career: He/She might excel in technical fields linked with electrical and radio industries. Most are hard working, even driven in his chosen field. Many choose careers like Astrology. Here are some occupations that he/she might consider such as Academic, Adventure travel guide, Advisor, Astronaut, Consultant, Entrepreneur, Inventor, Humanitarian work, Market researcher and Senator.

Health: His/Her lungs are particularly sensitive to cigarette smoke and air pollution.

Ideal Partner: The most suitable match for Aquarians is Leo, Capricorns.

Compatible Zodiac Signs: Gemini and Libra

Incompatible Zodiac Signs: Scorpio, Cancer and Sagittarius

Greatest Overall Compatibility: Gemini, Libra

Best for Marriage and Partnerships: Leo, Capricorns

Aquarius likes: Fun with friends, helping others, fighting for causes, intellectual conversation, and a good listener

Aquarius dislikes: Limitations, broken promises, being lonely, dull or boring situations, people who disagree with them

BUSINESS ASTROLOGY FOR AQUARIUS

Enthusiastic and hard working, often these individuals take much stress to make business successful. Under pressure and anxiety they become seriously rude and offensive. For a successful business they should learn to cope up with stress and have control on temper. Business booster Gem is Pearl.

Aquarius Compatibility with Aries

Aquarius & Aries Sexual Intimacy and Compatibility: Usually it is both. Their signs go well together in general and they support each other easily, since they both have a lot of energy to follow one another. Still, when it comes to their sexual and intimate relations, they could lack emotion. Aries is a passionate sign with lot of warm, creative emotions. This is a relationship that could bring out their worst nature and simply emphasize that they are a sign ruled by Mars, a cold, unemotional sexual hunter. While this can be really exciting to both of them, it will not be very fulfilling, because they both need to feel loved. There is an excess of masculinity and energy that could lead to very turbulent relations. Their roles are easy to understand with Aries giving energy and stamina to their Aquarius partner, and Aquarius giving crazy ideas and widening horizons of their Aries. Sexual contact between signs of Aries and Aquarius can be really stressful or extremely exciting. This is very fun at the beginning of their relationship, but after a while, it might get tiresome for there are not enough ideas to cover the emotional emptiness they could encounter. 65%

Aquarius & Aries Trust: This doesn't mean they will be faithful to their Aries partner forever, but they would think it is fair to keep an open relationship and tell them about their indiscretions. Unfortunately, Aries is ruled by Mars and needs to be the only one in the world that their partner ever lays eyes on. This could turn them into an angry, possessive person who obsesses about the movements of their partner. When we are discussing matters of trust between them that don't include other people, it is safe to say that they don't have a problem. They both simply don't understand why they would lie when there are so many interesting truths to discover. Trust is an important issue for Aries and Aquarius can understand that. They need to be free to speak their mind and accept that they will never avoid conflict, but that it can be used in a constructive way to better understand each other and strengthen their relationship. 85%

Aquarius & Aries Communication and Intellectuality: Aquarius partner will recognize this, laugh and shake their entire world. It is unimaginable to Aries, always moving straight, for someone to have such an open mind, going back and front, having new revelations every day and never losing energy for new, different topics. Aries could find an idol in their Aquarius partner and full-heartedly enter any dialogue because they are excited about what they might discover and how their perspective would change. On the other hand, Aquarius enjoys this role in their partner's life due to their ego issues with the Sun positioned at this sign. They will share their thoughts with their partner, trying to be as interesting as possible. Aquarius is motivated by their Aries partner and enjoys making tiny jokes at their expense. It is important for Aries not to take things personally when it comes to Aquarius humour and they might have a lot of fun together. Because of their strong natures, filled with energy, they could fight most of the time. Their conversations can be so exciting that many people would like to jump in. Aries is often kind of serious and asks for their boundaries to be respected. In most cases, Aquarius will not stand for ridiculous conflicts and will build a brick wall somewhere between them if needed. Still, they usually tear it down at the end of the day, for they cherish each other the way they are after all. 90%

Aquarius & Aries Emotions: We wouldn't exactly say that Aries is patient, so you can imagine the problem that could appear. From the perspective of an Aries, their partner is cold, distant and has no intention of opening their heart for them. Aquarius sees things differently and tries to stay rational at all times. In order for Aquarius to awaken their emotional nature, it usually takes a partner with enough flexibility and patience to get there. When Aries starts asking for the show of emotion, the true problem surfaces, for Aquarius might have shown how they feel the entire time, but no one would guess what they were showing. 40%

Aquarius & Aries Capricorn Values: They both value freedom by first impulse. But in time, Aries realizes that they don't really value freedom that much when they see it at work. In fact, they would often change everything in their lives only to take away the freedom from their Aquarius partner. This is not a conscious need, but Aries can be like a spoiled child wanting things (and people) all for themselves. Although they can share a great conversation, their values go their separate ways as soon as they touch the subject of freedom. So with Aquarius changing direction as the wind and never changing their nature, Aries can find themselves truly unhappy for they want someone to share everything with, not only what the wind carries in. 30%

Aquarius & Aries Shared Activities: With so much energy, their only mistake would be to stay at home and not share a chance to get all that energy out of their systems. Unless Aquarius suggests something truly unacceptable to their Aries partner, they will have an abundance of possibilities when it comes to their shared activities. 9%

Conclusion: They are not two brutes who let their relationship fade as soon as their passion does, but the distant examining look of Aquarius can take out the emotion out of it. Aries partner needs to be relaxed by their significant other, so they can melt down and show their true, warm emotional nature. In this relationship, they would have a distant partner that basically supports their primal, instinctive nature. Although it is nice to think that the point of each relationship is for partners to accept each other as they are, in this case that would take away every chance for an Aries to grow through togetherness and learn about their emotional nature. This is something they will never be satisfied with. Still, every relationship with Aquarius can surprise us as much as any individual Aquarius could. With them as a partner, there is always room for an enlightening scenario that leaves all things to free will. In case they decide to share their lives together, they should have a screaming room they could individually visit once in a while. This is a couple that lacks tenderness. This would probably do the trick. And about that lack of emotion, they could just put in a lot of physical tenderness to begin with and let things go from there. 68%

Aquarius Compatibility with Taurus

Aquarius & Taurus Sexual Intimacy and Compatibility: In most cases, they are not even attracted to each other and think of each other as boring or crazy, depending on the situation. However, they could really help each other blossom if they opened up for the possibility of unusual sexual encounters. If the tenderness of Taurus is projected on their independent, distant Aquarius partner, their creative and motivating side would awake, giving energy and speed to the productive gentle side of Taurus. Imagine the sex life they could have, different from each other, two outcasts, if they only shared enough respect and emotion. The slow, tender and smooth nature of Taurus will be ridiculously annoyed by the changeable and unusual nature of Aquarius. They will rarely get this far, for they seem to be looking for different things in a relationship to begin with. Taurus would like to have a secure, unbreakable partnership and Aquarius wants to be free of any attachment to this world, let alone emotional relationships. It is not easy for them to mend these differences or keep them out of their sex life, because they wouldn't feel like themselves in a relationship with disregard of their primal needs. 15%

Aquarius & Taurus Trust: Aquarius doesn't really understand the attitude Taurus has and least of all their fear of not being good enough. Guilt and self-criticism is the most difficult trait of Taurus, and one Aquarius is free from, finding it obsolete. This strict Aquarius opinion will scare Taurus to the point where they feel it is impossible to tell how they feel. This will end in a circle of lies and mistrust that cannot be repaired. There seems to be no flexibility in an Aquarius partner, although they tend to show a nature that is so open for people's differences. If Taurus wasn't so stressed out by their Aquarius partner, they might decide to be true and honest. In order to build the subtle trust, Taurus needs to be brave and stop thinking about the consequences of everything they say, while Aquarius needs to let go of their righteous attitude and be careful about the way their Taurus partner feels in their presence. 20%

Aquarius & Taurus Communication and Intellectuality: The sign of Taurus brings Uranus to its fall and all of those bright ideas Aquarius has, seem to go through the sieve of reality given by Taurus. This wouldn't be a problem per se, but sometimes the narrow-minded Taurus doesn't exactly see the true possibilities of the material world and can bring down their Aquarius partner to the point where they don't see how any of their dreams is possible. If Taurus shows understanding for their partner's need to fly, they could actually help them materialize what they have dreamed about. This doesn't happen often, for Aquarius rarely finds Taurus as a person to talk to, slow and boring with a "small town" attitude that inhibits the progress of our civilization. Their differences are hard to reconcile and when they fall in love, every little thing could become a huge problem and a reason for both of them to think about ending the relationship. If Taurus wants a white picket fence, Aquarius wants a condo on 67th floor. As a contact of Earth and Air elements, they can be so far apart that they can't find anything to talk about. If Taurus yearns for compassion, Aquarius doesn't care about opinions of others. If Taurus wants to go by foot, Aquarius wants to buy a plane ticket. In general, they can find that they aren't exactly made for each other; unless they both have enough flexibility to understand the ultimate difference in others, and enough openness to do things they don't care about just to see if they like them in the end. 10%

Aquarius & Taurus Emotions: Taurus will rarely fall in love with an Aquarius due to the fact that they don't recognize their ruler in a good context. Aquarius is distant enough as it is, and without excitement some other signs might offer, they will not exactly feel the electricity of being in love with an unmovable Taurus. However, they both might get tricked by the middle ground between them. They are lucky that the sign of Aquarius lifts up spirits of Venus, or they wouldn't really stand a chance. If Taurus sees the stable, Saturn side of Aquarius, and Aquarius recognizes the inner child in their Taurus partner, they could discover that they do belong together, even though this goes against the odds. 15%

Aquarius & Taurus Capricorn Values: One of them wants to be tied down, and the other wants to fly. There is really not much they can do, but accept the differences of their goals and natures, for there is truth and good in both approaches to life. While Taurus values material things and grounded behaviour, Aquarius values freedom in any shape and form. They can find certain things to value together, if they put their minds into creating them through Aquarius field of ideas and Taurus' practical realizations. 1%

Aquarius & Taurus Shared Activities: After the recourses of Taurus have been spent, there is not much else they will want to do together. By "recourses" we don't necessarily speak of money, but the overall energy for action. Taurus will gladly visit a strange place they have never been to, but after this, they will want to come home and have a nice dinner. Aquarius doesn't have this need and wants to always be on the move. They could be taken care of through the efforts of their Taurus partner, if they had enough patience to keep them well fed, dressed in clean clothes and took care of household activities. For a short time there could be a number of beautiful things they could do together. This compromise is rare to find, because the emotional satisfaction Taurus partner will get in return is not enough. 5%

Conclusion: There is a strange similarity and connection between their rulers and although very challenging, this is a relationship where both partners could fall in love with each other, over and over again, every single day. They are ruled by Venus and Uranus, both planets rotating in a direction opposite to the direction of other planets. They are two outcasts, different and standing out together, they understand that East can be where west is, and vice versa. Taurus and Aquarius are people from two different worlds. They understand diversity, change of direction and the excitement of love. However, they will rarely get to the point to understand each other because of their excessive need for peace (Taurus) and excitement (Aquarius). What a strange pair these signs are. With such an obvious opportunity for electric love, they go around it and search for something else. 11%

Aquarius Compatibility with Gemini

Aquarius & Gemini Sexual Intimacy and Compatibility: They don't need to get naked to have a sexual experience, although they will want to be naked all the time to set themselves free from all the human restrictions represented through clothes. They will get lost on their way to somewhere and have sex there or somewhere else. But who cares when they are in search for kindred spirits and want to have a good time while at it. They will both be aroused by the intellectual side of their relationship and if they are to be satisfied, they have to consider each other intelligent. Neither Gemini nor Aquarius will ever be in a serious relationship with someone who is, in their opinion, stupid. Even something that they would call an "insignificant sexual encounter" has to be with someone with enough wit and something to say. They can have sex anywhere and none of them would care. Gemini is a bit childish and can be ashamed in certain situations, but when Aquarius takes over, Gemini will realize that there is no limit to their freedom of expression. Gemini and Aquarius could probably have sex by simple verbal stimulation. These partners will try everything, communicate excessively and learn quickly about each other's body and the way to satisfy one another. Still, their relationship could lack emotion and true physical intimacy. This could lead to them pulling apart, often not aware that they both need something else in their partner. 85%

Aquarius & Gemini Trust: Aquarius finds lying ridiculous and Gemini will usually feel free enough not to lie. On the other hand, Aquarius understands one's need for privacy, for this is a sign where Neptune is exalted. They will both probably have this ultimate trust for their partner and are rarely deceived because of their premise to give and receive freedom as an absolute priority. Trust is a strange thing for this couple. We should emphasize that they will trust each other. None of them will have any satisfaction in storytelling or lying when there are so many interesting things to talk about with their weirdo partner, and so little to share that will be judged. 90%

Aquarius & Gemini Communication and Intellectuality: They stimulate each other's mind to such a point that they fire arguments they weren't aware existed in their thoughts. While Gemini will probably be fascinated by the belief system of Aquarius, always so rational and humane, Aquarius will have an opportunity to relieve some of their ego problem with their Gemini partner. The mutable quality of Gemini will allow them to adapt to some of those rigid Aquarian attitudes and opinions, even if they disagree. Gemini does have this mellow nature that understands the flow of the social touch with other people, and will rarely fight for their beliefs with someone they feel really close to. This is a good thing for their everyday life, but in general, this can present a problem because the authentic personality of Gemini could be shushed until they are not sure who they are anymore, once again. When Gemini and Aquarius engage in an intellectual debate, they are fun for everyone to watch. It is important for them to have enough flexibility for one another, however different their premises might be. Still, it is best if they share the same basic life philosophy, which they usually do, or they could get distant and lose interest in each other. As two Air representatives they find that communication is the solution to any problem, but aren't aware how far from Earth they might get with their ideas unrealized and their goals unreached because of too much talk, and too little action. 99%

Aquarius & Gemini Emotions: The unstable nature of Gemini can make them change their mind or their emotional state on a daily basis, and if they don't feel good in a relationship, they will set themselves free without over thinking the reasons why they had to do so. Aquarius is always in a rush to set themselves free from anyone or anything, so a breakup wouldn't really be something strange in their world. In most cases, their rational, mental natures will complement each other in an exciting way, but there is not much emotion to be built in the core of their relationship. We could say that Gemini and Aquarius understand each other perfectly when it comes to their emotions. Usually this is true, but that doesn't mean this is what they both need. It seems that both of these partners need to find someone a bit warmer in order to feel things more deeply and light their passionate hearts. They will much more often become friends than lovers, even if they were attracted to each other when they first met. 40%

Aquarius & Gemini Capricorn Values: However, Aquarius can be very passionate about their humane beliefs and will often support them strongly. This is something Gemini can understand but rarely supports. Suffice it to say that they both value intellect. The rest is just something that other signs worry about. Because of the fact that Aquarius partner values equality of the people as much as their own freedom, this can be their point of separation, even though Gemini partner does not really disagree. 95%

Aquarius & Gemini Shared Activities: This is not always the case because some of the Aquarius' weirdness can be a show off, designed by their need to mend their turbulent Sun. It is exciting for them both to enter this relationship, Aquarius to surprise and Gemini to follow them wherever they go. Their main activity to be shared is movement. Aquarius is the only sign capable to really surprise Gemini. They are so different than everyone else and represent a step that Gemini should climb if they want their life to be unbelievable. They could drive thousands of miles just to find a specific ice cream or for no reason at all. Mostly they can do anything together, from travelling and clubbing, to reading labels and instructions on the use of different kitchenware. 99%

Conclusion: When you look at things this way, you could say that there is no better match for them than the fabulous Aquarius. Aquarius needs someone to understand their grandiose ideas and discuss each one with them, and also someone who doesn't make them feel inhibited. Gemini needs a partner who doesn't bore them or make them feel inhibited. However, they could find themselves in a relationship that doesn't have enough emotion and compassion, and this is certain to surface as soon as the first disturbing thing happens in the life of one of these partners. They need to work on their emotional base and their non-verbal understanding if they want their relationship to last. 85%

Aquarius Compatibility with Cancer

Aquarius & Cancer Sexual Intimacy and Compatibility: Aquarius, on the other hand, is known as an innovator, someone to make the change, but in fact, they are a fixed sign, pretty set in their ways and as a paradox like unchangeable. When they engage in sexual activity, Cancer could be so stressed that they will have to set those boundaries and Aquarius will not be able to make the needed change to be gentler to their Cancer partner. There is too much energy in Aquarius that needs to be grounded through their physical activity and this includes sex. Cancer doesn't really understand this and is convinced that in sexual relations with someone you love only emotions should be shared. If Aquarius finds a way to slow down and not force anything on their partner, and if Cancer allows their rational mind to take over for some of the time they spend together, they might share an exciting sexual experience. A sexual relationship between Cancer and Aquarius can be stressful for both partners. Although Cancer is considered the most sensitive sign of the zodiac, governed by the Moon, they can be quite rough and distant when they feel the need to set strong boundaries. Cancer will bring emotions and tenderness to their sex life and Aquarius won't ever let boring routine take over. If they compromise on experimenting and emotional exchange, they could even start having fun. 1%

Aquarius & Cancer Trust: With Aquarius, they might feel stressed to share things and this could present an either way issue when it comes to trust. The liberal nature of Aquarius could seem crazy to a Cancer, and their partner's honesty about their craziness won't help the inner feeling of mistrust for their possible actions. Cancer is usually loyal and honest, except in situations when they are scared of the aggressive reaction of their loved one, or of hurting them badly. It is a complicated thing for them, because none of them wants to lie, but still they don't seem to trust the future they might share. 35%

Aquarius & Cancer Communication and Intellectuality: They could make grand ideas come true, especially those that need a lot of people involved on their way to become real. However, they might have trouble talking to one another in the same tone or understanding each other in the first place. Cancer is ruled by the Moon, the fastest heavenly body in the sky, but they are not fast to recognize what hides behind Aquarius' words. There is difficulty for Aquarius to express their inner state and this is something Cancer has trouble understanding. The best beginning of their relationship is guaranteed if Aquarius sees their Cancer partner as a weird human being that needs to be examined. Cancer and Aquarius are able to join forces in intellectual activity. The mind of a Cancer is sensitive enough to pay attention to details and interpersonal relationships when Aquarius fails to do so. This will allow them both enough space to get to know each other well, and this could influence all other areas of their relationship. If this happens, Aquarius will approach those strange activities Cancer needs as if they weren't ordinary at all. After all, not everyone can drink a morning coffee in total silence with their partner and enjoy this silence as much as these partners can. 55%

Aquarius & Cancer Emotions: That homey, cosy feeling Cancer needs can be deeply disturbed by the rebellious Air sign of Aquarius. They will bring stress and too much information in their life, and speed that cannot be handled by a subtle state of deep empathy Cancer has to live with daily. The way they show love is very different, but it can be wonderfully focused on their kids and the family they build if they get to this point. There is no sign in the zodiac predestined for a family life such as Cancer. In a relationship with Aquarius, they would take over the most of everyday activities and responsibilities. In return, their children would get a childhood without boundaries and a life of free choices that no other couple can give. This is a consequence of the difference between them and the tolerance they have to build in order to stay together. When they do fall in love, they will not be so quick to end their relationship. The unconventional nature of Aquarius interferes with Cancer's need to stay in a peaceful environment, and this is something they will find hardest to reconcile. Aquarius will approach it as a kind of challenge and understand the stability and love they get from this partner. Cancer will realize that they have never been this free to actually be themselves instead of living in a symbiotic relationship they are easily sucked into. Once they form a strong bond, it will be very hard for both of them to let it go. 50%

Aquarius & Cancer Capricorn Values: There is a difference between their worlds that might seem impossible to overcome, but if they hold on to their love of distances and travel or if they learn together, they could easily get over the fact that their values on other things differ so much. Cancer values knowledge almost as much as Aquarius values information. This is a fine connection between their worlds and if it is nurtured it could be just enough for them not to be set apart by other values they hold on to. Cancer does value stability, intimacy and family, while Aquarius values their freedom, intellect and new technology. 10%

Aquarius & Cancer Shared Activities: The key activity that could truly connect them is travel. Although Cancer seems homey and unmovable, this is not exactly true. Due to the fact that it is a sign of Jupiter's exaltation, they will want to travel far. While Cancer will want to stay at home, go for a picnic in the park or to a furniture store, Aquarius will look for the highest skyscraper, wish for a new laptop and read anything that falls into their hands. Aquarius will always want to board a plane and it would be perfect if they could parachute to a location where Cancer would safely land in a Boeing 747. 25%

Conclusion: However, the link between them can actually be wonderful when found, and they could open up such interesting new perspectives for one another if this happens. They both want to learn new things and could travel far if a strong base is made at home, so Cancer can remain peaceful. We could say that Cancer and Aquarius are not your usual happy couple in most cases. Their relationship can be too stressful for Cancer partner and the lack of intimacy will most probably tear them apart. For this couple to move in a positive direction, Aquarius needs to understand how unusual their partner is, and try to experiment on being homey while having fun. Cancer will have to take over the main set of responsibilities to hold on to the idea of their home as a base from which they can move wherever they want. In the end, Cancer might discover an unbelievable joy of freedom and Aquarius might develop closeness. If these partners can be silent together, sipping on their morning coffee, this is in most cases the first step to success. 31%

Aquarius Compatibility with Leo

Aquarius & Leo Sexual Intimacy and Compatibility: Leo is the king of the entire zodiac, and Aquarius seems to be there to bring down the king and fight for independence. Imagine the attraction and the passion between two such strong individuals, lying on the axis of Sun's rule and detriment. Their sex life is a struggle, a fight and an incredible experience for both. Liberating and yet warm and passionate, sensual but still interesting. When they find true emotion, Aquarius might actually end up respecting the king. The beauty of their sex life is in things they can learn about their bodies, their confidence and the way they look at the act of sex. The attraction is always great in relationships of opposing signs, and it is probably the greatest in a relationship of Leo and Aquarius. Through the struggle of insecurities and forced liberation, these are two partners to form a strong connection by a simple act of gravitation that the Sun has over Uranus. 99%

Aquarius & Leo Trust: However, these signs represent the axis of Neptune's exaltation and fall, and they will almost always have the challenge of trust and the search of truth in their relationship. Although they could find incredible understanding and freedom for both partners, usually when they separate they realize how little they have actually known about each other and how little trust they shared in the first place. Everything seems clear in a relationship between Leo and Aquarius when we look at it from a distance. 75%

Aquarius & Leo Communication and Intellectuality: If they end up fighting for the same cause, they could turn down entire governments and use their incredible force to change anything in the world. To get there, these two would have to stop the battle they have with each other, because energy can be scattered on their unnecessary fight for dominance in a relationship. Leo is a sign ruled by the Sun and has the ability to give clarity to any situation. No matter how confused they might be or how lost they may sound, if you take a closer look to the time spent with them, you will see that they've brought clarity in your life. Both Leo and Aquarius are heroes in their own way. While Aquarius is reaching for heroism, looking for ways to set free from repression, Leo was born a hero and sometimes doesn't even know it. Aquarius, on the other hand, understands the necessity of change and they seem to carry around a spark to ignite and excite any possible situation that they find worthy. This can be irritating to many, especially Leo, but in fact it is a necessity of liberation we all carry within. 90%

Aquarius & Leo Emotions: The warmth pours out from the centre of their being and one has to be blind or senseless not to pick up the signals. Aquarius can hide their emotions much better and often has trouble expressing and acknowledging how they feel. Leo is exactly what Aquarius needs to find love. It is a strange thing how they find each other, on the grounds of their former relationships, to liberate and shine as if they have been searching for one another for many lifetimes. It is a good thing that Leo's warm emotional nature will melt even the coldest of hearts and there will be no safer place for Aquarius to share their love than in these fiery arms. While Leo is the Sun, Aquarius is a lightning and it usually comes out on a rainy day. This is exactly what they need to understand that there is a time for both of them to shine and they don't endanger each other's chance to do so. When Leo falls in love, the entire world can feel it. The only thing that can endanger their emotional relationship is their everlasting ego battle and they should both pay attention not to be too proud to let go to love. 99%

Aquarius & Leo Capricorn Values: Although they will not agree on many other things, this is the one that could connect them strongly, because they are both such strong individuals in the eyes of each other. The deepest value they share is the value of individuality. Someone with a strong character, who knows exactly what she or he wants, cannot stay unnoticed by Leo or Aquarius. 80%

Aquarius & Leo Shared Activities: The need Leo has to show off might be a bit disturbed by Aquarius' tendency to show what others don't want to see. From a different perspective, this should help Leo feel more free and confident, although it might not seem so in the beginning. They will both like to show off, each in their own way, and it is only important for them to set the territory for both partners to be expressed. While Leo shines, Aquarius likes everything shiny. Just like the Sun and the lightning don't go together, Aquarius should take over on a rainy day, in a depressive crowd or in places where they both feel as if they would drown. This is where Leo needs to give in and let their partner rule the sky if they are planning to keep the relationship going. There will be enough regular, shiny chances and days with no clouds for Leo to rule. 90%

Conclusion: Imagine what these partners could do together if they let each other lead the way when the territory of their rule is in front of them. They both need to learn to let go of the image they have about themselves and about each other, or they won't get very far stuck in their unnecessary ego battle. Warm and cold, hearted and smart, nuclear gravitation and vacuum in space, it cannot be easy to mend their differences or form a stable, loving relationship. Signs of Leo and Aquarius combined represent the ultimate creativity, famous scientific discoveries, the first man in an airplane and the first man on the Moon. The best thing they could do is find a cause they will support together. This would give them a focus on the outer world and allow them to deepen the inner emotional world of their relationship while fighting outside of it. 89%

Aquarius Compatibility with Virgo

Aquarius & Virgo Sexual Intimacy and Compatibility: Their natures find it very hard to support each other, they are both intellectual but in a completely different way, and they will probably ruin any chance of a good sexual relationship by over thinking everything, each of them in their own direction. Aquarius really holds on to their spontaneity, but who's to say what does "spontaneity" really mean? They are rational when it comes to sex, and yet spontaneous? That sounds strange, doesn't it? The truth is, they choose when to be spontaneous and their intellectual strength often gives them the image of spontaneity because they've seen the result faster than other, "not-so-spontaneous" people. This could ruin their sex life, because Virgo is one of those people and will usually think long and hard about starting a sexual relationship. This is in no way an easy sexual relationship and unless some strong support is provided by their natal charts, Virgo and Aquarius will rarely be attracted to each other enough to start a sexual relationship at all. Over thinking is a true turn-off for Aquarius, although they do it too just in a quicker pace, and they will rarely find Virgo's analysis sexy in any way. Shy, thoughtful, sensitive Virgo will have trouble understanding their nude, weird and often too fast Aquarius partner. It is almost certain that none of them will have enough patience to build their sex life with someone so different from what they need. 1%

Aquarius & Virgo Trust: Although Virgo can be tough on trust, with Aquarius it is obvious even to Virgo that a lack of trust would lead nowhere. Still, there is a great chance they will drift apart, even when their first contact is passionate and strong, because they could simply start feeling that both of them need someone different. Their rational natures usually connect them in a trusting relationship, because they both find it stupid to lie or not trust their partner. If they want to hold on to their mutual trust, they need to keep their relationship fresh and accept each other for exactly who they are. 50%

Aquarius & Virgo Communication and Intellectuality: As a fixed sign, Aquarius is usually set in their ways and this can be difficult to handle for someone so willing to sacrifice their happiness, such as Virgo. Their communication should mostly be good and their topics similar. They will share interest in many things and usually be excited about similar details. However, they belong to the most different elements of all – to Earth and Air. As an Earth sign, Virgo can be very slow, too thorough and rarely inspired enough for Aquarius to even feel the need to share their ideas with them. On the other hand, Aquarius can seem unrealistic or even crazy to their Virgo partner. The best way for Virgo and Aquarius to function and be satisfied with their relationship, is to take each other seriously enough. Virgo is ruled by Mercury and has a mutable quality to it, which gives it this changeable, moveable and adaptable nature. Aquarius has no trouble dealing with Mercury, in most situations, and will most certainly like this adaptable Virgo quality. With Virgo's attention to detail, any plan for Aquarius' brilliant mind to come to life is possible. Their communication might be tricky and the lack of compassion Aquarius suffers from will sometimes hurt Virgo, but they still have a great opportunity to combine their minds and form a universal intelligence capable of creating anything at all. 40%

Aquarius & Virgo Emotions: Virgo is a sign of health and our daily routine, and its representatives can often be obsessed by their every meal or every check up they've had with their doctor. Aquarius will in most cases avoid doctors at all costs and the fact that they exalt Neptune will mostly turn them to all sorts of alternative medicine, rather than anything typical that Virgo might hold on to. The entire emotional world of their relationship could come down to Virgo worrying for their irresponsible Aquarius partner, and the lack of gratitude they might get in return. In fact, Aquarius usually doesn't need to be taken care of in this way. The emotional rollercoaster Aquarius gladly offers is something Virgo will probably despise. In case they do fall in love, they will have to deal with a constant fight for freedom and routine. This is a complicated emotional relationship because the worrying of Virgo degrades the personality of Aquarius and the best of intentions could have damaging consequences. The biggest problem in the relationship of Virgo and Aquarius is in the fact they both heavily rely on their rational mind. This leaves no room for the joy of seduction, love and satisfaction, and usually they both need a partner with more warmth, life or emotion to them so they could both be happier. 35%

Aquarius & Virgo Capricorn Values: They could motivate each other to develop their intellectual strengths and hold on to this asset if other things in their relationship aren't that good. Other things they value aren't really similar and while Virgo would always choose practicality, Aquarius would choose the unknown and a not so understandable reality. Virgo and Aquarius will both value intelligence and a clear mind most of all. This doesn't mean they will find the same people, actions or thoughts intelligent and they could often have opposite opinions on someone in their surroundings. 30%

Aquarius & Virgo Shared Activities: Their taste in many things can be almost the same, because the same attention to detail Virgo cherishes so much, makes some people great in their art and this is what fascinates Aquarius. The work of any artist with a great mind could connect them and they could easily be interested in similar shows, galleries and plays. However, Virgo is too cautious and predictable and most of the time they will have trouble fitting in that Aquarius' too exciting, unpredictable world. 25%

Conclusion: Imagine how incredibly irresponsible, chaotic and unrealistic Aquarius looks to them. Their strongest meeting point is in their rationality and communication, and this can be used to overcome many problems that their differences result in. Unfortunately, in most cases they will not have enough chemistry to start a relationship, let alone stay in a sexually satisfying one for very long. Virgo can represent everything that Aquarius runs from like practical, worried about health and earthly things, down to Earth, cleaning obsessed maniac. If they take each other seriously, they might create incredible things together, as their great minds merge. 30%

Aquarius Compatibility with Libra

Aquarius & Libra Sexual Intimacy and Compatibility: They will either be afraid to be judged and seem too asexual, or have a need to show it "in spite" of everyone's opinion and this can be quite repulsive for their partner. When it comes to sexuality, Aquarius has an entirely different approach like they simply don't care about anyone's opinion. They live their life in a constant search for freedom from any taboos or restrictions, and this will help Libra forget about other people, at least for some time. Aquarius can truly help Libra express their sexuality. The problems with Libra's Sun bring too much concern for opinions of other people, so representatives of this sing go to extremes when it comes to the way they show their sexuality. Although their sex life can be quite liberating for Libra, it can also be a bit challenging for Aquarius because they will be the one who has to fight against Libra's need to fit in. However, as two Air signs, they will both tend to be free to express their sexual desires to each other. They will like to experiment, learn about each other and their own inner desires and communicate with ease. Their sexual relations should be a strong pillar of their entire relationship, although they will usually think of their verbal ways to get along as the most important for their bond. 90%

Aquarius & Libra Trust: Their insecurities coincide very well, and they will usually help one another move through them, but the trust between them needs to be built, it is not implied. Both of these signs like to be attractive to different people and they should find a way to communicate this need in the right way. Because of their righteous natures they could trust each other without exception, if they were only that sure of themselves. The problem can arise when Libra starts to get attached and becomes emotionally dependent on their partner. This is not something Aquarius will easily deal with and it could damage the trust of both partners, in each other and their entire relationship. 85%

Aquarius & Libra Communication and Intellectuality: They are both stiff in their convictions and won't easily change their mind once they are set on it. Aquarius won't have such a good time waiting for Libra to make any decision in their life, no more than Libra will enjoy the spontaneous, unpredictable nature of Aquarius. Since they both rely on communication, they will have a lot to talk about and will usually find a language to solve all of their issues. Still, it won't be easy for them to reconcile some of the differences in their approach to things. Libra and Aquarius both have certain images to maintain. Libra wants to look nice and act nice for others, while Aquarius wants to go in the opposite direction of everyone, sometimes even if there is no reason to do so. Libra is indecisive but once they make a decision, they will rather stick to it than salvage their relationship, even if it is a simple meal in question. Aquarius will have a tendency to walk away as soon as they feel pressured into anything, even if it is that same meal. So basically, they could have an enormous problem about lunch if we talk about extremes. Still, they usually understand each other well on usual, daily things and have similar convictions that help them handle big changes in life well. In time, as they get to know each other, their personalities will get along better and they could realize that they respect each other to a point that is unreachable with anybody else. 40%

Aquarius & Libra Emotions: This is something they will get in a relationship with Aquarius and it could help them both build a very strong emotional bond. They have strangely different goals in life, but if they harmonize them, their emotional bond should be very strong and develop much faster than we would anticipate. The biggest obstacle that could present in their way is marriage, at some point in their relationship. Libra is a sign ruled by Venus and this makes them emotional in a way, but we should remember that it is also a sign of Saturn's exaltation and detachment is something that makes them feel very good. Libra is a sign that represents marriage and finds it very meaningful as the institution Saturn would support. Aquarius might think of it as obsolete, even run from it and they will probably enter it only for practical reasons. It is important not to make pressure to any of the partners when this point in their relationship arises, or they might both feel repulsed and angry, leading to unnecessary conflicts and even the end of their relationship. 80%

Aquarius & Libra Capricorn Values: This could represent a big problem in their relationship, and Libra partner could seem clingy and not at all independent, while Aquarius might seem like an uncontrollable lunatic who would do anything to destroy all relationships in the world. As much as Libra will value togetherness, Aquarius will value solitude. They both value communication and intellectual strengths enough to talk about their needs and desires, and this should help them overcome their differences. 50%

Aquarius & Libra Shared Activities: If something can launch Aquarius into the orbit it's the lack of spontaneity and Libra can sometimes be the opposite of spontaneous. Aquarius will want to do anything, really, for as long as their life doesn't fall into a boring routine. Libra will have trouble deciding what they want to do and this could drive their partner crazy. These partners could end up in a relationship in which only Aquarius pulls the strings and Libra follows. This wouldn't really be a good solution for any of the partners, and Aquarius will have to learn to show some patients in order for respect between them to remain intact. 65%

Conclusion: It can be quite difficult for their troubled Suns to get along and they will often have difficulty adjusting to each other's character and finding deep respect for one another. There is a strong understanding between a Libra and an Aquarius partner due to their shared element of Air. The best cure for any problem in their relationship is usually in time, but with Aquarius' need for spontaneity they often won't last long enough for time to mend what gets broken. Whatever their story, they will have a lot of exciting things to live through together and if they fall in love, it would be a shame for a couple such as this one, not to give their relationship a try however it might end. 68%

Aquarius Compatibility with Scorpio

Aquarius & Scorpio Sexual Intimacy and Compatibility: These signs combined represent the ultimate sexual freedom, a place with no restrictions or taboos. They are a combination of Water and Air, of emotion and information, all combined in a strong scent of attraction. If they get tied to each other and break up, they could end up hating each other and despising everything they've shared in their sex life. It is very difficult for these partners to find a balance of passion, emotion and rational thinking. Contact between a Scorpio and an Aquarius can be truly intense. As squaring signs, they should have a very troublesome contact, but the sign of Scorpio exalts the ruler of Aquarius, Uranus. While Scorpio's sexuality is hungry, deeply emotional and pervasive, Aquarius wants to be free of any boundaries and emotion, and will have real trouble being with a possessive partner. Their sex life can be like a battle arena, or like a wonderland, depending on the flexibility of both of them and the depth of emotions they share. As two fixed signs, they will most certainly have trouble changing their natures and adjusting to a partner that is too different from them. 40%

Aquarius & Scorpio Trust: As soon as Scorpio starts to assume that Aquarius should be tamer and belong to them in a loving relationship, it will result in a forceful rebellion and the counterattack of their partner. Being the two honest and straightforward individuals such as Scorpio and Aquarius don't have any problem to trust each other. The problem here shows its face when they get too close. Things could really get out of control if any sort of manipulation takes place, and unspoken tendencies might tear them apart in a matter of minutes. 1%

Aquarius & Scorpio Communication and Intellectuality: It is futile from their perspective, and although Scorpio likes to be in control of everything their partner does, it will be refreshing to talk to someone who says unusual things. The biggest quality of their relationship is an incredible connection of depth and width in only one couple. They will both have trouble understanding our society as it is, and have certain similar perspectives on anything out of the ordinary. Scorpio exalts Aquarius' ruler and this is why their relationship is the possibility for both of them to grow. Not only will Scorpio adore the intellectual strength of their partner, but they will also help them understand the way their ideas might be realized through a feeling of ultimate possibility. For as long as they don't give in to their stubborn, unmovable modes, these partners could have great conversations about all strange topics they can imagine. None of them will want to have small talk or discuss their day at work. The weakest link in their relationship is their respect for each other, combined with their static natures. We would think that both of these signs are in connection to change and they couldn't possibly be static, but in fact, they are static in their way of change, and their biggest challenge is to stop for a minute and treasure what they've found in each other. 50%

Aquarius & Scorpio Emotions: If love happens between them, the most typical scenario is for Scorpio to fall into an obsessive mess of feelings towards their uninterested Aquarius partner. It takes a lot of work and commitment to reach the emotional core of Aquarius, and it is impossible to get there without spontaneity and trust. Scorpio can be spontaneous in situations that are free of emotions, but will rarely let their love for someone be a part of a maybe, yes, maybe, or no, swing controlled by their partner. Aquarius will rarely tolerate or be with someone who tries to make them be more stable and down to earth, or anyone who quenches their desire to be free. As soon as they feel obligated to do anything, they will start pulling away and any emotion that might have been developing will suddenly be covered by the fear of commitment and the rut of everyday life. If they want to reach emotional balance, Scorpio has to be untied, realize that their partner will never belong to them and that they are free to leave anytime. They will have to understand that this relationship might end tomorrow and there is nothing they can do about it but accept it. On the other hand, Aquarius will have to confront their emotional depth and be ready to make certain changes in their approach to romantic relationships, so they can steadily feel understand Scorpio's emotional nature. 1%

Aquarius & Scorpio Capricorn Values: Unfortunately, most of the other things they would value in their partner are completely different. They will both value excitement and change and this will be a strong meeting point for their characters. While Aquarius values free spirit, communication and independence, Scorpio values commitment, sex and deep emotional connection. 30%

Aquarius & Scorpio Shared Activities: They will both like to take risks of any kind and their best date could be anything from parachute jumping to a night out in a casino. For as long as they stay out of their ego battle, they could find many things to do together. The best way for them to spend some quality time together is in some sort of intellectual activities and competitions, because this would allow them to manifest their possible hostility in a healthy way. 60%

Conclusion: This would be a bit extreme though. The truth is, Scorpio is the sign of Uranus' exaltation and as such, it adores Aquarius in a way. In most cases, Scorpio partner will show their affection obsessively, but this might actually feel good for Aquarius. When we look at the sign of Aquarius, we will see that it exalts Neptune, the ruler of a Water sign of Pisces, and all of our assumptions on their lack of emotionality will drown in their ultimate love. Someone might say that this is a karmic relationship, that these partners were enemies in one of their previous lives and that they could fight until one of them falls dead. The fact is they are both in a way outcasts and rebels. While Scorpio represents all of our emotions we don't want to deal with, Aquarius represents the way of thinking most of us are not ready for. It is best to look at them as announcers of change, for this is exactly what they will bring into each other's lives. 30%

Aquarius Compatibility with Sagittarius

Aquarius & Sagittarius Sexual Intimacy and Compatibility: Their sexual relationship will be very fun, because they both like to experiment and learn new things. Their communication will usually give them both so much satisfaction that sometimes they both almost won't even need the act of sex in order to get satisfied. Although their sexual connection can be very satisfying for both partners, they could have trouble creating intimacy. Sagittarius partner will bring just enough warmth in their relationship, but the mutable quality of their sign will make them easily turn their focus to something else, while Aquarius partner still holds on to the same things. Aquarius acts in a way Sagittarius thinks and this is quite an asset in their sex life. Their attraction can be strong, especially when a Sagittarius partner is at a crossroads in their life and need confirmation of their freedom and sexuality. They will both understand the necessity of change and incorporate it in their sex life. Still, the emotional bond and consequentially the intimacy between them could get weak and strong, on and off, too often for both of them to see each other as perfect partners. 80%

Aquarius & Sagittarius Trust: With them both knowing these things about each other, they could easily start questioning if they should trust one another or not. Although they both find their relationships very dependent on the level of freedom they have, this is probably something they won't be able to give to each other when they decide to commit to their romantic relationship. They will sometimes know each other's minds too well for them to create trust from the sense of absolute freedom. Sagittarius can be a sign prone to infidelity and Aquarius likes to be free to be available. The best remedy for the lack of trust in any of the partners is for both of them to realize they their relationship is just something out of the ordinary, casual and free from any restraints. 60%

Aquarius & Sagittarius Communication and Intellectuality: When they finally point in the right direction and choose to speak of something that awakens Aquarian interest, their conversations will become incredible. Both of these partners are rational and give a lot of attention to their chain of thoughts. Both of them are fast enough in coming to different conclusions. The contact between them will spark their need for intellectual sparring and they could end up in some great debates. When Sagittarius and Aquarius find a mutual interest, it becomes the infinite source of new topics, information and could even change their life philosophies. Aquarius partner can remain distant for a long time and Sagittarius might feel like a little child, talking excessively about uninteresting topics and trying to make a connection. When they share a love for something, they will talk about it passionately, excessively and find new ideas and solutions to incorporate in their approach to this subject. The speed of Aquarius mixed with the passionate state of constant belief of Sagittarius, could make their relationship one of the most productive in the entire zodiac. 99%

Aquarius & Sagittarius Emotions: It is a good thing that Sagittarius is so changeable, or they would have trouble keeping up with their Aquarius partner. Another good thing in this contact is the rational nature of both signs and their focus on mental processes. This will allow them to communicate about their emotions, whatever they are, without any sense of guilt or emotional pressure. It is hard to set the scale for emotions in this relationship. None of these partners is that emotional on the surface, although Sagittarius can fall in and out of love quite often. When their emotions start to build, it will take a long time before they are stable and both partners certain of their feelings for each other. Sagittarius will change their mind many times, probably going from one extreme to another because that is what they're inspired to do by their Aquarius partner. On the other hand, Aquarius will need to form attachment first and then wait for the certainty of their partner's love. 85%

Aquarius & Sagittarius Capricorn Values: As a sign of Neptune's exaltation, Aquarius has a special approach to honesty, and for a Sagittarius honesty is one of the things they value most. There are so many things they would agree on, starting from the usual like value of freedom, and moving on to their own qualities and expectations. They will both value wideness of one's mind, the optimism and the faith behind the brains, intelligence and vision. 90%

Aquarius & Sagittarius Shared Activities: They could have trouble reconciling their approach to religion and any religious activities could be the source of problems in their relationship, because their whole individual belief systems could be at stake. Still, in most situations they could have a lot of fun wherever they go. There will be activities that Sagittarius will want to commit to, while Aquarius will find them silly or even stupid. Sagittarius will easily put on a smile and follow any idea Aquarius has, for as long as it makes sense or they have something to learn from it. There will probably be no place strange enough for these two to discover their shared interests once again. 85%

Conclusion: They will get together when it is time for both of them to go through a change in their lives or leave a partner they feel restricted with. Their relationship is often a shiny beacon to everyone around them because it gives priority to the future and brings hope of a better time. A relationship between a Sagittarius and an Aquarius partner might seem like a same sex friendship to other people and whatever they might think of this, this is the type of relationship both of these partners might need. The main challenge of Sagittarius and Aquarius lies in their rational natures. Although their minds will have a wonderful relationship, they could have trouble reaching real intimacy and closeness. They both need to slow down and ask themselves how they feel before they end up in a heartless bond they find solace in as they run away from the world. 83%

Aquarius Compatibility with Capricorn

Aquarius & Capricorn Sexual Intimacy and Compatibility: The main problem in their sex life will be their different pace, and this is mostly caused by the difference in their elements. As an Earth sign, Capricorn is slow and thorough. A representative of this sign will rarely jump into a sexual relationship without attraction and respect for another person, and will want to give their best performance when sex finally happens. It would be easy to make a simple assumption that Capricorn is traditional and restricting, while Aquarius is the opposite, but they are both traditionally ruled by the same planet and it would be silly not to understand their similarities. Aquarius is an Air sign and this does make them kind of flaky and unreliable, although they are ruled by Saturn, the master of reliability. They will want things spontaneous and fast, without much thinking and as relaxed as possible. It is very rare for an Aquarius to have patience to wait for Capricorn to make a detailed plan, and this will be a great turn off for Capricorn because they don't like anything done in haste and the heat of the moment, especially when it comes to sex. They can both be very passionate when with the right partner, but the starting point in their approaches is usually simply too different to work out. Fortunately, the respect they will have for each other could make them become very good friends and if they manage to find the right way to communicate, they might even build a quality sexual relationship on a foundation of friendship. 5%

Aquarius & Capricorn Trust: Their ideas of trust are different though, and it will be hard for both of them to accept each other's natures the way they were meant to be. They don't trust one another. One of them has stone cold convictions and the need to never be wrong or make a mistake of any kind, while the other has no fear of confrontation and values the image of truth among the human race too much to have the desire to lie. The lack of trust between them is not really the lack of trust they have in each other, but in the possibility this kind of relationship will work out. 80%

Aquarius & Capricorn Communication and Intellectuality: They have this distant, cold, silent agreement that they are both worth each other's respect and this can seem terrifying at times as they distance from each other in order to remain in this agreement. They would rather be in no relationship than look at each other in a different way, and because of this they could form a beautiful, lasting friendship. An intellectual relationship between a Capricorn and an Aquarius can be kind of painful to watch when you are, for example, a Taurus or a Cancer. It is important to remember how different they are. Although there is no real freedom for Aquarius if the rules of their preceding sign aren't followed, it is not easy for them to understand each other's way of living. For as long as they hold on to shared interests and their mutual love for the seriousness of their bond, they will be able to maintain the image of a strong intellectual bond that makes them both satisfied. 65%

Aquarius & Capricorn Emotions: This is exactly why they are often described as selfish, taking care only of their own needs. It is not easily accepted by the more spiritual signs that someone gives so much attention to earthly things such as money, or any material or career oriented value. Aquarius belongs to the element of Air and it is a sign of ultimate overall faith, different than any religion or rule that a human might have created. They need heavenly love, someone to share all of their ideas with, so they can float together on a cloud to never land. The emotional side of a relationship between a Capricorn and an Aquarius is something really strange. In general interpretations they are both supposed to be unemotional, detached and closed up for emotional interaction with other people. However, this is not exactly the root of their lack of emotional connection. As an Earth sign, Capricorn needs emotions to be shown in a physical, practical way. They will not care much about money, food or even sex, if their elevated spirit gets wings and they get yet another chance to dream. In order to form an emotional bond, both Capricorn and Aquarius will have to accept the other side of reality. 5%

Aquarius & Capricorn Values: It is sometimes easy to forget that the sign of Capricorn precedes Aquarius and that there would be no sense of liberation without enough pressure. Capricorn values boundaries and Aquarius values freedom. Capricorn and Aquarius can't ever possibly be in a loving relationship. Their roles are intertwined in a strange way and they could find themselves valuing the exact same things if they dig into their personalities a bit deeper. To start with, they both value consistency and loyalty, and they will both consider all of the standard humane evaluations of people necessary. They will choose their relationships on the same grounds when it comes to long term commitments and rarely allow others to control them. 40%

Aquarius & Capricorn Shared Activities: Capricorn exalts Mars and knows exactly where to drain it and Aquarius doesn't really know what to do with it anyway. Even though they won't often choose the same activities, this should help them follow each other for long enough to determine what they both might like. None of these signs lacks energy. It is only necessary for Aquarius partner no to force, insist and speed up, and Capricorn needs not to inhibit, restrict and deny. 25%

Conclusion: Their most challenging point in a relationship is their emotional contact. If they are to stay together, Capricorn partner will have to separate from the ground, just a little, and Aquarius will have to come a bit closer to Earth. Capricorn and Aquarius might not find each other that interesting to begin with. Both of these sings are traditionally ruled by Saturn, but their roles in the zodiac are entirely different. They need to meet in the middle for Capricorn will be able to help Aquarius materialize their ideas, and Aquarius to be able to help Capricorn make the needed change in their life and turn to something new. 37%

Aquarius Compatibility with Aquarius

Aquarius & Aquarius Sexual Intimacy and Compatibility: They will fulfil each other's fantasies without any repression, and easily find a language in which to connect their strange sexualities. Both of these partners will have trouble fitting in the usual stereotypes of sexuality and will rarely understand the usual taboos and restrictions other people tend to impose. Two Aquarius partners can have a very interesting sexual relationship, full of excitements and experimentation. They could find an obstacle in emotional bonding that could present itself through a general lack of intimacy in their sex life. This will not quench the desire they will feel when they are truly attracted to one another, but they could be too detached for any of the partners to give enough warmth or focus to their relationship. This is why as soon as the first excitement and attraction start to fade, and the need for true emotional contact emerges, they might have trouble staying together. However, if they are both looking for an occasional fling, this might be the best contact of signs in the entire zodiac to have one. There will be no better understanding for the sense of freedom and the need for the lack of intimacy as these two might have in certain conditions. Unfortunately, this is usually not enough to support their future together and they will both probably need someone they could love more and who would care about them in a different way. 50%

Aquarius & Aquarius Trust: They silently understand each other's thoughts on a mental level that no other sign can reach. There is always an interesting bond between two representatives of the sign of Aquarius. When they begin their relationship, their trust will be built on a foundation of freedom, so there will really be no reason for either of them to lie. If any of these partners becomes possessive, they will both be going against their beliefs if they decide to stay in this kind of relationship. 99%

Aquarius & Aquarius Communication and Intellectuality: Whatever their relationship might be like, when they talk, they will most certainly have an electrical connection and ideas will fly from side to side as if the conversation was their playground. Aquarius tells to another might be something that only the two of them understand. This is a conversation we would all like to be a part of. Aquarius is a sign of God's voice and the image of the thoughts and reasoning of our higher selves. The problem in their intellectual contact can show up because of their ego issues and this is almost inevitable with two such strong individuals. The Sun represents our self-esteem, respect and individuality, and it is in its detriment in the sign of Aquarius. This could be the reason of a very unpleasant conflict between their personalities, for both of these partners have the need to set strong boundaries and easily get lost in their extreme individualities. The best cure for this situation is for each of them to slow down, breathe in, and ask themselves – do they really give each other enough freedom to be who they are? 75%

Aquarius & Aquarius Emotions: When it comes to a romantic relationship, two Aquarius partners could feel a lot for each other, but are more likely to perceive each other as friends. If they do fall in love, as representatives of a fixed sign, they will stay together for a long time although none of them really cares for how long the relationship will last. Aquarius isn't that emotional in a typical sense, but this doesn't mean they are not emotional at all. Their love is in a way shared on a group of people and although it is not that romantic, it is very important to them as the foundation of their entire belief system. The beauty of their love is in its detachment, however strange that may sound. Since they know it will never be found with another partner, they will only get closer as soon as true emotions start to show. 40%

Aquarius & Aquarius Values: As two rebellious, opinionated people, they need similar upbringing if they are to understand each other. It would be funny to assume that two members of the same sign have different values. But when we look at the sign of Aquarius we see that what they value is closely connected to their home and the way they were raised. The good thing is that they both value their freedom to extremes and this will connect them even when there are many other things they disagree on. 80%

Aquarius & Aquarius Shared Activities: They could go anywhere and do anything, for as long as it is interesting enough, educating enough or exciting enough. Maybe they are a bit selfish in the eyes of other people, but they won't really think about it when they are together. The best thing about a relationship between two Aquarius partners is that they can share extreme activities that other people don't understand. So if one of them is into paragliding and the other into climbing Mount Everest, why wouldn't they do these things together, one by one until all of their wishes come true? 99%

Conclusion: When you study Astrology, one of the first things you will learn is that Aquarius is the sign of divorces, breakups and setting free from regimes and relationships. It can be strange to imagine an Aquarius in love with another Aquarius. As a sign that carries opposition within, they are often not easy to be with for any sign of the zodiac, but this is exactly something both of them could understand in each other. If they truly respect each other, there is a great chance they will learn to understand other things in each other's lives too. As crazy as it may sound, these two partners have a great chance of staying together because they will know each other better than anyone else could. However, they are rarely that attracted to each other, and even when they are, it is very difficult for them to form a deep emotional bond. When they do, well... the sky isn't the limit. 74%

Aquarius Compatibility with Pisces

Aquarius & Pisces Sexual Intimacy and Compatibility: There is a strong link between these two signs, and in their sexual relationship, things will most certainly never get boring. At first glance, they don't exactly go well together, one of them romantic, looking for their perfect love, while the other distant, looking for ways to set themselves free from all emotion. The sign of Aquarius exalts Neptune, the ruler of Pisces. Their sex life can be quite amazing if Pisces don't get too attached and find a way to keep their distance until their partner shows emotion. As a mutable sign, Pisces have an understanding for constant change and the exhilaration and the excitement of the act of sex. Aquarius will happily follow, with a little less enthusiasm because they are, after all, rational. The beauty of their sex life could be in creativity, a game of emotion and the everlasting questioning that will bring even more excitement and emotion to the entire relationship. Unfortunately, in many cases Pisces just want their emotions flow and they will end the relationship, rather than deal with constant disappointments. The best chance for a satisfying sex life between an Aquarius and a Pisces partner is in a scenario where Aquarius already had some emotions to share, before their relationship even started. They need a good starting point and the ability of Aquarius to show emotion from time to time in a way their partner will understand it. 50%

Aquarius & Pisces Trust: Aquarian nature can be a bit aggressive from time to time and their personal rebellious needs don't actually help Pisces feel secure enough to share intimate thoughts. The potential emotional dependency of Pisces can make Aquarius partner give in to their perfect lying skills in order to feel freer. Trust is the most important issue for this couple and it can go from one extreme to another. Depending on the state of their intimacy, they could end up covered in lies or completely free of them. The only way they can build unconditional trust is to use their ability to feel each other's true core. If Pisces understand the soft side of Aquarius, the one that lies far beneath the surface, they won't run from telling the truth. When intimacy is found, Aquarius will finally be able to stop running away from commitment and the problem with the lack of freedom will be automatically solved. 50%

Aquarius & Pisces Communication and Intellectuality: The sign of Pisces should be able to ground Aquarius partner's ideas because their element is closer to earth. However, they often get lost on their way to do anything real, especially when they don't feel it is in the path of their own mission. The understanding between them doesn't always go very deep and although they can have a lot of fun together, they will have different approaches to the same things that interest them. They can dream well together, but unfortunately they will hardly ever make any dreams come true. The lack of reality in their relationship could hurt them both and they might not even know where the problem hides, while they feel frustrated by their relationship. For example, if they start talking about religion, they will end up in a philosophy battle that has no real value. By the end of it, Pisces will feel bad because they even tried to rationalize their faith, and Aquarius will feel like they have been talking to a foggy image of something resembling an opinion. 35%

Aquarius & Pisces Emotions: The distance of their partner will most certainly exalt them and create a special emotional rollercoaster for Pisces that could lead straight to disappointment. They could make a safe emotional environment for each other only if the initiative comes from Aquarius partner alone. Pisces would have to be completely silent, uninvolved, feminine and reactive. Aquarius is distant and impersonal while Pisces can't wait to get tied to someone from their fairytale love. Their relationship could be exhausting for Pisces partner if they try too hard to find the response for the feelings they might have. As soon as idealization appears, the fine balance will be shaken and the sense of freedom for Aquarius will be disturbed. 1%

Aquarius & Pisces Capricorn Values: They will both value freedom of any kind, love for humankind, excitement, change, inspiration and their ideas and dreams. For Aquarius, this love for humankind would mean absolute justice, equality and freedom of speech. The link between these signs simply doesn't let them get too far from one another and their values will mostly be the same, but entirely different in their realization. For Pisces, it would mean the eternal sound of the ocean as a blessing bringing us here. If we apply this significant character difference to all other things they value together, we will see that they will need a lot of deep understanding to mend their differences. 50%

Aquarius & Pisces Shared Activities: Pisces will gladly visit an art show, but why not make it a modern one so Aquarius could be interested as much? Their problem is in the perception of romance and disagreements on things couples should do together. Pisces partner will want unusual, surprising and exciting things, but followed with romance, physical pleasure and deep emotional understanding. It is a good thing that the sign of Pisces is always open for change, because Aquarius will gladly provide some. Their interests can be quite similar and with certain compromises they could find a lot to do together. Aquarius partner, on the other hand, will want inspiring conversations, intellectual stimulation and preferably some extreme activities included. 40%

Conclusion: It is not easy to create the fairytale version of this contact, but once they find the emotional balance and the one, core truth to each other, they will have no problem keeping their fairytale alive, day after day. As all neighbouring signs, Aquarius and Pisces don't necessarily have the best understanding of each other's personalities. However, the sign of Aquarius exalts Neptune, the ruler of Pisces, and this gives them a strong bond through the planet of all magic. 38%

PART-12
PISCES HOROSCOPE

(February 19th to March 20th):

Who possesses a gentle, compassionate, sensitive and spiritual nature?
Who are friendly and respond to suffering, which others encounter?
Element: Water
Quality: Mutable
Ruler: Neptune
Stone: Yellow Sapphire, Topaz & Coral.
Lucky Gemstone: Cat's eye, which is to be worn as a ring in ring   finger
Colours: Turquoise, Mauve, Lilac, Purple, Violet, Sea green
Lucky Colours: Pale Green & Turquoise
Lucky Numbers: 1, 2, 3, 4, 6, 7
Day: Thursday, Monday, Friday
Lucky Day: Friday
Lucky Flowers: Water Lily, White Poppy & Jonquil
General Insights: Pisces are very friendly, so they often find themselves in a company of very different people. Pisces are selfless; they are always willing to help others, without hoping to get anything back. Pisces is a Water sign and as such this zodiac sign is characterized by empathy and expressed emotional capacity. Their ruling planet is Neptune, so Pisces are more intuitive than others and have an artistic talent. Neptune is connected to music, so Pisces reveal music preferences in the earliest stages of life. They are generous, compassionate and extremely faithful and caring. People born under the Pisces sign have an intuitive understanding of the life cycle and thus achieve the best emotional relationship with other beings. Pisces-born is known by their wisdom, but under the influence of Uranus, Pisces sometimes can take the role of a martyr, in order to catch the attention. Pisces are never judgmental and always

forgiving. They are also known to be most tolerant of all the zodiac signs.

Personal Quality: The Piscean is charity, anxious, self-sacrificing, gentle, patient, malleable nature and has strong intuitive powers. He/She has superb observation, concentration while listening and good grasping power. He/She has instinct for nature, beauty, travelling, luxury and pleasure. He/She is good in subordinate positions and heads of small business. He/She is the kindest and most charitable of all the signs. He/She will make many sacrifices for other people. He/She lives life in lonely. As a lover, he/she is faithful and love to dabble in the art of sexual fantasy. He/She has many generous qualities and is friendly, good-natured, kind and compassionate, sensitive to the feelings of those around them, and respond with the utmost sympathy and tact to any suffering. He/She gets ancestral properties but he/she wishes to make money by his/her own efforts. Horseback riding, dancing, skating, swimming or sailing are favoured activities.

Positive Quality: He/She is versatile, compassionate, artistic, intuitive, gentle, wise, musical, and intuitive and has quick understanding. He/She observes and listens well, and are receptive to new ideas and atmospheres. He/She readily adapts to change.

Negative Quality: His/Her dominant keyword is "I believe", fearful, overly trusting, sad, desire to escape reality, and can be a victim or a martyr. His/Her nature tends to be too otherworldly for the practical purposes. He/She also dislikes disciple and confinement. The nine-to-five life is not for him/her.

Relationship: He/She tends to bond romantically well with Aquarians. He/She is never egotistical in personal relationships and gives more. He/She can be loving and affectionate partners for life.

Health Concerns: He/She can be threatened by anaemia, boils, ulcers and other skin diseases, especially inflammation of the eyelids, gout, inflammation, heavy periods and foot disorders and lameness. He/She is prone to all kinds of allergies and crippling headaches.

Ideal Partner: Aquarians are the best match for Pisceans and two tend to bond romantically.
Compatible Zodiac Signs: Taurus and Cancer
Incompatible Zodiac Signs: Aries, Gemini, and Leo
Greatest Overall Compatibility: Cancer, Scorpio
Best for Marriage and Partnerships: Virgo, Aquarians
Pisces likes: Being alone, sleeping, music, romance, visual media, swimming, spiritual themes
Pisces dislikes: Know-it-all, being criticized, the past coming back to haunt, cruelty of any kind

BUSINESS ASTROLOGY FOR PISCES

Driven by fascination and enthusiasm there business venture is often a mix bag of imagination, creativity and profits. But in this process they often miss reality. For a successful business they should plan better and remain grounded. Business booster Gem is Opal.

Pisces Compatibility with Aries

Pisces & Aries Sexual Intimacy and Compatibility: It is hard for them to bond, as much as it is hard for all of us to transcend, go beyond our physical body and be one with the Universe. With that said, it is understandable how difficult it is for their sexual natures to accept one another. Aries stands for instinctive sex. The sign of Pisces stands for orgasm. Although Aries cares about their orgasm, they will not make an art out of it. Pisces would rather satisfy themselves than be with someone who doesn't understand the art of orgasms. When they end up together, it can be torture for both, because they just don't understand what each of them needs. Aries would even have some success in understanding the need for tenderness and physical touch, but what Pisces want is like an unreachable wonderland that no one needs. In fact, they simply don't understand what it is they need. Aries looks like an inexperienced child to their Pisces partner, and although this can open the door for Pisces to enter this relationship, it does not feel that good when they realize that this is not about to change. Aries and Pisces are two signs that really have trouble connecting. The beginning of all things lies at 0° of Aries and their end at 29°

of Pisces. Their connection is like a "little death" making room for all that is new, untamed and inexperienced. If they are both open enough to find their intimate language, their sex life has to be weird and kinky if they want to succeed. Pisces will feel suffocated in anything ordinary and less satisfying than what they know they deserve, while Aries is usually not very interested in sharing emotions all night long and waking up in the afternoon. 20%

Pisces & Aries Trust: It is hard to open up when you don't trust your partner, so Pisces will stay in their little world for as long as they can, only to avoid being hurt and lied to. With Aries holding their head high, their attractive, straightforward attitude and their libido, it is not easy for sensitive Pisces not to pick up those signals emitted all around. This will immediately give effect to the degree of their confidence. Aries will see their partner's world as a phony, unclear image that there is no need for, and find their Pisces partner shady and unworthy of their trust. 1%

Pisces & Aries Communication and Intellectuality: Although they are interested in entirely different things most of the time, they are still connected as neighbouring signs and have a way of leaning on each other. Through their relationship they need to learn about their own weaknesses and how to mend them to be complete. It is not exactly as they complete each other, but the effect they have on each other can be like the correct medicine. Aries has a tendency not to look behind, question the past, or be too sharp and fast for their relationships. They could also have an ego with a shotgun, waiting for any potential partner to pass by and kill their desire to even think about dating an Aries, let alone be serious about a relationship with them. Pisces are sensitive enough to explain to Aries how they should soften up but keep their boundaries strong. Aries and Pisces could find many things to talk about if they open up for each other's support and advice. Pisces represent a dream land of Aries and they are able to show them that they could actually have a mission and a higher purpose, instead of just chasing through life. In return, Aries partner will help their Pisces partner find their grounding. They will not be that gentle about it, that is guaranteed, but could be realistic just enough to show Pisces how important it is to have initiative and build something you dream about in the real world. If they start their intimate relationship on these foundations, they could easily discover their middle ground for other segments of their relationship. In case they are not so open to change and are not in search of someone to help them create, they will hardly share many topics they both find interesting. 70%

Pisces & Aries Emotions: While Mars, the ruler of Aries, is covered in rust, a red colour desert with volcanoes, canyons and weather, Neptune is a blue gas giant, cold, whipped by winds and much farther from the Sun. This is exactly how their emotions differ. Those that Aries cherish most are well defined, strong, protected, and colour in a colour of passion. Their emotional worlds are like two different planets that rule their signs like Mars and Neptune. Pisces on the other hand, have a windy and changeable emotional world, colour blue like the colour of sadness and vision, and are easily cooled down as soon as they feel disappointment. 5%

Pisces & Aries Values: The core of these values is different for the two of them. Aries representatives will value them because of that sense of strength, power and because of the role of that one and only hero, smarter and braver than everyone else. It is strange how they both value honesty and have such trust barriers when they get together. When they get involved, trust becomes something like a sole purpose of their entire relationship. They will also both like fairytale heroes and value the usual pride, chastity and bravery scenario. Pisces value them for their ideals, happy endings and those utopian relationships between those few worthy men and women. 35%

Pisces & Aries Shared Activities: Yes, Pisces partner knows that sports are healthy, but they need to attach them to something from their infinite world. So water sports are fine, because of all the secrets of the water, the view of the ocean, being underwater and contemplating on the purpose of life, or a dive in the pool. Walk in the forest can be beautiful because they can hear the birds, trees saying "hello" and wait for two owls to rest on their shoulders. They could share a walk in the forest, or engage in water sports.

Other activities that Aries would gladly take on are not "spiritual" enough for Pisces. Pisces always need to have "a second perspective" and this can seem crazy to their Aries partner. In the world of Aries, things are really so much simpler and if they want to enjoy something, they will simply go and enjoy. In the same manner, they would run when they run, practice when they practice and watch the ocean when they watch the ocean. 40%

Conclusion: Pisces is ruled by Neptune, in charge of our entire aura and our permeability for outside stimuli. Since they are both responsible for our border with the outside world, it is hard to say which partner should loosen up and make it possible for them to come close. This is a relationship disturbed mostly by the lack of trust and the ability of both parties to open up to their partner. Aries is ruled by Mars, the planet that rules our first chakra, responsible for our ability to set good boundaries. Their only chance of a happy ending is if Aries partner dives in and their Pisces partner wakes up. 20%

Pisces Compatibility with Taurus

Pisces & Taurus Sexual Intimacy and Compatibility: This is a place where Venus is exalted, magical, mysterious and unbelievably satisfying for Taurus' ruler. They have the ability to get lost in each other, make their dreams come true and satisfy each other by pure existence. When it comes to sex, Taurus can easily end up in a rut if their partner isn't inspiring or creative enough. They don't even care, for as long as their emotional needs are met and their physical body respected. Pisces on the other hand, get lost in sexual experiences, and can even find them toxic if their impressions on other people are unrealistic. Taurus and Pisces are both all about pleasure. Taurus represents the art of love making, tenderness and sensuality. The sign of Pisces is a culmination of a sexual encounter like orgasm. When they meet the right Taurus partner, they can be intrigued and relieved by their nature, for what they see is actually what they get. Because of the emotional nature of the sign of Pisces and their deep sense of purpose, Taurus will feel loved to the point of getting lost in the sexuality of their partner. They will both pay very little attention to their own pleasure because of all those feelings guiding them. This is almost always a giving relationship where both partners are equally satisfied when it comes to sex. 99%

Pisces & Taurus Trust: The beauty of their contact is in the fact that when together, they both lose their need to hide and let their emotions grow with ease. The sign of Pisces is a sign of mutable quality, and they can unexpectedly change, without a clear reason. If this happens, Taurus will know that the trust is lost, however their relationship seemed just a couple of minutes ago. It is a deep sense of broken intimacy, lost in the Pisces' need for emotional exhilaration. Basically, when Pisces partner gets bored, they start to think of excuses and lies, before they even realize that their relationship is over. Because of the Pisces' tendency to enter each relationship with an idealistic approach, there is a great chance they will open up to their Taurus partner as soon as they realize how stable and secure they seem. It is up to Taurus to understand the flakiness of their partner. When they do, they can either accept the situation and fight for love, or end the relationship and move on. 80%

Pisces & Taurus Communication and Intellectuality: Subtlety of Pisces is something truly inspiring for Taurus and they will feel the need to get to know every detail in their partner's behaviour. Both of these signs are not very talkative and Pisces even lead Mercury to its fall. This is why they really need to form a strong emotional bond and listen to each other through very little words. That field of talent and creative energy that Pisces carries along goes well with Taurus' need for beauty. Unfortunately, they can get lost in the world of Pisces and really lose that grip they have on reality. Taurus and Pisces probably won't have the need to talk much. Instead, they will understand each other through all types of nonverbal communication, curious about each other's next movement. At first, this will be like a drug, an addiction, something they have been waiting for their entire life. As time goes by, the feeling will not be this good, for they will lose touch with themselves and have a feeling they don't know who they are anymore. The most important thing for Taurus in a relationship with Pisces is to stand their ground and hold on to their common sense, practicality and their usual need to live in reality. 95%

Pisces & Taurus Emotions: Taurus will feel, for however long, like the centre of someone's world, loved and cherished to the point of unbearable beauty. If this feeling goes on, they could stay in a beautiful relationship for a very long time. As soon as Pisces partner feels this beautiful emotion dying down, they will make a spontaneous manoeuvre to distance themselves from their Taurus partner. The funny thing is that in most cases Taurus won't be hurt at all. That simple feel of inadequacy will be enough for both of them to let distance take its course. Taurus and Pisces have a magical emotional connection. For as long as Pisces don't change their mind and swim off, their relationship should be filled with love and wonder. With Pisces exalting Venus, the ruler of Taurus, this is not only love but adoration. Even though Taurus has a tendency to get emotionally bound to their partner, their potential separation from Pisces will be as coming back to reality more than a devastating event. 99%

Pisces & Taurus Values: Their values differ a lot, but the one they share is incomparable to others like love. Taurus is turned to a material reality and Pisces to an emotional one. No other sign of the zodiac can truly understand the way these two value love, especially when they are in love with each other. 85%

Pisces & Taurus Shared Activities: This can lead to a lack of understanding when it comes to the way they want to spend their time together. At first, they will enjoy the same things, but Pisces partner will get bored very soon if the scenery doesn't change. It is not like them to stay in one place for too long. When Taurus is found in a beautiful situation, they will want to stay in it forever, holding on to the first image even when the beauty of it fades. The main problem these signs will run into is in the fact that Taurus is a fixed sign and Pisces mutable. This will slow down all movement and could really annoy their Pisces partner. 70%

Conclusion: Taurus will give their Pisces partner a chance to connect to the real world, showing them how to ground their creativity, while Pisces will lift up Taurus and make them a bit softer and more flexible. They seem to be on a mission of convincing them that true love exists. This is a relationship based on love and full of it while it lasts. They both crave romance and beauty in their lives, and will do anything that is needed to keep the beauty going between them. When their relationship is over, they will both know it instantly and very often a conversation about a breakup would be redundant. If they savour their trust and nurture the beauty of love they share, their relationship can last and be as inspiring as a dream coming true. 88%

Pisces Compatibility with Gemini

Pisces & Gemini Sexual Intimacy and Compatibility: They can be attracted to each other due to the fact that they are ruled by Mercury and Jupiter, the same planets that rule their opposing signs. Still, there is a big chance they won't even recognize each other as sexual beings or keep a distance from each other if they do. Gemini has a lot of creative potential, but isn't exactly in search for their one and only true love in order to have sex. Pisces, on the other hand, exalt Venus, and they only want to have sex with the love of their life, unless they've been disappointed too many times. It is a good thing there is so much creativity to Gemini's approach to sex, or they would really have difficulty making any sort of intimate connection with Pisces. If they meet after these numerous disappointments, Gemini will not find Pisces very attractive, for they will no longer have any childish energy or charm. If their sex life is supposed to be functional, both of them will have to find a way to be a bit more grounded than they normally are. Gemini will have to realize the truth behind their own emotional nature and give in to true intimacy, while Pisces will have to accept the differences of their partner instead of searching for a soul mate with predefined qualities. 15%

Pisces & Gemini Trust: They have completely different ways of dealing with their emotional relationships and their issues with self-image, and they will think of many different ways to bend the truth when they are together. Unfortunately, when any of them lies, they won't have much success.

Gemini is too smart to be lied to by their Pisces partner, and Pisces sense their Gemini partner's state too well to not realize when they're not telling the truth. Trust is already a weak spot in almost every relationship these two can have and when they get together, there is a chance they will have no idea how to create any trust between them. Basically they both dip into each other's unconscious and see each other in the way that none of them looks at themselves. 1%

Pisces & Gemini Communication and Intellectuality: Gemini can decide to make a joke and Pisces will laugh without really thinking about it. Pisces will then say something to poke their Gemini and Gemini will laugh without thinking about it. It is as if they never really listen to each other and sink into a strange pool of superficial relationships and small talk. If they start discussing their deep thoughts and feelings, they might end up in a conflict that none of them anticipated. We could say that they idealize each other, but only to a point of recognition. Neither Gemini nor Pisces will think of each other as their one true love unless they really are each other's one true love. So they will have this image of each other that is deviated from reality, due to the fact they don't really listen to each other. There is always a fairytale or two to share between them and some fun to have when they go out. They will laugh together, but this is a strange connection with a lack of real communication. The only possible way for them to have a conversation with depth is in a situation in which they have absolute emotional intimacy, such as family members usually have.20%

Pisces & Gemini Emotions: When they do fall in love, they are rarely on the same frequency and often only one of them has true emotions for the other. Gemini is one of the most rational signs in the zodiac and Pisces is the cureless romantic and one of the most emotional signs. They represent ideal candidates for an unreturned love scenario and can be a nuisance to everyone around them if they end up in a relationship with no emotional balance. 1%

Pisces & Gemini Values: In general, they will both hold on to what they know best and Gemini will value intellectual strength and won't get very disturbed by dishonesty for as long as their image of a relationship isn't disturbed. Pisces will value their partner's reliability and trust is very high on their list of priorities. Still, there is one thing they will share, hidden in the fact that they both value someone's ability to create. They both value what they stand for and although Gemini values someone to listen to them and love them unconditionally, this is not the same as the passionate love Pisces partner wants to have. Even though this comes out of different views on creation, it can bond them in the act of creativity. Pisces would provide talent and inspiration and Gemini their resourcefulness and practicality. 5%

Pisces & Gemini Shared Activities: When we speak of Gemini and Pisces, we have to keep in mind that both signs are mutable. Although their interests might differ greatly, they could find activities to share due to their mutual need for movement of any kind. Pisces will normally dream about movement rather than actually move and this is exactly what their Gemini could teach them like how to make the first step. 15%

Conclusion: They are both usually positive enough to have a superficial enjoyable relationship and go well together at large social gatherings. They could both forget to call each other when they agreed to, and they can both change their opinions in two seconds, but they simply don't share the same goals. Gemini and Pisces are squaring signs that often don't have that much in common. As a strongly mental and a strongly emotional sign, their lack of understanding can be hurtful for Pisces and sometimes for both of them. If they do fall in love and start a romantic relationship, chances are they will not last very long. However, there is a beauty in the creative side of this relationship and if Gemini decides to truly listen to Pisces, they could help them use their talent in a constructive way. In most situations Pisces will just drain the energy out of their Gemini partner, especially if they end up in their fragile, needy mode that some other signs could understand much better than Gemini. If they are to succeed in their persistence to be together, they should work together and socialize a lot. The most important thing for both of them in this relationship is to reach for their emotional cores and give in to true intimacy, or they will never manage to communicate. 10%

Pisces Compatibility with Cancer

Pisces & Cancer Sexual Intimacy and Compatibility: Cancer will bring intimacy into their sex life and the meaning behind the act. They will nurture their partner and care about their pleasure, giving them a stable and a safe approach to a healthy sex life. Pisces will bring in change, creativity, inspiration and probably a lot of sensuality due to the fact that this is the sign that exalts Venus. The beauty of this connection is in the emotion they share and the way they cherish each other and respect each other's sensitivity. Their main problem might arise because Cancer can be somewhat traditional when it comes to sex and Pisces partner doesn't really understand this. Cancer and Pisces are almost always brought together by a romantic love. Their sexual connection is usually primarily emotional. Pisces partner might seem a bit weird and kinky to Cancer, but they should have a feel for each other, strong enough for both of them to enrich their sexual relationship with their own quality. Pisces' need to connect and feel love is larger than any sort of rule humankind might have made for love. However, in most cases they will be tender enough to inspire their Cancer partner to let go of their rigid attitudes and shame, and give in to the beauty of sexual exchange of emotions. 85%

Pisces & Cancer Trust: This could be recognized as pressure to some point and this could lead to Pisces partner getting scared. When Pisces get scared, they somehow fail to tell the truth even on silly things in their life, because they feel the need to distance themselves from any pressure they might feel. It is a good thing that Cancer is usually not aggressive or pushy, or they could easily get dishonesty from Pisces as a response to their tendency to create intimacy and a happy home at any cost. Pisces don't really understand marriage except as a part of a fairytale ending or because of all that lace, and Cancer will usually want a wedding as a crowning of a loving relationship. It is a good thing that Cancer understands this and easily separates lies from intimacy. Whatever the situation, they will both probably be patient enough to have just enough trust in one another for their relationship to work out. 70%

Pisces & Cancer Communication and Intellectuality: Usually they communicate just fine, but there are situations in which they could float away on an idea made out of words. Cancer is looking for someone with clarity on the use and the practicality of everything they mention. Pisces is everything but focused on practicality in most everyday situations. If Pisces partner learns to be more silent, relying on their feelings, and starts fighting for what they wish for, they could sweep their Cancer off their feet. As changeable as Pisces are, they always have something to talk about. This can be inspiring or irritating to Cancer who would maybe rather deal with "real information". Unfortunately, their entire relationship will not last very long if only words are spoken but deeds don't follow. Cancer's opposing sign is after all – Capricorn, and they need a partner able to constructively use things, situations and emotions. 85%

Pisces & Cancer Emotions: Everything that seems easygoing and positive might have a hidden negative note in the Pisces world, and Cancer feels rather than listens, which makes them a perfect companion for someone like Pisces. When they sense this deep understanding, Pisces partner will return the favour by absolute tenderness and finally open up to their Cancer partner. Cancer can understand the sensitive nature of their Pisces partner better than anyone else. When they find this shared point of intimacy where true emotions are shared, this will affect all other segments of their relationship and be a fuel for it to have a fairytale ending. 99%

Pisces & Cancer Values: It is often said that Pisces idealize partners and different things in life, but in fact they get depressed when there is no magic and perfect beauty surrounding them. This is where the difference in their character really comes to focus. As much as they will both value being loved and cared for, Cancer will value a stable emotional situation and a cosy home to come to, while Pisces will probably value any chance for an emotional rollercoaster more. If their day to day life with a Cancer partner becomes anything similar to a boring routine, they will find a way to run off, find a lover or create any sort of truly exciting circumstances. 25%

Pisces & Cancer Shared Activities: At the beginning, this may seem like a great arrangement, but in time, Pisces might want too much activity for what Cancer partner really needs. This wouldn't be much of a problem if they would say this to their partner without fear of any one of them getting hurt. When they meet and start their relationship, they will probably have a lot of things to do together. A relationship with a Pisces partner is always exciting and inspirational, and Cancer will give it strength, stability and roots. If they start bending the truth, Cancer will feel their trust beginning to fade and this could begin a series on problems between them, that could have been easily avoided. 70%

Conclusion: Their main challenge is hidden in the changeable nature of the sign of Pisces, not because it is there, but because they might fear to show it. Their biggest problem lies in the fact that they give priority to different types of love in their life. If passion and sensual, sexual love isn't there, Pisces will rarely be satisfied with the love they get from their family, and Cancer would find a life without a family nest very depressing. As two Water signs, Cancer and Pisces connect through emotions, usually as soon as they lay eyes on each other. This is one of the typical combinations of zodiac signs for love at first sight. A fine balance needs to be made between excitement and stability, and they could be one of the most wonderful couples of the zodiac – Cancer inspired and Pisces with a feel of home. 72%

Pisces Compatibility with Leo

Pisces & Leo Sexual Intimacy and Compatibility: In return, Leo will think of Pisces as weak and unrealistic, completely separated from their own desires and the strength of their body or emotions. The truth is, they can both be incredible lovers but they will rarely discover this together. Their roles and characters seem to be too different for them to find a way to coexist in a satisfying sexual relationship. The main problem of their relationship is in the fact that the sign of Leo is a sign of the fall of Pisces' ruler, Neptune. In a practical sense, this means that Leo will burst the bubble of Pisces and endanger their sensitivity, idealism and go against their beliefs. This will ruin the romance between them and make it impossible for them to find any magic while they are together. Leo's openness and directness will make Pisces feel ashamed and rushed, and their sex life could be delayed indefinitely until Pisces partner feels secure enough to get naked. Because of differences in their approach to sex, Leo will in most cases seem like an insensitive brute, unless Pisces start understanding their emotional depth even though it is so different from theirs. It is incredible how two signs that represent love, can be so wrong for each other. Leo will seem like a brute, caring selfishly about their own needs, incapable of forming an intimate relationship with anyone, let alone Pisces. Although this is not true, it might be the obvious reality to Pisces if they end up in a relationship with Leo Partner. The best way for these partners to find a language that can sustain their sex life is by building emotional trust first and worrying about sexual satisfaction later. 1%

Pisces & Leo Trust: Pisces is the sign ruled by Neptune, and Leo brings it to its fall. This is a difficult combination of signs to reach a point of mutual trust. They will both seem dishonest to one another, not because they lie, but because their characters seem unreal. Neptune is a planet of all deceit and mistrust in the world, making things around us seem foggy, unclear and fake. Leo will think of Pisces as if they were always on drugs, and Pisces will feel sorry for Leo and their lack of faith. 1%

Pisces & Leo Communication and Intellectuality: They could share many interests due to the creative power of both signs. Pisces will easily give inspiration to Leo, but the problem is in the way Leo might use it. The best way for them to create a safe surrounding for both partners, is to stick to the subject they are individually interested in. Ideals Pisces have could be shattered by the approach of Leo if they get too close to one another. Leo is a warm sign, very passionate about their doings and their desires. Pisces will rarely show the same initiative to realize any of their dreams and this is their greatest difference. Since Leo always shines a light on our virtues and shortcomings, they will not miss a chance to show their Pisces partner how unrealistic they are.

This could help Pisces build a more realistic approach, but it could also affect their confidence and hurt them through a difficult perception of the world. 35%

Pisces & Leo Emotions: They will rarely fight for anything, convinced that perfection doesn't need fighting for and that real treasures are spontaneous and free of conflict. The middle point for these partners is in their realization that not everything needs to be won, as much as not everything should be uninfluenced. Although true emotions are supposed to develop without difficulty, sometimes life is testing us to see if we really care. Both of these individuals are extremely emotional, each in their own way. The Fire element Leo belongs to, makes them passionate and gives them the need to fight for their loved one and their emotions. Pisces is a Water sign, and much more passive, showing their passion through the flow of emotion. This is not something that happens all the time, and sometimes things need to be let go because they don't belong to us, and we don't belong to them. 15%

Pisces & Leo Values: Pisces partner understands the necessity of lies, but still lives for clarity of the mind and the realization of their true inner Self. A great link between their worlds of values is in Leo's heroic nature that seems to have roots in a fairytale of Pisces. It is interesting how much both of these partners will value clarity and honesty. As much as they will both value their individual set of beliefs, they will be able to find middle ground in the grandiose character of Leo and the idealizing nature of Pisces. 20%

Pisces & Leo Shared Activities: Although Leo is a Fire sign, always ready to start something new, they will like to stick to their routine and show themselves in all the usual places every day. Pisces want to be invisible and they will change the scenery often in order for people not to recognize them. As a fixed and a mutable sign, they will have trouble synchronizing their need for changes and new activities. Although they could share some interests and have activities they would like to share, they will rarely stick to the same place and same actions together for very long. 10%

Conclusion: The problem isn't in their element or their quality, as much as it is in their connection through the fall of Neptune, the ruler of Pisces. If they get attracted to each other, they will be subjected to the risk of great damage to their beliefs, their inner faith and usually succumb to mutual disrespect because of a simple lack of understanding. Leo and Pisces seem to be put on this Earth to spread entirely different kinds of love. The beauty of their relationship could be developed through the fairytale approach of Pisces, if they build the heroic image of their Leo partner to the point in which other differences between them fade. 14%

Pisces Compatibility with Virgo

Pisces & Virgo Sexual Intimacy and Compatibility: These partners have a task to find the place of physical intimacy in which they will both be relaxed to be exactly who they are.

Virgo partner will usually be shy, trying to show their sexuality through rational behaviour, and Pisces will see right through this. On the other hand, Pisces will fear close physical connection with another person, and this will be practically dismissed by Virgo. As they both learn that they cannot hide who they are, they will have no choice but to set themselves free from any fear and shame, giving in to the wonderful sexual experience Venus has to offer. Virgo and Pisces are opposing signs and their attraction is very strong. Since they also represent the axis of Venus' fall and exaltation, we can conclude that their relationship always has a lesson on Venus to teach. This is a couple that will never have instinctive sex, however passionate they might get. Virgo's analytical mind wouldn't allow for them to act "like animals", and this is something that Pisces will find so humanlike and attractive. Virgo will mostly be attracted by the purity of sex with Pisces, who truly approach it as an act of love, free from prejudice and following their inner feeling, wherever it leads. 99%

Pisces & Virgo Trust: In order to have a healthy relationship in which they both trust each other, they will both need to be secure and confident enough to be honest. Both of these partners will easily give in to dishonesty, although their convictions are the opposite of their behaviour. It can be torture for both of them if any one of the partners tells a lie.

Virgo can have some serious trust issues that Pisces won't actually help them overcome. Fortunately, they are both aware that some secrecy might even spark their relationship and give it more passion. This is why they will usually get over small intimacy outbreaks in order to trust each other on a higher level. 65%

Pisces & Virgo Communication and Intellectuality: The mutable quality of their signs will allow them to jump from topic to topic, both of them staying interested in the flow and the outcome of their conversations. The best person to pull Virgo out of their obsessive analysis is Pisces, with their smile and their wider picture. Pisces will give their Virgo partner faith, teach them how belief can form reality and help them be free from too much caution and fear from failure. Virgo often has this inner battle in which nothing they know, think or do is good or valuable enough. Pisces are able to inspire and find value in everything in life, and those insecurities and emotional problems of Virgo may seem like something needless that damages the self-esteem of everyone around them. They will do their best to help their partner reach the point of inner security in which they understand their worth. In return, Virgo will help Pisces reach the actual materialization of their incredible talents. They complement each other best through communication and intellectual stimulation. When they start their relationship, they are bound to realize how similar they actually are, even though they seem so different. They might do so through nagging and constant criticism, but in the end, Pisces will have many things to be thankful for. Not only does this relationship represent the axis of Venus' exaltation and fall, but it also represents the axis of the exaltation and fall of Mercury. As much as Virgo has trouble with Venus, Pisces have trouble with practical Mercury and their mind can send mixed signals making them lost and confused. Virgo will help them build an inner sense of intellectual security, in return for their emotional one. 85%

Pisces & Virgo Emotions: It is important for them not to build too much expectation around this idea of perfection, for no one can meet these requirements in real life. If they go astray, losing touch with their partner's true personality, they will easily get disappointed and have the need to end their relationship. With two mutable signs everything moves fast and changes are inevitable. There is no other sign in the zodiac that can awaken the emotional depths of Virgo better than Pisces. Understanding between them could reach points of perfection and this is something that both of them will probably be unable to find with anyone else. If they want to remain in a stable relationship, they need to find a fine balance between rationality and emotion, reality and dreams and love each other as if they were perfect just the way they are. 95%

Pisces & Virgo Values: This is their meeting point and it can make them divine, or constantly dissatisfied with the need to change everything about their partner. Both of these partners will value flexibility, someone's ability to adapt and change, and they will most certainly value the love they get from their partner. They share a great love of perfection. As much as Virgo will value one's perfect mind, Pisces will value a perfect emotion. Their differences in approach to one's beliefs and convictions might be huge, and acceptance between them needs to be unconditional. 75%

Pisces & Virgo Shared Activities: They are after all ruled by Mercury, and have this need to see and feel everything that this Earth has to offer. Once they get into the magical world of Pisces, they could discover the beauty of life they were fully unaware of. Virgo can be grumpy about doing anything from a fairytale of Pisces, but they will follow out of curiosity. When these partners are together, they make each other feel like anything is possible, for Pisces understand endless possibilities as much as Virgo makes things come true, understanding reality better than many others. 99%

Conclusion: This makes them partners with greatest challenges and the greatest potential for love in the entire zodiac. They need to find a fine balance of rationality and emotions, each one individually and together through their relationship. In many cases this is not a couple that will last very long, as their mutable quality makes them changeable enough to disregard the entire relationship quickly if they aren't satisfied. Virgo and Pisces represent the axis of the exaltation and fall of both Venus and Mercury. They need to realize that perfection they seek might not be presented in the form they expect. If they stay together for long enough to understand the benefits of their contact, they might discover that the love between them is the only true love they could find in this lifetime. 86%

Pisces Compatibility with Libra

Pisces & Libra Sexual Intimacy and Compatibility: Since they touch each other through this sensual, loving planet, they might find true sexual satisfaction together. They can both be selfless lovers, caring more about the satisfaction of their partner than their own. Tenderness shouldn't be an issue here for none of these partners will like too much aggression and roughness anyway. They could discover many different sexual preferences that they weren't aware of before, through a connection of very different natures. If they weren't connected by Venus, it would be very difficult for them to form a relationship on any kind, for their approach to life and sexuality is completely different. Libra partner wants someone strong, passionate and confident, while Pisces partner wants someone gentle, compassionate and aware of their feelings. Air and Water, Libra and Pisces seem to have almost nothing in common. However, we shouldn't forget their connection through Venus, the ruler of Libra, exalted in Pisces. Libra will want their sexual experiences fast and exciting, and Pisces will want them slow and sensual. The main issue of speed is usually overcome by the quickly changeable nature of Pisces, except in cases when they are too shy to jump into a sexual relation with someone as openly sexual as Libra. 50%

Pisces & Libra Trust: That sparkly, always in love, childish, flirty nature of Pisces will be a huge turn off for Libra, who will not be able to trust someone who openly shows their interest in other people. They will not understand each other well enough to share much trust. The need Libra has to be accepted and liked by other people will be wrongly interpreted by Pisces, for they don't understand how can someone's self-confidence be that low. The only way for these partners to remain in a trusting relationship is to approach it casually and build their understanding and trust from zero, as if they have never had any relationships prior to this one. 1%

Pisces & Libra Communication and Intellectuality: Unfortunately, they will often want to help them grow and change who they are so they could be "happier". The problem here is in the fact that Libra doesn't know or try to understand what would make Pisces truly happy. This is some basic disrespect and it could ruin the foundation of their entire relationship. The main challenge here is to remain in a respectful bond, however crazy Pisces might seem to their Libra partner, or however stiff and boring Libra might seem to Pisces. In many cases, Libra will cherish the optimism and the childish naïve nature of their Pisces partner. Their communication can be inspiring for as long as they don't try to change each other, explain to each other how they should think or feel in certain situations, or even worse, teach each other how to behave. Pisces partner can be quite direct and spontaneous and this might endanger the image Libra is trying to maintain in the eyes of others. If they find each other's actions inspiring in any possible way, they might reach a point in which they will truly communicate without judgment. 5%

Pisces & Libra Emotions: It will be a rare occurrence and their rationalities will weigh them down, for Libra does exalt Saturn, and Pisces are ruled by Jupiter. Their minds will probably be filled with all sorts of irrelevant information until they decide that it is simply too hard for them to be together. If they pass the point of disrespect and set strong boundaries, nurturing each other's individuality and self-sufficiency, they might reach the point of emotional interaction. As if they didn't have enough challenges, this will usually happen at a different time. Their emotional connection is mostly presented through the adoration of Venus in the sign of Pisces. This is an everlasting love waiting to happen, and the kind that could be born between these partners in case they both overcome their ego. Libra will rush into an emotional show off, realizing that they have found love, but Pisces will not feel any love until all the dust settles and they have the time to close their eyes and feel. If their timing is off, they will probably end their relationship on good terms, not expecting much from each other in the future. 15%

Pisces & Libra Values: This could give them a strong foundation for a sexual relationship if they are attracted to each other in the first place. The rest of their characters will differ greatly and while Libra will value consistency and stability, Pisces will value spontaneity and one's ability to follow their heart. They will both value love and this is something that will connect them over everything else. With Venus at focus, they will both need to be loved and cherish those who know how to show it, hot to enjoy life, food and sex, and how to make their loved ones satisfied. If they both believe that they have a mission here on Earth, and it happens that their missions cross paths, they could inspire each other to fight for what they value most – their names in the stars. 60%

Pisces & Libra Shared Activities: The problem here is that by the time Libra decides if they want to do something or not, Pisces will change their minds five times, not sure if this is the best thing for them, for their relationship, for their life's purpose and mission, etc. The questioning of Libra will raise suspicions in Pisces with painful ease. They will manage to find activities they will both like, and if nothing else works they can always turn to any sort of art. Indecisive Libra is exactly what their Pisces partner doesn't need in order not to feel completely lost in life and all of their chosen activities. The real question is that they will want to do anything while they are together. They could drain so much strength and confidence out of each other just trying to make a plan for the night. 40%

Conclusion: They could have real trouble adjusting to their partner's speed, and the mutable quality of Pisces often won't help them open up any faster to build a relationship in the pace that would fit their Libra partner. Both Libra and Pisces can selflessly be interested in the satisfaction of their partner, and this should help them stay on the good side of their relationship whatever happens between them. Libra and Pisces have a meeting point in the beauty of Venus. Still, they perceive it in two different ways and they will often not respect each other enough to find the beauty of Venus in one another. If they move past the disrespect and the unrealistic expectations from each other's personalities, they might find that they share real love. 29%

Pisces Compatibility with Scorpio

Pisces & Scorpio Sexual Intimacy and Compatibility: Scorpio is a sign that represents sex, as well as sexual repression, and depending on the upbringing and previous sexual experiences, they can be a bit rough on their sensitive Pisces partner. On the other hand, Pisces is a sign of orgasms, strange sexual experiences and all of the sexual weirdness. If they understand the emotional depth of Scorpio, they might be much more resilient than we would assume. The biggest challenge for these partners is their relation to Venus, the planet of sensual physical satisfaction. Scorpio doesn't care for Venus very much, leading it to its detriment, while Pisces adore it through exaltation. As two Water signs, both Scorpio and Pisces find it very important for emotions to be the most intense part of their sexual experiences. This can be very unfortunate if Scorpio dismisses this emotional need of Pisces to be satisfied and loved at the same time. If Scorpio partner is aware of their animal nature and instinctive sexual desires and in any touch with their feminine side ready to show it, Pisces will easily find a way to blend in their sexual world. 70%

Pisces & Scorpio Trust: Suspicious Scorpio can easily become a clingy, control freak in a relationship with Pisces. However, they will both be in search of their one, perfect love and this should bind them with certain honesty. As soon as one of them is cheated on or disappointed, their relationship should end, because none of these partners can handle the tainted image of love. Trust between them will be maintained for as long as Pisces have an idealistic approach, doing everything for their one true love. When their image clears and they realize who they are with and what their relationship looks like, it might become very difficult for them to stay in loop with Scorpio's expectation of honesty. 65%

Pisces & Scorpio Communication and Intellectuality: It will be almost impossible to have a healthy conversation in which there will be no hurt, distance or anger. They will rarely fight, for Pisces partner usually has no reason to fight with anyone, but they could have a lot of misunderstandings that lead to their separation pretty quick. If Scorpio partner is tender enough and Pisces partner possesses the needed boundaries, their communication can be pretty exciting and magical. Both of these signs are linked to different types of magic, and they will both be interested in the "behind the scenes" view on everything that surrounds them. As they start to communicate and get out of their silent zones, they could easily get carried away in topics most signs wouldn't understand. The possible problems in communication between Scorpio and Pisces are either the roughness of Scorpio or excessive sensitivity of Pisces. The emotional approach to everything in their lives will help them understand each other when it comes to rational choices, too. The most superficial experiences will become something incredible to talk about, and the truth behind everything in life will be mesmerizing. They should hold on to the fascination with each other, instead of giving in to their weaknesses. 90%

Pisces & Scorpio Emotions: Each river flows into the ocean or the sea, and this reflects the emotional connection between these signs in the best possible way. Pisces partner will have the ability to disperse the intensity of emotion from their Scorpio partner. This will allow them both to breathe more easily, for as long as they don't cross the line and endanger the part of this depth that is loved by Scorpio. If any sign other than Scorpio is capable of understanding Scorpio, it is Pisces. There is an emotional depth to Scorpio that not everyone is ready to face and Pisces are ready to face anything in the field of emotions. The sign of Pisces represents our oceans and seas, while Scorpio represents rivers. This is a special connection in which Scorpio partner needs to focus their emotions and Pisces partner needs to give them a purpose. However difficult and dark they both might get, they will share a deep emotional understanding that should be followed to see where their relationship will lead. 99%

Pisces & Scorpio Values: Scorpio is a sign ruled by Mars and there is always a certain admiration for chivalry. Pisces represent all the fairytales in which a prince became a hero and married a beautiful girl. Their mutual love for a connection with emotions and the depth of their emotional connection, will give them just enough shared values to hold on to. As much as Scorpio values someone's strength of character and depth, Pisces will value sensitivity and width. Pisces partner has a mission to teach Scorpio how to reach their fairytale through chivalry, and they should both stay focused on creating their shared dreamland, royalty or not. 75%

Pisces & Scorpio Shared Activities: If Scorpio gets tied to their Pisces partner, this might become tiresome for both of them, for too much scattered activity of Pisces can be irritating for focused Scorpio and the obsessive nature of Scorpio might weigh Pisces down. Still, they will have enough energy to follow each other and it should be easy for them to find shared interests. When it comes to activities they could share, they will probably be inseparable whatever they do. The main problem with the time they spend together could be the unconscious negativity of Scorpio partner. It might endanger the positive, happy image of the world Pisces want to carry around and this could push Pisces partner away if their emotional connection is not strong enough to keep them together. 85%

Conclusion: Scorpio and Pisces relationship will probably give them both new insights on emotional possibilities. They will both easily get carried away into an image of a fairytale love, and this image could keep them together for a very long time, even if they are both not that happy. As two Water signs, they will rely on their emotional judgments and understand this about each other, creating true intimacy. The challenge here is for the nature of Scorpio not to obsess and suffocate their changeable partner, and for Pisces to stop running away from negative emotions. 81%

Pisces Compatibility with Sagittarius

Pisces & Sagittarius Sexual Intimacy and Compatibility: Their sex life will have ups and downs, excitements and disappointments, too many expectations and a lot of surprises. The best thing about their relationship is the positivity both partners share, and a lot of laughter and fun they will share in their sex life. Unfortunately, the level of intimacy will rarely be satisfying for any of these partners. Since they are both ruled by Jupiter, they will be faced with their rational natures and their convictions. If they ever manage to end up in a physical relationship, they will have a lot of fun. As two mutable signs, there will be no end to their creativity and changes in positions, scenery and levels of commitment and intimacy. The main reason why their sexual relationship rarely comes true is over thinking of both partners. Sagittarius will wait for a grand emotion, grand gesture or any sort of passionate initiative from Pisces, while Pisces will wait for all of the pieces of the puzzle to fit in their perfect position. In most cases, neither of these things will happen and they won't move further from a platonic relationship. 30%

Pisces & Sagittarius Trust: Sagittarius partner is too passionate and loves to have a lot of options when it comes to relationships. They will rarely settle down with anyone who lacks a strong decision to win them over. Pisces, on the other hand, will be too sensitive while trying to show their imaginary strength. They could end up following their Sagittarius partner in adventures they are not ready for, sad because of the lack of emotional understanding from their partner and ready to open up to someone else. It will be very difficult for Sagittarius and Pisces to trust each other, but they will probably accept it as a perfectly normal thing. This is the beauty of Jupiter's rule like everything makes sense in their relationship. Regardless of the purpose of each little thing they do, they will often have twisted expectations from each other and this will lead to unintentional dishonesty. 1%

Pisces & Sagittarius Communication and Intellectuality: There is no way to determine how long their relationship will last, and unless supported by fixed signs in their personal charts, they will rarely stay in it for long. Sagittarius is ruled by Jupiter, and traditionally so is Pisces. This is the biggest planet in the Solar system and as such, it has a great influence on the personality of these signs. They will share the same optimism, the same vision and pretty much the same delusions. These partners will be linked through extremely beneficent influences and they will most certainly share the same sense of humour, operate at the same speed and learn a lot from each other for however long they are together. Jupiter is a planet of knowledge, and they will be fascinated by the unknown they can share with each other. In time, they will realize what their differences are in the most unusual way. When we rule out emotional and physical sides of their relationship, a Sagittarius and a Pisces partner will be best friends, almost inseparable, for a while. Sagittarius is a sign of convictions and will be more rational and reliable than their Pisces partner. At some point Sagittarius will start to form a distance because of expectations that haven't been met and the irresponsible, detached behaviour of their Pisces partner. In return, Pisces will have a simple feeling that this is no longer where they want to be. Both of them might never understand why, but they will simply separate with no ill intentions, and probably not much anger or hurt. The beginning of their separation lies within disrespect of each other's convictions and personalities. 85%

Pisces & Sagittarius Emotions: The relationship they will build through communication and understanding of each other's worlds will awaken emotions through excitement and the unexpected. They will laugh with each other with open hearts and share wonderful emotions for as long as they are in the beginning of their relationship. As soon as any problems start to arise, they will both feel their emotions fade, as if the entire relationship was superficial. Whatever the circumstances, it is important for both of them to remember that there is nothing superficial about this contact. Their relationship will be an emotional rollercoaster for both partners, but it will rarely last very long. The learning process and the beauty of their entire relationship shouldn't be forgotten, but kept as a base for all of their future relationships. They love each other in a strange way, idealizing each other, getting disappointed, choosing to stay apart even when they wish to be together. This is a complicated emotional contact because both partners easily fall in love, and the deepening of their relationship can make them both be swept off their feet. If the relationship ends in a disrespectful way, they could both lose a bit of their faith in love. 30%

Pisces & Sagittarius Values: If they connect through deep love, they will overcome this with ease and emotions they share will make Sagittarius understand their partner. In general, they will agree on many things. The biggest difference here is the value Pisces partner gives to emotions, for Sagittarius often doesn't really understand that approach. They will value each other's utopias, people with good hearts, knowledge, wide perspective and travels. They will understand each other's sense of not belonging and share the sense of a higher power. 65%

Pisces & Sagittarius Shared Activities: It is like they are both hungry for happiness in a dark world they have stumbled upon, and when they meet someone this fascinating it would be a shame to miss the opportunity to enjoy. For as long as they are in a fascination phase of their relationship, they will be inseparable. Both of them have the need to grab everything that is offered and leave nothing joyful unused, unsaid and left for tomorrow. In time they will realize that although they share the same need for movement and changes of scenery, they might not need the same contents in their lives and Sagittarius will turn to physical activity, philosophy and travel, while Pisces will usually go back to creative work and the pursuit of love. 90%

Conclusion: Once they get close to each other, their process of learning will begin and both partners will be fascinated by each other, thinking that their relationship could never end. This is a relationship of two kindred spirits that often doesn't last very long. At first, it will be challenging for them to leave the platonic zone and start building a physical relationship.

They will easily idealize each other, think of their relationship as the perfect love, but this infatuation won't last very long because of their changeable natures. The fact is their relationship represents a moment in time when they have both deserved to smile. For as long as it lasts and they are happy, it will be cherished by both of them. 50%

Pisces Compatibility with Capricorn

Pisces & Capricorn Sexual Intimacy and Compatibility: Their sexual relationship is a contact of two powerful individuals, one of them extremely strict and rational, and another flexible and emotional but confident about their beliefs. Differences between them will create a strong attraction, almost as if they were opposing signs. The sex life these partners can share is unexplainable when their characters are superficially observed. But in a different way than a Cancer, Pisces can reach emotional depth of Capricorn by a simple feel. This is not a matter of compassion, but a matter of their deep inner truths. Capricorn isn't unemotional, however obvious their coldness might be, and Pisces can be quite rational, even though they seem lost in emotion. They can awaken the best qualities in each other and share strong intimacy through deep emotionally-rational understanding. There is probably no better way for a Capricorn to relax, than to enter a relationship with a Pisces partner. Their sex life will move in a strangely spontaneous way. Capricorn will be inspired to let go and open up to their partner, while Pisces will easily ground their affections and find a way to show them through physical contact. Capricorn will feel more casual and Pisces will start to gather themselves up. If they stay together for long enough, they could make a perfect blend of stability, trust and emotional excitement. 99%

Pisces & Capricorn Trust: This is not always the case, and the possible roughness of Capricorn can sometime induce Pisces to tell a lie or two. On the other hand, if Capricorn is closed up and unreachable, trust will be disturbed by a simple fact that they don't know each other that well at all. They will often understand each other well enough to respect their relationship and keep it clear of dishonesty. The beauty of their relationship is in their approach to trust when they realize who they are dealing with. Since both of the partners don't trust the world to open up easily, they will have to earn it, day by day, from each other. This might seem like a game or a competition to win over the trust of their partner. 85%

Pisces & Capricorn Communication and Intellectuality: If anyone can inspire a Capricorn, it is their Pisces partner. They can both care a lot about communication in their relationship, but their bond will allow them to stop talking and start listening. They are both shy in a way, and in order to get to know each other, they will have to be very careful to pay attention to one another. It is a good thing they will both want to do so, intrigued by their partner's nature, excited about getting to know each other in depth. The problem in their communication can arise when Capricorn gives in to their rigid opinions and beliefs. The sign of Capricorn brings Jupiter to its fall, and Jupiter is traditionally the ruler of Pisces. This could truly endanger their entire relationship, for the strict and rational nature of Capricorn can damage the faith and the convictions of Pisces through simple disbelief. Pisces partner will rarely give up on their belief system, since this is what they live for, but they could question it and feel lonely because of the lack of understanding from their partner. "What will be, will be" might become "I decide what will be" and this can suffocate the spontaneity and the inspiration that Pisces partner carries within. 65%

Pisces & Capricorn Emotions: This doesn't mean any of them will actually change their ways, even though they will be the best they can be. These partners share a deep emotional bond that can be built for years until it truly blossoms. The most wonderful side of their relationship is in the expectance of emotion, the constant growth and their ability to bring out the best in each other. Capricorn might remain grumpy, while Pisces might remain flaky and unreliable. When they are too set in their ways, they could end up truly annoying each other as Capricorn quenches the beliefs and the magic of Pisces and Pisces disappoints reliable, earthly Capricorn. 90%

Pisces & Capricorn Values: Capricorn will surprisingly value one's ability to be in touch with their emotions and have a clear vision of a positive outcome. Unfortunately, they will have a real problem in approaching the use of beliefs and emotions in everyday life, and while Capricorn will value coolheaded thinking and one's ability to be rational, Pisces will value the opposite. They will sometimes be too different, Pisces dreaming of a perfect love, Capricorn knowing it is impossible, Pisces imagining a God with a golden beard, Capricorn believing in this moment, or visiting a church because of tradition rather than belief. There is a certain consistency in their way of approaching their values. It is strange how someone like Pisces can value stable emotions that much, but they help them to finally rest their mind and their heart when they are in a stable emotional relationship. It won't be easy but they might value each other enough to overcome their differences. 45%

Pisces & Capricorn Shared Activities: Even though their interests are usually very different, Capricorn will be inspired to get inside the world of Pisces, as much as Pisces will want to solve the equation inside their Capricorn partner. When they begin their relationship, they will want to do everything together. In time, their activities will separate, as they realize that things Capricorn wants to do are boring to Pisces, while things Pisces like to do, drive Capricorn crazy – with no plan, no intent or any kind of seeming usefulness. Still, they will both have a need to hold on to tradition, Capricorn for respect of tradition itself, and Pisces for romantic reasons, and this should help them build enough shared activities as time passes. 70%

Conclusion: If someone like Capricorn can be pulled into a crazy love story, exciting and unpredictable, this must be done by Pisces. In return, Capricorn will offer their Pisces partner stability, peace and some rest from their usual emotional tornadoes. There is a fine way in which Capricorn can help Pisces be more realistic and practical, while feeling more cheerful and optimistic themselves. A relationship between Capricorn and Pisces tells a story about possibilities of inspiration. There are challenges in their contact, mainly represented through their love of Jupiter. It might be hard for them to reconcile their different approaches to religion, faith and their different belief systems. To overcome this, it is best if they both ask themselves – does their belief system work? And does the one of their partner also work? If they understand answers to these questions, they might find enough respect to leave each other's Jupiter intact. 76%

Pisces Compatibility with Aquarius

Pisces & Aquarius Sexual Intimacy and Compatibility: At first glance, they don't exactly go well together, one of them romantic, looking for their perfect love, while the other distant, looking for ways to set themselves free from all emotion. Still, their sex life can be quite amazing if Pisces don't get too attached and find a way to keep their distance until their partner shows emotion. As a mutable sign, Pisces have an understanding for constant change and the exhilaration and the excitement of the act of sex. Aquarius will happily follow, with a little less enthusiasm because they are, after all, rational. The beauty of their sex life could be in creativity, a game of emotion and the everlasting questioning that will bring even more excitement and emotion to the entire relationship. Unfortunately, in many cases Pisces just want their emotions flow and they will end the relationship, rather than deal with constant disappointments. The sign of Aquarius exalts Neptune, the ruler of Pisces. There is a strong link between these two signs, and in their sexual relationship, things will most certainly never get boring. The best chance for a satisfying sex life between an Aquarius and a Pisces partner is in a scenario where Aquarius already had some emotions to share, before their relationship even started. They need a good starting point and the ability of Aquarius to show emotion from time to time in a way their partner will understand it. 50%

Pisces & Aquarius Trust: Depending on the state of their intimacy, they could end up covered in lies or completely free of them. Aquarian nature can be a bit aggressive from time to time and their personal rebellious needs don't actually help Pisces feel secure enough to share intimate thoughts. The potential emotional dependency of Pisces can make Aquarius partner give in to their perfect lying skills in order to feel freer. Trust is the most important issue for this couple and it can go from one extreme to another. The only way they can build unconditional trust is to use their ability to feel each other's true core. If Pisces understand the soft side of Aquarius, the one that lies far beneath the surface, they won't run from telling the truth. When intimacy is found, Aquarius will finally be able to stop running away from commitment and the problem with the lack of freedom will be automatically solved. 50%

Pisces & Aquarius Communication and Intellectuality: The sign of Pisces should be able to ground Aquarius partner's ideas because their element is closer to earth. However, they often get lost on their way to do anything real, especially when they don't feel it is in the path of their own mission. The understanding between them doesn't always go very deep and although they can have a lot of fun together, they will have different approaches to the same things that interest them. For example, if they start talking about religion, they will end up in a philosophy battle that has no real value. They can dream well together, but unfortunately they will hardly ever make any dreams come true. The lack of reality in their relationship could hurt them both and they might not even know where the problem hides, while they feel frustrated by their relationship. By the end of it, Pisces will feel bad because they even tried to rationalize their faith, and Aquarius will feel like they have been talking to a foggy image of something resembling an opinion. 35%

Pisces & Aquarius Emotions: Their relationship could be exhausting for Pisces partner if they try too hard to find the response for the feelings they might have. The distance of their partner will most certainly exalt them and create a special emotional rollercoaster for Pisces that could lead straight to disappointment. They could make a safe emotional environment for each other only if the initiative comes from Aquarius partner alone. Pisces would have to be completely silent, uninvolved, feminine and reactive. Aquarius is distant and impersonal while Pisces can't wait to get tied to someone from their fairytale love. As soon as idealization appears, the fine balance will be shaken and the sense of freedom for Aquarius will be disturbed. 1%

Pisces & Aquarius Values: They will both value freedom of any kind, love for humankind, excitement, change, inspiration and their ideas and dreams. For Aquarius, this love for humankind would mean absolute justice, equality and freedom of speech. The link between these signs simply doesn't let them get too far from one another and their values will mostly be the same, but entirely different in their realization. For Pisces, it would mean the eternal sound of the ocean as a blessing bringing us here. If we apply this significant character difference to all other things they value together, we will see that they will need a lot of deep understanding to mend their differences. 50%

Pisces & Aquarius Shared Activities: Their problem is in the perception of romance and disagreements on things couples should do together. Pisces partner will want unusual, surprising and exciting things, but followed with romance, physical pleasure and deep emotional understanding. It is a good thing that the sign of Pisces is always open for change, because Aquarius will gladly provide some. Their interests can be quite similar and with certain compromises they could find a lot to do together. Pisces will gladly visit an art show, but why not make it a modern one so Aquarius could be interested as much? Aquarius partner, on the other hand, will want inspiring conversations, intellectual stimulation and preferably some extreme activities included. 40%

Conclusion: However, the sign of Aquarius exalts Neptune, the ruler of Pisces, and this gives them a strong bond through the planet of all magic. As all neighbouring signs, Aquarius and Pisces don't necessarily have the best understanding of each other's personalities. It is not easy to create the fairytale version of this contact, but once they find the emotional balance and the one, core truth to each other, they will have no problem keeping their fairytale alive, day after day. 38%

Pisces Compatibility with Pisces

Pisces & Pisces Sexual Intimacy and Compatibility: This is not a consequence of a lack of initiative, but rather their own need to stay in the bubble of emotional perfection and their fear of bursting this bubble by forming a physical relationship that requires dealing with physical imperfections. When they begin a physical relationship, they could both be too cautious on choices of sexual activity they will suggest to each other. Their relationship might strangely inhibit them both, because of the possibility of unrealistic expectations and the fear of being let down. At first glance, two Pisces representatives might become a perfect couple, but when we scratch beneath the surface, we might see that they have real trouble getting close to each other. Their sex life can be magical, but in many cases they will not even get to the physical contact, keeping their relationship senselessly platonic. Fortunately, their mutable natures will in most cases allow them to progress and make enough adaptations and changes for their sex life to work. When they do, they will find the exact amount of tenderness and sexual freedom they both need, and understand each other's needs before they are spoken. 75%

Pisces & Pisces Trust: Their main problem is in the fact that they know each other too well. They can both recognize their own unstable and unreliable nature in their partner, so instead of building trust and changing them both for the better, they will easily get caught in a circle of attempts to be honest and dishonest, without the need for their flaky nature to change. Trust is a very difficult subject when two Pisces representatives begin a romantic relationship. The best way for them to create a safe and trustful atmosphere is in a lot of meaningful communication that they both usually find obsolete. 50%

Pisces & Pisces Communication and Intellectuality: They will both have their own image of what is important for their partner to know about them. It is a good thing they will have such a strong feeling on each other's point, because they would probably never meet each other at all if there was no emotional connection between them. They will have a tendency not to move from a certain point, both of them intensely focused on the idea of love, rather than actual activities. Pisces is the sign of Mercury's fall and these individuals will often be too closed up or lost to have good practical communication. Although they will share their dreams with one another, and probably inspire each other in many ways, it will be difficult for them to have discussions on ongoing things in their lives. This will easily lead them to a place where there is really nothing else to share and talk about especially if they don't share the same group of friends, or have other joined activities that they can discuss on a daily basis. If they lead separate lives, they could end up in a relationship where they simply don't know each other and crave for closeness with someone else. 60%

Pisces & Pisces Emotions: The sign of Pisces has a great responsibility to exalt the planet that is linked to all sides of a loving relationship. Emotional contact between two Pisces partners will rarely materialize, because both partners are in search for someone who needs inspiration, and the two of them don't need this from their partner. When they fall in love, this is a fairytale romance and their emotional contact is something that no other sign can reach. Two of the representatives of Venus' exaltation in the same relationship, are love multiplied. Their tenderness and the way they nurture their emotions toward each other, will be a true inspiration for everyone around them. Pisces is the sign that exalts Venus, the planet of love. Not only does Venus rule the sign of Libra, speaking of our relationships, but it is also the ruler of Taurus and represents physical pleasures and satisfaction of the physical body. There is no better partner to understand the emotional nature of a Pisces partner, than another Pisces. Their mutable quality will show through emotional changes and apparent inconsistency, but in truth, they will know exactly when to separate and when to be together in order for their love to remain exciting and beautiful. 99%

Pisces & Pisces Values: They will both value talent, one's ability to stay true, and the flexibility someone has in the outer world. However, two Pisces partners might realize that they don't share the same values when they are together.

With their sign exalting Venus, it is difficult to speak about their values while they are in a relationship. The problem here is in the fact that they are too alike and they will wake each other's deficiencies through mere existence. This will lead to misunderstandings and the conviction of both partners that they don't value the same things. Not because their values differ that much, but because their priorities might be different. 80%

Pisces & Pisces Shared Activities: Their relationship will be satisfying to all their senses, and they will spend many romantic nights, as well as exciting days, in each other's arms. That is, if they ever manage to meet. While one of them might appear at the agreed point in time, the other will get lost, miss a turn, or show up at their previous meeting point. When they finally manage to find each other, with many hours of delay, they might move in opposite directions and be unaware of that due to the fact they didn't talk to begin with. They will most certainly have a lot of fun. Two Pisces partners cannot ever be bored. In general, their shared activities are dependent very much on their physical contact, because if they have it, at least they could hold hands not to get lost. 75%

Conclusion: They don't need this from each other, because they already inspire themselves. When romantic love happens between them, they might have an actual fairytale story, the one with unicorns, rainbows, and an everlasting love. Two Pisces partners will have trouble trusting each other. Their changeable natures will shift their relationship all the time, and only if they share enough love, they might be able to handle the changes and stay together. In most cases, they will not fall in love, because of their inner need to inspire their partner and help them grow. 73%

CHAPTER-4
MOON SIGN
(Janma-Rashi)

The waxing (rising) Moon is a very auspicious planet, capable of causing Neechabhanga of other planets by his mere aspect. The position of Moon in a Sign is at the time of birth is called the Moon-Sign or Birth Rashi or Rashi. Example: If the Moon is in the Sign of Mesh (Aries), the Moon-Sign or the Birth Rashi or Rashi (Janma Rashi) is Mesh. It is also called the Moon-Sign Chart, in which Moon in the first House. Accordingly, the distance of the house of all the planets from the Moon is accessed, which is essential to predict the effects of the Maha Dasa and Antar Dasa of planet.

4.1 Aries Rashi (Moon in Aries) - (Aswini, Bharani, Kritika 1st quarter):

Name starts with phonetic (Chu, Hey, Cho, La, Li, Lu, Ley, Lo, Ae): The Moon in Aries is not congruent to her nature. He/She is restless, eyes, inflicted with diseases, unfaithful, gives pleasures to his wife, fears of drowning into water, hard working and is full of tranquillity in his old age. He/She is adventurous and is too changeable, moody, whimsical and flirtatious. He/She is likely to meet with all sorts of disappointments and even disillusion.

4.2 Taurus Rashi (Moon in Taurus) - (Kritika last 3 quarters, Rohini, Mrigasira 1st half):

Name starts with phonetic (Ee, U, Aye, Oh, Va, Ve, Vo, Vay, Vo): The Moon in Taurus is a natural domicile. He/She is charitable, pious, virtuous, wealthy, and full of radiance, good health and long lived. He/She has determination, loyal friend and is emotionally very strong and seldom changes his/her mind. He/She attracts opposite sex for strong romance. He/She takes up occupations of real estate, property, art, design, jewellery and business. He/She is an ambitious, selfish, has a goal for a luxurious home, plenty of money and intellectual but has a very practical, astute, shrewd ability to judge the average conditions in life.

4.3 Gemini Rashi (Moon in Gemini) - (Mrigasira last half, Ardra, Punarvasu first 3 quarters):

Name starts with phonetic (Ka, Ke, Koo, Gha, Jna, Cha, Kay, Ko, Haa): The moon in Gemini is considered weak. He/She has a melodious voice, talks sweetly, is kind hearted, very lusty and is prone to throat diseases, famous, wealthy, fair complexioned, tall, clever, genius, of firm resolution, efficient in work, and remains judicious in every situation. He/She thrives on communication. He/She prefers job in media, travel as well as sales. He/She chooses his/her partners. He/She is highly imaginative, educated, and is both, a good teacher and a sharp student. He/She cannot limit to one activity and business at a time and would do well with a strong & practical partner. He/She will be entrusting the partner with most of the decision-making in business. He/She is very successful as news person, advertising agency, writers, authors or any other creative field. He/She is likely to make plenty of money, and enjoy great popularity. He/She is good in the business world. He/She will retain a youthful look and behaviour and succeeds at "staying young forever."

4.4 Cancer Rashi (Moon in Cancer) - (Punarvasu last quarter, Pushya, Aslesha):

Name starts with phonetic (He, Hoo, Hey, Ho, Daa, Dee, Doo, Dey, Do): The Moon is considered royal in her own Cancer. He/She is wealthy, has patience, serves his teacher, very clever, lives in a foreign land, keeps good company, and has a high degree of intelligence. He/She is sympathetic, kind, compassionate and sensitive to others' feelings. He/She is with excellent memory, fond of home and parents, peaceful, gentle, affectionate, and romantic. He/She may get delayed marriage. He/She may be excellent artists, musicians, and poets and home life is very important to him/her. As a parent, he/she is quite nurturing and lavishes loved ones. He/She does not hesitate to use tears or a self-sacrificial attitude to get his/her point across. Real estate is an especially good area to invest in for him/her. He/She would also do well in a business run from home and has natural tendency to put on weight.

4.5 Leo Rashi (Moon in Leo) - (Magha, Poorvaphalguni, Uttaraphalguni 1st quarter):

Name starts with phonetic (Ma, Me, Moo, May, Moo, Ta, Tee, Too, Tay): The Moon in Leo is the natural domicile of the Sun. He/She forgives easily, loves to travel, likes to eat non-vegetarian food, is full of fear, keeps good company, is humble, has excessive anger, devoted to his parents and achieves fame. He/She is self-sacrificing, generous, conservative, discriminating, encouraging, romantic, optimistic, brilliant robust, strong, and decisive and a natural leader and exudes energy and drive. He/She loves excitement and action. He/She may sacrifice everything in the cause of righteousness and justice. He/She makes others dependent on him. He/She cannot be convinced against his/her will, nor be swayed against emotions. He/She is strong-minded and determined. He/She is warm, loving and outgoing. He/She doesn't settle for less than what he/she wants and is too much commanding. He/She will be smothered with love and care, and may be henpeck.

4.6 Virgo Rashi (Moon in Virgo) - (Uttaraphalguni last 3 quarters, Hasta, Chitra 1st half):

Name starts with phonetic (Too, Pa, Pee, Pu, Sha, Na, Tha, Pay, Poe): The Moon in Virgo is unimpeded mind, passion and flesh. The keyword is "criticism". He/She is a sensualist, respects the virtuous people, religious, clever, charitable, poet, follower of the Vedas, lover of humanity, interested in dance and music, likes to travel, and is troubled by his/her Spouse. He/She has no confusion in the mind, strong social conscience, good communication and is sociable, logical, back-seat drivers and clear-headed. He/She has a strong personal code of conduct and set high standards. He/She frequently takes nursing, dietetics, teaching, and secretarial work as careers. He/She is considered to be cold-blooded and overly ambitious. He/She is very conscious of keeping fit, both physically and mentally. He/She also enjoys taking part in national politics or social issues. He/She makes well, stable business partners and a successful professional. He/She is grave, sexless people with no sexual curiosity, and doesn't understand the meaning of sex. He/She is exceedingly active, and will put more energy into house cleaning, attention to business, and personal doctoring than any other type of person.

4.7 Libra Rashi (Moon in Libra) - (Chitra last half, Swati, Visakha first 3 quarters):

Name starts with phonetic (Ra, Ree, Ru, Ray, Tha, Thee, Thoo, They): The Moon in Libra is the best positions. The keyword is "decision". The moon in this position gives artistic temperament, creative ability, good mental understanding, but no executive ability. He/She gets angry unnecessarily, talks sweetly, has restless eyes, mixed fortunes, authority inside the house but powerless outside, a devotee and likes to travel. He/She is charming, creative, and diplomatic. He/She may be romantically amorous, notoriously fickle and wavering in romance. He/She is financially motivated, can be reckless, careless, and/or squandering. He/She is known for charm and social grace, and presents an image of total balance and harmony, and has problem of the kidneys and allergies. He/She takes professions as law, architecture, politics, the arts, and even homemaking. He/She is attracted to very gracious partner and finds a good one in life, although it may not be the first one. He/She is voluptuous, deceptive in habits, with a voracious physical appetite. Usually he/she is so pretty and has so much charm that marriage is a foregone conclusion.

558

4.8 Scorpio Rashi (Moon in Scorpio) - (Vishakha last quarter, Anuradha, Jyeshta):

Name starts with phonetic (Tho, Na, Nee, Noo, Ney, No, Yaa, Yee, Yoo): The Moon in Scorpio is favourable position. The keyword is "Ulterior Motivation". He/She is a traveller from his childhood, has yellow eyes, lusty, proud, behaves roughly with his relatives, acquires wealth through hard work, and is wicked towards his mother. He/She is intelligent, cold, sensual, emotional, materialistic and secretive. He/She can become superb occultists and astrologers. The position is favourable for jobs in medicine, surgery, chemistry and investigative work. He/She can be very possessive, jealous, very cruel and vindictive. He/She never forgets a wrong that someone has done, and will plot and plan for years or decades, if necessary, to seek revenge. He/She is intensely emotional but projects a perfectly cool exterior at all times. His/Her frenzied desires are never satisfied within home. He/She is constantly seeking outside satisfaction. There are few women who have a good deal of scandal running through the life.

4.9 Sagittarius Rashi (Moon in Sagittarius) - (Moola, Poorvashada, Uttarashada 1st quarter):

Name starts with phonetic (Yey, Yo, Ba, Bee, Bu, Dha, Pha, Dha, Bay): The Moon position in Sagittarius is risk-taking. The keyword is "enthusiasm". He/She is pious, wealthy and virtuous, has a loving nature, knowledge of fine arts, likes drawing and painting, has a wife full of good qualities, sweet-talker, has a heavy physique and in some rare cases, a destroyer of his family. He/She is eternal students with an urge to higher education, impulsive, blunt and outspoken, magnetic and forceful, actively philosophical and believes in justice and fair play and helps out anyone who is in need. He/She likes astrology and prophecy. He/She hates anything hidden or secret. He/She has a strong sense in gambling and believes that he/she simply cannot lose and therefore take unconsidered risks. If it is necessary to terminate a relationship, he/she does so quickly and cleanly without looking back. He/She knows something better is waiting just around the next corner. He/She is great spenders and enjoys a jolly, pagan sort of social life, uninhabited and full of romantic interest. He/She is incurable opposite sex partners' chasers and has put a great deal of enthusiasm into his/her dashing love affairs.

4.10 Capricorn Rashi (Moon in Capricorn) - (Uttarashada last 3 quarters, Sravana, Dhanishta 1st half):

Name starts with phonetic (Bo, JA, Je, Ju, Jay, Jo, gha, Ga, Gee): Moon in Capricorn is one of the least desirable positions. The key word is "Management". He/She has values in his family, is under the influence of his wife, scholar, undertakes charitable work, respects his mother, is wealthy, has obedient servants, is kind hearted, has a large family, and lot of worries also. He/She lacks sympathy, and has innate selfishness, self-preoccupation, dignity, tenacity and a realistic vision of the world. He/She has defeated ambitions and dreams, misfortunes, occupational and financial troubles, credit difficulties and all sorts of other misfortunes. He/She will have positions of executive, administrative, public and organizational positions and commercial pursuits. He/She has a natural desire to rise to a position of power and fame and is willing to work hard for accomplishments. He/She is not fortunate enough to achieve fame and fortune and becomes terribly frustrated and may even develop ill health as a result. He/She is very selfish, cautious, and thrifty and lays own aims and ambitions.

4.11 Aquarius Rashi (Moon in Aquarius) - (Dhanishta last, Satabhisha, Purvabhadra 1 to 3 quarters):

Name starts with phonetic (Goo, Gay, Sa, See, So, Say, Da): The Moon in Aquarius is fixed. The keyword is "disinterestedness". He/She is lazy, owns the most expensive vehicles, wealthy, is blessed with beautiful eyes, and has a simple nature. He acquires wealth and knowledge, achieves fame on account of his virtuosity and kindness, is fearless and enjoys his wealth. He/She is idealistic, caring for the global village, a little detached to home and paradoxically. He/She possesses integrity and honesty, and is not likely to ask for help when in trouble, but ready to help others. He/She is well liked and have strong religious and philosophical instincts coupled to a humanitarian urge. He/She possesses absence of jealousy and possessiveness, and favours all forms of humanitarian, political, and educational pursuits, exploration in all fields, authorship, and astrology too. He/She has a tendency to gossip and spread rumours. He/She loves the business and professional world, and has plenty of patience, kindness, intelligence and understanding.

4.12 Pisces Rashi (Moon in Pisces) – (Poorvabhadra last quarter, Uttarabhadra, Revati):

Name starts with phonetic (Dee, Du, Tam, De, Do, Cha, Che): The Moon in Pisces is not favourable position. The key word is "anxiety". He/She is brave, talks cleverly, but often has excessive anger, loved by his family, a devotee, a very fast walker, efficient in charity and knowledge, virtuous, sacrificing and receives affection and love from his friends and family members. He/She has strong creativity, powerful imagination and impressionability. He/She is natural worshippers of beauty, very loyal to friends and is more inclined to romantic attitudes. He/She succeeds best in intuitive judgement, discretion, assiduity, and detailed work. He/She does well as entertainers, dealers with liquids of all sorts, promoters, seafarers, and detectives. He/She has strong intuitive and psychic qualities. He/She should generally follow his intuition, and cares a great deal about others and seek to serve society as a whole in own way. He/She gives love freely to others, and may get deceived by others. He/She may feel like rejecting the world altogether. His/Her life probably will not be one of the Cinderella or Prince Charming of his/her dreams. He/She is not afraid to work hard and is subjected to wild swings in mood. He/She can give pure, unselfish, transforming love and compassion.

He/She can often find satisfaction and relief in religion and art.

CHAPTER-5
NAMING VEHICLE & CHILDREN

Give your Child a Correct Name: The name carries much value in life. Once, a person of around 30 years of age visited an astrologer telling that he has a lot of personal as well as financial problems. He has a dull and sombre face without lustre because of Saturn (Sani). Astrologer asks his name. His first name is Mandeswara Rao but changed to Ravi Kumar as per advice of an astrologer. Second astrologer is shocked. Ravi Kumar means "son of the planet Sun", which is none other than Saturn. Then second astrologer realizes that the person is completely under the clutches of Lord Saturn, and it is not so easy to come out of the influence of Lord Saturn.

It is very difficult for anyone to select a name filled with complete positive vibrations and may not be able to change his name for good or bad at middle age. But we can give a better name to our kids. There are 3 steps to choose a better name, filled with positive vibrations, as advocated by Indian Astrology and Numerology combined.

Step I: Set the positive vibration in the name starting with phonetic according to the birth star (Nakshatra) of the child. Astrology prescribes the starting sound to be used in a name depending on the Birth Star (Nakshatra) and or the Rashi (Moon- Sign) of the child, as shown below in Table:

Table: Naming Letter as per Birth Star or the Rashi (Moon- Sign)

Naming Letter as per Moon- Sign		
Moon- Sign	Nakshatra	Naming Letter

Aries Rashi	Ashwini 1,2,3,4 quarters	Chu, Che, Cho, La
	Bharani 1,2,3,4 quarters	Li, Lu, Le, Lo
	Krittika 1st part or quarter	Aa
Vrishabha	Krittika 2,3,4 quarters	Ee, Vu, Ae
	Rohini 1,2,3,4 quarters	Vo, Va, Vi, Vu
	Mrigashira 1 and 2 quarters	Ve, Vo
Gemini Rashi	Mrigashira 3, 4 quarters	Ka, Ki
	Aridra 1,2,3,4 quarters	Ku, Kham, Jna, Cha
	Punarvasu 1,2,3 quarters	Ke, Ko, Ha
Cancer Rashi	Punarvasu 4 quarter	Hi
	Pushya 1,2,3, 4 quarters	Hu, He, Ho, Da
	Ashlesha 1,2,3,4 quarters	Di, Du, De, Do
Leo Rashi	Makha 1,2,3,4 quarters	Ma, Mi, Mu, Me
	Purva Phalguni 1,2,3,4 quarters	Mo, Ta, Ti, Tu
	Uttara Phalguni 1 quarter	Te
Virgo Rashi	Uttara Phalguni 2,3, 4 quarters	To, Pa, Pi
	Hasta 1,2,3,4 quarters	Pu, Sha, Na, Dha
	Chitra 1 and 2 quarters	Pe, Po
Libra Rashi	Chitra 3, 4 quarters	Ra, Ri
	Swati 1,2,3,4 quarters	Ru, Re, Ro, Tha
	Visakha 1,2,3 quarters	Thi, Thu, The
Scorpio Rashi	Visakha 4th quarter	Tho
	Anuradha 1,2,3,4 quarters	Na, Ni, Nu, Ne
	Jyeshtha 1,2,3,4 quarters	No, Ya, Yi, Yu
Sagittarius	Moola 1,2,3,4 quarters	Ye, Yo, Bha, Bhi

Rashi	Purva Ashada 1,2,3,4 quarters	Bhu, Bha, Dha, Pha
	Uttara Ashada 1st quarter	Bhe
Capricorn Rashi	Uttara Ashada 2,3,4 quarters	Bho, Ja, Ji
	Shravana 1,2,3,4 quarters	Khi, Khu, Khe, Kho
	Dhanishta 1, 2 quarters	Ga, Gi
Aquarius Rashi	Dhanishta 3,4 quarters	Gu, Ge
	Shathabisha 1,2,3,4 quarters	Go, Sa, Si, Su
	P. Bhadrapada 1,2,3 quarters	Se, So, Dha
Pisces Rashi	P. Bhadrapada 4th quarter only	Dhi
	U. Bhadrapada 1,2,3,4 quarters	Dhu, Tha, Jha, Da/Tra
	Revati 1,2,3,4 quarters	Dhe, Dho, Cha, Chi

Note: If we could not find a name starting with the sound mentioned against the birth star, then we should use the sounds of the other star in the same Moon-Sign group. Example: Purva Phalguni born child name should be with the sounds: Mo, Ta, Ti, Tu or the star Makha sound: Ma, Mi, Mu, Me or the star uttar Phalguni sound of the same Moon-Sign sound: Te, To, Pa, Pi. Similarly Revati born child name should be with the sounds: Dhe, Dho, Cha, Chi or for the same Moon-Sign sound: Dhi, Dhu, Tha, Jha, Da/Tra.

Step II: Let us now set a harmonious vibration between the name and the birth date Number according to Numerology. Remember, every letter is assigned a numerical value as shown below:

Table: Letters with assigned Numerical Value for writing Vehicle Number

A, I, J, Q, Y	1

B, K, R	2
C, G, L, S	3
D, M, T	4
E, H, N, X	5
U, V, W	6
O, Z	7
F, P	8

Calculation: Now, take the name of the child, and find the total numerical value of the name using the above digits. Example: RAVI BABU is the name chosen.

Name	Value	Compound Number	Single Number
RAVI	2 + 1+ 6 + 1	10	1
BABU	2 + 1 + 2 + 6	11	2

Name compound number: 10 + 11 = 21 and name's Single Number Value: 1 + 2 = 3.

Now, take the birth date of the child, and make it a single digit Number. Example: the child is born on 27th April, 2006. Take only the birth date number, i.e., 27 = 2 + 7 = 9. The following numbers are having harmonious vibrations:

Table: Digits with assigned Harmonious Vibrations of Single Numerical Value of Date of Birth

Name's Single Number	Harmonics to Name's Number
1	1,2,9,3
2	2,1,5
3	3,1,2,9
4	3,6,2

5	5,1,6
6	6,5,8
7	7,1,2,9
8	5,6
9	9,1,2,3

If we observe the name's Single Numerical Value of RAVI BABU, it is 3 and his birthday number is 9. From the Table above, Number 9 is harmonic for 3. Hence we can give RAVI BABU name to the child who was born on 27th. Note: Once, the child is named as RAVI BABU; we should take care to call him with this name only. If others call him, RAVI only, then his name single number is 1, which is a harmonic to his birth number 3. Similarly, if he is called BABU only then also his name single number is 2, which is a harmonic to 3. So even if he is called RAVI BABU or RAVI or BABU, there is no problem.
Step III (Check with defined Favourable Compound Number): According to Numerology, the following compound numbers have good vibrations (others are bad): 10, 15, 17, 19, 21, 23, 24, 27, 32, 33, 36, 37, 41, 42, 45, 46 and 50. The compound number 'RAVI BABU' is 21, which is falling in the above favourable numbers. So we can confirm the name RAVI BABU is the best suitable name for the child.
Note: If the name compound number is more than 52, then add the digits again. For example, take 53; add 5 + 3 = 8. Take it as 8 only. Similarly, 54 are equal to 9, 55 = 10 and 56 = 11. By following the above steps, we can avoid any obstacles and chaos in the child's future and the child will become a happy and successful person.
Give your Vehicle a Correct Number: People often say that they got good luck after purchasing the vehicle and most of the people throw huge amount of money to get a lucky number for their new vehicle. In India, many people are seeking number '9999' for their vehicle. There is a high demand for this number than any one else. People who choose this number would be shocked if they know that this number is prone to accidents more often than any other

number. Yet, they choose such numbers out of knowledge, ignorance and misguidance.

Alphabet	Numerology No.
a, i, j, q, y	1
b, k, r	2
c, g, l, s	3
d, m, t	4
e, h, n, x	5
u, v, w	6
o, z	7
f, p	8

Example: Take a vehicle number AP 9 UL 9902 to understand the Numerology behind Vehicle Number. Numerology assigns numerical digits for each alphabet as shown in the table. Calculate the Vehicle Number representing each alphabet in the form of digits. Add all the numeric digits of the number to make a single digit. Thus, it is A P 9 U L 9 9 0 2. 1 + 8 + 9 + 6 + 3 + 9 + 9 + 0 + 2 = 47 = 4 + 7 = 11 = 1 + 1 = 2. So, the Vehicle Single Digit (VSD) number is 2. Now see the positive and negative traits of this number in the table below:

VSD No.	Planet	Positive traits	Negative traits
1	Sun	Good for works with government and forestry.	You appear to be egoist.
2	Moon	Good for short trips and picnics.	Emotional driving like sudden changes in the speed of vehicle.
3	Jupiter	Good for political, administrative and advisory works.	You are proud.

4	Rahu	Sudden monetary gains.	Wrong decisions and separation.
5	Mercury	Good for any commercial activity.	You are talkative and appear to be childish.
6	Venus	Good for artistic works and cinema-industry.	Too much indulgence in liquors and ladies.
7	Ketu	Spiritual thinking and pilgrimages.	Material benefits become less.
8	Saturn	Mobilizing masses and unions.	More work and less profits.
9	Mars	Quick decisions and dynamic actions.	You drive too fast.
0		A turning point in your life.	May be sudden downfall.

Since the major influencing digit in this vehicle number is 2 from the table, we can understand that you would enjoy short trips and picnics. Hence your car will be favourable for trips with your family. The negative part is you would drive your vehicle emotionally; sometimes very fast and sometimes too slow and you would be engrossed in changing thoughts. This may lead to minor accidents. Similarly, take another vehicle number 'UP 12 BD 2204'. This gives 6 + 8 + 1 + 2 + 2 + 4 + 2 + 2 + 0 + 4 = 31 = 4. So, the Vehicle Single Digit (VSD) is 4. This digit 4 shows separation, which means the person, may be separated from his family (death) or he may lose a limb.

Lucky number for the vehicle: There are 3 simple principles that we follow to set a lucky number for the new vehicle. Step 1: The number should be harmonious with the vehicle purchaser birth date number. For this purpose, take the date of birth of the purchaser and make it a single digit. Example: If he is born on 16th of some month, his Birth Single Digit (BSD) will be: 1+ 6 = 7. From the table below, we can see that the Vehicle Single Digit (VSD) can be 7 or 2. Thus, suppose the vehicle number is 'AP 20 Y 1202' = 1 + 8 + 2 + 0 + 1 + 1 + 2 + 0 + 2 = 17 = 8.

Birth single digit (BSD)	Favourable Vehicle Single Digit number (VSD)	Avoid/Unfavourable Vehicle Single Digit number (VSD)
1	1, 3, 9	6, 8
2	2, 7	8, 9
3	3, 1, 9	5, 6
4	4, 3, 6, 8	1, 2
5	5, 6	2, 9
6	6, 5, 8	1, 9
7	7, 2	6, 8
8	8, 5, 6	1, 2
9	9, 1, 3	5, 8

Step 2: Safe-guard the vehicle number from harmful influences. The vehicle number should not sum up to the single digits shown in the last column of the preceding table, i.e. the Unfavourable Vehicle Single Digit number (VSD). Example: If the Birth date Single Digit (BSD) is 7, then avoid 6 and 8 as the vehicle single digit number. Thus, the vehicle number AP 20 Y 1202 = 8 becomes unfavourable.

Step 3: Avoid Always the combination of accident-prone digits. Numerology identifies 9, 8 and 4 as highly dangerous numbers with respect to accident. Try to avoid too many 9 or 8 or 4 in the vehicle number as well as their combinations, such as, the number like 'AP 9 D 9908' because it will be very harmful even though the sum or VSD Number becomes 3. There are three 9's and one 8 which are 100% accident prone digits.

Note: The digit '0' is considered a wheel of fortune and it represents the ups and downs in life. Single zero will not affect in any way. So, if a zero comes in the vehicle number, there is no problem. But avoid repetition of zeros in the vehicle number too.

CHAPTER-6
ZODIAC SIGN ELEMENT

Every Zodiac Sign falls into one of four elements. There are four Elements, such as earth represents common sense; fire represents action, air represents thinking and communication skills, and water represents the ability to feel and intuitively know. Each Element is assigned to each sign depending to their orientation in the zodiac. Many astrologers consider the element of each of the planets when determining which of the elements may be more significant in a horoscope.

Fire (Aries, Leo, and Sagittarius):

Fire is active and masculine. People of the Fire element are outgoing, quite moral, very creative, courageous, passionate, impulsive, hot, dynamic, progressive, action oriented, and direct. Their essence is spirit. They are enthusiastic, optimistic, confident, naive, self-centred, open, confronting, loyal, tactless, impatient, honest, trusting, and independent and feel free.

Earth (Taurus, Virgo, and Capricorn):

Earth is a receptive, feminine sign. People are practical, cautious, and pragmatic approach to life and build solid, 'real' material success, i. e. car, home, career success and have long range planning and strong determination to succeed. They are safe/secure, suspicious, sensual, organised, dependable, introvert, and efficient and strong survival instinct.

Air (Gemini, Libra, and Aquarius):

They are active, curious, idealistic, unemotional, conceptual, devoid of feeling, good to communicate, social, objective, impersonal, distant, masculine, intellectual, changeable, and impractical, good speech, and natural communicators, extroverted, social, charming, and logical and air has least obvious bad qualities, theoretical, abstract, needs to socialise, needs to share ideas.. The lack of Air Element in a native birth chart indicates difficulty in the expression of that person. Communication of ideas and the ability to conceptualise may prove difficult.

Water (Cancer, Scorpio, and Pisces):

People of Water element dissolve everything in them coolly. They take the shape of who they are with, and are quite emotional, sustaining, emotional, sensitive, imaginative, protecting, compassionate, caring, artistic, moody, soulful, subconscious, irrational, introverted, but strong/powerful, vulnerable to hurt, intimate, defensive, psychic, past, suffering, suspicious initially, self-contained, picks up impressions and associated with healing.

Zodiac Sign Element Strength:

The method for evaluating the strength of an element in a birth chart is to assign a value of 4 to the element associated with the Sun; the Moon element is assigned a value of 3; Mercury, Venus, and Mars sign elements are assigned a value of 2 each, and Jupiter and Saturn each have a value of 1. Uranus, Neptune and Pluto are disregarded because their element is more societal affecting large groups of individuals born during a period. Using this approach, if as many as 8 points are concentrated in one element, it is considered "Preponderance" in that element. If we get less than 8 points with this approach in one element, it is considered "Absence" in that element.

Example: In one chart, the planets are placed as is given below: For Sun in Aquarius, an Air sign, we assign = 4 Points. For Moon in Libra, an Air sign, we assign = 3 Points.

For Mercury in Aquarius, an Air sign, we assign = 2 Points. For Venus in Capricorn, an Earth sign, we assign = 2 Points. For Mars in Sagittarius, a Fire sign, we assign = 2 Points. For Jupiter in Pisces, a Water sign, we assign = 1 Points. For Saturn in Fire, a cardinal sign, we assign = 1 Points.

Thus, we have Element strength, such as, 9 in the Air Element, 2 in the Earth Element s, and 3 in the Fire Element and 1 in Water Element in this chart. This shows a preponderance of Air element in this chart. The preponderance reading of Air element would be appropriate in the above chart. The preponderance readings of all elements are given below:

A Preponderance of the Fire element:

A preponderance of Fire Element indicates high spirits, great faith in self, enthusiasm, direct, honesty, intensely assertive, most daring, individualistic, active and self-expressive, good natured, fun loving, natural leader, having a good time than on material possessions, big egos. He believes so strongly in his own powers and abilities that he overlooks and frequently fails to take advantage of the talents and abilities of others. He tries to do it all himself and don't delegate well. He is constantly "out front" or "on stage", such as an Artist and they need to be recognized and admired for his attainment and accomplishments. Appreciation is more important than money in his estimation. Nothing hurts him more than being ignored. The fire sign sense of honesty is straightforward and often child-like. Thus, he believes everyone is, like himself, an open book.

A Preponderance of the Earth Element:

A preponderance of Earth Element indicates cautious, conventional, dependable but quite responsible, methodical, organizer, a builder, and a hard-worker. It provides the skills and attitude necessary to succeed readily in the world of business and never gamble or take unnecessary chances. They understand the reality of a situation and value, reliable and steadfast. They are dependent, diligence and a pragmatic, no-nonsense approach to life. Lack of ideas or imagination, dullness, rigid, conservatism, extreme materialism, and blind adherence to rules and regulations are their potential faults.

A Preponderance of the Air element:

The preponderance of Air Element suggests a strong emphasis on thought, ideas and intellectual and they communicate and express ideas with mental agility and become the impractical dreamers, constantly thinking, people-oriented, but more inclined toward the group than the individual. Your interests are varied, and you're apt to be a life-long student.

A Preponderance of the Water element:

The preponderance of Water Element indicates close emotional relationships, romantic, sentimental, affectionate, secure bond with partner, communicate best in non-verbal ways; emotionally, psychically, or through forms as art, dance music, poetry and photography. They have a natural feel and sense for the arts and are apt to let the heart rule the head, highly impractical and impressionable.

CHAPTER-7 ZODIAC SIGN IN HOUSE

The word zodiac is a loose translation of the Greek phrase for circle of animals. The zodiac is an imaginary line in space mapping the path of the rising and setting sun. This line follows the equator superimposed out onto the constellations. In the circle surrounding the earth, different distant stars signal the beginning of different signs. The zodiac circle is divided into twelve of these sections. Each section is known as a sign and is named for the constellation near which the sign begins in the zodiac. In Hellenistic, Vedic, Medieval and Renaissance astrology each house is ruled by the planet that rules the sign on its cusp. Example: Mars is ruler of the 1st house because it rules Aries, the first sign; Mercury rules the 3rd house because it rules Gemini, the 3rd sign; etc. The first house in a chart takes on the characteristics of Aries – it is where an astrologer gets information about your ego and ability to move forward in life. In a solar chart, the zodiac sign following your birth sign is your second house. The second house tells the characteristics of the 2nd sign, Taurus. It is the house of wealth and values. And so on. In mixing a sign with a house, there are myriad possibilities for an astrologer to consider. The solar chart of a Taurus would place Gemini in the 2nd house, indicating that the Bull will communicate his or her values directly and loves to discuss financial minutia with trusted friends and colleagues. When you get an astrology reading, each house provides vital information about you to your reader. When your astrologer is looking at your chart's houses, what follows here are the areas of your life from which your individual makeup can be surmised:

House	House Characteristics	Related Sign	Ruled By
First	House of the Self, Appearance & Ego Drive	Aries	Mars
Second	House of Personal Values and Finances	Taurus	Venus
Third	House of Communication & Local Culture	Gemini	Mercury
Fourth	House of Identity & Roots	Cancer	The Moon
Fifth	House of Children, Creativity & Gambling	Leo	The Sun
Sixth	House of Work & Ritual	Virgo	Mercury
Seventh	House of Partnerships	Libra	Venus
Eighth	House of Sex, Death & Other People's Money	Scorpio	Mars & Pluto
Ninth	House of Adventure, Travel & Learning	Sagittarius	Jupiter
Tenth	House of Career & Public Persona	Capricorn	Saturn
Eleventh	House of Gatherings & Crowds	Aquarius	Saturn & Uranus
Twelfth	House of Mystery and Intuition	Pisces	Jupiter & Neptune

PART-1
ZODIAC SIGNS IN 1ST HOUSE
(LAGNA/BODY):

Aries (Mesha) Lagna:

He/She is proud, wealthy, having excessive anger, dependent on others. He/She will be medium height, white complexion, smiling face, clever, and lean. He/She suffers from abdominal problems. He/She relishes helping the poor and has faith in God. He/She thinks very high but implements little. He/She inherits huge paternal property. He/She is prone to drowning and accidents. He/She is selfish and forgets a person who is no longer of use. He/She is prone to be cheated by friends and partners. He/She will improve in life and will earn money and believes in donation. Even born in medium income group, he/she earns good money of his/her own. He/She has a personality that is positive, aggressive, and competitive. He/She has leadership qualities, is bold, and empowered with more physical strength. He/She is best in sports, games, trekking, summer camp, and any other outdoor activities.

Sun: Surya is L – 5 measured from Mesha lagna, Mesha Chandra, Mesha Navamsha and so is very benefic planet for Mesha Lagna and is a friend of Mars (Lagnesh). Sun in Aries ascendant is exalted. If Surya is well-disposed and strong, Surya brings good luck, highly intelligent performance, and a royal position to enjoy all aspects in life, winning by speculation, good children and great pride by children. Surya will provide public roles, leadership responsibilities in politics and contests for power (including elections), theatrical performance, games, entertainments, royalty, courtly life, literary arts, creativity of children and avocation with youth group activities.

Moon: Chandra is L-4 measured from Mesha lagna, Mesha Chandra, Mesha Navamsha and so is neutral planet for Mesha Lagna and is a friend of Mars (the lord of the Ascendant). If Chandra is well-disposed and strong, Chandra will provide secure foundations, schooling, and all kind of pleasures, good childhood home, shelters and house properties, vehicles of all kinds, patriotism and love of the homeland.

Mars: Kuja (Mangal) is L-1 & L-8 measured from Mesha lagna, Mesha Chandra, Mesha Navamsha and so is very benefic planet as Lagnesh. If Mangal is well-disposed and strong, it makes the native very physically strong and competitive for winning competitions through physical skill, recognition in physical appearance, a competitive businessman and social rewards to him/her for his strong physical body engaged in healing service, such as, massage therapy, psychotherapy, Hatha yoga instructors unless Kuja is oppressed by Shani. Mangala will make a skilled and active sports enthusiast, a hunter, an athlete, dancer, surgeon or dentists and police officers. If Kuja occupies Makar in Vikrama (10th) Bhava, he/she marks excellence in business and a propensity to create self-made wealth. He will be exalted in the field of sales, marketing, administration, meetings, teamwork, colleagues, siblings and mental tasking. If there is a Parivartana Yoga with Shani rising in Scorpio (once in every 29 years a batch of these people are born), the native has tremendous self-made wealth through determined hard work, dominating aggressiveness, and sharp business judgment.

Mercury: Mercury is L-3 & L-6, lord of two bad houses the 3rd and 6th, measured from Mesha lagna, Mesha Chandra, Mesha Navamsha. Mercury is the most malefic planet for Mesha lagna and is unfriendly with Mars (the lord of the Ascendant). Budha and Kuja is adversarial planet to each other. Mercury create problem for the native, particularly for siblings (due to natural karaka of 3rd house), planner, accountant, organizer and media worker and so they feel victimized and disturbed. Budha is especially a pernicious planet for Mesha lagna. Mesha natives feel frustration when Budha occupies Mesha. Budha can even signify a divorce.

Jupiter: Guru is L-9 & L-12 measured from Mesha lagna, Mesha Chandra, Mesha Navamsha. On account of his being lord of the 9th (a Trine), Jupiter is auspicious for Aries Ascendant and is friendly to Lagnesh, Mars. Guru supports both public and private religious activities. Guru casts his auspicious aspect upon three Bhava and thus four of the twelve Sthana gets benefit from the graha-Drishti of the all-round good planet (L-9). Wise and expansive Guru controls good bhava-9 and bhava-12 (the house of loss of identity and loss of body).

Venus: Sukra is L-2 & L-7 measured from Mesha lagna, Mesha Chandra, Mesha Navamsha. Venus is the malefic planet for Mesha lagna and is unfriendly with Mars (the lord of the Ascendant). The male native will resist divorce and avoid breaking business partnerships even when conditions are very inhospitable.

Saturn: Shani is L-10 & L-11 measured from Mesha lagna, Mesha Chandra, Mesha Navamsha. Shani is the malefic planet due to Badhakesha (L-11) for Mesha lagna. If Sani occupies Lagna, His debilitation Sign (Aries), the native will suffer isolation and his identity is oppressed by herd mentality or crowd movements and the native has fear of entering into new connections with people and ideas. Saturn in 2nd house (Taurus), the house of his very intimate friend, Venus, will give position and income as Shani is lord of 10 and 11, more especially if that lord Venus is strong or if there is the aspect of Jupiter. Saturn will certainly cause some wasteful expenditure and some trouble in the family like loss of children. Saturn is in the 10th in his own sign, Capricorn, will give rise to Sasa Yoga and so native will command good servants, but his character will be questionable. He will be head of a village or a town or even a king, will covet others' riches and will be wicked in disposition.

Rahu: Rahu is co-lord of Kumbha, L-11, "gains", measured from Mesha lagna, Mesha Chandra, Mesha Navamsha.

Rahu's effects for Mesha lagna depend significantly on Rahu's Bhava, Rashi, and Drishti. Rahu amplifies the effect of any graha who are sharing Rahu's house. Rahu magnifies the effects of the lord of His occupied Rashi. Rahu and Ketu are said to give positive results always in the 3rd, 6th, and 11th Bhava from the lagna or from Chandra. As per BPHS; Rahu-Ketu are exalted in Vrishabha-Vrischika. Rahu and Ketu will give good results when Mangala and Sukra are well-disposed. The Exalted Rahu for Mesha lagna becomes a virulent Maraka in Dhana Bhava. Vimshotari periods of Rahu are associated with malignant diseases of Sukra (STD, sugar-related, drug-addictions).

Ketu: Ketu is co-lord of Vrischika L-8 "open secrets", measured from Mesha lagna, Mesha Chandra, Mesha Navamsha. Ketu effects for Mesha lagna depend significantly on Ketu Bhava, Rashi, and Drishti. Ketu amplifies the effect of any graha who is sharing Ketu house. Rahu and Ketu will give good results when Mangala and Sukra are well-disposed.

- A Jupiter-Saturn association doesn't confer a Rajayoga
- A Mars-Venus combination can be both favourable and fatal
- A Mercury-Mars combination in Virgo gives the native skin diseases, eruptions and wounds.
- An Aries native will have fear of diseases (especially smallpox), weapons and injuries
- A Sun-Jupiter-Venus combination in Capricorn helps the native enjoy dips in holy rivers like the Ganga
- A Sun-Mercury-Venus combination in Aquarius is a fortunate combination, with all three conferring riches during their Dasa and Bhukti
- A Sun-Moon combination confers Rajayoga and Saturn in Libra
- If Mars conjoins Jupiter in Cancer, he becomes a Yogakaraka
- If Mars conjoins Mercury, death by brain disease occurs in his Dasa and Bhukti
- If the above combination occurs in Gemini, there isn't any yoga
- Jupiter becomes a Maraka if he occupies Capricorn

- Jupiter in Aquarius won't automatically give good results in his Dasa
- Mars will prove a benefic during his Dasa if he's in Leo
- Mars, if in conjunction with Sun and Venus in Scorpio, will confer some fame
- Mars, when conjoined with Jupiter and Venus in Taurus, becomes a Yogakaraka
- Sun will become a Yogakaraka when aspect by Jupiter; t he same can't be said for Venus when she is aspect by Jupiter
- The native will possess self-earned wealth if Mars and Venus are in Libra
- There's a special yoga produced when Mars, Sun and Jupiter are in Sagittarius, with Venus
- This is the only lagna where it's great if the 2nd lord is in the 12th; this isn't the case with any other Lagnas
- Venus confers a Rajayoga if he's in lagna with the Sun and un-aspect by Jupiter
- Venus is a Maraka
- When Sun is in lagna and Moon in Cancer, the native enjoys a Rajayoga

Taurus (Vrishabha) Lagna:

He/She is a pleasant talker, scholar and loves all. He/She is tall, luxurious, clean-hearted, strong built, good personality, whitish complexion, smiling face, clever and attractive personality. He/She improves in life and earns money and believes in donation. He/She is prone to Litigation or imprisonment because of personal or property disputes. He/She is financially well off, earning much more. He/She is educated and enjoys a happy married life with educated and glorious progeny. He/She has differences with near relatives and is thus socially unpopular. He/She is prone to accidents. He/She has a tendency toward being heavy by both bone structure and the self-indulgence. Lord Krishna, Mata Amritanand Mayi and also Shri Basaveshwara were born in Taurus, but in Rohini Nakshatra. The second Drekana of Taurus gives the skill of fine arts, music and dance to the native.

Jupiter, Venus and Moon are malefic. Saturn and Sun are auspicious. Saturn will cause Raja Yoga. Mercury is somewhat inauspicious. The Jupiter group (Jupiter, Moon and Venus) and Mars will inflict death.

Sun: Surya is L-4 measured from Vrishabha lagna, Vrishabha Chandra, Vrishabha Navamsha and therefore is a benefic planet due to Kendradhipati dosha for the Taurus ascendant. If Sun is strong in the Kundali, there will be great pride to him in childhood in home, family care, shelters, vehicles, schools, access to social and emotional security, including education, vehicles and higher degree, license, exam pass, school-teaching and parenting.

Moon: Chandra is L-3 measured from Vrishabha lagna, Vrishabha Chandra, Vrishabha Navamsha and is a malefic planet for the Taurus ascendant. As L-3, lord of 8th from 8th, Chandra is a bit decisive and has dicey influence and extreme fluctuations caused by tsunami-quality churning of the wheel of death and rebirth. Sibling relationships are strengthened but mother conditions are prone to sudden upheaval, and extreme emotional upset. If Chandra is associated with Rahu or Ketu, the mother's life is certainly subject to considerable psycho-mental volatility. Mother will have her own business.

Mars: Kuja is L-7+ L-12 measured from Vrishabha lagna, Vrishabha Chandra, Vrishabha Navamsha and is a benefic planet due to Kendradhipati dosha for the Taurus ascendant. Presuming Vital and competitive Kuja is strong in the horoscope, it controls good alliances and agreements. The spouse is naturally competitive, active, but aggressive. Spouse will work hard and play sports hard. Sleep is less peaceful and he takes extreme measures to ensure his sleep due to mental restlessness and bed pleasures are sexually (Kuja) activated. Kuja dosha is less harmful for Vrishabha lagna because Kuja aspects to swakshetra Vrischika Rashi in 7th house, which gives the spouse more strong personality rather than harming marriage relations.

Mercury: Budha is L-2 + L-5 measured from Vrishabha lagna, Vrishabha Chandra, Vrishabha Navamsha. Mercury is lord of the 2nd, the house of wealth and the 5th, a trine, and so is a very auspicious or benefic planet for the Taurus

Ascendant. Presuming Budha is strong in the horoscope, it empowers him to talk, explain, analyse, articulate, announce, memory, money, historical knowledge, beautiful face, voice, language, speech and song, all matters of bhava-2 and children, politics, fashion, charisma, games and gambling, winning prizes, awards and trophies, fortune and luck, creative arts & literature, divine Intelligence, speculation , prestige, fame, fashion, theatre centre-stage roles, celebrations and self-expression, all matters of bhava-5. There are four main wealth houses, such as, Bhava-2 is for banked savings; Bhava-5 is wealth through gambling or speculative wins; Bhava-9 is wealth in the form of assets provided for acts of faith; and Bhava-11 is earned income wealth. Budha becomes a highly auspicious as L-2 and L-5 for Taurus Ascendant and, if strong, gives stored wealth, conversational and sexual power to Vrishabha Lagna native. The sources of the wealth will be as per the Bhava and Rashi of Budha position. Example: USA Pres-29, Warren G. Harding of Vrishabha lagna had publications and announcements of Mithuna, Bhava-2. Budha is L-2 for Vrisha lagna and therefore Budha may function as "Maraka" in the Vimshotari Dasha period of Budha. However, the other Maraka for Vrishabha lagna is Mangala, a natural aggressor Maraka. .

Jupiter: Guru is L-8+ L-11 measured from Vrishabha lagna, Vrishabha Chandra, Vrishabha Navamsha and is most malefic planet for the Taurus ascendant. Apart from that, the lord of the Ascendant Venus is not a friend of Jupiter. But, if Jupiter is in the Ascendant, 2nd, 4th, 5th, 9th, 10th or 11th (in his own sign) and strong, the native will have financial gains, abundant income, gains of wealth, secret or privileged information; generous and liberal friends and wisdom teachings. However, eighth lord (L-8) Guru in the Karmasthana (10th) cause very bad trouble in career.

Venus: Sukra is L-6+ L-1 measured from Vrishabha lagna, Vrishabha Chandra, Vrishabha Navamsha and is an auspicious or benefic planet for the Taurus ascendant even though He is also lord of Ari Bhava (6th). If there are dignified graha in 6th, the native goes to the carrier of medicine. If L-6 Sukra occupies Ari Bhava, one becomes an agent of addiction to sweet foods, candies, ladies, alcohol, and numbing drugs. The first wife may have health problems (6th), and has well-developed social personality (1), but typically argumentative mentality. It is said that a graha which rules two Bhava of a nativity will give the strongest results for its Moolatrikona Rashi. Shula's Moolatrikona Rashi is Tula (6th) for Vrishabha. If Sukra occupies bhava-6 in Tula Rashi, there will be success in cosmetic or reconstructive medicine, import-export business and work in foreign lands, decorator and artistic designer.

Saturn: Shani is L-9 + L-10 measured from Vrishabha lagna, Vrishabha Chandra, Vrishabha Navamsha and Shani is benefic planet and Yogakaraka (Most Beneficial) for Vrishabha Lagna. One of the best characteristics of Vrishabha lagna is that three Bhava will receive the auspicious Drishti and one Bhava will have position of Shani as dharma-karma-adhipati. Three houses thus get benefit from the Drishti of the all good ninth lord and the pretty-darned-good 10th lord. Shani as L-10 brings plenty of public responsibilities, and performing as public leadership and pursue profitable personal business. Example: OSHO Rajneesh, Yogakaraka (L-9+L-10) Shani is in Bhava-8 provides leadership via hidden knowledge or secret money. UK- Queen Victoria I, Yogakaraka (L-9+L-10) Shani in 11 provides leadership via market places and social networks. USA Pres-29, USA Pres-31, Herbert Hoover, Yogakaraka (L-9+L-10) Shani in 10th provides leadership via executive roles, ordering and regulating behaviours, imposition of policy and protocol in hierarchical organizations. Career: Shani is lord of both the karma Bhava for leadership roles and the dharma Bhava for loss-of-impetus (12th from10th) in leadership roles. Therefore, (L-10) Shani gives a series of gains and losses in positions. It is rare for Vrishabha native to obtain one leadership role and hold it for extended periods. Rather, he gets financially comfortable and respectable position holding basic decency and remarkably steady level of luxury so long Sukra is well-disposed. There is likely not too much ambition to "get to the top or highest post" but rather a pleasant, competent, lawful routine regularity of professional position that allows respect without too much pushing for recognition. The most productive period for public reputation and social recognition occurs during Shani Mahadasha; but each Shani Bhukti is also a step toward his longer-term credibility in management and executive roles. If the Shani Mahadasha occurs during middle age then his career becomes quite dignified. There is no need to fight for competition or push against nature, but slowly and steadily, one rises into visible positions of social responsibility that are much appreciated and generally well-paid. Second husband: In a female nativity, Shani is Significator of the second husband, who will be well-known or possessed of a high public dignity and responsibility in one's community (10) and may be involved in being a father, moral guide, and possibly an ethnic priest.

Rahu: Rahu is co-lord of Kumbha (bhava-10) measured from Vrishabha lagna, Vrishabha Chandra, Vrishabha Navamsha. Rahu's effects for Vrishabha lagna depend significantly on Rahu's Bhava, Rashi, and incoming Drishti. Rahu amplifies the effect of any graha who share Rahu's house and also the lord of His occupied Rashi. Rahu and Ketu are said to give positive results in the 3rd, 6th, and 11th Bhava counted from the lagna or counted from Chandra. Since R-K is exalted in Vrishabha-Vrischika, R-K will give good results when Mangala and Sukra are well-disposed. If Shani is strong, then Rahu may be empowered to aid matters of professional recognition and acquisition and development of respected social positions. If Rahu is in bhava-2 in Mithuna, Rahu Periods are notable for acquisitive uprising like a rising of cobra from the snake-charmers basket.

Ketu: Ketu is co-lord of Vrischika (bhava-7) measured from Vrishabha lagna, Vrishabha Chandra, Vrishabha Navamsha. Ketu's effects for Vrishabha lagna depend significantly on Ketu's Bhava, Rashi, and Drishti. Ketu amplifies the effect of any graha who share Ketu's house and also the effects of the lord of His occupied Rashi. Ketu is co-lord of Vrischika (bhava-7) measured from Vrishabha lagna, Vrishabha Chandra, Vrishabha Navamsha. Ketu's effects for Vrishabha lagna depend significantly on Ketu's Bhava, Rashi, and Drishti. Ketu amplifies the effect of any graha who share

Ketu's house and also the effects of the lord of His occupied Rashi.

- A Jupiter-Mercury association causes Dhana yoga
- If Mars and Venus are in lagna, and Jupiter is in Capricorn, both Mercury's and Jupiter's Dasa will be fortunate
- If Mercury and Venus are in lagna and Jupiter is in Scorpio, Mercury Dasa will be fortunate
- If Moon and Venus are in Libra, with Jupiter and Mercury in Pisces, Jupiter Dasa will cause a dhana yoga
- If Saturn, Mercury and Mars are in Capricorn, and Rahu in Pisces, the native enjoy dips in holy rivers like the Ganga during the Dasa of Mars and Rahu
- If there's a Mars-Jupiter combination in Capricorn or Rahu is in Aquarius, the native enjoy dips in holy rivers like the Ganga
- Jupiter's Dasa will give mixed results whereas Mars' Dasa gives wealth. If Mercury is in a quadrant, the native is blessed with happiness during its Dasa
- Lots of wealth are indicated in the Venus Dasa
- Mars, when in 7th, is a benefic. If Sun and Jupiter are conjoined in Pisces, long life is indicated
- Moon in lagna isn't very auspicious, and will afflict other dhana yoga present in the chart
- Moon is capable of producing yoga when posited in Leo and aspect by either Jupiter or Mercury
- Neither is Saturn a Yogakaraka, nor do Sun and Mercury confer any fortunes if they're in lagna
- The above dhana yoga is destroyed if either planet is aspect by Mars
- The native becomes very fortunate if Moon is in the lagna of any sign other than Taurus
- There will be debts during Mercury's Dasa if there's an association between Mars, Jupiter and Mercury

Gemini (Mithuna) Lagna:

He/She is proud, loves his friends, charitable, and wealthy. He is annihilator of his enemies and progresses slowly in life. He/She will be medium height, whitish complexion and a faithful friend. He/She is sweet-voiced, jolly and humorous. He/She is well-wisher for everyone and gets cooperation from parents. He/She seldom seeks help and doesn't work under any one. He/She has his own successful business set-up - big or small. He/She is considerate to subordinate and weaker people. He/She has long life, lean body. He/She improves in life; earns money and believes in hard working. He/She, even, born in medium income group, earns money of his/he own. He/She does not get help from his/her spouse. 3rd Drekana of Gemini blesses him/her with fine art, music and dance.

Mars, Jupiter and Sun are malefic, while Venus is the only auspicious Planet. The Conjunct of Jupiter with Saturn is similar to that for Aries Lagna. Moon is the prime killer, but it is dependent on her association.

Sun: Surya is L - 3 measured from Mithuna lagna, Mithuna Chandra, Mithuna Navamsha. Surya is malefic for Mithuna Lagna unless Surya is in his own sign or exalted in the horoscope. Surya is also an inimical of Mercury (the lord of the Ascendant). However, Surya is natural tanupati (Karaka of Ascendant). Surya in bhava-3 (Simha), in own sign, makes a magnificent writer or a local politician and places him in centre stage in business administration, writing skills, publishing, announcements; planning, scheduling & reporting; meetings and teamwork and favours communications with respect to sales, marketing, advertising, media production (films, internet visual arts, music,), short-journey services (holiday and business travel agent), seminars and conferences, editing, technical and skills working with the hands like painting, scribe or transcription, tool-using, type-setting, inks and literary press. However, Surya in the dushthamsha supports ego-identification. If Surya is not placed in a dushthamsha 6-8-12 from lagna, Mithuna is capable of making a central figure on the team work like movie script-writing Surya in 4 (Kanya), indicates a remarkable writer and a teacher. Surya in 10 (Meena), makes a capable writer and a social activist leader like the scientist, Einstein's leadership skills, that he was asked to accept the role of prime minister of Israel but he declined. Surya in bhava-6 makes one a big dramatist, actor or model, but provides break in contract from the industry due to mental disorder. Surya in bhava-12 makes one a dramatist, model, communicator but provides a severe dissolution of engagement from media industry productions due to mental breakdowns. Surya in 8-Makara, makes a splendid writer and publisher, but provides sudden upheaval or series of catastrophic mental upheavals leading to suicide due to suspected bi-polar disorder or similar like Mr. Monroe.

Moon: Chandra is L - 2 measured from Mithuna lagna, Mithuna Chandra, Mithuna Navamsha and so Chandra is neutral and Maraka for Mithuna Lagna unless he is in his own sign in the horoscope. Chandra (L-2) is an inimical of Mercury (the lord of the Ascendant) and Shukra (natural dhana-pati) and is in "double trouble". Therefore, Moon is highly fluctuating and becomes a conservationist in His role as lord of the house of the Treasury. If Moon is in bhava-6 (Vrischika), one has to leave his country due to some force measures and can become ill. If Chandra is yuti with Ketu, the native's family may be perfectly stable and organized, but he will not connect emotionally with others and may lack empathy and provides a feeling to suicide action. However, the Mithuna native with a strong and healthy Moon may enjoy a vibrant and deeply supportive relationship to the family history along with luxury and (Laxami) wealth. If Chandra is well disposed the native will eat and drink delicious nourishment. If Chandra is poorly disposed, matters of the mouth including tongue, teeth, and voice may become problematic. Chandra is a Maraka for the Mithuna lagna; Guru-Chandra and Chandra-Guru periods are both candidate triggers for separation from the fleshly body, i. e. death. According to BPHS chapter- 43, the period of Chandra-Guru can be fatal.

Mars: Kuja is L-6+ L-11 measured from Mithuna lagna, Mithuna Chandra, Mithuna Navamsha and so Kuja is most malefic for Mithuna Lagna unless he is in his own sign in the horoscope. Kuja is also an inimical of Mercury (the lord of the Ascendant and 6th house) and Shani (natural labhpati-11th house) and so is in "triple trouble". As lord of 6th and 11th (6th-from-6th), Kuja is a highly adversarial (malefic) graha for Mithuna lagna. Kuja is by far the most difficult graha (malefic) for Mithuna natives. The placement of Mangala will determine the effects like debt, disease, divorce, gains of marketplace, income and friendship networks. Periods of Mangala are likely to encourage marital conflict, personal illness or injury, even though Kuja might be well-placed in horoscope. The best profession for Mithuna lagna native is an advocate or an attorney, and if Mangala is strong in Kendra, he will be very successful in handling even the most aggressive accusations against one's legal clients. L-6 Kuja in bhava-1, bhava-4, or bhava-12 provides very challenges for marriage due to the L-6 (a natural malefic) and due to Kuja Dosha by Drishti upon Yuvati Bhava. Example: Bill Gates is the most acutely sufferer of his health and harmony in life due to Kuja in bhava-4. Female, in particular, tends to female-illnesses such as breast and reproductive cancers due to Mars in 4th.

Mercury: Budha is L-1+ L- 4 measured from Mithuna lagna, Mithuna Chandra, Mithuna Navamsha. Despite His Kendradhipati dosha, Lagnesh Budha is a highly auspicious benefic graha for Mithuna lagna. If Budha is well placed, the native is well indoctrinated and he easily completes examinations, obtains licenses, and earns diplomas, good capabilities in business, in writing, as well as skill in acquiring property and vehicles, good early childhood home, has good ability to talk, instruct, explain, analyse, articulate, announce and has luxurious life, personality, physical body appearance, competitions, innovation, muscular movement, vitality and birth lands and properties (esp. marine coastal), buildings, vehicles, property management, stewardship, schooling, diplomas, examinations, and licensing. Lagnesh Budha in bhava-10 provides a public figure. Lagnesh Budha in bhava-4, the Exaltation state, provides property and vehicle ownership, the securing of shelters, and matters of global education.

Jupiter: Guru is L-7+ L-10 measured from Mithuna lagna, Mithuna Chandra, Mithuna Navamsha and so Guru is malefic for Mithuna Lagna due to Kendra-adhipati dosha unless he is in his own sign or exalted in the horoscope. If Jupiter is in the Ascendant, 2nd (sign of exaltation), 4th, 5th. 7th (own sign giving rise to Hamsa yoga), 9th, 10th (his own Sign giving rise to Hamsa yoga) or 11th, he is considered benefic and will give good results. Brihaspati is the "bad boy" badhesha for Mithuna lagna and gives hindrance, harm, harassment. L-7 Brihaspati can generate troublesome experience in relationships. Even if Guru is well-disposed, it can provide fall from leadership power and loss of collegial relationships. Marriage and children are also affected by the badhesha function of Guru and falls below expectation. The marriage and children have not realized their growth potential.

Venus: Sukra is L-5+ L-12 measured from Mithuna lagna, Mithuna Chandra, Mithuna Navamsha. Being the lord of the 5th, a trine, Venus is an auspicious or benefic planet for this Ascendant. Moreover, Venus is a friend of Mercury, the lord of the Ascendant. Presuming Sukra strong indicates possessiveness, pleasures, satisfaction, treasures-and-pleasures, luxuries (Laxami), possession of children and political power (5) and all types of sweets. Sukra rules bhava-5 (romance) and Guru rules the Yuvati Bhava, suggesting multiplicity of alliances. Sukra being L-5 and L-12 indicates getting indulge in clandestine, private sensual and extramarital romance. Shula's agency promotes reproduction and pleasures from theatre-politics-fashion, as well as pleasures of the more private variety association with the bedroom and long-term foreign residence. If Sukra is well placed there is much happiness from children from the first wife living in a foreign land.

Saturn: Shani is L-8 and L-9 measured from Mithuna lagna, Mithuna Chandra, Mithuna Navamsha. Saturn is lord of the 9th, a trine, and Mercury, the lord of this Ascendant, is a friend of Saturn too. So, Saturn is a very auspicious or benefic planet for Mithuna lagna. Presuming Shani strong, it gives wisdom after age of 60; the L-8 Shani begins to give remarkably valuable wisdom about the cycle of deaths and rebirths.

Rahu: Rahu is co-lord of Kumbha, L-9, Bhagya Bhava, measured from Mithuna lagna, Mithuna Chandra, Mithuna Navamsha. Rahu becomes a beneficial and Bhagya amplifier for Mithuna lagna. Rahu is said to give positive results in the 3rd, 6th, and 11th Bhava from the lagna or from Chandra.

R-K are exalted in Vrishabha-Vrischika as per BPHS, R-K will give good results when Mangala and Sukra are well-disposed. However, Rahu being in any Rashi of Maraka, Guru is dangerous for Mithuna lagna.

Ketu: Ketu is L- 6 + L – 8 measured from Mithuna lagna, Mithuna Chandra, Mithuna Navamsha. Ketu is temporal co-lord of Ari bhava-6 (11th from 8th), Vrischika. Ketu is the natural co-regulator of bhava-8 (Makar). Ketu effects depend significantly on Ketu Bhava, Rashi, and Drishti. Ketu does not connect the effect of any graha sharing. Ketu scatters the effects of His planetary lord. Even if Ketu is auspicious, the nature of litigation and lawsuits (6) shall certainly get "scattered".

- The placement of Sun and Mercury in Leo proves fortunate in Mercury's Dasa
- A Venus-Moon-Mars conjunction in Cancer results in wealth during the Venus Dasa
- There will be mixed results during Saturn Dasa if Mars is in Cancer, with Moon and Saturn in Capricorn
- With the above planetary placement, Mars confers wealth during his Dasa
- I f Mars and Saturn are in Cancer and Moon in Capricorn, the native's prosperity and wealth get mostly destroyed during the Dasa of Mars and Saturn
- Despite being the lord of 2nd, Moon isn't a Maraka
- When Saturn is in Aquarius and there's a Mars-Moon conjunction in Aries, a powerful dhana yoga results
- I f Saturn and Jupiter are in Aquarius, the native enjoy dips in holy rivers like the Ganga during the Dasa of Jupiter and Saturn
- When Mercury is in Aries, misunderstandings with elder brother are indicated

Cancer (Karaka) Lagna:

He/She is religious, handsome, long, has good personality, whitish complexion, and donating. He/She is good-looking, rich and famous. He/She respects his elders and teachers. He/She is prone to head injury during childhood. He/She excels his business away from birth place. He/She is prone to be cheated by partners and close relatives. He/She gains from business abroad in white coloured items. He/She is intelligent and heads an organisation or society. People flock to him/her for advice. He/She may undergo political imprisonment. He/She, even, born in medium income group, earns money of his own. He/She does not enjoy his life due to hardship. His family life is not happy and always difference of opinions between husband and wife. He/She is very protective of those who are close to him/her. He/She is affectionate, emotional, home loving and lovely in their approach.

Venus and Mercury are malefic; Mars, Jupiter and Moon are auspicious. Mars is capable of conferring a full-fledged Yoga and giving auspicious effects. Saturn and Sun are killers and give effects, according to their associations.

Sun: Surya is L-2 (the house of wealth) for Karka lagna measured from Cancer (Karaka) Ascendant, Karaka Chandra, Karaka Navamsha and Sun is also a friend of Moon, the lord of Cancer Ascendant. Sun is a neutral planet but inclined to give benefic results to native. The second house is also a Maraka house (house of death), so Sun is strong Markesh or Maraka planet for cancer Lagna. He is more beneficial, if he occupies his own house or is exalted in the horoscope. Presuming Sun strong, Karka native will have public roles and professional leadership responsibilities related to speech, song, language, story-telling, memory and historic conservation.

Moon: Chandra is L-1 for Karka lagna measured from Cancer (Karaka) Ascendant, Karaka Chandra, and Karaka Navamsha. Lagnesh Moon is very beneficial or benefic planet for the Cancer Ascendant. Lagnesh Chandra provides emotional sensitivity and deepens the natural connection to the parents. Native is comfortable with boating, fishing, and life near the sea shore. If Chandra is strong or exalted, in own sign, in Parivartana with his house lord, or in Kendra, the Karka lagna native will enjoy worldly success. Guru, Sukra, and other benefic will contribute and amplify the Moon powers to enhance talent and capability. If Chandra occupies Vrishabha (Taurus) in Labh Bhava, one has many friends and good business intuition. Female friends will provide many valuable opportunities. Exalted Chandra in Bhava-11 gives connections, networks, goals, accomplishments, and the marketplace suggests that this native will be well-known in his community for traits of beauty, artistic sense, and financial stability, love of fine foods and drinks, and celebrations. This person has a wonderful team spirit and will be well-liked even if other planets cause distress. If Chandra occupies Vrischika (Scorpio), His sign of debilitation in Putra Bhava, the individual may find that his creativity and self-expression is in the dark side. If Lagnesh Soma in the 8th house [2nd from 7th], the first spouse will bring the wealth for him.

Mars: Kuja L-5+ L-10 measured from Cancer (Karaka) Ascendant, Karaka Chandra, and Karaka Navamsha. Mars being lord of the 5th a trine (Trikona) and 10th house quadrant (Kendra) is a very auspicious or benefic planet, Yoga Karaka and is friend of Moon, the lord of Ascendant. Presuming Mars strong, he will be blessed with children, intelligence, good fortune, name, fame, honours and success in his professional career. Yogakaraka Kuja has the power to bring fame, winnings, leadership roles, and public dignity for him. He is more beneficial, if he occupies his own house or is exalted in the horoscope. Debilitated Kuja residing in radical lagna usually gets neechcha Bhanga and becomes very beneficial for public recognition (10) through vigorous, self-defending competition in politics and entertainment (5) and makes him a famous sea captain, industrialist, politician, entertainer, or adventurer. Mars in Bhava-2, maybe he is a highly energized public (10) who can suit the facts of the world to one's own financial (2) purposes like India-PM Indira Gandhi, USA Pres-43, G. W. Bush or financier Donald Trump.

Mercury: Budha is L-3 + L-12 measured from Cancer (Karaka) Ascendant, Karaka Chandra, and Karaka Navamsha. Mercury is lord of two inauspicious houses, the 3rd and the 12th, so he is malefic planet unless he is in his own sign in the 3rd and 12th. Moreover, Moon, the lord of this Ascendant, is an enemy of Mercury too. If Budha is strong, he tends toward shallow, short-term thinking, unless

Budha is quite well disposed or there are good occupants in Bhava-12. He is especially successful in publications business such as bookstore or media advertising. However the Karka native is not a philosopher but rather one's mentality is looking for short-term advantage (3) and opportunity for bedroom sanctuary (12) or indulgence of articulated fantasies (12).

Jupiter: Guru is L-6+ L-9 measured from Cancer (Karaka) Ascendant, Karaka Chandra, and Karaka Navamsha. Being lord of the Bhagya Sthana (9th), a Trine, Jupiter is considered to be an auspicious or benefic planet for Cancer (Karaka) Ascendant. However Guru as L-6 can function as a divorce agent. If the Jupiter is in the Ascendant, he will be exalted and in 9th, he will be in his own sign and these dispositions give rise to powerful Raja Yoga. He is more beneficial, if he occupies his own house or is exalted in the horoscope. If Guru is yuti Lagnesh Chandra, he is prone to mix sincere religion with deep corruption

Guru in Bhava-6 makes him a sincere humanist, a wonderful teacher especially of the underprivileged, but the circumstances of his school or temple will be full of conflict, illness, and debt. Vimshotari periods of will be an end of agreement phase in relationships and degradation of the professional status due to spiritual priorities. During a Guru Bhukti, he will be nourished by attending religious services, practicing in your own home, travelling to holy places, enjoying the company of priests and wise persons, and also taking continuing education classes in your profession. Guru period will be excellent for global travel and he finds himself working less and travelling more. It is a splendid time for travelling abroad with those like-minded.

Venus: Sukra is L-4+ L-11 measured from Cancer (Karaka) Ascendant, Karaka Chandra, and Karaka Navamsha. Venus is lord of the 4th (Kendradhipati Dosha) and 11th houses (inauspicious house), and accordingly Venus is not an auspicious but a malefic planet for Cancer Ascendant. Moreover, Moon, the lord of this Ascendant, is an enemy of Venus. Still we feel that the 4th and the 11th are auspicious houses and so Sukra is by and large a beneficial graha for Karka lagna, if he occupies his own house in 4th and 11th or is exalted in the horoscope. Money and Finance: If Sukra (karaka for wealth as L-11) is well placed and strong, it provides a special opportunity to him for financial gain of wealth via market place gains and L-4 for the wealth via owned properties, shelters, vehicles in Vimshotari periods of Sukra. He will get well-educated or land-holding (4) female/business partner having socially and aesthetically well-connected. His partner will be a good housewife having strong interest in the children's education and home-loving and the couple may enjoy an active social network. Sukra brings him wealth and pleasure through large social networks and gainful market relationships. Seva via large organizations and providing financial support and alliances can empower wealth to him.

Saturn: Shani is L-7+ L-8 measured from Cancer (Karaka) Ascendant, Karaka Chandra, and Karaka Navamsha. The 7th is a death inflicting (Maraka) house. Saturn is very malefic planet for Cancer (Karaka) Ascendant, still can give good results, if he occupies his own house or is exalted in the horoscope. Shani provides burdensome or difficult marriage-unions, such as, spouse tending to be older, or hailing from a lower class than him. The partner's character may be rigid and inflexible, with little room for growth or change in the marriage behaviours. Shani in Bhava-6 or Bhava-9 encourages religious sectarianism, great sense of duty in providing service to the sick and poor. Uchchamsha Shani or the highly disciplined and non-sexual Shani-7, if gains Drik-bala often transform the marriage into a celibate spiritual practice.

Rahu: Rahu is co-lord of the dushthamsha Bhava-8 (Kumbha) measured from Cancer (Karaka) Ascendant, Karaka Chandra, and Karaka Navamsha, therefore Rahu is quite inauspicious or malefic planet unless he occupies his own house or is exalted in the horoscope. Rahu and Ketu are said to give positive results in the 3rd, 6th, and 11th Bhava from the lagna or from Chandra. In addition, some authorities posit that since R-K are exalted in Vrishabha-Vrischika (per BPHS), R-K will give good results when Mangala and Sukra are well-disposed.

Ketu: Ketu is co-lord of Bhava-5 (Vrischika) measured from Cancer (Karaka) Ascendant, Karaka Chandra, and Karaka Navamsha, therefore Ketu becomes a beneficial or benefic graha for the Karka native. Ketu effects for Karkatva lagna depend significantly on Ketu Bhava, Rashi, and Drishti. Ketu amplifies the effect of any graha who are sharing Ketu's house; and also Ketu magnifies the effects of the lord of His occupied Rashi. Remedial Rant for Karaka Lagna Ketu- (Cat's Eye - Vaidurya):

- While Mercury proves fruitful, Jupiter doesn't cause any yoga
- Mars is a Yogakaraka, and even more so if posited in either of his own houses
- Venus can also be a benefic, but only if he's posited either in Leo or Gemini
- If Mars, Jupiter and Moon are in Leo, with Sun and Venus in Scorpio, the native will become wealthy and fortunate
- If Mercury and Venus are in Scorpio, Mercury confers fortune during its Dasa
- If Mercury, Venus and Moon are in Gemini, Jupiter in Cancer and Sun in Aries, the result is a Maha Rajayoga and will definitely make the native a king
- If Sun and Mars are in Aries, the native will always be wealthy. Death results during the Jupiter Dasa
- If Mercury and Venus are in Gemini, Venus Dasa will prove fortunate
- A Jupiter-Moon conjunction in lagna will make the native famous and fortunate
- When Moon is in lagna, Rajayogas are produced when either Mars is in Capricorn, Saturn is in Libra, or Sun is in Aries
- When there's a combination of Sun and Mercury in lagna, Venus in Libra, and Moon-Mars-Jupiter in Taurus, the native becomes bankrupt during the Sun Dasa, while other Dasa will be good
- If Mercury and Jupiter are in Taurus, and Saturn and Rahu are in Scorpio, the native enjoy dips in holy rivers like the Ganga during the Rahu Dasa

Leo (Simha) Lagna:

He/She is annihilator of his enemies, has few children, is tall, strong built, good personality, whitish complexion, and smiling face, clever and has attractive personality. He/She is efficient, undertakes tough tasks, and is hard-hearted and always successful. He/She is self-dependent and doesn't trust others. He/She spends as quickly as he earns. He/She crushes his/her enemy, is religious and donating. He/She doesn't forget or forgive his enemy and takes revenge. He/She has differences with father. His wife is long-lived but he/she keeps quarrelling with his/her. He/She is a devoted friend who will remember and repay a kindness. He/She has royal tastes and a sense of luxury. He/She is angry but vents the anger quickly. He/She goes to any authority to prove that he/she is right.

Mercury, Venus and Saturn are malefic. Auspicious effects will be given by Mars, Jupiter and Sun. Jupiter's Conjunct with Venus (though, respectively, Kona and Kendra Lords) will not produce auspicious results. Saturn and Moon are killers, who will give effects, according to their associations.

Sun: Surya is L-1 measured from Cancer (Karaka) Ascendant, Karaka Chandra, Karaka Navamsha and is an enormously positive graha with the potential to bring great power and fame in politics and theatre, great pride in one's physical appearance and the products of one's unique social personality. If Sun is strong, it is very good for Politics and makes leaders like, Emperor Napoleon Bonaparte (Sun in Simha/1); Richard Nixon, (Sun in Dhanusha/5); G. H.W. Bush 41 (Sun in Vrishabha/10); JFK, Jr. had he lived (Sun in Vrischika/4). Sun favours literary and dramatic arts, theatre. Surya is karaka for career visibility and public regard, and as L-1 for Simha lagna, it provides public roles and professional leadership responsibilities. Career for Simha lagna tends toward spectacular rise and fall. The Simha politician who starts high office on a "landslide" election win may end one's term of office only a few years later, as an object of public contempt.

Moon: Chandra is L-12 measured from Cancer (Karaka) Ascendant, Karaka Chandra, and Karaka Navamsha. If

Chandra is well, the native has well-being. If Chandra is disadvantaged, the native will need to compensate and cope in order to gain the essential levels of security and protection needed in this lifetime. L-12 Chandra provides emotional need for residence in distant lands near the sea-coast of distant shores on foreign beaches and on long ocean-going journeys. The marriage may be problematic one due to Shani as lord of 7). Simha typically seeks an elder spouse or one senior in business rank and one profitably connected in the marketplace. If Moon (L-12) is in yuvati bhava (Kumbha-Chandra for the Simha lagna),

He is irrepressibly lustful. He receives money from his spouse and gains renown in distant lands. Associations and partners are social outcasts or foreigners involved in shady deals."

Mars: Kuja is L-4+ L-9 measured from Cancer (Karaka) Ascendant, Karaka Chandra, and Karaka Navamsha. Kuja is L-4+L-9 and so is Yoga Karaka for Simha lagna. Kuja periods will bring high educational accomplishment. For the Simha lagna with Kumbha Chandra, Sukra and Mangala create a special pair which controls 3-4 and 9-10. When both Sukra and Mangala are strong, the native is much empowered according to their location. Mangala is Yogakaraka as L-4+L-9 from Simha lagna and L-3+L-10 from Chandra. Sukra is Yogakaraka. Behari says that Mars in Leo is eminent in its own fiery element and destroys everything unwanted in life and actively seeks spiritual experiences. He has to sacrifice material pleasure for spiritual realization, either by denial of marital happiness or absence of social associations or loss of family inheritance or abandonment from one's siblings. He stands completely on his own. Accidents, surgery, problems with the blood or diseases connected with the urinary tract are the worldly trials which impel you toward a higher understanding.

Mercury: Budha is L-2+ L-11 measured from Cancer (Karaka) Ascendant, Karaka Chandra, and Karaka Navamsha. If Budha is strong and well placed, native wants to talk, instruct, explain, and announce and has hoardings, accumulation, acquisition, money, collections of records, libraries, historical knowledge, beauty, face, voice, speech and singing ability. He gains income, profit by marketplace associations, networks, financial systems, connections, community and large assemblies. Budha becomes a highly auspicious as L-2 / L-11 and becomes a wealth graha for Simha lagna. The sources of the wealth may be found in the Bhava and Rashi of Budha, such as, Paramahamsa Yogananda material wealth arose primarily from sales of his extraordinary books due to Budha + Lagnesh Surya in 5, creative literature and in Dhanusha for humanistic religion; George H. W. Bush -41 has material wealth derived mainly from his government positions and his ownership of natural resources (oil + gas, also cattle + land) due to Budha + Lagnesh Surya in 10 for government positions; and in Vrishabha for natural resources. For Simha lagna, presuming that Budha is fairly well disposed, periods of L-2+L-11 Budha generates considerable wealth via family-of-origin hoarded assets and knowledge-values, as well as earned income from the marketplace of goods and services. Buddha's materialistic behaviour may increase wealth, solicit death, strengthen a second marriage, increase aptitude for languages, improve relations with the family history, intensify the articulator powers of speech and song, and improve the capacity to store and acquire virtually any type of items of value, from food to furniture to art collections to currency money. If well-disposed+ L-2 + L-11 calculating Budha = an excellent graha for the financial professional. Bhukti periods of Maraka L-2 (Budha) can be fatal. Example: Simha-lagna USA Pres-37, Richard Nixon died of a debilitating stroke age 81, Ketu-Budha period. Simha-lagna JFK, Jr. (son of the USA Pres-35, JFK) died of airplane crash into the ocean age 39, Shani-Budha period. Simha-lagna India-PM 1984-89 Rajiv Gandhi died of injuries from a bomb explosion age 46, Rahu-Budha period.

Jupiter: Guru is L-5 + L-8 measured from Cancer (Karaka) Ascendant, Karaka Chandra, and Karaka Navamsha. Jupiter is lord of the 5th and 8th houses. Being lord of 5th house, a trine, Jupiter is considered as an auspicious planet for this Ascendant.

Venus: Sukra is L-3 + L-10 measured from Cancer (Karaka) Ascendant, Karaka Chandra, and Karaka Navamsha. Sukra is mutual enemy of Lagnesh Surya. Sukra is L-3 (self-owned-commercial business) + L-10 (governance and public-interest leadership) and provides natural capability to lead and govern, either by self-earned wealth (3) or by manual craft such as writing, drawing, planning behaviours. The native switches gracefully between the public sector and private business, earning wealth from both as one capitalizes on commercial skills when in governing roles and upon government contacts when in private business. Shani regulates 7 being 10th-from-10th and presuming Shani is reasonably well disposed, Simha are often capable bureaucrats with a pragmatic Shani instinct for step-wise careful climbing ups the ranks of the target hierarchy. He may be engaged in scientific systems consulting, computerized governance. Politicians = excellent for socially diverse partnerships + wife's help with large-scale community fundraising wives. Sukra is strength for the Simha lagna and enhances success in writing, publications, announcements; commercial verbiage of sales, marketing, advertising; relationships with siblings and team-mates; and ensemble work in any venue. If exalted Chandra or exalted Rahu occupy Karma Bhava, the career flourish in matters of women, fashion, luxury, and the social-emotional experience of partnering in human relationships. For the Simha lagna with Kumbha Chandra, Sukra and Mangala create a special pair which controls 3-4 and 9-10. When both Sukra and Mangala are strong, the native is much empowered according to their location. Mangala = Yogakaraka L-4+L-9 from Simha lagna + L-3+L-10 from Chandra. Sukra = Yogakaraka L-4+L-9 from Chandra + L-3+L-10 from Simha lagna.

Saturn: Shani L-6 + L-7 measured from Cancer (Karaka) Ascendant, Karaka Chandra, and Karaka Navamsha. Periods of Shani provides ill health and faltering of the will. Shani makes trouble for Simha native's marriage. L-7 Shani and Surya, lord of Simha, are bitter enemies. L-6 Shani is an especially difficult graha for Simha native as it is L-6 from both radix lagna (material enemies) and Chandra lagna (emotional resistance to agreement). Death: Physical separation of the soul from the flesh is triggered mainly by the Vimshotari period of the "Maraka" lords of bhava-2 or bhava-7, which for Simha lagna are Budha and Shani. Other Maraka are Mangala and Rahu-Ketu giving the effect of a Maraka graha ruler. The Shani Drishti to Ayur-bhava (8) or even strong Shani in 8 will deliver a fairly long lifespan, otherwise the periods of Shani or Budha will be the death timing for Leo Lagna native. Examples: India-PM 1984-89 Rajiv Gandhi left his body in Rahu/Budha period. JFK, Jr. left his body in Shani/Budha period.

Rahu: Rahu is co-lord of bhava-7, Kumbha measured from Cancer (Karaka) Ascendant, Karaka Chandra, and Karaka Navamsha. Yugasthana is not a trine angle from Simha lagna. Rahu's effects for Kumbha lagna depend significantly on Rahu's Bhava, Rashi, and incoming Drishti. Rahu amplifies the effect of any graha who are sharing Rahu's house; and also Rahu magnifies the effects of the lord of His occupied Rashi. Rahu and Ketu are said to give positive results in the 3rd, 6th, and 11th Bhava from the lagna or from Chandra. In addition, some authorities posit that since R-K are exalted in Vrishabha-Vrischika (per BPHS), R-K will give good results when Mangala and Sukra are well-disposed. Numerous other schemes for evaluating the elusive aprakasha graha also exist.

Ketu: Ketu is L-4, Scorpio measured from Cancer (Karaka) Ascendant, Karaka Chandra, and Karaka Navamsha. Ketu's effects depend significantly on Ketu's Bhava, Rashi, and Drishti. Chid karaka Ketu does not connect the effect of any graha sharing Ketu's house. Ketu scatters the effects of His planetary lord. Ketu = temporal co-lord of bhava-4 = 9th-from-8th = Vrischika. Ketu is the natural co-regulator of bhava-8.

- Sun-Mercury-Mars or Sun-Jupiter -Mercury conjunctions indicate wealth, whereas only a Sun-Mercury combination indicates moderate fortunes
- A Jupiter-Venus combine causes destruction of other Yogas
- Venus can't produce yoga. However, he becomes a benefic when in Libra but a malefic when in Taurus
- If there's a Sun-Mercury-Mars conjunction in lagna, the Mercury Dasa brings in much wealth

- The Saturn Dasa will be fortunate if there's a Saturn-Mars combination in Cancer

Virgo (Kanya) Lagna:

He/She is endowed with beauty and has a good fortune, medium height, broad chest, whitish complexion, smiling face, clever, very fast in doing the job and is selfish and harm too much out of his selfishness. His/Her young age is very happy. He/She is a successful in politics because he/she has something inside and speaks something else in public. Nobody can measure his political capability. He/She is never crude or coarse. He/She prefers the role of researcher, observer, critic, or teacher. He/She is fault finding type and hypercritical. He/She feels proud in finding the fault and drawback in others.

Mars, Jupiter and Moon are malefic, while Mercury and Venus are auspicious. Venus Conjunct with Mercury will produce Yoga. Venus is a killer as well. Sun's role will depend on his association.

Sun: Surya is L-12 measured from Virgo (Kanya) Ascendant, Kanya Chandra, and Kanya Navamsha and so is malefic planet for this ascendant unless Sun is in his own sign in the 12th house or exalted in the horoscope. Because Ravi = L-12, the native may have a less dramatic personality or leadership responsibilities.

Moon: Chandra L-11 measured from Virgo (Kanya) Ascendant, Kanya Chandra, and Kanya Navamsha. Moon the lord of the 11th and so he is malefic Planet unless Moon is in his own sign in the 11th house or exalted in the horoscope.

Presuming Chandra is well placed and strong, he may gain from ocean and its products and elder sibling especially a sister, cousin, or a close-in-age aunt, can be most helpful as a mentor in this life.

Mars: Kuja L-3 + L-8 measured from Virgo (Kanya) Ascendant, Kanya Chandra, and Kanya Navamsha and so is a great functional malefic or highly inauspicious for the Virgo Ascendant unless Mars is in his own sign in the 3rd or 8th house or exalted in the horoscope. He has an overtly sexual, animal instinctive, L-8 (Kuja), and has a tendency to get involved in impulsive sexual relationships with girlfriend, a team-mate or office-mate or neighbour and can get into trouble that may threaten his marriage. Mangala for Kanya lagna becomes a karaka for the "office marriage type relationship or sexualized connection" and the native spends more time in office getting involved with a team-mate than with his own spouse. His communication style (L-3) is overly aggressive to the point of being self-destructive (L-8) unless Mangala or Budha are very nicely disposed. He can have typically quite sudden death involving blood disease, severe injury, or results of conflict. From 1, 2, and 5 Bhava, Mangala casts Drishti upon Randhra Bhava and gives sudden and possibly violent death from causes given by Mangala Rashi. Example: Kanya lagna, Mangala in bhava-1 (Kanya) gives intestinal disease, argument, iatrogenic causes, particularly mixed drugs, poison, spoiled food, often from a hostelry or jail. Kanya lagna, Mangala in bhava-2 (Tula) gives kidney disease, adrenaline surge, disease of external genitals, sexually transmitted disease. Kanya lagna, Mangala in bhava-5 (Makar) gives bone disease, skeletal disorder, and multiple broken bones. Kanya lagna, Mangala in bhava-7, death is swift and direct for the first and second spouses. Mangala-Mesha-8: circumstances of death involve weapons, engines that propel moving vehicles (car crash), or both like JFK, in a moving car when killed by rifle bullet.

Mercury: Budha is L-1+ L-10 measured from Virgo (Kanya) Ascendant, Kanya Chandra, Kanya Navamsha and so is a great functional benefic for the Virgo Ascendant in the horoscope. Presuming Budha (L-1) is strong and well placed, he has very good personality & physical body, vitality, contests and competitions ability, innovation and fashion, professional or social dignity: leadership roles, position and attributes of distinction, rank and reputation, promotions, government bureaucrats, policies and procedures, legislators and laws matters. Budha (L-10), is a booster for a communicative career in administrative, information-intensive, and policy specialties. As L-10, Budha for Kanya lagna is a Significator of career, public reputation, and social status. The native tends to favour professions in the verbal, media, explanatory, or communicative arts, with a special affinity for careers that involve writing. Budha as lord of 10th Navamsha for a Kanya native indicates a great facility in sales, marketing, advertising, public relations, or any of the communicative-administrative activities of commercial business. The person may be an extraordinary sales professional if Budha is well disposed. Jupiter: Guru is L-4 + L-7 measured from Virgo (Kanya) Ascendant, Kanya Chandra, and Kanya Navamsha. Counted from Mithuna lagna or Kanya lagna, Brihaspati is the "bad boy" badhesha who gives hindrance, harm, harassment. As lord of two Kendra, Guru acquires Kendradhipati dosha and so he is a great functional malefic for the Virgo Ascendant unless Jupiter is in his own sign in the 4th or 7th house or exalted in the horoscope. Presuming Guru is strong and well placed, he will have good home life having compatible spouse (unless Guru is damaged). Otherwise L-7 Brihaspati can generate troublesome and failed expectations and fewer fulfilments in relationships to both personal and professional, marriage, Children, grandchildren and marriage and children are affected by the badhesha function of Guru for Kanya Lagna. There is a general life feeling overall for most Mithuna-lagna and Kanya-lagna natives that marriage and usually children too have not realized their growth potential. The spouse's career (10th from 7th = 4) is affected by badhesha Guru and during Guru Periods it too will experience unexpected setbacks, hindrance, and harm due to spousal misplaced confidence. However, Guru periods helps in significant educational completion, such as learning a new trade skill or a professional diploma, presuming Guru is healthy and there is no severe resistance in house-4 or house-7.

Venus: Sukra is L-2 + L-9 measured from Virgo (Kanya) Ascendant, Kanya Chandra, and Kanya Navamsha. Venus, being lord of the 2nd house (house of wealth) and the 9th (Bhagya), a trine, is a very auspicious planet for the Virgo Ascendant. Sukra becomes a highly auspicious L-2 / L-9 wealth graha for Kanya lagna. Presuming Venus is strong and well placed, he will enjoy the company of women, wealth (banked savings assets-2, gambling or speculative win assets-5, or earned income assets-11), knowledge as well as world travel, religion and his wife will have financial skills and an international lifestyle. As L-9 Sukra favours artistic and musical talents for the grandchildren and accomplishments in arts and music for students of the guru. Sukra is also a Maraka graha who will effect a saturation of pleasure, when the time has come.

Saturn: Shani is L-5 + L-6 measured from Virgo (Kanya) Ascendant, Kanya Chandra, and Kanya Navamsha. Saturn is lord of the 5th (a trine) and the 6th houses for the Virgo Ascendant and so Saturn is an absolutely auspicious (benefic) planet for this Ascendant. However, Shani will bring conflict in marriage, losses from lawsuits, health issues, and financial strain from loan debt. Periods of L-5 Shani -or- Guru -or- L-9 Sukra will produce children, eventually - unless Shani occupies house-5 or house-9 in an unfavourable sign or house-5 is otherwise damaged. Shani = the L-6 can offer particular difficulties of chronic illness and exacerbation of resistance to keeping agreements in contractual relationships, during Shani periods. Rahu: Rahu is co-lord of maha-dusthamsha bhava-6 (Kumbha) measured from Virgo (Kanya) Ascendant, Kanya Chandra, and Kanya Navamsha and so Rahu is a malefic planet except Rahu is in his own sign in the 6th house (Kumbha) or exalted in the horoscope. Rahu is a malefic planet except for careers in medical clinical practice and human services ministries, e.g. social work, druggist, police, divorce law, bankruptcy, criminal attorney. Rahu effects for Kanya lagna depend significantly on Rahu Bhava, Rashi, and Drishti. Rahu effects for Kumbha lagna depend significantly on Rahu Bhava, Rashi, and incoming Drishti. Rahu amplifies the effect of any graha who are sharing Rahu house; and also Rahu magnifies the effects of the lord of His occupied Rashi. Rahu and Ketu are said to give positive results in the 3rd, 6th, and 11th Bhava from the lagna or from Chandra. In addition, some authorities posit that since R-K are exalted in Vrishabha-Vrischika (per BPHS), R-K will give good results when Mangala and Sukra are well-disposed.

Ketu: Ketu is co-lord of bhava-3 (Vrischika) dushthamsha measured from Virgo (Kanya) Ascendant, Kanya Chandra, and Kanya Navamsha and so Ketu is a malefic planet except Ketu is in his own sign in the 3rd house (Kumbha) or exalted in the horoscope. Ketu effects depend significantly on Ketu Bhava, Rashi, and Drishti. Chid karaka Ketu does not connect the effect of any graha sharing Ketu house. Ketu scatters the effects of His planetary lord. Ketu is temporal co-lord of bhava-3 (8th-from-8th – Vrischika). He will study yogic literature, engage in penance, and establish a new school of thought. Ketu will bring cataclysmic changes to his life, resulting from the distrust, humiliation, and animosity generated by colleagues and associates. After attaining great proficiency in trade, he will be forced to alter your life pattern. After having confronted opposition from several quarters, he will gain equanimity and spend time more of less like a recluse. He will seldom open his inner feelings and ideas to others, and will experience a great metamorphosis in life.

- A Sun-Moon-Venus association will bring in lots of wealth during the Sun Dasa
- The native loses wealth during the Venus Dasa. Mixed results can be expected during the Moon Dasa
- When Moon and Venus are in Pisces, with Jupiter in Cancer and Sun in Aries, during the Jupiter and Venus Dasa, the native will have multiple wives, who will all be alive. Virgo natives possess women of high rank
- A Jupiter-Venus combine in Sagittarius produces fortunes during their Dasa and Bhukti
- If Saturn is in Cancer, its Dasa will be fruitful

Libra (Tula) Lagna:

He/She is a scholar, earns his livelihood by virtuous means, wealthy and is respected by everybody. He/She is tall, fair complexion and healthy. He/She is sweet-voiced and benevolent. He/She doesn't stick to one profession and keeps spending too much on research. He/She enjoys little reputation at home but is reputed outside. He/She excels in occupations related to iron. He/She has problem with brothers/sisters. He/She is talkative, is not hard-working and depends on fate, and thus leads insecure life. He/She is unlucky for father's business and gives setback at the age of

12. Initially he/she begins with service but later settles down in own business. He/She leads happy married life with kids.

Jupiter, Sun and Mars are malefic. Auspicious are Saturn and Mercury. Moon and Mercury will cause Raja Yoga. Mars is a killer. Jupiter and other malefic will also acquire a disposition to inflict death. Venus is neutral.

Sun: Surya is L-11 as measured from Tula lagna, Tula Chandra, or partner-effects from Tula Navamsha. Ravi (L-11) is enemy of Lagnesh Sukra and also enemy of Shani the natural 11th lord and so is malefic planet and Double-trouble maker for Tula lagna unless Surya is well placed like in own Sign or is in exaltation in the Kundali. Surya selfishness or autocratic behaviour causes trouble in relationship to marketplace earnings, friendly networks, the elder sibling or father's brothers.

Moon: Chandra is L-10 measured from Tula lagna, Tula Chandra, or Tula Navamsha. Chandra L-10 is enemy of Lagnesh Sukra and enemy of Shani the natural 11th lord and is double-trouble maker. Presuming Chandra is well placed and strong, it will provides leadership capabilities and style, a natural capability to lead and govern, to make decisions which provide social order and lawfulness with special emphasis upon matters of home and homeland, folkways customs and ways one's people live upon the land, property ownership, houses and schools, farms and fisheries, parenting and protection of the weak, physical-emotional security and grounded in a fixed place and provides him with servants and employees who bring good fortune and also good relationships with social workers, followers, employees, and members of the marginalized classes including refugees, criminals, the dispossessed, chronically ill and addicted persons, and will form the basis of a creative strategy for his leadership. He helps the underclass and the underserved classes who will contribute to his positive public reputation and empowerment to make decisions affecting the downstream like Mahatma Gandhi's career similarly was based in his creative solutions to the chronic exploitation (6) and disenfranchisement of the Indian masses and Adolph Hitler who famously created from the massive discontent of the German working classes who felt victimized by the Treaty of Versailles, after WWI. Mahadasha of Chandra may signify an exceptionally strong engagement with public organizations such as church-temple, children's school, or charitable group for a person with an overall public-service oriented nativity. His mother will have a strong career in karakatwa of the Bhava of Chandra and she is role model in his developing professional identity.

Mars: Kuja is L-2 + L-7 measured from Tula lagna or Tula Chandra; partner-effects from Tula Navamsha. Mars is the lord of two death inflicting houses (Maraka houses), the 2nd and the 7th and also is an enemy of Venus, the lord this Ascendant and so is malefic planet, unless it occupies its own sign or sign of exaltation. Presuming Mars is well placed or in bhava-2 and strong, it can energize the Treasury (2) of money and knowledge. Mars in 7th can very good marriage alliances, and make him advisers such as physicians, attorneys, brokers. The negotiation skills are fuelled and sharpened like a fierce warrior's sword. Mangala periods indicate fleshly death for the mother, father, and other family members in his early age. Mangala can signify the end of a romantic relationship or a cutting of family ties during its period. L-2 Mangala facilitates a split with a family and brings him to the dentist's chair, or may signal a new food diet. As lord of both 4th-from-11th and 9th-from-11th, Mars is a highly auspicious Yogakaraka graha in regard to his ability to set and attain his goals. When Mangala occupies Kumbha, it engages him and his life-partner toward earnings (L-7 in 11th from swakshetra). If Mangala in a Rashi of Shani in 5, it suggests miscarriage or therapeutic abortion of at least one conception; however if Guru is strong there may be other live births when Mangala is not the time lord. There may be a still-birth, which will cause much sorrow. He may be cruel, ruthlessly striving to attain the objects of desire and causing terror to his adversaries.

Mercury: Budha is L-9 + L-12 measured from Tula lagna or Tula Chandra; partner-effects from Tula Navamsha. Mercury is Raja Yoga Karaka for Libra Lagna. Mercury is lord of the 9th, the house of Bhagya, (a trine), and is a friend of the ascendant, Venus and will prove very beneficial. Budha as lord of 12th from 10th drains energy away from public leadership roles so that the native can study philosophy and deliver wisdom teachings to world. Budha gives a negative influence on career, and a strong Budha is often the signal of a weak career for Thula-1. If Budha periods occur during his young age or at adult years, it gives unfavourable results career-wise. Good results begin to accrue in regard to the native's ability to communicate in positions of leadership and dominion, but only with the passage of time. Budha as lord of both 3rd from 7th and 6th from 7th is a major trouble-maker in human partnerships, and is a general miscreant in marriage. Budha may excite internal argumentation that can damage partnership negotiations and harm an otherwise good marriage.

Jupiter: Guru is L-3 (Malefic Bhava) + L-6 (Malefic Bhava) measured from Tula lagna or Tula Chandra; partner-effects from Tula Navamsha and is unfriendly to Venus and is, therefore, not considered an auspicious planet for this Ascendant and is a problem maker. This combination can however help a self-owned business in religious trainings such as conferences and seminars, and this combination is also beneficial for academic publishing. Jupiter in this position activates his mind so much that his mental balance may be disturbed and he suffers from a mental disability due to L-3+L-6 Guru which makes awkward and uncomfortable to him. He sometimes face danger, such as, few children as daughters only or partner is barren or ailing, dispute or separation with partner or 2nd marriage, serious illness to the elder brother, poverty, indebtedness, imprisonment, loss in business, disrepute, leaving birth place are some of the unfavourable results.

Venus: Sukra is L-1 + L-8 measured from Tula lagna or Tula Chandra; partner-effects from Tula Navamsha. Sukra as Lagnesh is excellent and benefic planet for Tula lagna. As Lord of Lagna, Venus gives powerful results for Vaanija (the trader) and gains of income (11) in career. He earns an adequate income (unless there are very difficult graha in bhava-8) like others' monies, such as inheritance, insurance settlement, and in particular the assets of the spouse. Education and Properties: 5th-from-4th (Vrishabha) and 10th-from-4th (Tula), Sukra provides support in matters of education and property, acquisition of properties; and ascending to a respected social position, if Sukra is a well-placed. Sukra provide the education of his children, parenting or other culture-sustaining service that expresses individual intelligence. He may experience a burst of creativity and ingenious speculative behaviours in response to the need to sustain personal security. Sukra in 11 provides marketplace and profits, a very distinctive wealth and luxury from marketplace transactions within a network of buyers and profits. Swakshetra Sukra in Tula lagna provides the social personality embedded in the iconic material appearance and a cult of personality. Sukra in 9 provides affiliations with temples and universities; He is fundamentally oriented toward attractiveness, wealth, making agreements, living in harmonious balance with other humans, sensual pleasures from art, music, and the experience of beauty.

Saturn: Shani is L-4 + L-5 measured from Tula lagna or Tula Chandra; partner-effects from Tula Navamsha and so Shani are Yogakaraka for Tula Lagna Tula natives will successfully complete schooling and earn a diploma. Good children are born. There will be legacy educational diploma, license to practice a profession, scholarly performances, well-structured speculative ventures, political empowerments earned by accepting responsible duties, and lawful production of well-raised children. Shani is Yogakaraka for Tula lagna, but this fact does not make Shani an "easy results" graha. Examples: Partner of USA Pres-35 Jacqueline Kennedy has Yogakaraka Shani in 3, which provides great achievements in media communications. Mahatma Mohandas Gandhi has Yogakaraka Shani in 2 which provide great achievements in true knowledge and true wealth. USA Sec of State Hillary Clinton has Yogakaraka Shani in 10 which provide great achievements in leadership and governance (10). Tula makes plenty of mistakes in life, perhaps even more than average for an intelligent human being. However Yogakaraka Shani compensates for deficiencies in personal intelligence and as a result the native actually learns from one's mistakes.

Rahu: Rahu is co-lord of Bhagya bhava-5 measured from Tula lagna or Tula Chandra; partner-effects from Tula Navamsha and so Rahu a benefic planet and friendly to Venus and Saturn. Rahu is Raja Yoga Karaka for Libra Lagna. Rahu in Dhana-Bhava or Yuvati Bhava is associated with the end of life or death. Example: Tula lagna Jacqueline Kennedy died of lung cancer from lifetime cigarette smoking age 64, in Rahu-Chandra period. Chandra = L-10 but both Rahu + Chandra occupy the Makar Yuvati Bhava. As co-lord (with Shani) of Kumbha/5, Rahu gives passionate and unusual children. Due to austerity-karaka Sani ruler ship of the house of children, Tula natives typically have few offspring. However those few produced are complex, adventurous, ambitious, scientific, and passionate taboo-challengers by nature. Rahu and Ketu are said to give positive results in the 3rd, 6th, and 11th Bhava from the lagna or from Chandra. In addition, some authorities posit that since R-K are exalted in Vrishabha-Vrischika. R-K will give good results when Mangala and Sukra are well-disposed.

Ketu: Ketu is co-lord of Dhana Bhava, bhava-2 (Vrischika) measured from Tula lagna or Tula Chandra; partner-effects from Tula Navamsha and so is an auspicious or benefic planet for Tula lagna. Ketu amplifies the effect of any graha sharing Ketu house; and also Ketu scatters the effects of the lord of His occupied Rashi. Ketu contributes detachment and misfit qualities to the family identity. Ketu is beneficial to provide wealth and language, acquisition, knowledge of history, and relations with the family history, if the lord of bhava-2 and its occupants are auspicious. Ketu effects depend significantly on Ketu Bhava, Rashi, and Drishti. Special case: If Ketu in Vrishabha-8, it arouses supernatural powers, and his intuitive understanding of relationships will be surprising and will acquire much fame. Remedial Ratna for Tula lagna Ketu (Cat's Eye - Vaidurya): Vaidurya could be a beneficial gem in the case that Sukra and Mangala are both well disposed. Ketu, even if auspicious, shall certainly scatter the marital and other alliances.

- Saturn is a Yogakaraka. Though lord of 3rd and 6th, Jupiter is also capable of producing Yogas

- Mars doesn't become a maraca, although it lords the 2nd and 7th houses
- If Jupiter and Venus are associated or are aspect by Saturn and Mars, during the Jupiter or Venus Dasa and Bhukti, the native suffers from skin infections and wounds
- If Sun and Mercury are in Virgo and aspect by Saturn, the father will be fortunate
- If Mars is in association with Sun, Saturn and Mercury, it produces immense good
- If Sun, Mercury and Saturn are combined with either Moon or Mars, a Rajayoga is produced
- If there's a Sun-Venus-Mercury combination in lagna, the native gets very fortunate and wealthy
- If in lagna there are Mercury, Saturn and Venus, or the Moon and Mars in Aries, the Mercury Dasa will prove fruitful
- The presence of Mars and Mercury in Leo, Jupiter in Taurus and Saturn in Gemini creates a Rajayoga
- If Moon is in lagna with Jupiter either in Virgo or Pisces, the Saturn Dasa brings in wealth and fortune
- Venus becomes a maraca if he's in the ascendant
- If the Saturn is in lagna and Moon in Cancer, a Rajayoga results
- If Saturn, Jupiter, Mercury and Mars are in Aquarius, and Rahu is in Cancer, the native enjoy dips in holy rivers like the Ganga during the Rahu Dasa

Scorpio (Vrischika) Lagna:

He/She is wealthy and a scholar. He/She is tall, lean, either very rich or very poor. He/She lives away from home since very early in life and dominates over his family members as well as outsiders. He/She doesn't forget or forgive his enemy. He/She has his own business. He/She has financial problems up to 30 years but later he/she earns money and supports others. He/She cannot sit idle and is world-famous and heads an organisation or society. He/She is prone to injury during fight or accident. He/She gets married more than once. He/She will be rich, famous, popular and smart in love matter. He/She is successful in politics. He/She earns sufficient money in the life but not much savings. This Sign is not good for domestic happiness.

Venus, Mercury and Saturn are malefic. Jupiter and Moon are auspicious. Sun, as well as Moon is Yoga Karakas. Mars is neutral. Venus and other malefic acquire the quality of causing death.

Sun: Surya is L-10 measured from Vrischika lagna or Vrischika Chandra; Vrischika Navamsha and Sun is a friend of Mars the lord of Scorpio Ascendant. Therefore, Surya is benefic planet for Vrischika Lagna. L-10 Surya is a double career karaka. Presuming Ravi is well-disposed and strong; he is capable of channelling a remarkable amount of divine intelligence and accepts public roles and professional leadership responsibilities of a large corporations and government bureaucracies. When [Surya] the lord of the 10th house occupies the Navamsha of Jupiter, he finds his living with the help of Brahmins, deities, or through state favour, recitation of Puranas, studying Sastra, preaching or giving religious instructions, or lending money.

Moon: Chandra is L-9 measured from Vrischika lagna or Vrischika Chandra; Vrischika Navamsha and Chandra is a friend of Mars the lord of Scorpio Ascendant. Therefore, Moon will prove very beneficial or benefic planet for Vrischika lagna. Presuming Moon is well-disposed and strong, L-9 Chandra provides religious ritual and priestly duties, auspicious travel, spiritual advancement through religious fellowship and a lifelong emotional affinity for universities, seminaries, and teaching in temples. Moon will promote religious and charitable inclinations and purity of mind, brings good fortune and will prove good for the longevity of father.

Mars: Kuja is L-1+L-6 measured from Vrischika lagna or Vrischika Chandra; Vrischika Navamsha. Mars is lord of the Ascendant and also of the 6th and therefore, lordship of the Ascendant will prevail over the lordship of the sixth house and so Mars is beneficial or benefic planet Vrischika Ascendant. Kuja character in Rashi and Bhava will determine the field of action for the perpetual competition and combat which propel this highly vital native into life and work. He may be a ground-breaking researcher, a brilliant healer, a genius financier, or a fierce warrior and may hunt with physical weapons, or with the mind. Conflict in his life can become intensely psychological and even overtly violent due to the maha-dushtastana being ruled by an aggressive malefic graha. Deadly significations are intensified. It may give the death of a child (2nd-from-5th) or gains from confidential insider information (11th-from-8th) Kuja. If Kuja is exalted, in own sign, in Paravatamsha with his lord, or in Kendra, he will enjoy worldly success. The stronger Kuja traits provide the strong the native's performance in his specialty. Example: if Kuja occupies Makar in Sahaja Bhava, he marks excellence in business and a propensity to create self-made wealth. He will be exalted in the house of sales, marketing, administration, meetings, teamwork, colleagues/siblings, and mental tasking. When Kuja = uchcha and in bhava-3, the native flourishes in all varieties of commerce and business administration, with a persistently competitive thrust toward domination and control. If there is also a Parivartana yoga with Kuja lord Shani rising in Scorpio (every 29 years a batch of these people are born) expect self-made wealth (3) through determined hard work, dominating aggressiveness, and sharp business judgment. By contrast, if Kuja occupies Kumbha, the sign of oppressive Shani, in Bandhu Bhava, the individual may find that the life force tends to implode into domestic disputes, conflict with educational system, and private scientific interests.

Mercury: Budha is L-8+L-11 measured from Vrischika lagna or Vrischika Chandra; Vrischika Navamsha. Budha and Mangala are enemies and therefore, Mercury is not considered an auspicious planet for this Ascendant.

Jupiter: Guru is L-2+L-5 measured from Vrischika lagna or Vrischika Chandra; Vrischika Navamsha. Guru is friend of Mars, lord of this Ascendant. Being lord of 5th house, a Trine, Jupiter is considered an auspicious or benefic planet for this Ascendant. L-2+L-5 Guru benefits him with family lineage and children and bring childbirth, publications of literary works, recognition of the value of one's intelligence, success in speculation, and wealth.

Venus: Shukra is L-7+L-12 measured from Vrischika lagna or Vrischika Chandra; Vrischika Navamsha. Venus is lord of the 7th (Kendradhipati dosha) and the 12th. Moreover, Mars, lord of this ascendant, is not a friend of Venus. Venus is malefic planet for this Ascendant. Sukra provides diplomatic agreements and meditation, prayer, a good deal of sleep, a beautiful architectural sort of imagination, and long separations due to residence in foreign lands. L-7+L-12 Sukra cause foreign travel and contractual partnerships of all kinds. Periods of Sukra are unfavourable for the affairs of one's children, as Sukra is L-3+L-8 vis-à-vis putra Bhava. Sukra is weakened by conjunction with Ketu or Rahu. Sukra + Rahu are a risk-taker who self-elevates via partnerships. Sukra+ Ketu is apathetic toward partnership and prone to unsatisfied abandonment of contractual unions. According to BPHS chapter-i 43, the period of Chandra-Sukra can be fatal.

Saturn: Shani is L-3+L-4 measured from Vrischika lagna or Vrischika Chandra; Vrischika Navamsha. Moreover, Mars, lord of this ascendant, is not a friend of Saturn. Saturn is considered as a neutral planet for this Ascendant. L-3 + L-4 Shani are good for practical education and social connections leading to increased social security. Periods of Shani give educational confirmation such as diplomas and certificates. Homes are gained and lost during Shani period. Favours all kinds of training which forms foundation for later accomplishments.

Rahu: Rahu is co-lord of bhava-4, Kumbha, therefore Rahu-Ratna (gomedha) may become a beneficial gem for matters of home-ownership, schooling in the ethnic culture, diploma completion, licensing and passing examinations, obtainment of vehicles and shelters, parents and care-taking including stewardship of the land and patriotic defence. However Rahu in either Rashi of Maraka Sukra is dangerous for Vrischika lagna. Even for the security-and-stability target, the Rahu-Ratna should be considered only if the lord of bhava-4 and its occupants are highly auspicious. Rahu and Ketu are said to give positive results in the 3rd, 6th, and 11th Bhava from the lagna or from Chandra. In addition, some authorities posit that since R-K are exalted in Vrishabha-Vrischika (per BPHS), R-K will give good results when Mangala and Sukra are well-disposed.

Ketu: Ketu is co-lord of Vrischika; therefore Ketu may be a beneficial or benefic planet. Ketu's effects for Vrischika lagna depend significantly on Ketu Bhava, Rashi, and Drishti. Ketu amplifies the effect of any graha who are sharing Ketu's house; and also Ketu magnifies the effects of the lord of His occupied Rashi.

- An association of Jupiter and Mercury brings in wealth
- If Jupiter is in Capricorn, the native will be charitable
- If Sun, Mercury and Venus are in Taurus, the Mercury Dasa confers lots of fame and power
- When Jupiter and Mercury are in Pisces with Moon in Virgo, the native will be fortunate
- If there's a Jupiter-Moon-Ketu combine in Cancer, the Ketu Dasa will be ordinary but the Jupiter Dasa excellent

Sagittarius (Dhanu) Lagna:

He/She is an expert in policy matters, religious, important person in his family, medium height, whitish complexion, smiling face, and attractive personality. He/She has strong faith in God, is vegetarian, simple living, believes the people very easily, and is a businessman. He/She does his work with well planning. He/She likes discipline, truth, justice, kindness and independence in his/her life. He/She takes everything and everyone for granted. His/Her reasoning powers are superb. He/She enjoys the good fortune of having thought patterns that remain young and fresh throughout life. He/She makes lots of promises but fail to maintain them. He/She has a great sense of fairness and adopts only fair means to handle the job.

Only Venus is inauspicious. Mars and Sun are auspicious. Sun and Mercury are capable of conferring Yoga. Saturn is a killer, Jupiter is neutral. Venus acquires killing powers.

Sun: Surya is L-9 measured from Dhanu lagna or Dhanu Chandra; Dhanu Navamsha. Sun is the lord of the 9th house, the house of Bhagya (fortune). Sun is also a friend of Jupiter, the lord of the Ascendant and so Surya is, therefore, a very suitable or benefic planet for natives of this Ascendant. His father is very philosophical, although strong-willed; and confident. He receives public recognition, and gets high professional leadership responsibilities within universities, temples (9) and world travelling religious or philosophical groups (9). Presuming a strong Surya, it produce a celebrity in the fields of religion, philosophy, moral teachings like ritual priesthood , university professor, specialties of sacred literature and wisdom teaching, temple-based education, religious missionary, preacher, moral talks or international travel pursuits. But Ravi is L-12 from 10th, so is a culprit for loss of professional dignity or "job loss" or other public indignities.

Moon: Chandra is L-8 measured from Dhanu lagna or Dhanu Chandra; Dhanu Navamsha. Moon is the lord of the 8th and so is a malefic planet. However, Moon is friendly with Lagnesh, Jupiter. Presuming psycho-emotional Chandra (L-8) is strong and posited well, he is the happiest (not always) in mysterious roles that express healing behaviours like psychotherapist, surgeon, psychic intuitive, money launderer, silent business partner, secret sexual partner, or as the holder of highly confidential information in secure positions such as executive secretary or military officer. Chandra's Radix and Navamsha Rashi and any Drishti to Chandra will show the native's specific style of pursuing security and emotional fulfilment. Example: Moon ruled by Sukra gives confidential roles in beauty and justice; ruled by Shani gives political and military roles handling government and corporate secrets; ruled by Surya a priestly role in communicating religious secrets, etc. Both he and his/her mother may suffer mysterious, medically not diagnosable illnesses which a psychic would know to have emotional causes. Chandra being L-8 provides sudden, apparently unprovoked eruptions of violence. secrets; disasters; upheavals

Mars: Kuja is L-5 / L-12 measured from Dhanu lagna or Dhanu Chandra; Dhanu Navamsha. Mars is the lord of the 5th, a trine house, and so Mars is auspicious or malefic planet for Dhanu lagna as lordship of the 5th a trine will prevail over the lordship of the 12th. If Mars is in his own sign in the 5th, it will be more benefic. Mars will blessed with children, good fortune, name and fame, performance arts, politics, brilliant acts of genius, as well as long-term foreign residence, contemplative arts, and dissolution of the physical identity.

Mercury: Budha is L-7 / L-10 measured from Dhanu lagna or Dhanu Chandra; Dhanu Navamsha. L-7 / L-10 Budha suffers from Kendradhipati dosha on account of his lordship of two Kendra and is so the Badhaka graha or malefic planet for Dhanu lagna or "harming" planet.

Jupiter: Guru is L-1 / L-4 measured from Dhanu lagna or Dhanu Chandra; Dhanu Navamsha. Jupiter is lord of the Ascendant and the 4th house and so is a benefic planet of this Ascendant. Wise and expansive Guru defines the physical body level of social personality identity, as well as the native's approach toward matters of education and property ownership and is a natural humanistic educator.

Venus: Sukra is L-6 + L-11 measured from Dhanu lagna or Dhanu Chandra; Dhanu Navamsha. Venus is lord of the 6th and 11th houses. Both lordships are inauspicious according to the principles of astrology and so is malefic or inauspicious planet for Dhanu lagna. Apart from that, Venus is an enemy of Jupiter, the lord of this Ascendant. L-6 Sukra has generally a problematic influence for the Dhanusha lagna. Sukra provides both conflict (6) and gainfulness (11). Sukra brings both the hostility of imbalanced relationships which produce enemies and the structured network of association through the marketplace of goods and ideas which creates a great circle of friends.

Due to Sukra inauspicious lordship he has quite a bit of trouble (especially during Sukra periods) with the range of Venus significations like partnerships and alliances of all kinds. He should in general avoid business partnerships, or at least be very careful to have the terms of agreement made perfectly. He is prone to divorce. Sukra is an especially difficult graha for Dhanusha nativities in which Sukra is L-6 from both radical lagna (material animosity) and Chandra lagna (emotional imbalance).

Saturn: Shani is L-2 / L-3 measured from Dhanu lagna or Dhanu Chandra; Dhanu Navamsha. Saturn is an enemy of Jupiter, the lord of this Ascendant. Saturn is a malefic planet for Dhanu lagna. Self-made wealth from bhava-3 depends on the disposition of Shani. If Saturn occupies the Ascendant at a person's birth, he will be poor, sickly, love stricken, very unclean, suffering from diseases during his childhood and indistinct in his speech. But an exception can be made if Saturn is posited in the Ascendant. Good results of such disposition of Saturn are described by Brihat Jatak. However, if any of the signs, such as, Sagittarius, Pisces, Aquarius, Capricorn and Libra is the Ascendant and Saturn occupy it at birth, the person concerned will be equal to a king, the headman of a village or the mayor of the city, a great scholar and will be handsome.

Rahu: Rahu is co-lord of Kumbha (bhava-3) measured from Dhanu lagna or Dhanu Chandra; Dhanu Navamsha. Rahu is malefic planet for Dhanu lagna. Rahu and Ketu are said to give positive results in the Upachaya 3rd, 6th, and 11th Bhava from the lagna or from Chandra. In addition, some authorities posit that since R-K are exalted in Vrishabha-Vrischika (per BPHS), R-K will give good results when Mangala and Sukra are well-disposed.

Ketu: Ketu is co-lord of Vrischika (bhava-12) measured from Dhanu lagna or Dhanu Chandra; Dhanu Navamsha. Therefore is a malefic planet for Dhanu lagna. Ketu effects for Dhanusha lagna depend significantly on Ketu Bhava, Rashi, and Drishti. Ketu amplifies the effect of any graha who are sharing Ketu house; and also Ketu magnifies the effects of the lord of His occupied Rashi.

- Saturn, when in Aries, confers prosperity during its Dasa
- This is the only ascendant for which Saturn in 11th confers yoga
- If Sun and Venus are in Leo with Saturn in Aquarius, the Saturn Dasa will bring in wealth

Capricorn (Makar) Lagna:

He/She is inclined towards evil deeds, is greedy and has many children, but hard working. He/She is tall, strong built, good personality, whitish complexion, clever and but selfish, changing his faces frequently as per situations, and very talkative. He/She works under someone and subjects to heavy ups & downs. He/She spends immediately what he/she earns in bad deeds. He/She dominates spouse and quarrels with other family members on that account. He/She stays away without information. He/She is abusive and short-tempered. He/She fails in business and has to go in for service. He/She is attached to mother and has differences with father. He/She is helpful to brothers and sisters and has more daughters. His/Her expense is more than earning and hence always faces shortage of money. The conjugal life is not happy and there is always difference of opinions between them.

Mars, Jupiter and Moon are malefic, Venus and Mercury are auspicious. Saturn will not be a killer on his own. Mars and other malefic will inflict death. Sun is neutral. Only Venus is capable of causing a superior Yoga.

Sun: Surya is L-8 measured from Makar lagna, Makar Chandra, Makar Navamsha.

Sun is lord of eighth (a very inauspicious Bhava) and a bitter enemy of Saturn, the lord of Capricorn Ascendant and so Sun is most malefic planet for this Ascendant. As a fierce enemy of Lagnesh Shani, ego-promoting Surya causes social awkwardness and psychological frustration for the careful, cautious, law-abiding Makar native. Sun gives a blazing hot sudden jolt and an unexpected upsurge of an inner clarion call to him. L-8 Sun manifest as personality disturbance, even anti-social behaviour. If Surya is weak by placement, it may generate secret or covert self-assertion behaviours like prostitution. If L-6 Budha (prostitution) + L-9 Budha (public religion) is exacerbated by sexuality-karaka L-7 (Chandra) in hidden Simha (Bhava-8), she will get indulge more likely in prostitution.

Moon: Chandra is L-7 measured from Makar lagna, Makar Chandra, and Makar Navamsha. Moon is the lord of the 7th (Kendradhipati dosha) and Moon is also an enemy of Saturn, and so Moon is a malefic planet for Makar lagna. The highly fluctuating and impressionable Chandra regulates contracts, agreements, alliances, and peer-to-peer relationships (bhava-7). Chandra's condition will determine the quality of counselling and advising relationships in the native's personal and professional life. After marriage, a man will find his wife to be his most influential advisor in all matters. For a woman, her own mother remains her closest counsel.

Excellent for women in legal and diplomatic practice (presuming Chandra is well disposed).

Mars: Kuja is L-4+ L-11 measured from Makar lagna, Makar Chandra, and Makar Navamsha. Mars is lord of the 4th and 11th. Mars is also an enemy of Saturn, and so Mars is a malefic planet for Makar lagna. Badhaka (L-11) provides trouble from the elder sibling, friendly association, the marketplace, the nature or method of setting goals, achievements and in profits.

If Shani occupies Vrischika or Mangala-Ketu are damaged in Kundali, he has considerable troubles with the network of friends and associates and may suffer isolation or the identity is oppressed by sudden upheavals in the network or selfish actions of a mentoring friend.

Mercury: Budha is L-6+ L-9 measured from Makar lagna, Makar Chandra, and Makar Navamsha. Lagnesh Shani is a good friend of Budha. Mercury is lord of the 6th and the 9th, a trine house, which is the Moola Trikona sign of Mercury and therefore is very auspicious. Venus and Mercury together are Yoga Karaka for Makar lagna. Budha is an especially strong influence in His natural domain bhava-6. As lord of bhava-6, Budha brings social conflict, exploitation, personal imbalance, mental resistance to marriage fidelity and a quick-witted avoidance of the terms of agreement. In particular, Budha can signify medical and mental issues with substance addiction and the psycho-physical consequences of sex (Budha) addictions. (Addiction indicates a behaviour which becomes so compulsory that it interferes with other necessary functions in work and family life.

As lord of bhava-9, Budha brings a mental engagement with religious principles and practices, affinity for sacred space, and personal priesthood. If Budha is well-disposed, the native has knowledge of the sacred scriptures of one's own religious tradition like Swami Vivekananda whose Budha in 1 had Parivartana Yoga with L-1 Shani and like Jeddu Krishnamurti whose Budha in 5 had Parivartana Yoga with L-5 Sukra. If Budha occupies a dualistic Rashi such as Mithuna or Meena, one may grasp the meanings offered by the sacred writings of traditions outside one's own, as well. If Budha is prominently placed, such as in lagna or with

Chandra, the native's personality and behaviour may appear contradictory or at least enigmatic.

Jupiter: Guru is L-3+ L-12 measured from Makar lagna, Makar Chandra, and Makar Navamsha. Lagnesh Shani is mutually neutral toward Guru. Jupiter will be lord of the 3rd and 12th Bhava, both inauspicious Bhava, and therefore Jupiter is not an auspicious planet and is malefic planet for this Ascendant. Guru L-12 is 6th from 7th, Badhaka or enemies of the marriage alliance.

Venus: Sukra is L-5 + L-10 measured from Makar lagna, Makar Chandra, and Makar Navamsha. Lagnesh Shani is a good friend of Sukra. L-5+L-10 Sukra is a powerful Yogakaraka for Makar lagna. (Sukra is also L-4+L-9 Yogakaraka for Kumbha lagna). Sukra Yogakaraka L-5 (politics) + L-10 (governance) has natural capability to lead and govern, either by royal entitlement (5) or by charismatic charm that facilitates successful democratic election. He is empowered to make decisions which provide social order and lawfulness like Barack Obama. Shukra gives excellent results in regard to matters of children, politics, speculative ventures, romance, professional respect, leadership roles, and public duties. He will enjoy the company of women. Depending on Shukra full character (Rashi, drishti, etc.) periods of Shukra are usually very beneficial for the Makara native. A good position of Shukra in the Makara lagna chart can compensate for many other difficulties. Shukra is best in any Kendra, but especially in lagna where it makes the person physically attractive and brings much happiness from profession and children.

Saturn: Shani L-1 + L-2 measured from Makara lagna, Makara Chandra, and Makara Navamsha. Shani is L-1, a trine house, and is Lagnesha and so a benefic planet. If Shani is well placed, he will lead a disciplined life and accrue stores of material wealth over time. The Maraka woman looking forward to her second marriage may consider wearing a beautiful blue sapphire on the longest finger of the left hand in order to encourage qualities of dignity, self-discipline, respect for order, and conventional behaviours in her forthcoming alliance with a second spouse.

Rahu: Rahu is L-2 (Kumbha) measured from Makara lagna, Makara Chandra, and Makara Navamsha. Rahu is co-lord of bhava-2 (Kumbha), therefore Rahu is a beneficial or benefic planet for Makara lagna and for matters of wealth acquisition, stock-piling and hoarding, learning in history and languages, second marriage, and expressions of voice and face such as song and appearance in story-telling imagery such as films. However Rahu in the Rashi of Maraka Chandra is dangerous for Kumbha lagna. Rahu and Ketu are said to give positive results in the 3rd, 6th, and 11th bhava from the lagna or from Chandra. In addition, some authorities posit that since R-K are exalted in Vrishabha-Vrischika (per BPHS), R-K will give good results when Mangala and Shukra are well-disposed.

Ketu: Ketu is L- 10 (Vrischika) measured from Makara lagna, Makara Chandra, and Makara Navamsha. Ketu is co-lord of bhava-10 Vrischika, and so Ketu is a benefic planet for Makara lagna. Ketu's effects for Makara lagna depend significantly on Ketu bhava, Rashi, and drishti. Ketu amplifies the effect of any graha who are sharing Ketu Bhava; and also Ketu magnifies the effects of the lord of His occupied Rashi.

- If Mercury is in Leo with Jupiter in lagna and aspect by Venus, long life is conf erred, but with poverty
- Venus in Taurus is good; if in Libra, though, he may not prove so fortunate
- If Venus and Mercury are in lagna with Moon in Taurus aspect by Jupiter, the native is certain to become an emperor. This is called maharaja yoga
- If Jupiter is in lagna with Venus and Mars in Scorpio, brothers will give the native riches during the Jupiter Dasa
- When Sun, Moon and Mercury are in lagna, with Mars and Venus in Sagittarius, the native will be replete with wealth
- A Saturn-Mercury combination in Virgo confers fortune. Rahu becomes a Yogakaraka if he's with Jupiter in Sagittarius
- The presence of Moon in Cancer and Mars in lagna causes a Rajayoga

Aquarius (Kumbha) Lagna:

He/She leads a happy and contented life. He/She will be well educated, gentle, peaceful, always ready to help others, having good thinking; tall, whitish complexion and attractive personality and straight forward in nature. He/She successfully tackles early age problems and heads for good time later. He/She would spend any amount of time and money to crush his/her enemy or achieve his/her aim. He/She likes to gossip and interact with women and has interest in astrology. He/She is financially well off and is prone to chest problems. He/She will be very hard working and faces difficulties in life. He/She may suffer with the stomach and heart diseases in old age. He/She is strong willed, detached and unyielding in nature. The child of this Sign is unpredictable regarding his/her behaviour and can change frequently himself/herself to any extent during his childhood.

Jupiter, Moon and Mars are malefic, while Venus and Saturn are auspicious. Venus is the only Planet that causes Raja Yoga. Jupiter, Sun and Mars are killers. Mercury gives meddling effects.

Sun: Surya L-7 measured from Kumbha lagna, Kumbha Chandra, and Kumbha Navamsha. Sun is L-7, a Maraka house (house of death), and is also an enemy of Saturn, the lord of the Ascendant, so Sun is malefic for Kumbha lagna. Ravi as L-7 controls the marriage house, therefore for Kumbha lagna the condition of Ravi in the nativity becomes a strong indicator of the character of the spouse. Career: The dominant indicator of career is lord of 10th Navamsha, and lord of 10th radix; however Surya = a lesser karaka for career and public recognition.

If L-7 Surya is strong, he undertakes career roles in consulting, advising, counselling, arbitration, mediation, and coordinated work such as partner dancing, partner skating, professional

Moon: Chandra is L-6 measured from Kumbha lagna, Kumbha Chandra, and Kumbha Navamsha and so Moon is the highly inauspicious agent and is a Problem creating planet. Chandra is enemy of Lagnesha Shani, and unfortunate lord of the maha-dusthamsha Ripu bhava, so Chandra is the most malefic planet for Kumbha Lagna. The divorce is a common occurrence for Kumbha, due to the L-6 Chandra. Moon as L-6, 12th-from-7th (alliances, agreements and trusts) indicates dissolution of contract, divorce, disagreement and distrust. Most acute results will be during Chandra Mahadasha or Vimshottari periods of graha in shad-ashtaka 6-8 angle to Chandra. He may feel often beset with jealousy and mood swings, digestive problems and body pain and financial problems due to indebtedness against taking of loans. The child may be accused of a parental strategy to control over a child being used as a servant. If Rahu in 6 or Rahu involved with Chandra, higher likelihood of criminal extremes such as human's trafficking. Marriage and Alliances: Kumbha lagna and Kumbha Chandra unbalances relationships between individual partners, between individuals and various groups, and between the individual and society. The Kumbha native feels that he does not belong to the society. Chandra occupies swakshetra-6, leading to an exceptional career in maternal-child medicine or the care of the service class. If Chandra is in the uchcha Vrishabha (bhava-4), he may become a happy and prosperous owner of a great many real-estate properties. One tends to get in and out of relationships very quickly, Chandra is an especially difficult graha for Kumbha Chandra = L-6 from both radix lagna (material animosity) and Chandra lagna (emotional imbalance). Kumbha natives are emotionally oriented toward providing service in exchange for security and protection.

Mars: Kuja is L-3 / L-10 measured from Kumbha lagna, Kumbha Chandra, and Kumbha Navamsha. Kuja as L-10 (Kendradhipati dosha) is reasonably auspicious at birth and is a benefic planet for Kumbha lagna. Vital and competitive Kuja controls daily mental process and leadership roles and he has strong administrative business inclinations.

Mercury: Budha is L-5 / L-8 measured from Kumbha lagna, Kumbha Chandra, and Kumbha Navamsha. Saturn is a friend of Mercury and on account of his lordship of a trine; Mercury is accepted mostly as an auspicious and is a benefic planet for Kumbha lagna. Well-disposed and strong Budha provides children, intelligence, occult, politics, charisma, games and gambling - winning prizes, awards and trophies, fortune and luck, creative arts, speculation, prestige, fame, self-expression and Theatre centre-stage roles. If Budha is not well disposed, he may develop blustering and boasting nature about knowing occult secrets, detecting hidden resources and making transformative changes.

Jupiter: Guru is L-2 + L-11 measured from Kumbha lagna, Kumbha Chandra, and Kumbha Navamsha. Jupiter is lord of the 2nd (Maraka) and the 11th (inauspicious house) and Saturn, the lord of this Ascendant, is neutral toward each other, however, Jupiter is accepted as an inauspicious and is a malefic planet for Kumbha lagna. However Jupiter is lord of the 2nd (Wealth house) and is posited in own house, it forms a wealth yoga. There are four main wealth houses: 2 = banked savings; 5 = gambling or speculative wins; 9 = assets provided to be used for acts of faith; and 11 = earned income. Wealth: Presuming Guru is fairly well disposed or in own sign, periods of L-2+L-11 Guru generate considerable wealth via family-of-origin hoarded assets and knowledge-values, as well as via earned income from the marketplace of goods and services.

Venus: Shukra is L-4+ L-9 measured from Kumbha lagna, Kumbha Chandra, and Kumbha Navamsha. L-4+L-9 Shukra are a powerful Yogakaraka for Kumbha lagna. As a natural benefic L-4+L-9, Shukra becomes a potent and positive Yogakaraka in control of schooling (4), attainments in arts and music (Shukra), property and vehicle ownership (4); relationships with priests and professors (9), and privilege of access to sacred teachings (9). A good position of Shukra in the Kumbha lagna chart can compensate for many other difficulties. Shukra is best in any Kendra, but especially in lagna where it makes the person physically attractive and brings much happiness from property ownership, interior decorating, architectural practice and also from global scholarly travels in the context of alliance-building and acquisition of higher knowledge.

If strong Shukra occupies bhava-5, he obtains the most attention from an admiring public who are entertained and amused by his drama (acting).

Saturn: Shani is L-1+ L-12 measured from Kumbha lagna, Kumbha Chandra, and Kumbha Navamsha. Shani as Lagnesha controls both identity and loss of identity and is the agent of acquisition of a great inner peace. He is prone toward meditation and pilgrimage. Even in middle age, undertaking long sojourn abroad during Shani period is easy for the Kumbha native due to the identity-dissolving effects of 12 combined with the new-clothes or new-attributes effect of 1.

Rahu: Rahu is co-lord of Kumbha; L-1 measured from Kumbha lagna, Kumbha Chandra, and Kumbha Navamsha and is friendly with Saturn, lord of Kumbha. Rahu is a benefic planet for Kumbha Lagna. Rahu effects depend significantly on Rahu bhava, Rashi, and incoming drishti. Rahu amplifies the effect of any graha who are sharing Rahu house; and also Rahu magnifies the effects of the lord of His occupied Rashi. Rahu and Ketu are said to give positive results in the 3rd, 6th, and 11th bhava from the lagna or from Chandra. In addition, some authorities posit that since R-K are exalted in Vrishabha-Vrischika (per BPHS), R-K will give good results when Mangala and Shukra are well-disposed.

Ketu: Ketu is L-8 + L-10, i. e. co-lord of 8th and 10th measured from Kumbha lagna, Kumbha Chandra, and Kumbha Navamsha. Ketu as L-10 (Kendradhipati dosha) is reasonably auspicious at birth and is a benefic planet for Kumbha lagna. Ketu effects depend significantly on Ketu bhava, Rashi, and drishti. Ketu UN-connects the effect of any graha sharing Ketu house. Ketu scatters the effects of His planetary lord. Ketu is the natural co-regulator of bhava-8

For both Aquarius and Leo ascendant natives, the mere association of the 9th and 10th lords doesn't confer any Rajayoga

- If Venus and Rahu are in lagna with Sun in Scorpio, yoga will be caused in the Dasa of Rahu and Jupiter
- If Sun and Mars are in Virgo, the native suffers during their Dasa. The Mercury Dasa will be better
- Jupiter in lagna with Saturn in Pisces will see mixed results during the Jupiter Dasa and an ordinary time during the Saturn Dasa

- If Saturn and Venus are in Sagittarius, the Venus Dasa proves fortunate
- A Sun-Mercury-Jupiter combine in Aries is good, especially during the Sun Dasa when the native enjoys power

Pisces (Meena) Lagna:

He/She is wealthy, educated, gentle, peaceful, religious, and always ready to help relatives, medium height, and beautiful curly hair and have self confidence. He/She is famous and heads an organisation or society. He/She studies very hard, but he/she is not a high scorer. He/She helps friends and serves society physically but without spending money. He/She works overtime to finish the work same day. He/She becomes favourable of family members and outsiders. He/She just cannot work under any one and leave his/her service very soon for own business. He/She loses temper beyond control but calms down very quickly. He/She is prone to cheating by partners and should better work alone. He/She is likely to break first marital relation (matured) or otherwise, be unhappy with spouse but happy with progeny. He/She may have great interest in writing, music and acting. He/She is likely to set high goals. Drug, alcohol and false promises attract him/her easily.

Saturn, Venus, Sun and Mercury are malefic. Mars and Moon are auspicious. Mars and Jupiter will cause Yoga. Though Mars is a killer, he will not kill the native (independently). Saturn and Mercury are killers.

Sun: Surya is L-6 measured from Meena lagna, Meena Chandra, and Meena Navamsha and is "The Problem creator". Sun is lord of the 6th, an inauspicious house and does not rules any other house and so Sun is malefic planet for Meena lagna. Ravi controls the most evil dushthamsha, bhava-6. Naturally, most of the significations of Ravi like father and father's figures become problematic. His father has been born into the service class (6). The father will be restless and discontent with his social position. He must be inordinately cautious of his own independent, egoistic instincts, since self-assertion tends quickly toward conflict.

Moon: Chandra is L-5, a Trine house, measured from Meena lagna, Meena Chandra, and Meena Navamsha, a very auspicious house and so Moon is a benefic planet for Meena lagna. He is the best emotionally satisfied in creative social roles that express individual intelligence and self-determination, and allow him to take speculative risks. He is the best in parenting, literary author, speculative investor, independent educator, and almost any political or entertainment role including "genius" innovative teaching styles. He loves children, and expects to profit by raising them. He is prone toward lifelong self-improvement and often involved in the consciousness industry and will achieve some degree of celebrity or notoriety.

Mars: Kuja is L-2 + L-9, a Trine house, measured from Meena lagna, Meena Chandra, and Meena Navamsha. Kuja is a money-maker, as He controls two important wealth/values houses out of four wealth houses like, 2, 5, 9, and 11. For Matsya Lagna, presuming that Mangala is fairly well disposed, Mangala generates considerable wealth via family-of-origin hoarded assets and knowledge-values, as well as via the higher philosophical education and world travel.

Mercury: Budha is Kendra lord L-4 + L-7, measured from Meena lagna, Meena Chandra, and Meena Navamsha. Here also on account of ownership of two Kendra Mercury suffers from Kendradhipati dosha and so Mercury is inauspicious or malefic planet for Meena lagna. Budha controls two of the most basic relationships, like parents (4) and spouse (7). Budha in 11th (Makar) and 12th (Aquarius) ruled by Shani (L-11+ L-12) will have his quiet and conservative early childhood home and marriage. Budha in 3rd (Vrishabha) and 8th (Libra) ruled by Sukra (L-3+L-8) will have a luxury-loving spouse and parents; however the pursuit of that luxury may become problematic. Budha in Simha ruled by the L-6 Surya will naturally have some struggles regarding personal sovereignty, both in the early childhood home and the marriage. If Surya is well-disposed, these conflicts can be managed.

Jupiter: Guru is L-1+L-10 measured from Meena lagna, Meena Chandra, and Meena Navamsha and is a superb career graha for Meena lagna. Guru is L-1, a Trine house, a very auspicious house and so is an auspicious or a benefic planet for Meena lagna. He benefits from a leadership responsibilities. Presuming that Guru is a healthy graha, he enjoys fairly high levels of social recognition and a favourable public reputation. Guru is not helpful for money, because of L-10 (12th from 11th and L-1 is (12th from 2nd). Guru typically raises the prestige and lowers the bank accounts.

Venus: Sukra is L-3 + L-8 measured from Meena lagna, Meena Chandra, and Meena Navamsha. Venus is an enemy of Jupiter, the lord of this Ascendant. Sukra is lord of 3rd and 8th, an Upachaya house and a dushthamsha (two inauspicious houses), and so Sukra is an inauspicious or malefic graha for the Meena lagna. Shukra is unfavourable for men, because Shukra describes his female partners. He may attract a female partner who is handicapped by her suspicious mind, particularly in regard to her husband's hidden monies or inheritance, and issues of marital fidelity. Harmonious relationships are hard to come by. In fact, Shukra is nearly as problematic for Meena lagna.

Saturn: Shani is L-11 + L-12 measured from Meena lagna, Meena Chandra, and Meena Navamsha. Shani is an enemy of Jupiter, the lord of this Ascendant. Shani is lord of 11th and 12th, an Upachaya house and a dushthamsha (two inauspicious houses), and so Shani is an inauspicious or malefic graha for the Meena lagna. Shani gives profit and takes it away. Shani creates loneliness, isolation, and a profound loss of identity. It is the great irony of the turning of the cycle of birth and death, that at the moment of accomplishment of a goal, the karma is completed, and one tends to lose interest in the activity. If Shani is favourably disposed, the L-12 Shani may create foreign travel, ashram retreat, healing hospitalization, and relaxing spas, rather than marital conflict or wandering.

Rahu: Rahu is co-lord of Kumbha (dushthamsha bhava-12) measured from Meena lagna, Meena Chandra, and Meena Navamsha. Rahu is an enemy of Jupiter, the lord of this Ascendant. Rahu is lord of 12th, a dushthamsha (inauspicious house), and so Rahu is an inauspicious or malefic graha for the Meena lagna. Rahu in either Rashi of Maraka Mangala is dangerous for Meena lagna. Rahu and Ketu are said to give positive results in the 3rd, 6th, and 11th bhava from the lagna or from Chandra. In addition, some authorities posit that since R-K are exalted in Vrishabha-Vrischika (per BPHS), R-K will give good results when Mangala and Shukra are well-disposed.

Ketu: Ketu is L-9; co-lord of Vrischika (bhava-9) measured from Meena lagna, Meena Chandra, and Meena Navamsha. Ketu is lord of 9th, a Trine house (an auspicious house), and so Ketu is an auspicious or benefic graha for the Meena lagna. Ketu amplifies the effect of any graha who are sharing Ketu house; and also Ketu magnifies the effects of the lord of His occupied Rashi.

- While Venus in 12th produces benefic results for all other ascendants, this isn't the case for either Pisces or Aquarius
- Saturn in Aquarius is good. However, if Moon is in Aquarius, the native suffers poverty (though there's an alleviation of this in Jupiter Dasa, Moon Bhukti)
- Jupiter in Cancer will bestow more daughters than sons
- If Moon is in Aries and Mars in Cancer, Moon's Dasa brings wealth
- If Moon and Mars are in Capricorn, Jupiter in Leo, Venus in Libra and Saturn in Scorpio, the native will be highly fortunate
- A Moon-Mercury-Mars combine in Capricorn is an indicator of wealth and conveyances
- If Saturn and Moon are in lagna, Mars in Capricorn and Venus in Leo, predict fortune for the native
- If Mercury, Jupiter, Moon and Mars are in lagna, the Dasa of these planets bring in immense fame, power and prosperity
- Jupiter posited in Sagittarius certainly causes a Rajayoga
- If Moon is in Taurus, Sun in Leo, Mercury in Virgo, Venus in Libra, Jupiter in Sagittarius, Saturn in Aquarius and Mars in Capricorn, there'll be much fortune to the native.

Even if one or two of these combinations are absent, there'll still the full effects of a Rajayoga

PART-2
ZODIAC SIGNS IN 2ND HOUSE
(DHAN/WEALTH)

Aries in 2nd Bhava: The Aries in the 2nd Bhava indicates the uncertainty about his/her earning. Sometimes he/she earns more money and sometimes very less money. He/She spends more than the earning on his/her show business and is sometimes harmed by his/her enemies. His/Her good luck starts after marriage.

Taurus in 2nd Bhava: The Taurus, in the 2nd Bhava, indicates his/her good earning. He faces ups and down in life and is harmed by his/her spouse or partners in the business. His/Her good luck starts after 18, 22, 24, 33 and 35 years of age. He/She has a strong drive to earn money; to build and hold financial worth and material possessions. Venus, the planet of love ruling the second house suggests a real love of money. He/She is good at business affairs. He/She is a natural for making and accumulating money.

Gemini in 2nd Bhava: The Gemini, in the 2nd Bhava, indicates his/her weakness in the earning. He/She wastes money in bad relation with the other woman/man. His/Her enemies sometimes harm him/her financially. His/Her good luck starts in business; the private services or in life insurance, small industries, electrical parts industry etc. The Gemini influence in the second house focuses the mind on material matters and on making money. He/She has an active interest in financial affairs. His/Her resourcefulness in accumulating money may result in holding more than one job simultaneously.

Cancer in 2nd Bhava: The Cancer, in the 2nd Bhava, indicates the less earning as compared to his/her hard work.

He/She is miser in nature and spends less than the earning and saves more. He/She faces ups and down in the business and is harmed by his/her partners in the business. His/Her good luck starts after 20, 26, 33, 44 and 54 years of age. The influence of Cancer in the second house shows that he/she is protective of his/her financial assets and possessions.

Leo in 2nd Bhava: The Leo, in the 2nd Bhava, indicates the uncertainty about his/her earning in his middle age. He/She spends very happy childhood and does not face any financial crisis in childhood. He spends more than the earning on his/her show business or in uncertain business and is sometimes harmed by his/her enemies. He/She has good luck always and finishes all his/her work successfully. He/She earns money through political works or government job. The Sun denotes his/her self-esteem, so his/her earning capacity may have a lot to do with his/her sense of self-worth.

Virgo in 2nd Bhava: The Virgo, in the 2nd Bhava, indicates he is financially strong. He/She earns money by hard work in business, especially ready-made shop or fancy store shop. Initially he/she faces financial difficulty due to not taking right time decision or due to hot temper. He/She appears very careful with financial affairs, but often he/she can become somewhat penny-wise and pound-foolish.

Libra in 2nd Bhava: The Libra, in the 2nd Bhava, indicates the uncertainty about his/her earning. Sometimes he/she earns more money and sometimes very less money by business. He/She spends more than the earning on his show business or luxuries life and is sometimes harmed by his/her bad habits. He/She has good luck in business like hotel or restaurants. There is balance and harmony in material affairs. This sign suggests the accumulation of possessions is highly dependent on ventures with a partner, normally the marriage partner.

Scorpio in 2nd Bhava: The Scorpio, in the 2nd Bhava, indicates the uncertainty about his/her earning. He/She spends more than the earning on his/her show business or on his/her big planning and is sometimes harmed by his/her friends or relatives. He/She has good luck in business like small industries rather than service.

Sagittarius in 2nd Bhava: The Sagittarius, in the 2nd Bhava, indicates very lucky and good earning. His/Her good luck starts after 24, 28, 33, 37, 48 and 55 years of age. He/She should be very careful while signing any paper. He/She is lucky in this regard. His/Her personality attracts financial success naturally. He/She is a risk taker who may get burned from time to time.

Capricorn in 2nd Bhava: The Capricorn, in the 2nd Bhava, indicates very lucky and god earning and is financially strong. He/She earns good money by business in mines or stone query as compared to service. He/She may serve in planning commission or plans making organization because he/she is very good in big planning. He/She has prudence and is practical in the handling of money. He/She has a 'poor' complex and does not know his/her earning potential and so he/she continues to live as though he/she had very little. When buying investments, he/she is inclined toward the blue chips and sure bets. He/She is very cautious and practical. He/She may reject luxury or expensive living.

Aquarius in 2nd Bhava: The Aquarius, in the 2nd Bhava, indicates very lucky and good earning and is financially strong. He/She earns good money by many source of income particularly by reporter, writing, publications or businesses as compared to service. He/She does not get benefits from brothers or relatives and gets harmed on good faith by the people. He/She is benefited in partnership and latter life is better than the middle age. He/She is not afraid to take chances financially, and he/she looks for unusual and inventive ways to invest.

Pisces in 2nd Bhava: The Pisces, in the 2nd Bhava, indicates very lucky and good earning and savings and is financially strong. He/She earns good money by many source of income particularly by doctor professions, by sale of medicines, by share business or small industries. He/She controls his expenditures and earns money like anything and does not have peace of mind in life but always busy in earning money. His/Her good luck starts after 24, 28, 33, 37, 48, 55 and 60 years of age. He/She has a sense of timing and intuition that may be an asset financially.

PART-3
ZODIAC SIGNS IN 3RD HOUSE (RELATIONS/SIBLING)

Aries in 3rd Bhava: The Aries, in the 3rd Bhava, indicates that he/she will be strong mussel and good built, strong built wider shoulders, good personality, very courageous. He/She earns money and believes in making the situations in his/her favours and obedience to his/her seniors. He/She will be talkative and artist and will save money in life. He/She has a strong need to communicate with others, and to communicate forcefully. His mind is active, alert and capable of making quick decisions. He/She is capable in expressing himself/herself. He/She is mentally competitive. He/She is an aggressive learner, always seeking new ideas and new knowledge.

Taurus in 3rd Bhava: The Taurus, in the 3rd Bhava, indicates that he/she will be strong built, good personality, very courageous. He/She earns money by writing, poetry, portrait making and other arts and believes in making the situations in his/her favours and admired by the family members. He/She will be artist. He/She will be able to save money in life. He/She is a slow learner, but once an idea is lodged in his/her brain, it's there to stay. In early education he/she may have been an indifferent student, but as he/she matures he/she accumulates a wealth of knowledge and understanding. He/She has a strong interest in the arts, especially music.

Gemini in 3rd Bhava: The Gemini, in the 3rd Bhava, indicates that he/she will be strong built, good personality, very lucky. He/She earns money and believes in making the situations in his/her favours and honoured by the government. He/She will enjoy the happiness of luxurious vehicles. He/She will be able to save money in life and has very happy family life and good respect and cooperation from spouse. He/She has quickness both in thought and speech. Mercury, ruling Gemini, is the planet of intellect, thought, speech, and wit. He/She is a good conversationalist, a fact collector, and a mentally stimulating person. He/She can be a good diplomat because he/she can agree with divergent views and present rational alternatives.

Cancer in 3rd Bhava: The cancer, in the 3rd Bhava, indicates that he/she will be strong built, good personality, very lucky. He/She earns sufficient money by good business and real estate or construction work and believes in making the situations in his/her favours by his/her noble nature and admired by the society. He/She will enjoy the happiness of friends. He will be able to save money in life and has very happy family life and is religious minded. He/She is very protective of brothers and sisters. He/She has close and emotional ties to the immediate family.

Leo in 3rd Bhava: The Leo, in the 3rd Bhava, indicates that he/she will be strong built, good personality and very courageous and has excellent imaginative power. He/She earns money by writings, poetry, and publication and believes in making the situations in his/her favours and has interest in the great music. He/She will be artist. He/She will face difficulties in education in childhood but finally gets good educations in life and be able to save money in life. His/Her powers of self-expression are outstanding.

Virgo in 3rd Bhava: The Virgo, in the 3rd Bhava, indicates that he/she will be good personality, has very good knowledge of Vedas. His/Her friends are helpful but not the family members. He/She will be short tempered. He/She will be suffering from inferiority complex on communication skills, especially in the early years. He/She can be overly critical of brothers, sisters or neighbours.

Libra in 3rd Bhava: The Libra, in the 3rd Bhava, indicates that he/she will be flicker minded and has relations with low status people. He/She will not be able to take fast decision but too much talkative and does mistake while talking for which he/she repents later on. He/She has differences in family life always. He/She gets along easily with family members because he/she dislikes conflict and argument.

Scorpio in 3rd Bhava: The Scorpio, in the 3rd Bhava, indicates that he/she will be flicker minded and has relations with low status people and is addict of bad habits. He/She will not be able to take fast decision. He/She has differences in family life always and live medium life. He/She is very angry man and does mistake in his/her angriness. He/She does not get help from brothers. There can be friction between him/her and family members, as well as friends and acquaintances.

Sagittarius in 3rd Bhava: The Sagittarius, in the 3rd Bhava, indicates that he/she will lose money in business and will not get success in business. He/She will be able to save money in life in military, police or other government service. He/She is very talkative and has a very cheerful outlook on life, natural exuberance and optimism. He/She has a natural ability to communicate, especially important issues such as politics, education, and religion. He/She has an executive type of mind.

Capricorn in 3rd Bhava: The Capricorn, in the 3rd Bhava, indicates that he/she will be strong mussel and good and strong built handsome body & good attractive face and personality. He/She is very lucky in respect of children and famous among friends and is religious minded. He/She has careful expression of thoughts and ideas. He/She refrains from writing or saying anything unless there is a reason for doing so. He/She has little capacity for small talk, and many people may find it difficult to communicate with him/her. In his early years, he/she may not have good education, but later he/she changes, and he/she is apt to turn to studies to attain his/her ambitions.

Aquarius in 3rd Bhava: The Aquarius, in the 3rd Bhava, indicates that he/she will be peaceful nature and has good relations with brothers but does not get help from them. He/She will get respect in the society and has interest in music. He/She has a sparkling intellect and an inventive mind. He/She is well ahead of time, and sometimes radical. He/She seeks education simply for the sake of learning instead of just for preparing to earn a living.

Pisces in 3rd Bhava: The Pisces, in the 3rd Bhava, indicates that he/she will be good built, handsome body & good attractive face and personality. He/She will be a wealthy man and lucky in respect of children and get full help from children in old age. He/She is famous in society, keeps everybody happy in the society and is religious minded and interested in religious work. In communicating, he/she can become over emotional and he/she may experience problems articulating, especially in his/her youth. He likes to be alone when he/she is performing any type of mental work.

PART-4
ZODIAC SIGNS IN 4TH HOUSE (BANDHU/PLEASURE):

Aries in 4th Bhava: The Aries, in the 4th Bhava, indicates that he/she will be having many cattle. He/She will have relations with many women simultaneously but still he/she enjoys the life happily in different ways and peacefully. He/She is more benefited by agriculture and business. He/She has an aggressive attitude toward the home and family. He/She always takes an active interest in the affairs of the home. He/She has a tendency to force issues and demand too much. The fourth house denotes activities in the latter part of life, the later years will be very active and more daring and outgoing, even youthful personality emerges, as he/she grows older. He/She is more assertive and physically active in the latter part of life.

Taurus in 4th Bhava: He/She is very lucky in respect of children who help him/her in old age and is religious minded. He/She does the social work in big ways. He/She is famous in society, has patient, peaceful, keeps everybody happy in the society and is religious minded and interested in religious work, celebrates worships in big ways. He/She is more benefited by worshipping the lord Shiva. He/She enjoys the happy family life. He/She has strong instincts to provide materially for his/her family. He/She has a very pleasant, easygoing home environment, and harmony, serenity and graciousness in his/her latter years.

Gemini in 4th Bhava: The Gemini, in the 4th Bhava, indicates that he/she will be handsome body & good attractive face and personality. He/She has medium luck in life. He/She is very sexy and happiest person and enjoys the sex in relation with the most beautiful women and spends lots of money on them and fashion and cosmetics. He/She has to work hard to earn money. He/She suffers a lot in his/her old age due to loss of heavy money. His/Her ties to home are not particularly strong.

Cancer in 4th Bhava: He/She will be the luckiest person and has many friends. He/She is very happiest person and enjoys the married life with the most beautiful good nature and fortunate spouse and spends lots of money on fashion and cosmetics. There is a deep attachment to family traditions and home relationships. He/She may be especially protective of his/her parents and assume a role of responsibility for them. He/She may depart his/her early shelter at a young age.

Leo in 4th Bhava: He is the happiest person and enjoys the married life with the most beautiful good nature and fortunate wife and spends lots of money on fashion and cosmetics. He/She is the most angry and irritated person. His/Her relation with the brothers and sisters are not good and does not get help from them. Children are not so happy with him/her. He/She has more female child than male child. He/She has properties and owns a piece of land and lives in a reflection of his/her ego. He/She is apt to be the one that 'rules the roost.' In either sex, this placement often shows the one who is the boss and makes the decisions in the family.

Virgo in 4th Bhava: He/She will be the luckiest person and most fortunate. He/She is very happiest person and enjoys the full and happiest married life with the most beautiful good nature and fortunate spouse and spends lots of money on fashion and cosmetics. He/She gets married in young age and has more male child than female child. He/She has sufficient money to enjoy the life. The childhood is difficult financially but becomes wealthy man after the age of 28 years and is the richest man at 36 years. He/She is learned, educated good nature person.

Libra in 4th Bhava: The Libra, in the 4th Bhava, indicates that he/she will be handsome body & good personality. He/She will be the successful businessman and will spread his business all-around and is most fortunate even though he/she will be poor in childhood. He/She is very happiest person and enjoys the full and happiest married life till old age. The childhood is difficult financially but becomes wealthy man after he/she starts working after marriage. He/She is educated, kind, peace loving and good nature person who helps others. He/She spends some of his/her wealth on religious matters and works and keeps at the distance from the bad people. He/She wants to own his/her home and possessions, free and clear.

Scorpio in 4th Bhava: The Scorpio, in the 4th Bhava, indicates that he/she will be very anxious and worried person; his/her mind will not be peaceful and will face many difficulties in life. He/She is afraid of enemy as they harm him/her frequently. Every time, he/she faces difficulties while starting any job but gets success at the end. He/She starts life from poor childhood but earns money as he/she grows older and saves money to enjoy the family life till old age. He/She has extremely strong feelings about the home and family life. He/She has a strong loyalty and protective tendency shown toward family members. He/She has a sense of royalty, splendour, and space in the home environment.

Sagittarius in 4th Bhava: The Scorpio, in the 4th Bhava, indicates that he/she will be fighting nature and always fighting with people and will be successful in war. He/She will earn money and gets success in business of lending money & interest as compared to service. He/She is involved in litigation and court cases. He/She gets success in military or police job and is his/her own fortune maker and enjoys a medium happy family life. He/She has a strong urge to control and direct family matters. His/Her latter portion of life will be very beneficial. He/She will get wealthier as grow older, both in material ways and in spiritual ways.

Capricorn in 4th Bhava: The Capricorn, in the 4th Bhava, indicates that he/she will be the owner of gardens and vegetations and serve in the same field. He/She has many good and helping friends and is fortunate in respect of friends. He/She has differences with wife but the children are helpful to him/her and beneficial to him. He/She has much responsibility in house of home and family. With the planet

Saturn ruling the home, issues of security can be of paramount importance. There is also ambition linked with this placement. He/She dominates his/her home with a practical and no nonsense outlook. He/She is a strong disciplined man, with strict, old-fashioned principles and a lofty code of ethics.

Aquarius in 4th Bhava: The Aquarius, in the 4th Bhava, indicates that he/she will be happiest person and enjoys the married life with the most beautiful, good nature, educated and fortunate spouse who is helpful in making the fortune in the life and spends lots of money on fashion and cosmetics. He/She gets wealth from father in-law. His/Her fortune starts after marriage. He/She has a strong demand for freedom in the affairs of the home. He/She is likely to find it necessary periodically to change his/her address, and he never likes the idea of moving. His/Her home environment is distinctive, perhaps even a bit unusual. He/She has many original ideas and wants the latest innovations as a part of his life style.

Pisces in 4th Bhava: The Pisces, in the 4th Bhava, indicates that he/she will be a captain on the ship or will be in service connected to water transport. He/She will be educated and peaceful person and will get respect in the society due to his/her good nature. He is educated, kind, peace loving and good nature person who helps others. He/She spends some of his/her wealth on religious matters and works and keeps at the distance from the bad people. He/She has sufficient money to enjoy the life. The childhood is difficult financially but becomes wealthy and enjoys the happiest family life. He/She has an emotional tie to the home. He/She is sentimental about his/her family and willing to make sacrifices for his/her loved ones. He/She has a strong need for domestic peace and seclusion.

PART-5
ZODIAC SIGNS IN 5TH HOUSE
(SANTAAN/CHILDREN)

Aries in 5th Bhava: The Aries, in the 5th Bhava, indicates that he/she will be very angry and foolish man and does mistake in haste for which he/she repent later. There is always difference of opinions among the husband, wife and children. He/she is always restless and can't take a decision on any matter immediately. He/she never minds spending his/her money for leisure time activities, whatever he/she may be. He/she loves the outdoors and a need to stay on the move. He/she has a sporting attitude that makes him/her fun to be around. Physical activity is necessary for his/her well being and happiness.

Taurus in 5th Bhava: The Taurus, in the 5th Bhava, indicates that he/she will have no issue or will have issue after late or the issue will be weak, unhealthy or mentally retarded. He/she will have better understanding, peaceful, kind and patient. He/she will earn money easily and sometimes from lottery or gambling. He/she will be marrying to a beautiful and fortunate spouse who will be taking all the care of the family. He/she has a very loving nature toward his/her offspring, and has very strong and fixed views regarding their behaviour and the proper upbringing. He/she has personal artistic talents, and he/she may have a natural artistic ability leading to self-expression in some form of the arts.

Gemini in 5th Bhava: The Gemini, in the 5th Bhava, indicates that he/she will have issue and will have all happiness from the issue who will be healthy and educated. He/she will have better courageous, rigid, strong will. He/she will be well educated even though faces difficulties during education. He/she gets angry soon and cooled down soon too. He/she has a cool and intellectual approach to romance. Gemini is the sign of mental energy and the fifth house denotes creativity. He/she is writers and otherwise talented people.

Cancer in 5th Bhava: The Cancer, in the 5th Bhava, indicates that he/she will have more daughters than son in number or the son may take birth at latter age or very late. However, he does not get happiness from the children but pain. He/she will be lazy and has faith on others easily. He/she is very protective of his/her family, and is very maternal toward offspring. His creative work may tend toward the artistic, especially theatre. He/She has good writing abilities and he/she can communicate more easily in writing than in speech.

Leo in 5th Bhava: He/she will have more daughters than son in number or the son may take birth at latter age or very late. He/she will be working away from the native place to earn money for the family. He/she will be courageous, rigid, strong will and hard working. He/She devotes fully to whatever creative activity has his/her interest at the moment. He/She identifies strongly with his/her children and he/she is very proud of his/her accomplishments. He/she is a born gambler and speculator, with more than his/her fair share of luck.

Virgo in 5th Bhava: The Virgo, in the 5th Bhava, indicates that he/she believes in unnecessary show business other than the reality, and does not have issue or does not get help or happiness from the issue. He/she take fast decision and misuse the power immediately after getting it. Relationships with offspring can be strained as he/she lacks the patience for properly disciplining them and understanding their needs.

Libra in 5th Bhava: The Libra, in the 5th Bhava, indicates that he/she will be a gentle man, peace loving, and effective personality, very handsome and educated. Romantically he/she is charming, but very inconsistent and fickle. He/She is lucky at love. Libra is naturally suited to the raising of children, although he/she is somewhat prone to spoiling them. He/She has many child-like qualities in his/her make-up.

Scorpio in 5th Bhava: The Scorpio, in the 5th Bhava, indicates that he/she will be a gentle man, peace loving, effective personality, very handsome, religious minded and educated. He/she will suffer from venereal disease. He/She will get happiness from the children. He/She has a good planning and completes the job taken in hand successfully. He/She has attachments and romantic encounter. He/She is concerned for his/her children, almost to an extreme degree.

Sagittarius in 5th Bhava: The Sagittarius, in the 5th Bhava, indicates that he/she will be interested in horse riding or horse driving like horse race man, or horse cart man. He/She respects the people and enjoys the full happiness from his/her children because all are obedient. He/She has a constant need to show off, and he/she enjoys the thrill of any adventure. He/She is a natural gambler who will speculate on just about anything. He/She has a good understanding of young people, and he/she can get along with them so well because he/she treats them with true respect as individuals.

Capricorn in 5th Bhava: The Capricorn, in the 5th Bhava, indicates that he/she will a bad man, which enjoys the bad work or bad politics, very clever to get job done by the enemies too, principle less. But, if she is a female, she will be a big Guru Matta and worship able by the people, knowledgeable of all the Vedas and religious books. He/She is a hard worker, good with details, and can make a scrupulous teacher and disciplinarian of young people. In romantic affairs, he/she may be prudish, and may even appear cold.

Aquarius in 5th Bhava: The Aquarius, in the 5th Bhava, indicates that he/she will a fixed mind, good in education and get all happiness from the children. He/She talks truth and help and gets respect from others. He/She is courageous and never gets afraid of problems in life and face with courage. He is cool and detached concerning romance. His/Her children are apt to display a rebellious nature, and they need to be taught discipline at an early age.

Pisces in 5th Bhava: The Pisces, in the 5th Bhava, indicates that he/she will be very sexy, popular in a female society and spends a lot for them, always smiling face, sentimental. He/She has differences with spouse and children and hence has medium level happiness from children. He/She has frustration with his/her offspring. He/She may sometimes be disappointed and disillusioned. He/She may dream about creating something so significant that he/she will be remembered for his/her efforts long after he/she is gone. Raising children can be difficult and confusing.

PART-6
ZODIAC SIGNS IN 6TH HOUSE (DUKHA/SADNESS)

Aries in 6th Bhava: The Aries, in the 6th Bhava, indicates that he/she will be bad and corrupt man, hard working but always complaining to someone and telling badly about others. He/She is a very hard working member of society. He/She may run into difficulties with co-workers and employees because of his/her critical nature.

Taurus in 6th Bhava: The Taurus, in the 6th Bhava, indicates that he/she will be a frustrated man; hard working but always having differences with family members and his/her children are aggressive type. Teaching is very attractive to him/her because he/she gets the time for his/her varied extracurricular activities.

Gemini in 6th Bhava: The Gemini, in the 6th Bhava, indicates that he/she will be a frustrated man; hard working but always having differences with family members and his/her children are aggressive and fighting nature. He/She does business rather than service but always loses money in life. He/She dislikes repetitious tasks. He/She is likely to succeed in fields related to scientific research, business or finance.

Cancer in 6th Bhava: The Cancer, in the 6th Bhava, indicates that he/she will not be taking rest in life and will not allow his/her relatives or subordinates to take rest. He/She will always be fighting with others and in very much love and affection with his/her children. As a supervisor, he/she is very understanding and concerned about the welfare of his/her employees.

Leo in 6th Bhava: The Leo, in the 6th Bhava, indicates that he/she will be angry-man, infighting and very sexy and will suffer financially due to entangle with other man/women. He/She will suffer and loose relation with spouse and children due to other man's/women's relation. He/She can lose himself/herself completely in work and service dominates over co-workers and subordinates. He/She has a feeling of authority where work and services are concerned.

Virgo in 6th Bhava: He/She will lose huge money due to contact and connection with low status/ level man/women. He/She seldom eats or drinks too excess and he/she is health conscious. History is especially interesting to him/her because of his/her ability to accumulate and relate details.

Libra in 6th Bhava: The Libra, in the 6th Bhava, indicates that he/she will be very rich man but will be selfish and will not believe in god. He/She is cooperative, tactful, and diplomatic in the work place.

Scorpio in 6th Bhava: The Scorpio, in the 6th Bhava, indicates that he/she will be sexy and luxurious. He/She will be hard working and will make money with hard work. He/She will not be so fortunate. He/She has commitment and seriousness about work and gets intensely involved in his/her job. His/Her work often involves matters of investigation such as journalism, research, laboratory science or psychology.

Sagittarius in 6th Bhava: The Sagittarius, in the 6th Bhava, indicates that he/she will be poor and has to work hard and struggle for bread and food whole life. He/She has a tendency to overwork and to over-extend himself/herself in his work. He/She eats or drink too much, or to have an extravagant taste in food and drink.

Capricorn in 6th Bhava: The Capricorn, in the 6th Bhava, indicates that he/she will be poor and has to work hard and struggle for bread and food whole life. His/Her expenditure is more than earning because of children is always sick. As a supervisor, he/she is a disciplinarian. Some Capricornia avoid work altogether because of disinterest in his/her job.

Aquarius in 6th Bhava: The Aquarius, in the 6th Bhava, indicates that he/she will work on ship or in navy. He/She is quarrelling with his/her boss and colleague. He/She has danger of drowning in water in childhood and also at the age of 41 years. He/She can function especially well within group situations.

Pisces in 6th Bhava: The Pisces, in the 6th Bhava, indicates that he/she will be infighting and so does not has good relations with children, spouse and other family members and so feel loneliness in whole life. He/She may be spending in litigation. He/She has excessive worry about the job.

PART-7
ZODIAC SIGNS IN 7TH HOUSE (JAYA/WIFE)

Aries in 7th Bhava: The Aries, in the 7th Bhava, indicates that his//her spouse will be cruel, angry-nature, rigid, very frequently developing bad relation with husband and habituated of making disturbed family life, even though she will be educated. His/Her partner is energetic, aggressive, sexy, innovative, and pioneering.

Taurus in 7th Bhava: The Taurus, in the 7th Bhava, indicates that his//her spouse will be very beautiful, intelligent, sweet-talking, good mannered, better-understanding and talkative but proudly of her beauty. He/She will be good in conjugal life. He/She will be a loyal marital partner. There is much stubbornness in both him/her and his/her partner.

Gemini in 7th Bhava: The Gemini, in the 7th Bhava, indicates that his//her spouse will be beautiful, intelligent, sweet-talking, good mannered, better-understanding and talkative. He/She will be knowledgeable of stitching, dance, music, good cooking. He/She will be fortunate but always unsatisfied in conjugal life.

Cancer in 7th Bhava: The Cancer, in the 7th Bhava, indicates that his//her spouse will be very beautiful, intelligent, attract others due to her good manner, imaginative, thinker, better understanding and excellent in beauty. He/She will be knowledgeable of stitching, dance, music, good cooking and will be fortunate. He/She will be sentimental and good in conjugal life.

Leo in 7th Bhava: The Leo, in the 7th Bhava, indicates that his//her spouse will be cruel, angry-nature, rigid, very frequently developing bad relation with others and stay separately even in the friend's party. He/She will be too selfish and will be attached too much to ornaments and money, even though he/she will be educated, intelligent and good thinker. He/She will have happy married life. His/Her partner is dynamic, dramatic, strong, and vital.

Virgo in 7th Bhava: The Virgo, in the 7th Bhava, indicates that his//her spouse will be very beautiful, intelligent, always helpful to his/her spouse in difficulty and painful time, good mannered, better understanding and excellent in beauty. He/She will have good conjugal life. Marital partner is hard working and effectual, and assumes most of the responsibility of helping him/her and takes care of his/her practical affairs.

Libra in 7th Bhava: The Libra, in the 7th Bhava, indicates that his//her spouse will be very beautiful, intelligent, attract others due to her excellent beauty, helping attitude, interested in religious work, donor and good mannered, imaginative, thinker, better understanding and excellent in beauty. He/She will be helpful to his/her spouse and will be fortunate. He/She will have good conjugal life.

Scorpio in 7th Bhava: The Libra, in the 7th Bhava, indicates that his//her spouse will be less educated or uneducated, does not know stitching or any ladies' art and not even interested in the same, very unfortunate in life since childhood, always facing difficulties in whatsoever work he/she undertake. He/She will be suffering from diseases like headache and stomach. His/Her mate may be inwardly powerful and dynamic.

Sagittarius in 7th Bhava: The Sagittarius, in the 7th Bhava, indicates that his//her spouse will be cruel, angry-nature, rigid, very frequently developing bad relation with others. She/He will be less educated and will have no interest in stitching or any other ladies' arts. He/She will have happy married life, fortunate spouse, who will bring monetary help from his/her parent and will give good children, but reluctance to be hemmed in formal partnership entrapments.

Capricorn in 7th Bhava: The Capricorn, in the 7th Bhava, indicates that his//her spouse will be cruel, angry-nature, rigid, very frequently developing bad relation with others and educated and will have no interest in stitching or any other ladies' arts. He/She will have happy married life. He/She is restricted by the duties of wedlock or of any partnership. He/She needs partners to be his/her nurturing support and a mate as the parent figure to give a solid base, be supportive and responsible.

Aquarius in 7th Bhava: The Aquarius, in the 7th Bhava, indicates that his//her spouse will be very beautiful, intelligent, always helpful to her husband in difficulties, hard working, ready to face strongly in difficulties, fearful of god, respecting elderly people, interested in sacred work. He/She will have good conjugal life and will give good children.

Pisces in 7th Bhava: The Pisces, in the 7th Bhava, indicates that his//her spouse will be beautiful, intelligent, helping attitude, interested in religious work, donor, good mannered, better understanding and very active in social work. He/She will be helpful to his/her spouse and will be fortunate. He/She will be good swimmer and have good conjugal life and deliver intelligent and good children. He/She is willing to make sacrifices to benefit the union. His/Her partner helps him/her expand his/her horizons to new fields.

PART-8
ZODIAC SIGNS IN 8TH HOUSE (MRITYU/DEATH)

Aries in 8th Bhava: The Aries, in the 8th Bhava, indicates that he/she will be mostly settled in foreign country. He/She will be suffering from the disease of talking or walking in sleep. He/She will be unhappy by remembering his/her past incidents of life. He/She will be rich and die in foreign country. In joint financial affairs, there can be conflict and disagreement about money. He/She is decisive in this regard. Sometimes such conflicts may result in litigation.

Taurus in 8th Bhava: The Taurus, in the 8th Bhava, indicates that he/she will be seriously suffering from the disease of cough and die due to serious cough congestion of respiratory system. He/She has a good head for business and a sense for sound investments. There is a tendency to marry for money or security, as well as for love. There may be ups and downs in his/her financial affairs, but eventually matters turn out profitably.

Gemini in 8th Bhava: The Gemini, in the 8th Bhava, indicates that he/she will be seriously sick of Prameha and Gurda Roga in old age. He/She will die due to fighting with enemies. He/She is full of ideas about how properly to handle joint finances.

Cancer in 8th Bhava: The Cancer, in the 8th Bhava, indicates that he/she will die due to drowning in water or in the house peacefully at old age. He/She rarely exposes emotional needs and ceases to be the go-getter that usually marks his/her style.

Leo in 8th Bhava: The Leo, in the 8th Bhava, indicates that he/she will die due to snake biting or due to fighting with thief He/She has a large capacity for romance. There is also a much creative energy directed toward business and large-scale investments. Moneymaking becomes a game for him/her to be enjoyed. This position usually promises living to a ripe old age.

Virgo in 8th Bhava: The Virgo, in the 8th Bhava, indicates that he/she will die due venereal diseases or due to consumption of poison. This sign suggests a restrained or constrained sex life.

Libra in 8th Bhava: The Libra, in the 8th Bhava, indicates that he/she will die due to hunger or due to fighting with shoulder or due to anger and blood pressure and has financial gain through marriage and partnerships. He/She gets just everything he/she really wants with a diplomatic approach.

Scorpio in 8th Bhava: The Scorpio, in the 8th Bhava, indicates that he/she will die due to skin diseases or due to consumption of poison.

Sagittarius in 8th Bhava: The Sagittarius, in the 8th Bhava, indicates that he/she will die due to drowning in water or in the house peacefully at old age among the family members. He/She has a natural flare for business and good fortune when it comes to money. He/She has good fortune to benefit from large-scale enterprises that grow and prosper. Jupiter, the ruling planet of Sagittarius, provides good fortune and abundance in the part of the chart it controls. He/She gets financial help throughout his/her life, due to Jupiter influence. He/She can usually get what he/she wants out of life and is good-humoured nature and generous, outgoing demeanour.

Capricorn in 8th Bhava: The Capricorn, in the 8th Bhava, indicates that he/she will die due to old age and worshipping god in the house peacefully among the family members. He/She does not like to borrow and being in debt. He/She would never stoop to cheating or deceptions in business dealing. He/She handles other people's money and does so with serious concern.

Aquarius in 8th Bhava: The Aquarius, in the 8th Bhava, indicates that he/she will die due to burning or due to respiratory problem or due to saviour wound problem.

He/She has a strong interest in the spiritual side of life. He/She is very unorthodox on sex, birth control and abortion.

Pisces in 8th Bhava: The Pisces, in the 8th Bhava, indicates that he/she will die due to lever problem or due to blood diseases in old age among the family members.

PART-9
ZODIAC SIGNS IN 9TH HOUSE
(BHAGYA/FORTUNE)

Aries in 9th Bhava: The Aries, in the 9th Bhava, indicates that he/she will be fortunate. He/She is always worried for money. His/Her expenses are more than the earning. He/She makes profit by purchase and sale of animals, agricultural lands and house properties, & product of religious work. He/She loves to travel.

Taurus in 9th Bhava: The Taurus, in the 9th Bhava, indicates that he/she will be fortunate at the age of 28 and 36 years. His/Her child hood is difficult and face problem in education but is successful in completing his/her education. He/She makes profit in business of fancy store but progress slowly in service. He/She earns wealth in second part of life and earns good name & fame and enjoys the life luxuriously. Venus makes travel enjoyable, and lives far from his/her native home.

Gemini in 9th Bhava: The Gemini, in the 9th Bhava, indicates that he/she will be fortunate at the age of 27 years, is simple, vegetarian, and gentleman and talks sensibly. He/She is successful in completing his education and helps poor people. He/She makes profit in business and thinks of always about business. He/She earns wealth in middle part of life and earns good name & fame and enjoys the life as a respected person. He/She especially enjoys travel mostly for observing different people and places.

Cancer in 9th Bhava: The Cancer, in the 9th Bhava, indicates that he/she will be simple, vegetarian, and gentleman and talks sensibly. He/She is successful in teacher, reporter, publisher and writer and helps poor people. He/She suffers from stomach problems and gastric. He/She takes too much time in taking the decisions and due to this, he/she suffers. He/She earns wealth but his/her middle part of life is full of struggles and loss of wealth.

Leo in 9th Bhava: The Leo, in the 9th Bhava, indicates that he/she will be totally against the religions and will be arguing against the religions. His child hood is difficult and face problem in education but is successful in completing his/her education. Whole life he/she will be struggling and will do hard work for survival. He/She earns wealth in second part of life and earns good name & fame and enjoys the life luxuriously. His/Her nature of job will be touring type.

Virgo in 9th Bhava: The Virgo, in the 9th Bhava, indicates that he/she will be luxurious at the young age and gentleman and talks sensibly. He/She earns wealth in middle part of life but fortune never favours him/her. He/She will be fortunate at the age of 23 years and enjoys his/her life at old age with grand children.

Libra in 9th Bhava: The Libra, in the 9th Bhava, indicates that he/she will be totally religious minded and will be arguing always in favour of the religions. His/Her child hood is difficult and face problem in education but is successful in completing his/her education. Whole life he/she will be struggling and will do hard work for survival. He/She will be more successful in business than service. He/She earns wealth from business and earns good name & fame in the society as a leader and enjoys the life luxuriously in leadership. He/She gets wealth from his/her in-law. He/She will be fortunate after the age of 24 years.

Scorpio in 9th Bhava: The Scorpio, in the 9th Bhava, indicates that he/she will be mean mind. His/Her child hood is difficult and face problem in education, will be struggling and will do hard work for survival. He/She will be more successful after the age of 28 years. He/She earns wealth after 28 year of age and earns good name & fame in the society and enjoys the life luxuriously. His/Her nature of job will be touring type.

Sagittarius in 9th Bhava: The Sagittarius, in the 9th Bhava, indicates that he/she will be clever, peaceful and gentle man. His/Her child hood is good and does not face problem in living. He/She will be struggling and will do hard work for survival at the middle of age. He/She will be more fortunate and successful after the age of 45 years and achieve very high position and wealth. He/She earns wealth after 45 year of age and earns good name & fame in the society and enjoys the life luxuriously. His/Her nature of job will be related to water like captain, sailor or any other service with irrigation department where he/she gets lot of success. He/She will have much more journey in life.

Capricorn in 9th Bhava: The Capricorn, in the 9th Bhava, indicates that he/she will be clever, peaceful and gentle man. His child hood is not good and faces problem in living. He/She will be struggling and will do hard work for his survival. He/She will be more fortunate and successful after the age of 45 years. He/She earns wealth after 45 year of age and achieves very high position and wealth and earns good name & fame in the society and enjoys the life luxuriously.

Aquarius in 9th Bhava: The Aquarius, in the 9th Bhava, indicates that he/she will be clever in politics, a leader. His/Her child hood is not good and faces many problems in living. He/She will be struggling in education and will do hard work for his/her survival. He/She will be more fortunate and successful after the age of 45 years and 36 years and achieve very high position and wealth and earns good name & fame in the society and enjoys the life luxuriously. He/She starts earning wealth after 28 year of age.

Pisces in 9th Bhava: The Pisces, in the 9th Bhava, indicates that he/she will be totally religious minded and child hood will be good but is successful in completing his/her education. He/She will be more successful in business than service, particularly in yellow metals business. He/She will be fortunate after the age of 27 years and earns wealth from business and earns good name & fame in the society as a leader and enjoys the life luxuriously in leadership.

PART-10
ZODIAC SIGNS IN
10TH HOUSE
(KARMA/PROFESSION)

Aries in 10th Bhava: The Aries, in the 10th Bhava, indicates that child hood is difficult and face problem in education but is successful in completing his/her education. Whole life he will be struggling and will be busy in more than one work and will do hard work for survival. He/She earns wealth in second part of life but does not earn good name & fame and does not enjoy the life luxuriously. He/She has tendency toward shyness and to disappear on the public stage, or even in public view.

Taurus in 10th Bhava: The Taurus, in the 10th Bhava, indicates that child hood is good but will spend more than his/her earning in his life. Whole lives he/she will be struggling and will be busy in earning and will not be able to save much. He/She will be patriotic to his/her father. He/She chooses career, such as art, theatre, and music.

Gemini in 10th Bhava: The Gemini, in the 10th Bhava, indicates that he/she will be patriotic to his/her father and respect and obey the elders and will be agriculturist and earn money from agriculture or real estate. He/She may earn more money from business rather than service.

Cancer in 10th Bhava: The Cancer, in the 10th Bhava, indicates that he/she will be child hood is good but will spend more than his/her earning in his/her life. Whole lives he/she will be busy in helping poor people and doing social work such as making temples and drinking water facilities for the people and will be busy in earning.

Leo in 10th Bhava: The Leo, in the 10th Bhava, indicates that child hood is difficult and face problem but is successful in completing his/her education. Whole life he/she will be struggling and will do hard work for survival. He/She earns wealth in second part of life but does not earn good name & fame and does not enjoy the life luxuriously. He/She does not get help from the family. He/She is likely to get a position of leadership and authority, and be admired and is happiest running his own business.

Virgo in 10th Bhava: The Virgo, in the 10th Bhava, indicates that he/she will be patriotic to his/her father but does not enjoy the life luxuriously. He/She will be self respected person and works to maintain his/her respect at any cost. He/She will not do buttering of any other person in life, even though he may lose something. He/She will earn good money from business rather than service. He/She finds employment in a large, well-established organization such as civil service, a church, or an educational institution.

Libra in 10th Bhava: The Libra, in the 10th Bhava, indicates that he/she will be patriotic to his/her father and respect and obey the elders and enjoy the life luxuriously. He/She will earn more money from business rather than service. Business will be lucky for him/her and will progress in life by business. He/She will be rich of his/her speech and fulfil whatever he/she will promise. He/She will reach to the highest position in the middle age. He/She will enter a business with a partner, or in cooperation with others. He/She will be an excellent administrator. His/Her public standing may rise well above that of his parent.

Scorpio in 10th Bhava: The Scorpio, in the 10th Bhava, indicates that he/she will be totally religious minded and get popularity and respect in the society and will be clean and honest. His/Her child hood is good but will spend more than his/her earning in his/her life in religious work. Whole lives he/she will be busy in helping poor people and doing social work and will be busy in earning and will not be able to save much. He/She will reach to the highest position in the service. He/She make his/her mark in the world. There has drive to succeed and is naturally able to impress others with the forcefulness of his/her mind.

Sagittarius in 10th Bhava: The Sagittarius, in the 10th Bhava, indicates that he/she will be patriotic to his father and respect and obey the elders and enjoy the life luxuriously. He/She will earn more money from business rather than service. Business will be lucky for him/her and will progress in life by business. He/She may face a lot of problem in service but will be hard working and makes everything favourable in life. He/She will get popularity and respect in the society. He/She has the abilities to be a leader and never hesitates using his/her influence to attain goals. A good deal of travel is likely to be associated with his/her career. He/She may have more than one career in his/her life.

Capricorn in 10th Bhava: The Capricorn, in the 10th Bhava, indicates that he/she will be totally religious minded and get popularity and respect in the society and will be clean and honest. His/Her childhood is good but will spend more than his/her earning in religious work. Whole lives he/she will be busy in helping poor people and doing social work. He/She will be able to make favourable conditions and will reach to the highest position in the middle age and people will be astonished with his/her rise. He/She has a very strong sense of duty, an attitude of dedication. The progress in the career may be slow, but it is consistent. He/She is capable of climbing to the top. He/She can build a solid public image. This is a very achievement oriented placement.

Aquarius in 10th Bhava: The Aquarius, in the 10th Bhava, indicates that he/she will be political minded and get popularity and respect in the society and will be talkative. His/Her child hood is good but will be busy in helping people. He/She has followers in politics. He/She makes favourable conditions and will reach to the highest position. He/She is a team player and function well as a part of the team. He/She may be involved in many humanitarian ventures.

Pisces in 10th Bhava: The Pisces, in the 10th Bhava, indicates that he/she will be patriotic to his father and respect and obey the elders and enjoy the life luxuriously. He/She will earn more money from service related to water. He/She may face a lot of problem in life but will be hard working and makes everything favourable in life. He/She will get popularity and respect in the society. He/She can make money by business too.

PART-11
ZODIAC SIGNS IN
11TH HOUSE
(GAIN/INCOME)

Aries in 11th Bhava: The Aries, in the 11th Bhava, indicates that he/she will be patriotic to his/her father and respects and obeys the elders and does not enjoy the life luxuriously. He/She will be self respected person and will not do buttering of any other person in life, even though he/she may loose something. He/She will be hard working person and does not get help from the father and family. Whole life he/she will be struggling and will be busy in more than one work and will do hard work for survival. He/She is fortunate after 28 years of age. He/She is active in associations, club or other such organizations. He/She attracts a wide and varied circle of friends. He/She makes the most of his contacts. Often, friends are the key to helping him/her attain his goals.

Taurus in 11th Bhava: The Taurus, in the 11th Bhava, indicates that he/she will belong to the medium family and does not enjoy the life luxuriously. He/She will be self respected person and work to maintain his/her respect at any cost. He/She will be hard working person and does not get help from the father and family. Whole life he/she will be struggling and will be busy in more than one work and will do hard work for survival. He/She will be known to the great personalities and will live in his/her surroundings. He/She gets benefited with the opposite sex. His luck starts from his/her middle age and reach to the highest post. He/She will be making money, accumulating a comfortable standard of living, and establishing a secure situation. His/Her interest is not just money and possessions for the sake of wealth.

Instead he/she wants to gain a sense of security, and status to overcome a basic insecurity. He/She has well-to-do friends, who may be called on to help him/her attain his/her goals.

Gemini in 11th Bhava: The Gemini, in the 11th Bhava, indicates that he/she will be hand to mouth since child hood. He/She will be self respected person and work to maintain his/her respect at any cost. He/She will be hard working person and does not get help from the family. Whole life he/she will be struggling and will be busy in more than one work and will do hard work for survival. He/She is fortunate in business as compared to service. Up to 42 years, his financial conditions are miserable. He/She will start earning from many sources after 42 years of age. He/She has many personal connections. He/She is naturally attracted to people who are witty, intelligent, and verbal. He/She has a good sense of humour and he can laugh, even at himself/herself. He/She is not a loner, and he/she needs constant, mentally compatible companionship to be happy and fulfilled.

Cancer in 11th Bhava: The cancer, in the 11th Bhava, indicates that he/she will be patriotic to his/her father and respects and obeys the elders and enjoys the life luxuriously. He/She will be self respected person and work to maintain his/her respect at any cost. He/She will be hard working person and does not get help from the father and family. Whole life he/she will be struggling and will be busy in more than one work. He/She will progress in the service as compared to business. He/She will be more attached to his/her family members due to his/her sentiment and he/she may sacrifice for his/her family or friends. He/She can earn money from lottery. He/She will have a successful life.

Leo in 11th Bhava: The Leo, in the 11th Bhava, indicates that he/she will be business minded and calculate everything in term of profit and loss before doing the work. He/She will be self respected person and work to maintain his/her respect at any cost. He/She will be hard working person and does not get help from the family. Whole life he/she will be calculating and will be busy in more than one work and will does hard work for survival. He/She is fortunate in business as compared to service. He/She will start earning wealth in business from many sources and makes lot of money. He/She is kind and respects everybody. He/She will never spend money on luxury and live a simple life. He/She has leadership qualities displayed within groups and organizations. He/She has a very deep need for friends and associations. He/She takes pride in his/her friends and associates; some may be rich and famous. He/She may tend to draw strength from his/her friends as if they fulfilled a special need in his/her life. He/She is always careful to dress in good taste, and notices what others are wearing as well. He/She has a very wide circle of friends, most of who have much respect for him/her.

Virgo in 11th Bhava: The Virgo, in the 11th Bhava, indicates that he/she will be business minded and calculate everything in term of profit and loss before doing the work. He/She will be far sighted and self respected person and work to maintain his/her respect at any cost. He/She will be hard working person and does not get help from the family rather he/she will be helping his brothers and other family members. Whole life he/she will be calculating and will be busy in more than one work and will do hard work for survival. He/She is fortunate in politics as compared to business and service. He/She will start earning from many sources and makes lot of money from politics. He/She is kind and respects everybody. He/She will never spend money on luxury and live a simple life. He/She has a readiness to serve close friends.

Libra in 11th Bhava: The Libra, in the 11th Bhava, indicates that he/she will be calm, quiet, intelligent and gentle person. He/She will be able to identify the right or wrong timing and to do the miracle even in the miserable conditions. Family and friends will be helpful to him/her. He/She has an amiable demeanour with friends and working in group situations. In this context, he is diplomatic and even handed. He/She is likely to be selected to head a group just because of he/she is acceptable to those with divergent interests. He/She has a highly social attitude. He/She really loves to be with friends. Often this sign denotes marriage to a friend of long standing.

Scorpio in 11th Bhava: The Scorpio, in the 11th Bhava, indicates that he/she will be so intelligent that he/she will speak anything suitable to the time, person and place. He/She will be able to do many things at a time and will be successful in all the works undertaken. He/She will be doing the business related to the land and spends a lot for this. He/She will be benefited by agriculture. He/She develops close relationships with intense and aggressive friends who are influential and powerful. He/She never accepts weak individuals as friends.

Sagittarius in 11th Bhava: The Sagittarius, in the 11th Bhava, indicates that he/she will be business minded and calculate everything in term of profit and loss before doing the work. He/She will be self respected person and work to maintain his/her respect at any cost. He/She will be hard working person. Whole life he/she will be calculating and will be busy in more than one work and will do hard work for making lot of money. He/She will have popularity among service class officers and managers and will make use of them to earn more and more money. He/She is fortunate in business as compared to service. He/She will start earning wealth in business from many sources and makes lot of money. He/She is kind and respects everybody. He/She will never spend money on luxury and live a simple life. He/She has a wide circle of friends and associates and enjoys such casual relationships.

Capricorn in 11th Bhava: The Cancer, in the 11th Bhava, indicates that he/she will be a self made person, active and intelligent to take right decision at the right time, right place. He/She will be self respected person and work to maintain his respect at any cost. He/She will be hard working person and get help from the family. He/She is fortunate in service as compared to business. He/She will start earning wealth in service and from many other sources and makes lot of money. He/She is kind and respects everybody. He/She will never talk lye and fulfil his/her promises at any cost.

Aquarius in 11th Bhava: The Aquarius, in the 11th Bhava, indicates that he/she will be a self-made person, active and intelligent to take right decision at the right time, right place. He/She will be self respected person and work to maintain his/her respect at any cost. He/She will be hard working person and does not get help from the family. His/Her childhood is very troublesome but as he/she grows up, he/she starts earning and his/her life is comfortable. He/She is fortunate in service as compared to business. He/She will start earning wealth in service and from many other sources and makes lot of money. He/She is kind and respects everybody. He/She will never talk lye and fulfil his/her promises at any cost. He/She spends money on luxury and lives a simple life. He/She is successful in politics too. His/Her friendships are intellectually motivated rather than sentimentally stimulated.

Pisces in 11th Bhava: The Pisces, in the 11th Bhava, indicates that he/she will be a rich man, self made person, active and intelligent to take right decision at the right time, right place. He/She will be self respected person and work to maintain his respect. He/She will be hard working person and get help from the family. He/She is fortunate and shins in a particular line such as singer or scientist or in arts and in service as compared to business. He/She will start earning wealth in service and from many other sources and makes lot of money. He/She is kind and respects everybody. He/She will never talk lye and fulfil his promises at any cost. He/She spends money on luxury and lives a simple life.

PART-12
ZODIAC SIGNS IN
12TH HOUSE
(LOSS/EXPENDITURE)

Aries in 12th Bhava: The Aries, in the 12th Bhava, indicates that he/she will enjoy the life luxuriously since childhood. He/She will earn more money from business rather than service in beginning but will become bankrupt in the middle age and there will be financial crisis. Business will be lucky for him/her in beginning or early age and will progress in life by business up to middle age. He/She may face a lot of problem in business at middle age but will be hard working and makes everything favourable in life. He/She will get popularity and respect in the society. His/Her eyesight will be weak and will get operated in the old age. Though he/she is very slow to anger, when he/she does cross the line, his/her temper can be irrational.

Taurus in 12th Bhava: The Taurus, in the 12th Bhava, indicates that he/she will be a rich man, self made person, active and intelligent to take right decision at the right time, right place. He/She will be self respected person and work to maintain his/he respect in the society. He/She will be hard working person. He/She is fortunate and shins in a particular line such as administrator or tourist or salesman or writer and in service as compared to business. He/She will start earning wealth at the age of 30 years and from many other sources and makes lot of money. He/She is kind and respects everybody. He/She will never talk lye and fulfil his/her promises at any cost. He/She does not spend money on luxury and lives a simple life. He/She has more worry about financial affairs than his/her happy-go-lucky demeanour would imply.

Gemini in 12th Bhava: The Gemini, in the 12th Bhava, indicates that he/she will be imaginative and sentimental type. He/She believes any person and get hurt and cheated in due course of time by that person, particularly by relative, brothers, sisters and others. He/She will never talk lye and fulfil his/her promises at any cost. He/She spends too much money on luxury and lives a luxurious life. Whole life he/she will be struggling and will be busy in more than one work and will do hard work for survival. He/She is fortunate in business as well as service. Throughout, his financial conditions are miserable. He/She will start earning from many sources and will spend too much money. His/Her eyesight will be weak and will get operated in the old age. He/She will find it very difficult to live with or near his brothers, sisters or other close relatives.

Cancer in 12th Bhava: The Cancer, in the 12th Bhava, indicates that he/she will be rigid and angry promise fulfiller type. He/She makes mistake due to his angriness. He/She will never talk lye and fulfil his/her promises at any cost. He/She spends too much money on luxury and lives a luxurious life. Whole life he/she will be struggling and will be busy in more than one work and will do hard work for survival. He/She is fortunate in business as well as service. Throughout, his/her financial conditions are miserable. He/She will start earning from many sources and will spend too much money. He/She believes any person and gets hurt and cheated in due course of time by that person, particularly by relative. He/She is a generous person. He/She gets hurt when those near and dear are beginning to take him for granted.

Leo in 12th Bhava: The Leo, in the 12th Bhava, indicates that he/she will be imaginative and sentimental calm and quite, kind and sweetly spoken type. He/She believes any person and gets hurt and cheated. He/She will never talk lye and fulfil his/her promises at any cost. He/She will not like to spend too much money on luxury and lives a simple life. Whole life he/she will be struggling and will be busy in more than one work and will do hard work for survival. He/She is fortunate in service rather than business. He/She will start earning from many sources and will not spend much money on decorative items. He/She is a power behind the scenes. He/She plays a back room manoeuvring role in matters. He/She is never one to blow his/her horn.

Virgo in 12th Bhava: The Virgo, in the 12th Bhava, indicates that he/she will be sentimental calm and quiet, and sweetly spoken type and successful businessman. He/She will be able to take decision very fast in any matter. He/She will be in business and find out many source of income in it. He/She will not like to spend too much money on luxury and lives a simple life and teach his/her children the same. Whole life he/she will be struggling and will be busy in more than one work and will do hard work for survival. He/She is fortunate in business rather than service. He/She will start earning from many sources and will not spend much money on luxury items. He/She may have to face litigation and charges and due to that, he may face problem. He/She will be self respected person and work to maintain his/her respect at any cost and will do the religious and social work.

Libra in 12th Bhava: The Libra, in the 12th Bhava, indicates that he/she will be a common person, self-made person, active and intelligent to take right decision at the right time, right place. He/She will be self respected person and work to maintain his/her respect. He/She will be hard working person and does not get help from the parent. He/She will be able to make any person in his/her favours and become popular in the society. He/She will have many friends and known personalities and will get benefited from them. He/She is fortunate and shins in service as compared to business. He/She will start earning in service and from many other sources. He/She is kind and respects everybody. He/She will never talk lye and fulfil his promises at any cost. He/She spends money on luxury and lives a simple life. He/She is a loner with some apprehensions regarding reliance on others. His/Her partnership arrangements, including the marriage, may be fated in some way and acting in collaboration with others, rarely works out well for him/her.

Scorpio in 12th Bhava: The Scorpio, in the 12th Bhava, indicates that he/she will be self made person, active, religious minded person and intelligent to take right decision at the right time, right place. He/She will be self respected person and work to maintain his/her respect in the society and will do the religious work. He/She will be hard working person. He/She is fortunate and shins in a particular line such as writer or publisher and in service as compared to business. He/She will start earning wealth at the age of 16 years and from many other sources. He/She is kind and respects everybody. He/She will never talk lye and fulfil his promises at any cost. He/She does not spend money on luxury and lives a simple life.

Sagittarius in 12th Bhava: The Sagittarius, in the 12th Bhava, indicates that he/she will be intelligent. He will be able to take decision very fast in any matter. He/She will be in business and find out many source of income in it. He/She will not like to spend too much money on luxury and lives a simple life. Whole life he/she will be struggling and will be busy in more than one work and will do hard work for survival. He/She will start earning from many sources and will not spend much money on luxury items. He/She will be self respected person and work to maintain his/her respect at any cost and will do the religious and social work and will be popular in the society. He/She will be able to make any person in his/her favour and become popular in the society. He/She will have many friends and known personalities and will get benefited from them.

Capricorn in 12th Bhava: The Capricorn, in the 12th Bhava, indicates that he/she will be a common person, self-made person, active and intelligent to take right decision at the right time, right place. His/Her childhood will be troublesome. He/She will be self respected person and work to maintain his/her respect. He/She will be hard working person and does not get monetary help from the parent. He/She will be able to make money by his/her own hard work and become popular in the society. He/She will start earning wealth at the age of 30 years and from many other sources. He/She will have many friends and known personalities and will get benefited from them. He/She is fortunate and shins in service as compared to business. He/She will start earning in service. He/She is kind and respects everybody. He/She will never talk lye and fulfil his promises at any cost. He/She spends money very cautiously and makes balance between income and expenditure and lives a simple life. He/She has subconscious feelings of limitation and inadequacy deeply ingrained in the psyche.

Aquarius in 12th Bhava: The Aquarius, in the 12th Bhava, indicates that he/she will be intelligent, a common person, self-made person, active, religious minded and intelligent to take right decision at the right time, right place. He/She will be self respected person and work to maintain his/her respect. He/She will be hard working person. He/She will be able to take decision very fast in any matter and will be in service and finds out many source of income in it. He/She will spend too much money and does not have control over it and lives a simple life. Whole life he/she will be struggling and will be busy in more than one work and will do hard work for survival. He/She will start earning from many sources and will not spend much money on luxury items. He/She will be self respected person and work to maintain his/her respect at any cost and will do the religious and social work and will be popular in the society. He/She will be able to make any person in his/her favour and become popular in the society. He/She will have many friends and known personalities and will get benefited from them.

Pisces in 12th Bhava: The Pisces, in the 12th Bhava, indicates that he/she will be a common person, self-made person, active and intelligent to take right decision at the right time, right place. His/Her childhood will be simple. He/She will be self respected person and work to maintain his/her respect. He/She will be hard working person and does not get monetary help from the parent. He/She will be able to make money by his/he own hard work and become popular in the society. He/She will start earning wealth at the age of 32 years and from many other sources. He/She will have many friends and known personalities and will get benefited from them. He/She is fortunate and shins in service as compared to business and reaches to the highest position by his/her hard work. He/She will start earning in service. He/She is kind and respects everybody and will never talk lye and fulfil his promises at any cost. He/She spends money very cautiously and makes bank balance and lives a simple life. He/She will be helpful to others.

: THANK YOU :